W9-DFE-408

Modeling Biological Systems

Principles and Applications

James W. Haefner
Utah State University

ITP® International Thomson Publishing
New York • Albany • Bonn • Boston • Cincinnati • Detroit • London • Madrid • Melbourne
Mexico City • Pacific Grove • Paris • San Francisco • Singapore • Tokyo • Toronto • Washington

Cover design: Andrea Meyer; emDASH inc.

Printed in the United States of America

Chapman & Hall
115 Fifth Avenue
New York, NY 10003

Chapman & Hall
2-6 Boundary Row
London SE1 8HN
England

Thomas Nelson Australia
102 Dodds Street
South Melbourne, 3205
Victoria, Australia

Chapman & Hall GmbH
Postfach 100 263
D-69442 Weinheim
Germany

International Thomson Editores
Campos Eliseos 385, Piso 7
Col. Polanco
11560 Mexico D.F
Mexico

International Thomson Publishing–Japan
Hirakawacho-cho Kyowa Building, 3F
1-2-1 Hirakawacho-cho
Chiyoda-ku, 102 Tokyo
Japan

International Thomson Publishing Asia
221 Henderson Road #05-10
Henderson Building
Singapore 0315

1 2 3 4 5 6 7 8 9 10 XXX 01 00 99 98 97 96

Library of Congress Cataloging-in-Publication Data

Haefner, James W.
Modeling biological systems : principles and applications / James W. Haefner
p. cm.
Includes bibliographical references and index.
ISBN 0-412-04201-0
1. Biological systems--Computer simulation. 2. Biological systems--Mathematical models. I. Title.
QH308.2.H325 1996
574'.01'13--dc20 96-7085
CIP

British Library Cataloguing in Publication Data available

To order this or any other Chapman & Hall book, please contact **International Thomson Publishing, 7625 Empire Drive, Florence, KY 41042.** Phone: (606) 525-6600 or 1-800-842-3636. Fax: (606) 525-7778. e-mail: order@chaphall.com.

For a complete listing of Chapman & Hall titles, send your request to **Chapman & Hall, Dept. BC, 115 Fifth Avenue, New York, NY 10003.**

For Ally —

and with many thanks
to my parents John and Dorothy

Contents

Preface

This book is intended as a text for a first course on creating and analyzing computer simulation models of biological systems. The expected audience for this book are students wishing to use dynamic models to interpret real data much as they would use standard statistical techniques. It is meant to provide both the essential principles as well as the details and equations applicable to a few particular systems and subdisciplines. Biological systems, however, encompass a vast, diverse array of topics and problems. This book discusses only a select number of these that I have found to be useful and interesting to biologists just beginning their appreciation of computer simulation. The examples chosen span classical mathematical models of well-studied systems to state-of-the-art topics such as cellular automata and artificial life. I have stressed the relationship between the models and the biology over mathematical analysis in order to give the reader a sense that mathematical models really are useful to biologists. In this light, I have sought examples that address fundamental and, I think, interesting biological questions. Almost all of the models are directly compared to quantitative data to provide at least a partial demonstration that some biological models can accurately predict.

As a result, I have generally kept the mathematical manipulations and requirements to a minimum. This is not a text in theoretical or mathematical biology; several of these already exist, and, being written by bonafide mathematicians, they have much to recommend them. The minimum mathematics needed for this book are statistics to the point of simple, single-variable linear regression, a small knowledge of probability distributions, and one semester of calculus.

The book is divided into two parts. The first, *Principles*, gives the basic steps that take a modeler from a biological question to a conceptual model to a quantitative specification of the system. The conversion of vague questions and ambiguous information into precise and quantitative mathematical forms is one with which biology students have the greatest difficulty. I have found that a set of heuristic "rules-of-thumb" applied to hypothetical situations is an effective teaching approach. Once these skills are mastered, the text describes techniques for constructing computer programs to solve the equations. Following this, methods to analyze computer output to answer the initial question are presented. These include equilibrium and stability analysis, sensitivity analysis, error analysis, and validation. The concepts developed in *Principles* apply to virtually any subject or question that can be addressed or formulated such that the answer can be gleaned from the dynamics of variables that describe the system (e.g., population size). Since the majority of biological theory is formulated in terms of differential equations, I stress techniques appropriate to "continuous systems" simulation.

The second part, *Applications*, is a series of chapters in which fundamental equations and problems from various biological disciplines are presented. Here I have tried to provide students and instructors with tools that will permit them to design their own course in biological modeling. By separating the details of subdisciplines from modeling fundamentals, I hope to provide a format in which a coherent portrait of the modeling enterprise can be obtained as well as background in modeling particular biological systems. Space, interest, and expertise have limited the suite of topics considered. Since my area of interest is ecology, I have perhaps stressed this field, but as most ecologists would admit, physiology and biochemistry are relevant fields. I include some fundamental equations and examples from these areas. The intent is not to give a comprehensive review of each topic; this is well beyond my expertise. Rather, I want to whet the students' appetites, providing enough background so that the references can be used in an intelligent manner and so that meaningful exercises can be attempted.

The process of modeling biological systems is certainly not a science, but neither is it as unconstrained as the creation of a work of pure art that is evaluated solely on its esthetic content. I prefer to analogize modeling with crafting a tool useful for human problem solving. To aid in the acquisition of this craft, I have provided problems and exercises for most of the chapters. Some of these require computer programming, and I have given an example using the C programming language. I believe C is rapidly becoming required for literacy in scientific computing. The very small amount presented in this book will give the reader a taste. A discussion of simulation languages and environments also provides access to other, relatively painless methods of implementing simulation models.

For the Instructors: There seem to be two methods for teaching quantitative and mathematical methods in biology: present a large number of models from many biological disciplines and expect the commonalities and principles to emerge on their own; or, present a set of modeling fundamentals extracted from general principles with relatively few examples and hope that students learn to apply the principles to new situations. Both methods have advantages and disadvantages; I like the latter approach, as the structure of the book suggests. Nevertheless, I have tried to accomodate both and I hope those of you favoring the former teaching style will find the book useable.

For the Students: At my university, I use this book in a course for seniors and new graduate students. It really is an introduction to the subject insofar as someone, somewhere, has already written an entire book on the subject of each chapter. If you are in a considerably earlier stage in your academic career and find the book approachable, consider yourself fortunate to be smart and to have had good teachers.

While the author cannot claim to be smart, he has been fortunate to have had good teachers over the years. It seems appropriate to mention three of them here not so much as to afix blame, but to recognize their contributions. Thanks to Charles Warren, Scot Overton, and George Innis. Finally, this book in whole and in part has been examined by a number of my friends, notable among them being Linda Abbott, Susan Durham, Laura Hartt, Upmanu Lall, Alice Lindahl, Keith Mott, Darcie Neff, Jim Powell, Kirk Steinhorst, and former students in my graduate classes. While their efforts were valiant, unintentional errors remain. Remember: *Never attribute to malice anything that can be attributed to stupidity.*

Modeling Biological Systems

Part I

PRINCIPLES

Chapter 1

Models of Systems

1.1 Systems, Models, and Modeling

'I want to understand everything,' said Miro. 'I want to know everything and put it all together to see what it means.'
'Excellent project,' she said. 'It will look very good on your resumé.'
— Card (1982)

WHEN thinking about systems, models, and understanding everything, it is good to begin with the famous parable of six blind men standing around an elephant. They are asked to identify the object before them which they cannot see. One man, feeling the elephant's leg, thinks he is touching a tree trunk. Another, grasping the elephant's trunk, thinks he is holding a snake. A third, standing near the moving ear, thinks it is a large, feathered fan. And so it goes for the other men touching the tusk, the side, and the tail of the elephant. Each man gave a different description of the same object, but none was correct.

Three fundamental lessons can be gleaned from this simple parable. First, in the real world, we don't know it's an elephant: there is no omniscient observer with special access to the truth. Imagine you are one of the blind men; now imagine yourself propounding the new "tree-trunk" theory to your fellow observers. Very likely, they are not amused. And this is a typical interaction among scientists. Second, all of the men collected basic data and generated an hypothesis consistent with the data. This activity, which is distinct from deduction or induction, is called *abduction* (attributed to Charles Peirce, see Hanson 1972). It is easy and natural for humans to practice abduction; as the parable suggests, it is an activity that occurs frequently in daily life. Third, abduction is not infallible. However it is accomplished, abduction is not a fail-safe method for discovering truth, beauty, or the meaning of life. Descriptions and hypotheses may vary in their quality or value. We must, therefore, go beyond the simple description, if we are to gain confidence that our initial perceptions were valuable.

This book describes some tools by which we may formally and quantitatively extend to specific predictions the qualitative descriptions abduced from observations on biological subjects

In essence, each blind man created a model (the description) of a system (the elephant). By these concepts we mean the following. A **model** is *a description of a system.* A **system** is *any collection of interrelated objects.* An **object** is *some elemental unit upon which observations can be made, but whose internal structure either does not exist or is ignored.* Finally, for completeness, a **description** is *a signal that can be decoded or interpreted by humans.* In short, systems are anything humans wish to discuss and models are one tool that facilitates the discussion.

Before discussing these definitions, consider an example. Suppose the system of interest is the set of students and the professor in a typical classroom situation. There are many potentially interesting relations between these objects, but let us focus on their spatial position at a moment in time. We could model this system by drawing a map of the objects based on some arbitrary coordinate system (e.g., Cartesian coordinates with origin in one corner of the room). This map then counts as a model because the objects and their relations (the system) are combined in a form that can be interpreted by humans. The relations between objects identified in this example are the spatial relations. Other relations could be used, for example, the relation *knows more than.* Thus, we could describe the system in the classroom by drawing arrows between objects to indicate that the object at the tail of the arrow knows more than the object to which the arrow points.

Although we can give particular examples, the definitions stated above are so general that they are nearly useless in normal discourse. Superficially, they imply that virtually everything is a system and that models are used and defined in every facet of human activity. For example, the simple declarative sentence "It is raining outside" counts as a model of a system composed of the atmosphere outside the walls of the building. Consequently, the definitions do not aid in defining and delineating our subject of interest. Nevertheless, the definitions make several points. First, modeling is a fundamental activity between humans: we use models to communicate a view of the world. (Indeed, this book is a model of modeling.) Second, any particular system with its specific objects and relations is defined, if not arbitrarily, then at least by some convention that may in the end be a matter of convenience.

Because of the generality in the definitions, we must narrow the class of models. We do this by identifying a class of uses to which models may be put. There are many possibilities: we use them to convince (e.g., use of analogy in a court room), delight (e.g., a painting or sculpture), inform (e.g., a map), and so on. However, it is the class of *scientific* uses that concerns us here and that will give us a framework for restricting the class

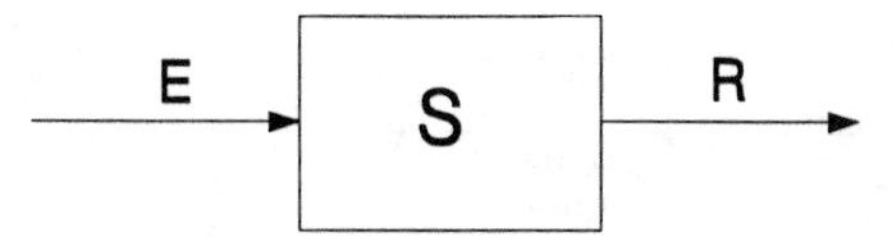

Figure 1.1: A general system represented as an input (E), a system object (S), and the output (R). (From Karplus 1977, Fig. 1a. © 1977 Simulation Councils, Inc. Reprinted with permission Simulation Councils, Inc., publisher.)

of models.

1.2 Uses of Scientific Models

There are three primary, technical uses of models in science:

- *Understanding* – of either a real, physical system or of a system of logic such as another scientific theory.
- *Prediction* – of the future or of some state that is currently unknown.
- *Control* – to constrain or manipulate a system to produce a desirable condition.

Karplus (1983) provides a simple conceptual framework of systems that defines these three uses of models. A system (Fig. 1.1) can be thought of as a black box (system object, **S**) with a single input (excitation, **E**), and a single output (response, **R**). Additional structure in the form of objects and relations could be provided within the box, but the idea is general, considering only a single object. The output is produced by the object's action on the input. For example, suppose **S** is a whole plant (not differentiated into parts), **E** is the amount of fertilizer added to the soil, and **R** is the amount of new growth.

This scheme permits a definition of the three uses of models (Fig. 1.2). Three general problems that humans face with respect to any discipline or body of knowledge are:

- *Synthesis* – use knowledge of inputs and outputs to infer system characteristics.
- *Analysis* – use knowledge of the parts and their stimuli to account for the observed responses.
- *Instrumentation* – design a system such that a specified output is the result of an input.

Models can be used in each of these problem areas and when they are, they allow us to predict, understand, and control systems.

There are also important secondary uses of scientific models that derive from the social characteristics of science:

1. Use as a conceptual framework for organizing or coordinating empirical research (e.g., designing experiments or sampling studies, allocating limited research dollars).

Type of Problem	Given	To Find	Uses of Models
Synthesis	E and R	S	Understand
Analysis	E and S	R	Predict
Instrumentation	S and R	E	Control

Figure 1.2: Knowledge needed for models of different uses. (From Karplus 1977, Fig. 1b. © 1977 Simulation Councils, Inc. Reprinted with permission Simulation Councils, Inc., publisher.)

2. Use as a mechanism to summarize or synthesize large quantities of data (e.g., a simple linear regression equation $y = mx + b$ to reduce all of the $x - y$ pairs of data to two parameters m and b).
3. Identify areas of ignorance, especially when defining relations between objects (e.g., Does species A eat species B?, Does Professor X know more than Student A?).
4. Provide "insight" to managers or planners (or others) by performing "what-if" simulations ("gaming").

1.3 Example: The Leaky Bucket

An example will help clarify some of these concepts. Consider a rectangular container of water that has a single hole in its bottom. Our problem is to determine how much time is required for all of the water to leak out. We know the following: the shape and dimensions of the container, the initial height and (therefore) volume of water in the container, and the diameter of the hole. We also note the common sense fact that immediately after puncturing a container of any fluid, the flow rate is initially high, but gradually decreases as the volume of fluid decreases.

We need a strategy to solve this problem. A little reflection will convince you that this is a rate problem. It is analogous to, but more complicated than, a problem such as: "How long will it take to travel 300 miles if we are moving at a constant speed of 30 miles per hour?" Fluid volume is analogous to the number of miles still to travel and flow rate is similar to travel speed. The main difference is that the fluid flow rate, unlike travel speed in our analogy, is not constant. Indeed, flow rate decreases with the volume of the fluid; it's as if our vehicle gradually slowed down as it neared its destination. Vehicles, fortunately, don't behave this way in real life, but liquids do. So, how to solve it? We cannot simply divide the volume by the initial flow, as we can with the travel problem. But, if we know the initial flow and how the flow depends on the volume, we can calculate the new volume after some short period of time, for example, 1 sec. With this new volume, we can calculate a new flow for the next interval of time and then another new volume. We can repeat this process until there is no more fluid in the container. We embody this strategy in a *recursive* equation for the volume of fluid (V):

$$V_{t+1} = V_t - \text{flow}_t, \tag{1.1}$$

where flow_t is the flow at time t over a fixed interval of time. When $t = 0$, the container is full and this is the starting condition (V_0). Applying the equation once gives us the new volume at $t = 1$. Applying the equation again using the new value of V_1 on the right-hand side of Eq. 1.1 gives us the volume at $t = 2$, and so on for as long as we wish.

The only remaining problem is to write an equation for the flow rate that portrays the effect of the current volume. There are many forms we could consider, but the theory of fluid dynamics helps out here in the form of *Torricelli's Law.* This law, derived from first principles of physics (Bernoulli's equation) and some assumptions about the fluid, relates the velocity of water at the hole (v) to the height of fluid above the hole (h_t):

$$v_t = \sqrt{2gh_t}, \tag{1.2}$$

where g is the acceleration due to gravity of a body near the earth's surface (980 cm/sec^2). Since v_t is a speed (length units over time units) and not a flow, we calculate the volume leaving in a unit period of time by multiplying the speed times the area of the hole:

$$\text{flow}_t = v_t \pi r^2, \tag{1.3}$$

where r is the radius of the hole.

Our Eq. 1.1 is now almost complete except for one minor problem. Equation 1.3 depends on the height of fluid above the hole and Eq. 1.1 tells us what the current volume is. To make the two equations compatible, we rewrite h_t in Eq. 1.2 in terms of the volume of fluid in the container. Assuming that the cross-section of the container is a square with sides of length b, the complete recursion equation is

$$V_{t+1} = V_t - \pi r^2 \sqrt{2g\frac{V_t}{b^2}}. \tag{1.4}$$

This equation constitutes the mathematical model of the system. It describes the behavior of the objects. In this simple problem, there is only a single object: the container of fluid. We describe it in terms of the amount of the fluid in the container. The environment of the object is everything other than the fluid in the container. Equation 1.4 has one dynamic variable (V) and three parameters that must be estimated: r, b, and g. The latter is a physical constant, so we need not repeat the measurements done by others. The other two parameters are simple length measurements that we obtain with a ruler.

Our use of Torricelli's Law in this situation has a number of problems that would be valuable to research and discuss, but we leave it as an exercise. Instead, we ask: How well does this model describe the physical system? The obvious approach to this question is to perform an experiment

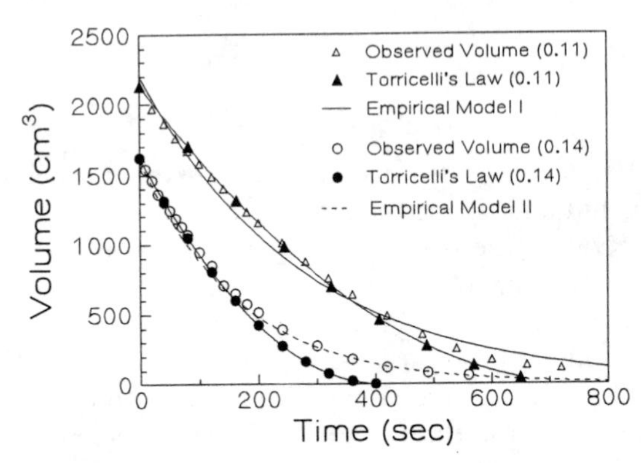

Figure 1.3: Comparison of predictions based on Torricelli's Law and an empirical model with observations of two containers with holes of radius 0.11 cm and 0.14 cm.

and compare the model output (its response) with observations. Figure 1.3 shows the dynamics of measured water volume in two experimental containers (plastic 1 gallon milk containers) with holes of different sizes. The observed values are plotted with model predictions based on the parameter values $g = 980$ cm/sec^2, $r_s = 0.11$ cm and $r_l = 0.14$ cm, and $b_s = 13.9$ cm and $b_l = 14.1$ cm.

There are clear discrepancies between Torricelli's Law and real leaky containers (Fig. 1.3). Part of the modeling process is to determine if some other model might do better. Since Torricelli's Law is based on solid physical principles, but derived with special assumptions, a natural alternative is a model that ignores physical principles and simply attempts to represent empirically the dynamics. One possibility is a model that hypothesizes that flow is a constant proportion of the current volume

$$V_{t+1} = V_t - fV_t, \tag{1.5}$$

where f is the fraction lost in each time step and is empirically determined for each hole radius. Since this model is simple, it has a mathematical solution based on the exponential function:

$$V_t = V_0 e^{-ft}$$

where V_0 is the estimated initial volume and f is an estimated, arbitrary shape parameter that summarizes the effect of holes of different sizes (small and large). For empirically fit f_s and f_l, the dynamics of this alternative model are plotted in Fig. 1.3.

The leaky bucket system is one in which there is a great deal of understanding and good data. We could probably use a model for product design if, for example, we wished to construct a device that emptied in a certain, specified time (a control problem).

This simple example illustrates most of the important concepts and problems of modeling biological systems that we address in this book. It shows the forms of some mathematical models for a given phenomenon. It also shows that models can be made to predict the same kind of data that experimentalists can observe and that these predictions will not match the data exactly. The discrepancy raises the problem of evaluating the relative quality of models. In this regard, we must answer such questions as: Which model is the best? Why did the model based on first principles perform poorly, particularly at low volumes? Could the theory be correct, but the data in error? What kind of error is made in the measurements? Of all these sources of error, which one is the most important? Which component of error should be studied most carefully in the future? We will discuss, in later chapters, tools for addressing these kinds of questions.

1.4 Classifications of Models

1.4.1 Forms of Models

Not all scientific models are precise, numerical, or quantitative. There are four forms:

1. *Conceptual* or *Verbal* – descriptions in a natural language.
2. *Diagrammatic* – graphical representations of the objects and relations (e.g., ecological "box-and-arrow" diagrams of energy flow, physiological diagrams of metabolic pathways such as the Krebs cycle).
3. *Physical* – a real, physical mock-up of a real system or object (either larger or smaller: a "tinker-toy" model of DNA or a scale model of an airplane for a wind tunnel).
4. *Formal* – mathematical (usually using algebraic or differential equations).

Our primary interest here will be in (2) and (4).

1.4.2 Mathematical Classification

The mathematical equation used to describe water flow (Eq. 1.4) is known as a recursive finite-difference equation. It is only one form that a model could take. To show the scope of the range of mathematical models that are potentially applicable to biological systems, we construct a simple classification of mathematical models. The basis of the classification is whether the mathematics incorporates (or not) a particular mathematical structure. In some cases, it is a matter of opinion whether the mathematics displays the character or not.

I. **Does the mathematics have an explicit representation of mechanistic processes?**
YES: *Process-oriented* or *mechanistic* models (e.g., Torricelli's Law, hydrology models using Newtonian physics and chemistry, or population dynamics models with details of reproductive physiology).

NO: *Descriptive* or *phenomenological* models (e.g., empirical leaky bucket model, Boyle's law relating temperature, pressure, and volume, or a density-independent population dynamics model with reproduction represented as a single parameter).

II. **Does the mathematics have an explicit representation of future system states or conditions?**
YES: *Dynamic* models (e.g., leaky bucket models).
NO: *Static* models (e.g., linear regression equation relating variables x and y).

III. **Does the mathematics represent time continuously?**
YES: *Continuous* models, time may take on any values (e.g., 3.3 sec).
NO: *Discrete* models, time is an integer only.

IV. **Does the mathematics have an explicit representation of space?**
YES: *Spatially heterogeneous* models (e.g., objects have a position in space, or occupy a finite region of space).
   A. Discrete: space is represented as cells or blocks, and each cell is represented as spatially homogeneous.
   B. Continuous: every point in space is different (e.g., diffusion equations in physics).

NO: *Spatially homogeneous* models (e.g., simple equations of population dynamics or enzyme kinetics, Torricelli's Law).

V. **Does the model allow random events?**
YES: *Stochastic* models (e.g., random temperature values may produce random changes in the intrinsic rate of increase in population dynamics models: $X_t = X_0 e^{r(N(0,1))t}$, where X is population size and r is rate of increase, which varies in time and is drawn from a normal distribution with mean 0.0 and variance 1.0 [$N(0, 1)$].
NO: *Deterministic* models (i.e., all parameters constant).

1.4.3 System Concept Classification

Based on the above classification, the model of the leaky container (Eq. 1.4) is a *deterministic, spatially homogeneous, discrete time, mechanistic, dynamic* model. This model is also an example of **compartment** models: models that describe the flow of physical material (e.g., water) between physical or biological storage compartments. While this is a very general conceptualization that applies to many biological modeling problems, there are many other biological applications for which differential or finite difference equations and compartment models are not the best representation.

There are three other broad classes of models that are appropriate to biological systems and to which the above classification also applies reasonably well. **Transport** models are those that transport material, energy, or momentum from point to point in physical space. They are similar to compartment models but use special mathematical structures (partial differential equations) and conservation principles. **Particle** models are those that follow the fate of individual particles moving in space (e.g., individual blood cells flowing through veins) or they may be individual organisms changing their condition (e.g., body size). **Finite state automata** are models that represent an object as being in only a few, finite number of *states* or conditions. For example, we might model weather dynamics as a system that has only *good*, *bad*, or *intermediate* weather quality. This is different from compartment models such as the leaky bucket, where the container could have any volume of fluid.

So, compartment models and differential or finite difference equations are not always appropriate, depending on our conceptualization of the system. Conversely, in other biological systems, differential equations may be a felicitous description, but the system should not be thought of as flows between compartments (e.g., movement of individual organisms over continuous two-dimensional space). The system conceptualizations mentioned are not mutually exclusive; a given model can contain elements of several or all of them. For example, a transport model of a pollutant in a river can contain a compartment model of the effects of the substance on the biota in the river. These distinctions will be made clearer when we present models based on alternative representations in later chapters.

1.5 Constraints on Model Structure

Models are used for many purposes, and the purpose influences the degree of system detail that is represented by the mathematics. For example, it may not be necessary for our purposes to provide an explicit spatial component in the model. In this case, a spatially homogeneous model suffices. Moreover, as we provide greater detail, the number of systems to which our model applies will decrease. For example, in a physiological model of blood flow, if we include a "gizzard" as one of the objects (compartments), then we have restricted the model to birds and it will not apply to mammals.

Levins (1966) has synthesized these trade-offs by identifying three properties of all models. No model can maximize all three simultaneously (but see Orzack and Sober 1993).

1. *Realism:* the degree to which model *structure* mimics the real world. In formal models that are realistic, the equations are correct, not just the model output. In physical models (e.g., a scale airplane) maximal physical detail is present (i.e., every rivet).
2. *Precision:* the accuracy of the model predictions (output). In precise models, the air flow around the scale model is exactly the same as that around the full-size plane. *Precision* is not used here in the statistical sense, which refers to the degree of variability of a set of measurements.

3. *Generality:* the number of systems and situations to which the model correctly applies. In physical models, a general scale airplane model applies to both a Piper Cub (small, single-engine aircraft) as well as a Boeing 747 (large, multiple-jet engine aircraft).

Each of these properties trades off against the other two. If a model contains great realism, it cannot also possess great generality, except at a level of description that is very imprecise. Since no model can simultaneously maximize all three, the uses to which the model is to be put will influence which is sacrificed to increase the other two. Prediction needs little generality, but great precision and (to a lesser extent) reality. Understanding implies the need for great generality and (to a lesser extent) reality, but precision is not necessarily important. Control needs great reality, but lesser amounts of precision (corrections can be made frequently) and even less generality. This conceptualization of models has recently been challenged; see the Exercises.

1.6 Some Terminology

In the chapters to follow, we will use a number of terms that need definition here (Table 1.1). Not all modelers will agree with these definitions, but they will help you read this book. Some of these terms will not be understandable as you read through the first time, but I hope their meaning will become clear as you learn more.

1.7 Misuses of Models: The Dark Side

When you have a hammer, you look for a nail.
When you have a good hammer, everything looks like a nail.

— Anonymous

A model, like a hammer, is a tool to solve a problem. It is possible to use a good hammer to insert a screw, but it isn't recommended. In the same way, a model may be inappropriately applied to a given system. Unfortunately, as the parable of the blind men illustrated, we often do not know if we have a nail or a screw. Inappropriate application of a model is pernicious in any form of model, but is especially misleading in quantitative models such as we will discuss, since the output of the models are numbers which often acquire a reality of their own. It is difficult to identify the source of the errors in these models.

There are many ways that models may be misapplied, but an important one is the application of quantitative models to areas of study in which there is great uncertainty in the data or to the degree that the underlying mechanisms are understood. Both Holling (1978) and Karplus (1977) have discussed this problem, and we synthesize their insights in Fig. 1.4. Holling noted that different scientific disciplines could be generally characterized

Table 1.1: A few more terms.

analytical model	*(n)* a mathematical model whose solution is not obtained by simulation or numerical approximation, but by purely mathematical argument or a model where mathematical properties (e.g., stability of equilibria) are achieved by mathematical argument
dynamic model	*(n)* a mathematical model that describes the changes over time of quantities representing the system objects (e.g., population sizes)
mathematical model	*(n)* a set of mathematical equations that describe a system
mathematical modeling	*(v)* the human activity of creating a set of mathematical equations that describe a system
model	(a) *(n)* a description of a system, (b) *(v)* the human activity of creating a description of a system
objective	*(n)* (a) the purpose for doing something, a goal, (b) a verbal statement that guides and constrains modeling, (c) a list including at least some of the following: objects and relations modeled, environment of the system modeled (influencing variables, objects not modeled), length of time that the model applies to the system, spatial and temporal scales of resolution, questions addressed of the model
simulate	*(v)* (a) to produce a solution to a simulation model, (b) to model
simulation	*(n)* (a) a set of one or more numbers that together constitute a numerical solution to a simulation model, (b) one run of a computer program that numerically solves a simulation model
simulation model	*(n)* a mathematical model whose solution is obtained by numerical approximation, usually involving computers; not an analytical model
solution	*(n)* (a) an answer to a problem, (b) a set of numbers whose values satisfy a mathematical equation (e.g., the roots to a polynomial equation)
system	*(n)* a collection of objects and relations between objects
system state	*(n)* the set of particular, numerical values of all system objects at a given time (e.g., grams carbon in all species in an ecosystem)
well-defined system	*(n)* the smallest set of objects and relations whose states (values) cannot be proved to be unnecessary to achieve the objectives of the model

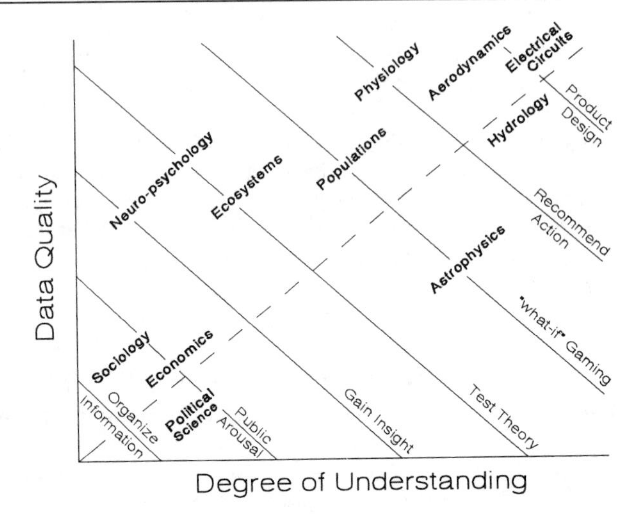

Figure 1.4: Appropriate uses of models related to degree of uncertainty in systems. Contour lines represent combinations of data quality and degree of understanding in disciplines for which models may be used as indicated on the lines. The dotted line is a continuum of uncertainty (lower left: much, upper right: little).

by two numbers: the precision and accuracy of the data upon which the discipline is based and the degree of mechanistic understanding. No doubt, these axes are not completely independent. There are not many sciences in which we have great understanding of the mechanisms, but very poor data, since usually we require good data in order to elucidate mechanisms. This scheme cannot be pushed too far for it is only intended to be a qualitative model.

Karplus (1977) viewed disciplines similarly but positioned them along a continuum from "black boxes" (poor data and shallow understanding) to "white boxes" (good data and deep understanding). This corresponds to a line in Holling's space from the origin to the upper right corner. Karplus went further and identified specific disciplines along this continuum. We can subjectively position some of these according to whether their place in the continuum is due to data quality or degree of understanding. Those disciplines that are black boxes should not use models for detailed, quantitative predictions, while white box disciplines can use models to design saleable products (e.g., electronic components, airplanes). Complete black box sciences should, at best, use models only to arouse public opinion. A notorious example is Jay Forrester's World Dynamics model which simulated the world's economic, social, political, and environmental systems in rather general terms and predicted a major population crash at about 2050 (Forrester 1971). This model was intended not to make accurate predictions, but to bring to the public's attention the need for better planning, particularly in the area of birth control. I have represented this qualitative assessment of model use in Fig. 1.4 by contour lines. The labels for model use do not apply to all disciplines. For example, it is hard to imagine what actions astrophysicists might recommend, much less the products they might design – but, then, one never knows.

The point of Fig. 1.4 is that discipline maturity dictates appropriate uses of the models. To use a model for a more rigorous purpose than appropriate for a discipline can be misleading, at best, or dangerous, at worst.

1.8 Exercises

1. Using your own discipline (e.g., ecology, biochemistry, natural resources, agroecosystem), draw a figure analogous to Fig. 1.4. Discuss cases where and how mathematical models may be misused.
2. Suppose a model attempted to integrate concepts and information from political, social, economic, and ecological systems. Would this model be more, or less, accurate as a single model in any one of the separate disciplines?
3. Based on the definition of a *well-defined system* in Table 1.1, what is an *ill-defined system* and why might it be undesirable?
4. In Fig. 1.3, the predictions of two models (Torricelli's Law and the exponential model) are compared to the data. Which is the "better" model? Why? Why does Torricelli's model deviate when the volumes are low?
5. Repeat the experiments for the leaky container with your own equipment. What are some of the problems in obtaining the data to test models? Were your measurements exact? How can you incorporate replicates?
6. Recently, S.H. Orzack and E. Sober have challenged Levins' trichotomy between model realism, precision, and generality. Read and discuss the original articles by Levins (1966), Orzack and Sober (1993), and the reply by Levins (1993). Specifically, do you agree with Orzack and Sober that the distinctions have no merit and that model robustness bears no relationship to model validity? Is Levins' reply that models are "relativistic" and must be evaluated in terms of their context relevant?

Chapter 2

The Modeling Process

2.1 Models Are Problems

WHEN we embark on a modeling project, we immediately have a problem. We want something that we don't have: a model. The *modeling process* is a semiformal set of rules that guides us through a solution to this problem. The rules are not mechanical instructions, not like a set of computer instructions we can step through one at a time and be guaranteed of arriving at the correct answer at the end. Modeling is real-world problem solving; it's hard and fraught with many opportunities for failure (or, if you're an optimist, opportunities for new insights). So, it is useful to begin by noting George Polya's four steps to solving mathematical problems (Polya 1973). Associated with each step is a question that we must answer. (1) *Understand* the problem (i.e., What is the *question*?) (2) *Devise a plan* for solving the problem (i.e., *How* do we solve it?) (3) *Execute* the plan (i.e., What is *an* answer?) (4) *Check* the correctness of the answer (i.e., Was it *right*?).

Certainly, these instructions are very general, perhaps only heuristically plausible, but they work on all problems, including the problem of producing a model. In this and the remaining chapters, we will see some more specific rules and tools that work in the more restricted domain of mathematical and computer models of biological systems.

As a problem to solve, then, the modeling process consists of the steps we take to *produce* a model, *implement* it in some formal language, *derive consequences* (predictions) from the model, and *evaluate* these in light of the use to which the model is to be put. Since the statement of the model inevitably requires making assumptions, comparing model consequences with observations is a major test of the adequacy of the assumptions to "explain" the observations. In its broadest form, then, modeling is the hypothetico-deductive approach to science and *vice versa* (Nagel 1961, Romesburg 1981). Here, we will describe this process in a way that emphasizes several important quantitative and computational procedures that are relevant to computer simulation.

2.2 Two Alternative Approaches

The classical description of the modeling process is shown in Fig. 2.1. This basic approach is presented in many texts (Shannon 1975, Spriet and Vansteenkiste 1982, Grant 1986). Its essential feature is that models should be constructed one at a time, and the quality of each is evaluated sequentially. Another model is not constructed until the current model is shown to be inadequate. For many biological systems, this is an appropriate methodology, but for others, a slightly modified view of this modeling process will be effective.

2.2.1 The Classical View

Objectives The beginning of the process is a statement of the objectives or purposes of the model. At this stage, we demonstrate our understanding of the problem (Green 1979). If we cannot give a clear statement of the reasons for building a model, then we do not understand the problem. If we do not understand the problem, then we are unlikely to discover the solution. Consequently, substantial detail should be provided in the statement of the objectives to answer the following questions:

- What is the system to be modeled?
- What are the major questions to be addressed by the model? (How will the model be applied?)
- What is the *stopping rule* for the modeling activity? (How good must the model be? To what will it be compared?)
- How will the model output be analyzed, summarized, and used?

Because of the importance of a clear statement of objectives, we will discuss this aspect of modeling in more detail later in this chapter. Here we note that the objective statement is a document that defines the reasons for producing the model in the first place. In cases of large, complicated modeling projects, it can ensure that the goal is well defined and achievable. Even when exploring theoretical concepts with small models, by answering the four questions above, the theoretician is forced to evaluate the scope and importance of the original questions.

Hypotheses The second stage is to translate the objectives and current knowledge of the system into a list of specific hypotheses. These are usually verbal statements. For example, a simple idea in population ecology is that crowding increases as numbers of individuals in the population increase and this, in turn, reduces the reproductive capacity of females. This can be qualitatively stated as: "increasing density decreases per capita

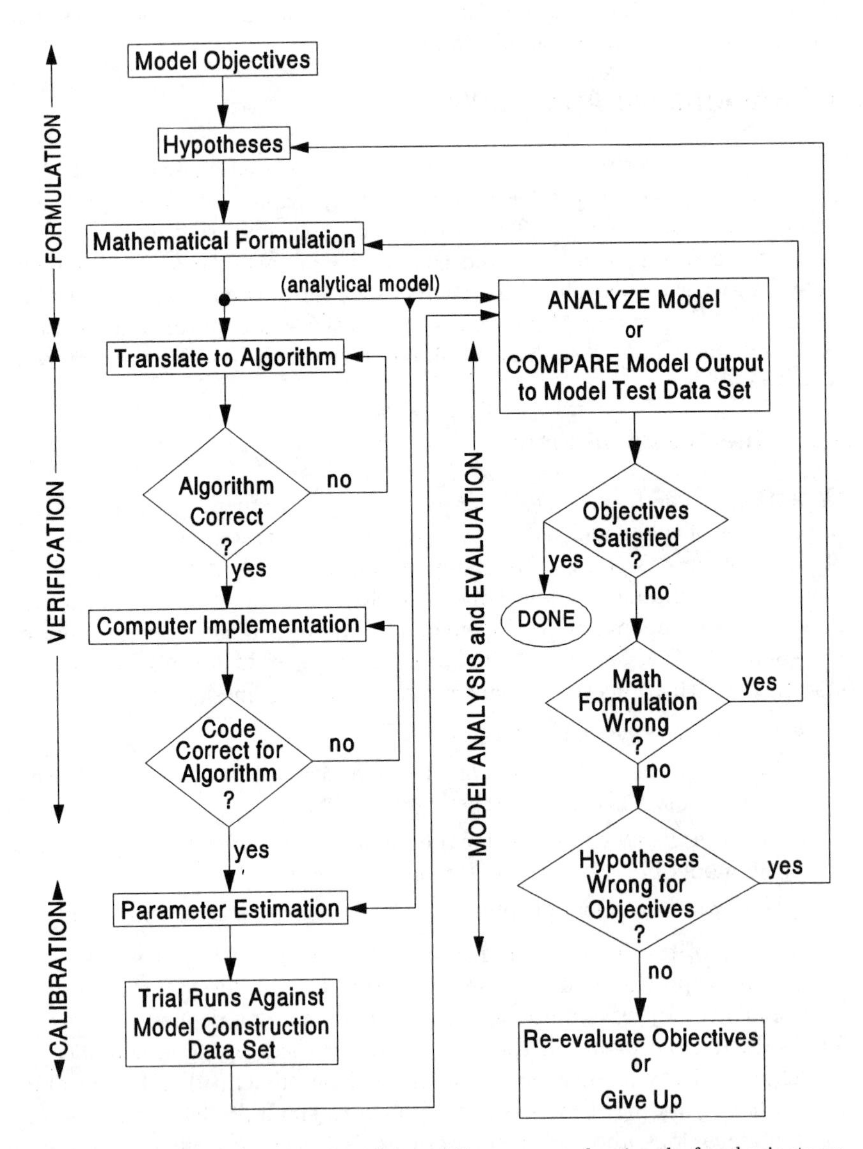

Figure 2.1: The classical approach to the modeling process, showing the four basic stages. In this approach, alternative models are developed sequentially, conditional on the failure of a previous model.

growth rates." Hypotheses may also use more quantitative relationships. For example, in simple models of blood circulation, the heart chambers expand as they fill with blood, but the rate of expansion decreases at large volumes because heart wall elasticity is limited. More quantitatively, we can say that the degree that chamber volume increases with a unit increase in blood volume decreases linearly as total volume increases. At this stage, we can also describe the complete model qualitatively with a graphical formalism that pictorially shows the objects modeled and their relations (e.g., flow of blood between organs). However it is accomplished, the function of this stage of modeling is to identify more fully the set of objects in the system and to bound the set of relations that connect the objects.

Mathematical Formulation Qualitative hypotheses must be converted into specific, quantitative relations that can be formulated with mathematical equations. In the third stage, the actual equations are defined. This corresponds to Polya's stage to *devise a plan* for solving the problem. This step uses the initial physical, chemical, and biological information available for model construction to derive and check the correctness of the equations we hope will describe the dynamic behavior of system objects. For many beginners, this is the most difficult and frustrating stage. It requires a certain level of mathematical sophistication, but more importantly, it requires that vague concepts and loose relations be made definite in the cold light of mathematics.

Verification Many mathematical models cannot be solved analytically, but can only be solved approximately using numerical techniques. Today, this means solving the equations using a digital computer. The fourth stage is a set of activities in which the equations are translated into computer code. At this stage, it is necessary to *verify* that the computer algorithms and code are correct for the mathematical relationships defined. Modeling projects that do not require numerical solution of the equations will replace this step with mathematical verification activities performed during the formulation stage. For example, in Chapter 1 we solved the leaky bucket model by a recursive equation. As we will see in Chapter 4, we could (and possibly should) have written the model as a differential equation. There are numerous numerical techniques for solving these equations (e.g., Runge-Kutta), and, depending on the nature of the equation, some methods are inappropriate. Thus, the choice of algorithm is important and can influence the predictions of the model. Similarly, for any algorithm, there are many different ways to write the computer code; some of these will be wrong. Models of biological systems can easily involve scores of dynamic variables and hundreds of parameters. This is especially common in models with explicit spatial processes. In writing a computer program to solve the equations, it is a nontrivial exercise to demonstrate that the computer output is correct. This is a concern of software engineering, and there are some basic programming procedures that can help in this regard (e.g., object-oriented programming).

Calibration After the model is correctly implemented on a computer, output can be produced. But before simulations can be performed, numerical values for the initial conditions (e.g., the starting volume of water in the leaky bucket) and constants in the equations must be specified. Calibration is the set of activities by which this is done; the basic problem involved is parameter estimation. Usually, this involves defining relations between observed quantities and the parameters so that statistical methods (e.g., linear regression) can be applied to produce the best estimates for the parameters (e.g., the slope and intercept of a straight line). These relations may require that specific laboratory experiments be performed. For example, in physiological models, one may wish to estimate the parameters for the quantitative effects of temperature on oxygen production in leaves. Laboratory measurements of oxygen at defined, controlled temperatures provide the necessary data. Often experiments cannot be performed, but uncontrolled observations over time are available (e.g., in ecological succession: plant biomass over several years). If this variable is an output of the model, some parameters can be estimated by curve fitting wherein the model is run repeatedly using different parameter values and compared to the same dynamic data set until a satisfactory fit is obtained. This stage is discussed in more detail in Chapter 7.

Analysis and Evaluation Once the model is calibrated, we can use it to produce the answer that our objectives specified. This corresponds to Polya's *execution* of the plan. For numerical models, this involves running a computer program and recording the numbers produced. This is primarily a mechanical exercise that can be automated to a great extent. For analytical models, execution may range from simple computations to complicated mathematical argument and theorem proving. This latter activity can require substantial creativity and may be the most difficult step in the process.

For both numerical and analytical models, the answer should be evaluated for its quality according to the objectives. It should be *checked* (Polya 1973) in some way. Often in purely theoretical studies where the primary objective is to "understand" the system, this involves, at most, only a qualitative comparison of model output and data. For example, in a theoretical plant succession model we may be satisfied if the model shows an initial increase in plant biomass followed by a decline, if this were the observed pattern. Ideally, however, we also desire models that are quantitatively correct as well. To establish this for a particular model, we need to *validate* (or *corroborate*) the model against independent data sets.

We have already noted the similarities between modeling and the hypothetico-deductive approach to scientific investigation. A component of this method is the doctrine of *falsification* (Popper 1968), which states that hypotheses cannot be proved, but only disproved (i.e., falsified). The same framework applies to models, since they are basically collections of hypotheses. Many modelers (e.g., Holling 1978, Hall and DeAngelis 1985) have adopted this view to the point of stating that the objective is to invalidate the model, that is, discover evidence that contradicts it, not evidence that supports it. There is much philosophical and logical weight behind this view; nevertheless, there is also a real psychological need to be able to point to a model, theory, or body of experiments and say: "We believe this is the way it is." On the one hand, logic permits only falsification; on the other, we desire positive statements that summarize our beliefs, if only at a moment in time. We need an approach that synthesizes these two different approaches. A candidate is proposed below that developments and tests *multiple working hypotheses* as well as the *resultant alternative models.*

If the model passes the validation criteria specified in the objectives, the project is complete. If it fails, then errors were made earlier in the modeling process and the hypotheses and/or mathematical formulations need to be revised. The entire process is repeated. Finally, depending on the objectives, further analyses of the model through computer simulation or mathematical analyses are performed. These topics are discussed in Chapters 8 and 9.

2.2.2 Problems with the Classical View

Many statisticians believe that for statistically rigorous hypothesis testing to occur, prior knowledge should not influence the test. (But the Bayesian school of statistical analysis disagrees, and this will be discussed in Chapter 8.) Therefore, sequential passes through the modeling process must use new data for validation. If only one independent data set is available, subsequent comparisons are only exercises in *curve fitting*, since the modeler has become familiar with the validation data during the development of the second and subsequent models. Thus, the major problem with the classical approach is that independent data sets necessary for validation are often difficult or expensive to obtain. A modification of the classical approach, based on multiple hypotheses and models, avoids this problem.

Multiple or alternative models are valuable for another reason. When we are uncertain about the correct equations to use (which we usually are), there is a danger that when we derive a model that we cannot reject, we will believe that this is a correct description. In fact, there may be many other models that would be equally likely to be validated as the one we chose. If we never create these models and their predictions, then we will never know if the original model was unique in its accuracy.

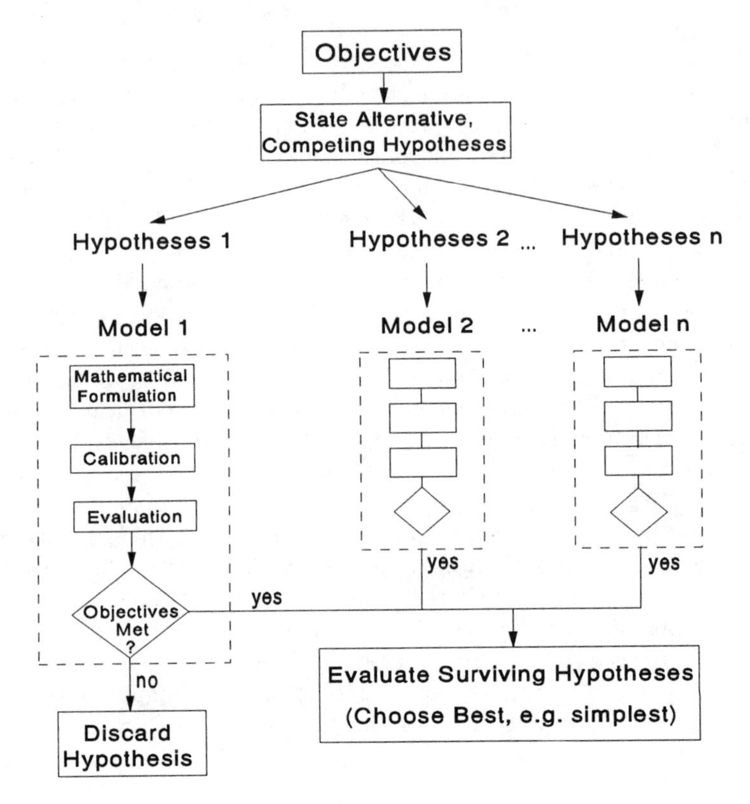

Figure 2.2: Another view of the modeling process, in which alternative hypotheses and models are developed and tested independently.

2.2.3 Multiple Working Hypotheses

> *A man who does not know one answer from another is as ignorant about the question as he can possibly be. The only state of greater ignorance is not to know the question.*
>
> — Tribus and McIrvine (1970)

An alternative to the sequential approach is a parallel approach that involves implementing and evaluating several different competing hypotheses and models simultaneously (Caswell 1976). This approach is diagrammed in Fig. 2.2. It is based on the ideas of statistical alternative hypotheses. Platt (1964) refers to these *multiple working hypotheses* as a component of *strong inference* and emphasizes the latter's value to incisive scientific analysis in all its forms (not just to modeling). Holling (1978) and his colleagues (e.g., Walters 1986) have also shown the practical wisdom of using this approach in developing models to assist the management of renewable resources.

Using this approach, we formulate several hypotheses and models each with separate computer implementation, verification, and calibration stages.

Every model is compared simultaneously (*in parallel*) to all of the validation data that are independent of data used to construct the model. The resulting comparisons are then independent and any models that survive the comparisons can be evaluated further with other quality criteria. A common auxiliary criterion is simplicity, which is the basis for the Principle of Parsimony or Occam's Razor. This approach presupposes that we can uniquely rank models from simplest to most complex, and this is not always so. Another criterion is the likelihood that one of the models is true; we will discuss this possibility in Chapter 8.

An example may make this clearer. Many species of seed-harvesting ants will exhibit mass recruitment of large numbers of foragers to rich resources (e.g., large insects or patches of seeds). Under other circumstances, ants forage individually, ignoring other ants and responding only to their local environment. The precise mechanisms required for these ants to perform these actions have not been determined, although experimental evidence indicates that they lay chemical trails and can remember previous successful foraging areas.

Suppose we wish to use a simulation model to explore the consequences of the foraging behavior of individual ants for seed consumption rates over the entire ant colony in order to evaluate the relative importance of different mechanisms. We can identify a number of possible candidates (Fig. 2.3; Haefner and Crist 1994): Random Walk (individual ants walk randomly and independently of other ants), Memory Only (individual ants remember previous successes but do not lay a pheromone trail), Pheromone Trail (ants lay a pheromone trail from the resource but do not use memory), Memory + Pheromone (ants use memory and pheromone trails), Omniscient (ants know the location of all the seeds). The first model serves as a null or random hypothesis in which no significant biological or social behavior is present. The last model represents a "super" ant and (presumably) defines the maximum rate of seed return to the nest. Together these models constitute a continuum of "ant intelligence." Since we can easily measure the colony's seed return rate in the field , the purpose of examining such a range of models is to determine where, along the continuum of models, the truth (i.e., real ants) lies. This addresses the question: "How smart does an ant have to be to forage in the way we observe?" We cannot definitively answer such questions with simulation models, but we can identify classes of models and hypotheses that are inadequate.

An important feature of this example, and one that should be used whenever possible, is the construction of a *base* model that incorporates as little of the biology as possible and yet still produces output that can be compared to observations. In this example, the base model eliminated all forms of communication between ants, but moved ants randomly so that they had the possibility of discovering seeds. Thus, the two extreme models, random and omniscient, bound the range of possible explanations.

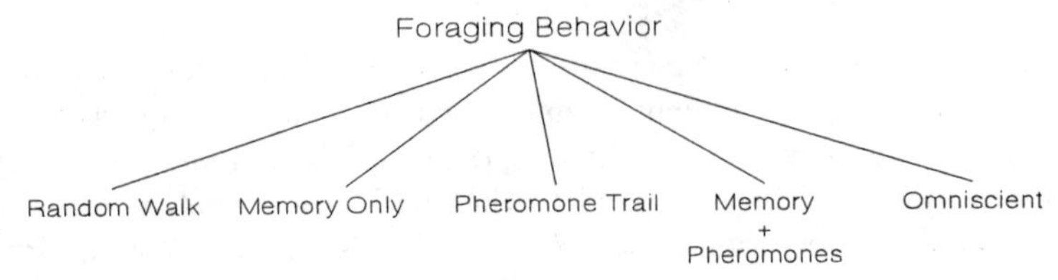

Figure 2.3: A family of competing hypotheses on the mechanisms used by ants to find seeds and recruit nestmates.

The base model concept is similar to a null or neutral model (Caswell 1975). The value of including these models is that they are simple explanations. However, we should not stop with these; as Albert Einstein is credited with saying: "a theory should be as simple as necessary, but no simpler." Or, to put it another way, simple models are good, but getting the right answer for the right reason is also good.

2.3 An Example: Population Doubling Time

We now summarize the idea of the modeling process applied to alternative models with a quantitative example. Suppose we wish to answer the question: When will the world's population double its current numbers? We identify the following objective statement.

Objective: Construct a description of the dynamics of the world's population such that the time when the population size is twice its starting value can be computed.

The above statement has the following desirable properties of an objective statement: (1) It defines the system of interest as the world's population without mention of spatial heterogeneity. (2) It defines the purpose of the model: determine the doubling time. (3) It indirectly identifies the analysis of the output to be used: a computation of the time at which the population is twice the initial condition. A major deficiency of the objective statement is that it does not mention validation criteria. We cannot tell from this statement when we should stop developing models.

To illustrate the idea of multiple working hypotheses, we will develop two models. One model assumes that per capita growth rate does not vary with increasing population size (density-independent growth) and the other assumes that the growth rate decreases linearly with population size (density-dependent growth). In addition to these assumptions, the two models share the following incomplete set of hypotheses.

1. Per capita growth rate is not influenced by any extrinsic variable (e.g., ozone, UV radiation, temperature).
2. The sex ratio is 1:1 (or we assume there is only a single sex).
3. There are no age differences among individuals (no age classes).

4. There are no geographical differences in growth rates (all countries and regions of the world are the same).

Our objective statement says that we intend to determine the doubling time by following the dynamics of the population. This suggests each of our mathematical models will implement the two hypotheses using equations that project population numbers forward in time. Recalling the Karplus (1977) **ESR** model of systems from Chapter 1 (Fig. 1.1), our problem is to write an equation for **S** that transforms the population numbers at time t into the population numbers at $t+1$. There are several kinds of mathematical equations we could use here, but for simplicity, we will use recursive finite difference equations (FDE), the same form of equation we used in the leaky bucket example of Chapter 1.

Our two hypotheses make two different assumptions: (1) the number of offspring produced per female (*per capita rate of increase*) is independent of (i.e., does not change with) the current numbers in the population, and (2) the per capita rate of increase decreases linearly with increasing numbers.

The FDE for the density-independent model (hypothesis 1) is familiar to ecology students:

$$N_{t+1} = N_t + rN_t, \tag{2.1}$$

where r is the intrinsic (or maximum) per capita growth rate of the population. In this model, the *per capita* rate of increase does not change with greater population numbers. [Note that the *absolute* rate of increase (rN_t) does change.] The per capita rate is constant and equals r, and the model asserts that the population increases each time step by a constant proportion (r) of the current population.

To *calibrate* this model we can solve the model for r:

$$r = \frac{N_{t+1} - N_t}{N_t}$$

and obtain population estimates over successive periods of time (N_0, N_1, $N_2, \ldots, N_t$) to compute r. These data would probably be taken from an historical data set, but could be from a field or laboratory experiment. To solve the equation and to predict numbers over time, we specify the numbers at time $t = 0$ (the initial conditions) and iterate Eq. 2.1 for $t = 0, 1, 2, \ldots, n$ time steps. This model produces the familiar exponential population increase over time (Fig. 2.4). Since the model output is population numbers over time, computing the doubling time is simply a matter of observing the time interval at which the predicted numbers are twice the initial numbers.

The alternative model is handled in a similar way. The equation is:

$$N_{t+1} = N_t + rN_t\left(1 - \frac{N_t}{K}\right), \tag{2.2}$$

where K is the carrying capacity of the population. The key aspect of

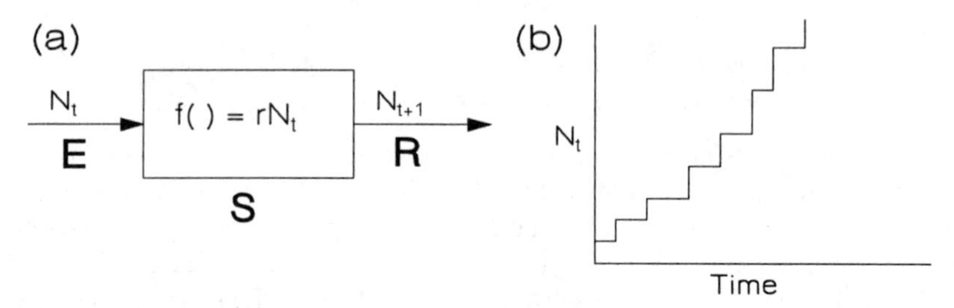

Figure 2.4: (a) The ESR scheme and (b) a typical dynamic trajectory for density-independent population growth using Eq. 2.1.

this equation is that the expression in parentheses depends on the current population numbers (N_t). This causes the numbers of offspring produced by each female to be reduced as population numbers increase. Although the mechanisms for this phenomenon are not described, they may be due to competition among females for food or child rearing costs. Notice that the relationship between population growth rate and this algebraic expression is similar to that between fluid flow rate and Torricelli's Law discussed in Chapter 1 (Eq. 1.4).

Equation 2.2 has two parameters that we calibrate by finding an expression involving r, K, and measureable quantities. Rearranging Eq. 2.2 to again form the realized per capita growth rate on the left-hand side yields:

$$\frac{N_{t+1} - N_t}{N_t} = r - \frac{r}{K} N_t.$$

This is a linear equation in which the left-hand side is the y-axis, or dependent variable, and N_t is the x-axis, or independent variable. We can use linear regression to obtain estimates of the intercept (r) and the slope ($\frac{r}{K}$) from which we can calculate K. The dynamics produced by this model are the classical sigmoidal or S-shaped curve of the *logistic* equation (Fig. 2.5). We will use the same approach to calculating the doubling time for this model as for the first model.

To this point, we have developed alternative hypotheses, their respective mathematical and computational formulations, and a strategy to answer the original question. The next step is to validate the models. Since the

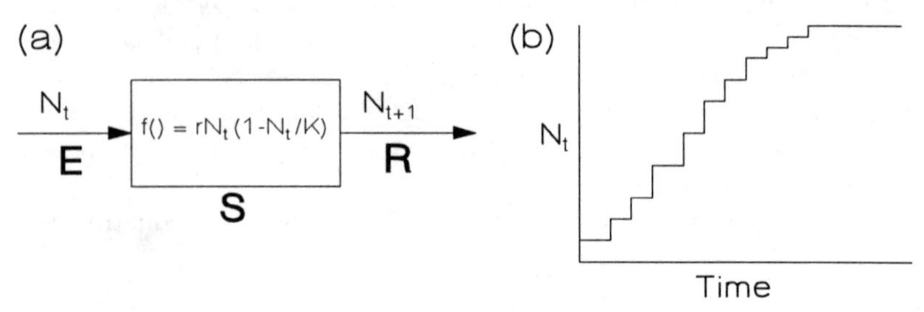

Figure 2.5: (a) The ESR scheme and (b) a typical dynamic trajectory for density-dependent population growth using Eq. 2.2.

model describes the world, we cannot realistically hope to find a similar, alternative system to study (not in this solar system, anyway). We might, however, validate the models by comparing each to an earlier historical record, one not used in the formulation of the model (e.g., from the period 1800–1850). This approach to validation makes some important assumptions about the nature of the system in the past and the present, but it is perhaps as good as we can expect when we cannot replicate the system.

After constructing both models and subjecting them to independent comparisons against the same data set, we may reach the conclusion that either none, one, or both of the models are inadequate to explain the data. Based on the results, we would choose between the two models, if possible (Walters 1986 and Chapter 8). Given that one or more of the models passed our validation test, we could then proceed to analyze the model by calculating the expected doubling time.

It is natural to ask: "Which alternative hypotheses should a modeling problem compare?" There is no general answer since it depends on the subdiscipline and the objectives of the model. Nevertheless, the two examples given (ants and populations) have something in common among their alternatives. Both examples have a *null* model: a model that hypothesizes that the observed dynamics are not caused by complicated biological processes. In the ant example, the simplest model was one in which seeds were returned to the nest as a result of random movement of individual ants: no interactions between ants were modeled. In the population model, the density-independent model assumed there were no interactions (e.g., competition) between individuals. In my usage here, a null model need not be completely random (e.g., density-independence), although we could have constructed such an alternative. So, it is a matter of degree how far removed from biology one wishes the alternatives to be, but at least one of the models should be as simple as possible; removing biological processes is one method of constructing simple models. As Richard Levins once said: "In order to understand complex systems, it is necessary to study something else instead" (Levins 1970). By this he meant not only models of the system, but also simple models. In the case of biological systems, this may mean models with little or no biological processes in them. The objectives statement should indicate the degree to which biological processes are to be removed from one of the alternatives.

The alternative modeling approach is not useful in all applications. In relatively mature disciplines such as physiology, in which either data quality is high or understanding is deep (Fig. 1.2), there will be less debate over the correct form of equations. At some point where a science matures from using models for "what-if gaming" to "recommending action," the equations become less debatable. In these systems, alternative models are less important. However, in the less mature disciplines such as ecology, especially where mechanisms are not understood, there is greater uncertainty, and the effects of using a particular set of equations need to be investigated with alternative models.

2.4 Model Objectives

Never weed your garden in the dark. — JWH

We have repeatedly referred to the objective statement and its role in constraining model structure. It is worthwhile to delve a little deeper into this concept and discuss the attributes of a good objective statement. A careful statement of the objectives of a model is important because it defines the problem to be solved and can, therefore, be used to devise the implementation and analysis of the model. The objective statement can also define the domain of applicability of the model. This latter use is important since it can reduce possible misuse of the model and help identify certain kinds of criticism as being directed not to the substance of the model, but to its objectives. These are and should always be held to be two different types of criticism. So, while model objectives do not always appear in print, they should be explicitly stated at some point.

Modelers do not agree on the content of a good statement of objectives, but Overton (1977) contains the most explicit rendition. He emphasizes that effective objectives are those that are stated as *goals* with *purposes.* For example, "Construct a model of photosynthesis [*goal*] to determine the effects of elevated UV light [*purpose*]." But beyond the purpose, an objective statement must provide the following information.

1. The objective *question(s).*
2. The *perturbations* and *stimuli* accommodated in the model.
3. The exact *system* and *environment* which the model addresses.
4. The temporal and spatial *scales* over which the system is to be described.
5. The temporal and spatial *scales of extrapolation and prediction.*
6. The *factual information* and *theoretical concepts* used in model construction (data, assumptions, sources, etc.).
7. The *criteria of validation* (empirical and theoretical).

To illustrate one of the best and most complete statements of model objectives, I give an extended quote from Innis (1978). The objective applies to a large, complex model, so this perhaps justifies the lengthy statement.

> The objective of this modeling activity was to develop a total-system model of the biomass dynamics for a grassland that, via parameter change, could be representative of the sites in the US/IBP [United States/International Biological Program] Grassland Biome network and with which there could be relatively easy interaction.

There are several key points in this objective that deserve elaboration. First, the term *total-system* model refers to the inclusion of abiotic, producer, consumer, decomposer, and nutrient subsystems. This requirement was imposed to assure that the modeling effort played the integrative role delegated to it ...

Second, *biomass dynamics* identifies our principal concern with carbon or energy flow through the system. Focus on biomass facilitated the comparison of model and data but turned out to be unfortunate because it is not conserved. The model, therefore, tracks carbon and converts it to biomass (and vice versa) in a number of places. We are concerned with dynamics as part of the general objective of the International Biological Program (IBP).

Third, *representative* expresses our desire to have the model apply, with minimal effort, to sites in the US/IBP Grassland Biome study. Changes of parameters are certainly necessary as these describe site characteristics (among other things). The representation was to depict "normal" dynamics as well as the response of the system to a variety of perturbations.

Finally, *relatively easy interaction* was a desideratum because of the role the effort was to play in program direction ...

This objective provides only the broadest guidelines to the modelers as to their respective functions. The purpose of the objective is to found the decision making processes that accompany model building. This involves clarification as to how many producers and consumers should be included, the amount of detail required in a representation of a producer, and whether a phosphorus, calcium, or lead model is required [i.e., resource management and research design]... In 1970 it was agreed that this objective would stand, with the first model addressing four specific questions:

1. What is the effect on net or gross primary production as the result of the following perturbations: (a) variations in the level and type of herbivory, (b) variations in temperature and precipitation or applied water, and (c) the addition of nitrogen or phosphorus?

2. How is the carrying capacity of a grassland affected by these perturbations?

3. Are the results of an appropriately driven model run consistent with field data taken in the Grassland Biome Program, and if not, why?

4. What are the changes in the composition of the producers as a result of these perturbations?

These questions were further specified with definitions of terms such as "variations," "level," and "type"; acceptance criteria were chosen.

This is a description of a whole ecosystem-level model, and it is quite possible that the reader will not appreciate the motives for or value of building these types of models. Nevertheless, it provides a reasonably clear statement of what the model is intended to do. Other disciplines may not require for publication such a self-conscious and direct statement, but, at some point, the modelers probably do.

2.5 Exercises

1. To what extent has Innis incorporated Overton's criteria for objectives statements?
2. How good was the objective statement of the "doubling time" model?
3. Using Innis' statement and Overton's criteria as guides, write an objective statement for the following problem: "How many cases of AIDS will occur in Utah in 1999?" Would the objectives change if the location had been San Francisco? Why or why not? What role does spatial scale of extrapolation play in this problem?
4. Write an objective statement for the leaky bucket problem of Chapter 1.
5. Write an objective statement for this problem: "What should be the best grazing pressure on the Foobar National Forest to simultaneously maximize cattle production and forest quality?"
6. We noted in the discussion of the model of the world's population that our abilities to validate the model were limited by our inability to replicate the system. Under what circumstances, if any, is it worth while to model systems that cannot be replicated or tested using rigorous statistical methods?
7. Read pages 10–13 in Reckhow and Chapra (1983b) and decide if there is a need to distinguish *validation* and *corroboration*.
8. Read an article in a current journal describing a model and critique the objective statement. In the models described in the chosen journal, how many discuss validation?

Chapter 3

Qualitative Model Formulation

3.1 How to Eat an Elephant

BUILDING a model is like eating an elephant: it's hard to know where to begin. As with almost all problems, it is helpful to break a big problem into smaller, more manageable pieces. We do this with model formulation (Fig. 2.1) by first creating a *qualitative* model and then converting this to a *quantitative* model (Chapter 4). Qualitative model formulation, then, is the conversion of an objective statement and a set of hypotheses and assumptions into an informal, conceptual model. This form does not contain explicit equations, but its purpose is to provide enough detail and structure so that a consistent set of equations can be written. The qualitative model does not uniquely determine the equations, but does indicate the minimal mathematical components needed. The purpose of a qualitative model is to provide a conceptual framework for the attainment of the objectives. The framework summarizes the modeler's current thinking concerning the number and identity of necessary system components (objects) and the relationships among them.

Qualitative model formulation is not always explicitly performed. If a modeling project is simple enough, elaborate plans for writing the equations are not necessary. Most of us do not need detailed instructions for getting out of bed in the morning. But with large models having many variables that interact in complicated ways among themselves and with the environment, it is easy to become confused. By providing an overview of the system, a qualitative version of the model can help reduce this confusion.

Qualitative models can take any form (except mathematical), but diagrams are the usual representation. Given our emphasis on differential equations and compartment models, three important diagrammatic schemes are: *block structure* diagrams (having origins in electrical engineering and analog computers), *Odum energy flow* diagrams (similar to block structure diagrams but based on energy flow within ecosystems), and *Forrester* diagrams (having origins in systems analysis and operations research). All

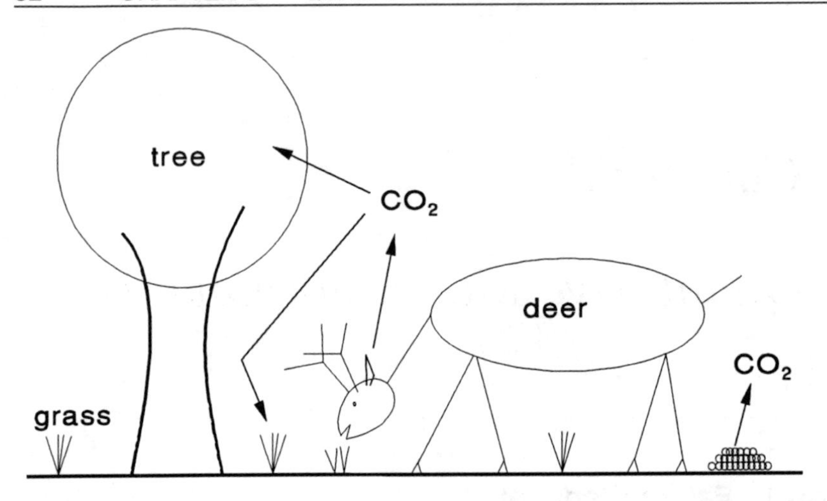

Figure 3.1: A simple ecosystem in which carbon moves among the labeled components.

three share the ability to represent systems as a set of objects and their interrelations. We will stress the latter here, but the interested reader can learn more of block structure diagrams in (Shannon 1975) and Odum energy diagrams in (Odum 1971).

3.2 Forrester Diagrams

Forrester diagrams (Forrester 1961) were invented by Jay Forrester, an MIT professor famous for work on early digital computer hardware and the simulation of social systems. Forrester diagrams are designed to represent any dynamic system in which a measurable quantity flows between system components.

Consider a simple ecosystem in which carbon flows between a population of grass and a population of deer (Fig. 3.1). Let us suppose that our objectives suggest that only deer and grass are interesting and that the grams of carbon in these two components are the relevant measures. Because of our simplification, we will not explicitly consider other components that may have large quantities of C (e.g., atmospheric CO_2 and excretion by deer). Consequently, two numbers (grams of carbon in grass and grams of carbon in deer) completely specify the condition of the system at a moment in time. By accepting this simple view of the ecosystem, we are stating that other variables or quantities are irrelevant and do not add to our knowledge of the system. For example, other consumers (e.g., insects), producers (e.g., the tree), or other variables (e.g., nitrogen) are not important. Moreover, these two numbers may change in time so that the condition of the system is dynamic. The exact nature of the temporal changes depends on the rates of flow of carbon into the grass component (growth) and into the deer population (grass consumption).

Figure 3.1 is a crude qualitative model in diagram form of the system, but since it makes specific reference to *deer* and *grass*, it has limited application to other systems. We want an abstraction of the basic concepts of *system components* and *material flows* to obtain a general tool for qualitative modeling of systems. Forrester diagrams are such an abstraction.

To understand the basis of the diagramming scheme, recall the general definition of a system: *a collection of objects and relations among them.* There are two kinds of objects: (1) those that are inside the system and are explicitly modeled and (2) those that are outside the system and are not modeled. The internal objects are called *state variables* and are those that, taken all together, characterize the condition or *state* of the system. In the example above, the state variables are *grass* and *deer.* These variables are dynamic and change their state over time. (See Caswell et al. 1972 for a more rigorous definition of state variable.)

The outside or external variables are either sources or sinks and are not modeled explicitly (i.e., no equations are written for these). For example, atmospheric CO_2 is both a source and a sink. It is a source because it represents an unmodeled pool of C that is an input to a state variable (grass). It is also a sink since gaseous CO_2 is a product of deer respiration.

Each state variable is described by its current level of the quantity of interest: the quantity in which units we measure the state of the variable (e.g., numbers of individuals, grams of carbon, temperature, etc). *Relations* between system objects have two forms: (1) the direction and rates of flow between the quantity of interest and the objects and (2) the influences of a variable (e.g., the quantity of interest) on the rates of flow.

Forrester diagrams are direct graphical representations of these concepts that permit easy translation to mathematical equations. They can be thought of as a graphical "language" with phrases that can be connected in certain prescribed ways. The graphical vocabulary items of the language are listed in Fig. 3.2 and are described below.

Objects System objects are the state variables of the system (called *levels* by Forrester). They are the primary system components whose values over time we wish to predict. They are dynamic quantities and are represented by a rectangular box (Fig. 3.2a). The box should contain a mnemonic name for the object and its unit of measurement. Many descriptions of models of this type refer to levels as *compartments*, and the type of models being represented by Forrester diagrams as *compartment models.*

Material Flows Flows are one manifestation of relations between system objects, which we will call a *flow relation.* A flow is represented as a solid arrow (Fig. 3.2b) and identifies the pathway over which the quantity of interest (e.g., grams of carbon) flows. In most models, the rate of flow is a dynamic quantity that is influenced by system components, and this rate is symbolized by a *control valve* (the "bow-tie") on the flow relation.

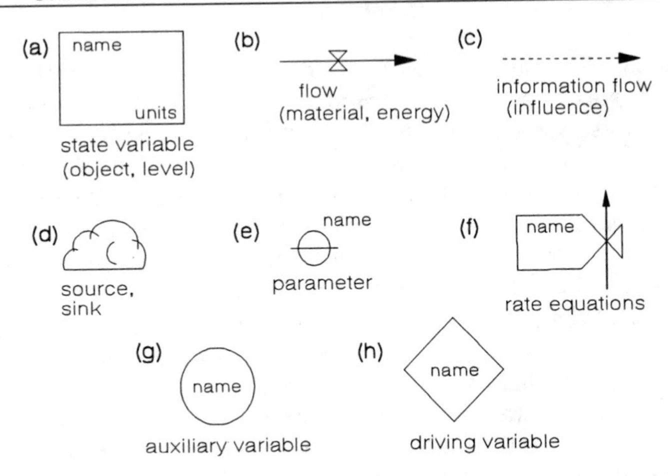

Figure 3.2: The basic components of a Forrester diagram.

Information Flow or Influences The second manifestation of relations between objects are the effects that the quantity of one object has on the rates of inputs to or outputs from another object (e.g., effects on growth rates). These are *control relations.* State variables affect the control valves of material flows of other state variables (including themselves). These influences are represented as *information transfers* (dotted arrows in Fig. 3.2c) connecting state variables and control valves. The tail of the arrow indicates the influencing component and the head of the arrow indicates the affected rate. Possible sources of information transfer are state variables, parameters, driving variables, and auxiliary variables or equations.

Sources and Sinks Objects that are defined to be outside the system of interest, but which are inputs to state variables or outputs from state variables, are represented as "clouds" (Fig. 3.2d). They are not state variables since they are not modeled explicitly and are not represented by dynamic equations. (Hence, they are nebulous and vague — traits well represented by clouds.) Sources or sinks cannot be involved in an information transfer. That is, they cannot alter a rate, nor can their condition be altered.

Parameters Constants in equations are noted in the diagrams by small circles with lines (Fig. 3.2e). They invariably are used as the tail of an information transfer, since their values influence flow rates and other equations within the model. Since they are constants, their values are not changed by an information transfer.

Rate Equations Total (or absolute) rates of input to, or output from, a state variable are described mathematically with *rate equations.* It is useful to identify and label these explicitly by modifying the control valve symbol (Fig. 3.2f). The equations usually describe information transfers from state variables and parameters.

Auxiliary Variables and Equations Auxiliary *variables* (large circles, Fig. 3.2g) are variables that are computed from an auxiliary *equation.* The auxiliary equation can be a function of other auxiliary variables, state variables, driving variables, and parameters. Auxiliary variables change over time because they depend on either (a) a state variable, (b) a driving variable that depends on time, or (c) an auxiliary variable that depends on a state variable or driving variable. Auxiliary variables are never constants, nor are they state variables.

Auxiliary variables are primarily used to simplify the writing of rate equations. In this use, they may be substituted into the equation, but they are isolated for clarity or computing efficiency (they may be used by several state variables). Consequently, they are often shown influencing rate equations. A secondary use is to convert, for output purposes, a state variable or another auxiliary variable.

Driving Variables Dynamic events that relate to variables that are not state variables (e.g., season or temperature in some models) are often used as *forcing functions.* These driving variables are represented as large diamonds (Fig. 3.2h). Driving variables may take as input only other driving variables. Usually, they have no inputs and time is assumed to be a component of the variable (e.g., temperature values on different days). Here are two examples when one driving variable may influence another: (1) A driving variable of time could influence a driving variable that specifies temperature over space. The temperature at depth (space) in a water column could be influenced by season (time): different temperatures at depth at different seasons. (2) A driving variable of time at one scale (slow) could be used to determine a variable that occurs at a faster time scale [e.g., season (a slow time-dependent driving variable)] can influence hourly temperature values (a fast time-dependent driving variable). The units of the driving variable (e.g., time, space) should be specified in the diagram.

3.3 Examples

As illustrations of this diagramming technique, we consider some simple examples.

3.3.1 Grass–Deer "Ecosystem"

Consider a system composed of grass and some deer that eat the grass (Fig. 3.1). For the sake of definiteness, we will make the following biological assumptions.

1. The per capita rate of growth of grass (g C produced per g C of existing grass) is constant. Therefore, the total growth will be the per capita rate times the total amount of C present.

2. The only loss to the quantity of C in the grass population is by deer consumption.

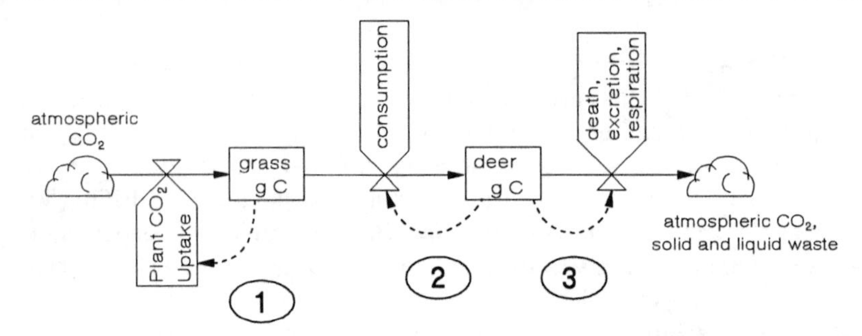

Figure 3.3: Forrester diagram for the grass–deer ecosystem. Solid arrows are pathways for C flow; dotted arrows represent relations between levels and input or output rates as hypothesized. (Numbered ellipses on information flows are not part of Forrester diagrams, but are used for explanatory purposes only.)

3. Deer compete with one another for grass so that, as the quantity of deer increases, each deer receives less C.
4. Deer excrete or respire a fixed proportion of their existing C as either atmospheric C or solid/liquid waste.

None of these hypotheses are detailed enough to allow us to uniquely define the equations, but they do permit us to draw the Forrester diagram in Fig. 3.3.

The assumptions indicated only two state variables: grass and deer. Therefore, there are only two boxes (levels) in the Forrester diagram. Also from our assumptions, there are only three flow relations: source to grass, grass to deer, and deer to sink. The diagram implies that any other flows are assumed to be unimportant to the objectives of the model. For example, we explicitly precluded C from flowing directly from grass back to the atmosphere or another sink. Information transfer 1 is a diagram of the concept that total grass growth depends on the amount of grass present. Information transfer 2 is similar, but we know from our verbal statement that deer are competing with one another, and grass is not competing (per capita rates are constant). Therefore, given the similarity of information transfers 1 and 2 (Fig. 3.3), it is clear that different hypothesized control relations can have the same Forrester diagram presentation. This implies that a single Forrester diagram can represent many different sets of hypotheses. Forrester diagrams do not uniquely determine the model equations. Information transfer 3 represents the effect of deer on the loss rate of C from the deer population. The verbal statement of this control relation is similar to that for grass growth rate, so the information transfer arrow is similar.

3.3.2 Population Growth with Explicit Birth and Death

To demonstrate the relation between diagrams and equations, the next example will start with an equation and produce a consistent diagram.

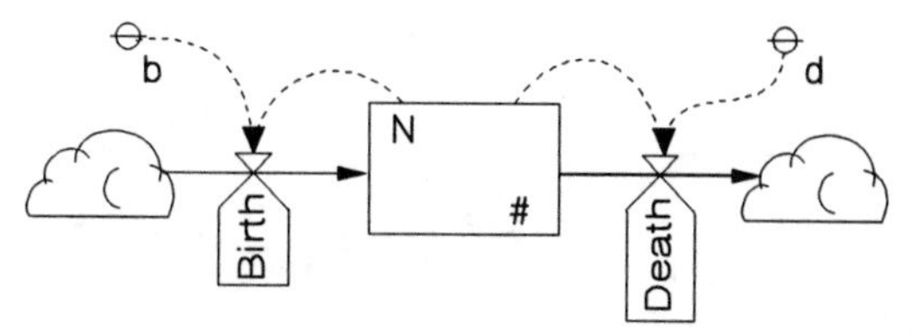

Figure 3.4: Forrester diagram for one form of the density-independent population growth model.

The classic, density-independent population model written as a finite difference equation (FDE) is $N_{t+1} = N_t + rN_t$, where r is the net per capita growth rate. Suppose we reparameterize it using the identity $r = b - d$, where b is the per capita birth rate, d is the per capita death rate, and both are positive quantities:

$$N_{t+1} = N_t + bN_t - dN_t. \tag{3.1}$$

Note first that there is a single state variable (N); therefore there will be a single box in the Forrester diagram. In general, there will be exactly as many boxes (levels) and FDEs as there are state variables. Second, note that Eq. 3.1 has two components of change: a positive value (bN_t) and a negative value ($-dN_t$). These correspond in Forrester diagrams as inputs to and outputs from a single state variable. Thus, for this form of the model, we have a Forrester diagram as shown in Fig. 3.4. Note the use of clouds (sinks and sources) to represent the origin of newborn individuals and the destination of dead individuals.

To illustrate the use of auxiliary variables and equations, consider the case where birth rates decrease linearly as numbers of individuals increase, but total death is a simple proportion of the population:

$$N_{t+1} = N_t + bN_t \underbrace{\left(1 - \frac{N_t}{K}\right)}_{R} - dN_t. \tag{3.2}$$

The second (middle) term of the right-hand side is the absolute rate of births in the population. The third term is the absolute rate of death. Birth rate is determined by a "reduction factor" that approaches zero as N approaches a constant K [i.e., $(1 - N/K) \to 0$ as $N \to K$]. Our modeling objectives might suggest that this is a particularly important quantity (e.g., we want to examine a range of functional forms, not just the linear one above). Consequently, we isolate that subexpression with a special symbol (R) and we treat it as an auxiliary variable. Figure 3.5 shows the Forrester diagram for this model. Note that it is similar in form to Fig. 3.4, but that we have used an auxiliary variable to represent the effect of density on the reduction factor. The "effective" per capita birth rate is bR, where b is the maximum per capita birth rate. Note that R is a function of N (state variable) and K (a parameter), so information transfer arrows

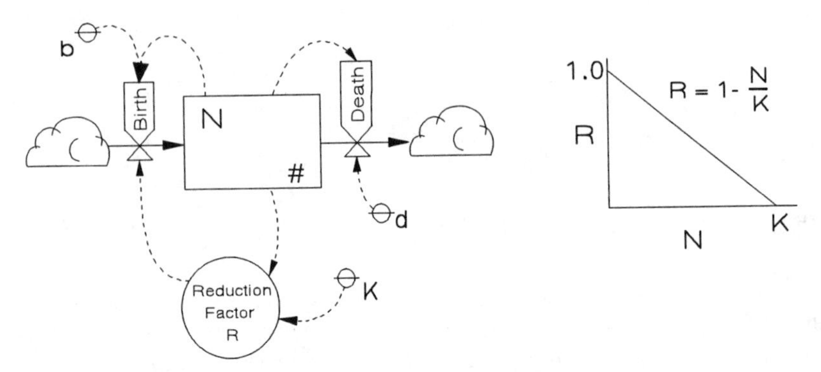

Figure 3.5: Forrester diagram for one form of density-dependent population growth model.

connect these entities with R.

It is somewhat a matter of taste to separate R and b. Alternatively, we could draw the diagram using a different auxiliary variable, perhaps called "effective per capita birth rate," corresponding to the variable $b(1 - N/K)$. This would require a minor modification of the control relations (information transfer arrows). Finally, it is possible to draw the Forrester diagram for Eq. 3.2 without any auxiliary variables; it depends on the emphases the diagrammer wishes to achieve.

3.3.3 Net Population Growth

The above models used explicit birth and death to show the relations between the parameters governing increases and decreases, and the input and output arrows in the diagrams. The typical presentation of these models subsumes birth and death into a net rate parameter r, which may be positive or negative. For these forms, the corresponding diagrams for the two models (Fig. 3.4 and Fig. 3.5) are shown in Fig. 3.6. Note the double-headed material flow arrows used to indicate that the parameter r controls both the inflow (away from source) as well as outflow (toward the sink). The single cloud serves a double purpose here as both sink and source.

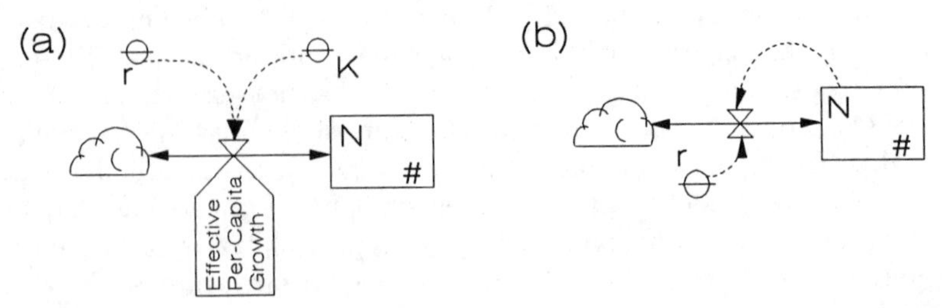

Figure 3.6: Forrester diagrams for density-dependent (a) and density-independent (b) growth using the normal parameterization.

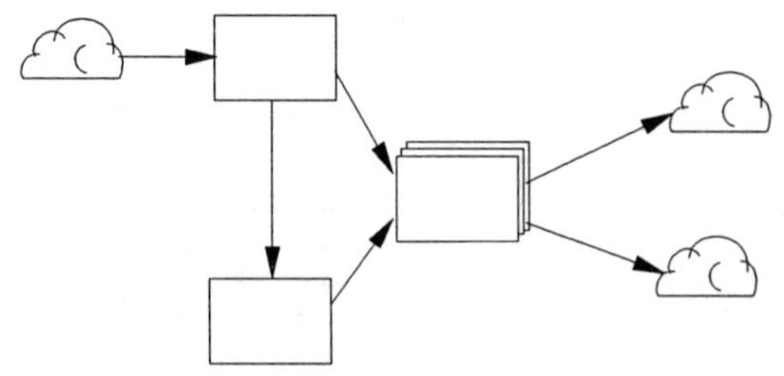

Figure 3.7: Forrester diagram showing multiple state variables. The set of three offset boxes represents three state variables all of which have the same relations (inputs and outputs) to other state variables in the system.

3.3.4 Multiple State Variables

It is often clearer to isolate different inputs and outputs to a state variable, even though they may be additive and could be lumped. This may be important if the controls on the different rates vary significantly, usually due to different parameters. This is diagrammed by multiple material flows into or out of a level.

When a model has more than one state variable (e.g., an ecosystem model with equations for plants, herbivores, and carnivores), then each object is represented by a box (level) that connects with the others according to the flow of material (energy) defined by the relations (i.e., foraging relationships). Figure 3.7 illustrates this for a simple case. The critical point for models of this type is that the units of state variables and the units of flow must agree. Some models have state variables that possess identical inputs and outputs (e.g., discrete soil layers in a water flow model); to simplify the diagram, these are shown as offset boxes (Fig. 3.7). A similar scheme can be used for auxiliary variables.

A more complicated case is illustrated in Fig. 3.8 for a simple agroecosystem model in which there are fertilization regimes, pests, and crop harvesting schedules. In this model, suppose the broad objective is to *determine the effects on profits of different schedules of fertilizer and pesticide applications to fields of alfalfa.* By "schedule," we mean the timing and amounts of applications. The major pests of alfalfa are weevils and aphids, but these are dynamic since pesticides will kill some of them. So, at least one state variable must represent the pests. We are also interested in the effects of fertilizer applications, but this also will be dynamic (it is applied at certain times and in variable amounts). Consequently, another state variable should be the soil nutrient pool. As we are primarily interested in the profits of farmers, we will need to know both the amounts of crops in the field and the amounts harvested.

Thus, the state variables are: nutrients, insect pests, field alfalfa, and harvested alfalfa. All of these must have common units, so for the sake of the example, we will assume that nitrogen is the limiting nutrient to be added and that all other state variables will be quantified in units of

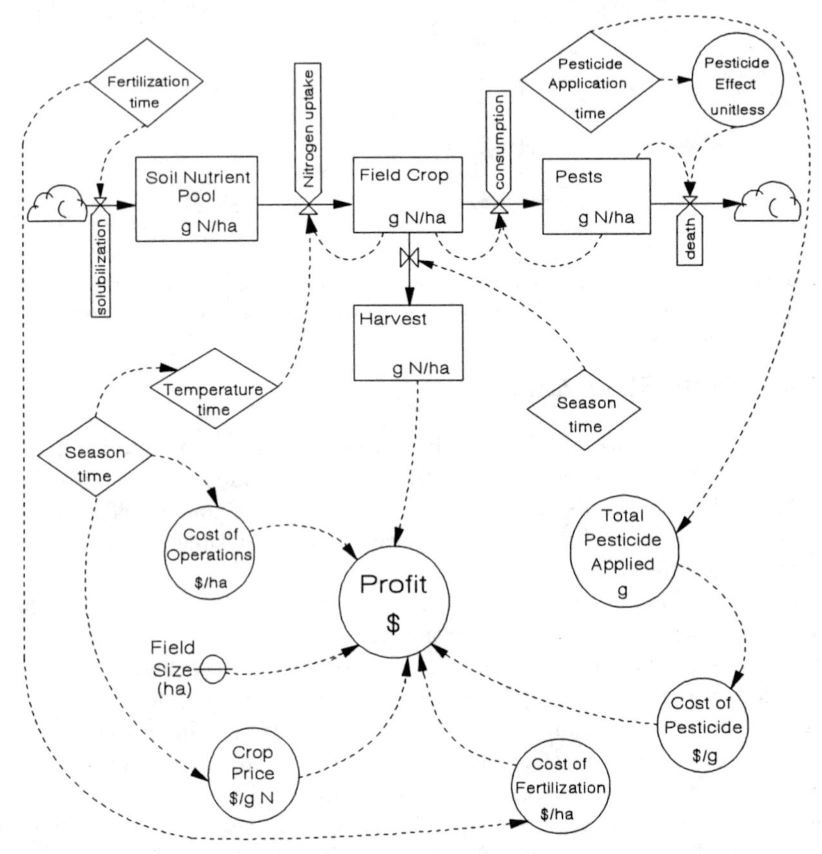

Figure 3.8: Forrester diagram for a hypothetical agroecosystem model showing multiple state variables of an agricultural system.

g N/hectare. These are not the most natural units by which to measure alfalfa and insect pests, but we can always use a conversion factor (auxiliary variable) to create other units.

The scheduling of management events such as pesticide application and fertilization is represented by driving variables, as are natural events such as season and temperature (Fig. 3.8). The objectives state that one of our primary interests is farmer **Profit**. Because we have chosen the dynamics to be stated in units of g N/ha and the units of profit are dollars, we need to convert from g N/ha harvested to dollars. To accomplish this, we use auxiliary variables such as **Fertilization Cost** ($/ha), **Field Size** (ha), **Alfalfa Price** ($/g N), and so on (Fig. 3.8).

The diagram is not complete because we have omitted the parameters, but without more specific hypotheses on the dynamics of the components it is difficult and not useful to add this facet of Forrester diagrams. The reader should study Fig. 3.8 so that the components (levels) and flows (material and information) are clear. In particular, it should be evident how a mathematical model based on this diagram will address the original objectives.

3.3.5 Multiple Flow Variables and Units

When different units on flow variables are modeled (e.g., g N and g C or blood pressure and blood oxygen in a physiological model), *parallel* models (or *multiple models*, Rideout 1991) must be used to avoid having "apples" flow into "oranges." The dynamics of many biological processes depend on several interacting variables. There are two broad applications of this concept in modeling: (1) the variables are at the same level of biological organization but may interact in their influence on the dynamics, or (2) the variables are at different levels of organization, but both are needed to address the model objectives.

Two variables (A and B) are on the same level of biological organization if all of the measurements that can logically be made on A can also be made on B, *and* there are no measurements that can be made on B that cannot be made on A. So, for example, two chemical molecules (CO_2 and H_2O) are on the same level because we can measure on both such things as molarity, boiling point, molecular weight, and so on. In contrast, an individual organism and a population of organisms are on different levels of organization since we can measure population growth rate on the population, but not on a single organism.

Variables that are on the same level of organization may interact to affect some biological process negatively (negative feedback), positively (synergism), or independently (substitutable). For example, the electrical potential across the membrane of a nerve cell is determined by the difference between the net charge inside the cell and the net charge outside the cell. Therefore, two variables that might be modeled and that interact negatively are positive ions exemplified by potassium (K^+) and negative ions such as chloride (Cl^-), since the net charge is the sum of positive and negative ions. In other situations, two different variables might complement each other and enhance the rates of change of biological processes [e.g., nerve cell activity and electrical potential and the different forms of positive ions: K^+ and sodium (Na^+)]. In still other systems, the two variables may influence dynamics independently, for example, grass species A and B may each increase deer growth rates by an equal amount.

In all of the above examples, it is conceivable (but not necessary) that a model would portray the dynamics of both quantities (K^+ and Na^+, or species A and B). In all three possibilities, if we wish to describe the dynamics of the affected process as influenced by the variables, then we must describe the dynamics of the individual variables and their effect on the process. Therefore, since the physical quantities cannot flow among themselves (i.e., g K cannot flow from a compartment containing g Na), we represent the separate dynamics as parallel models.

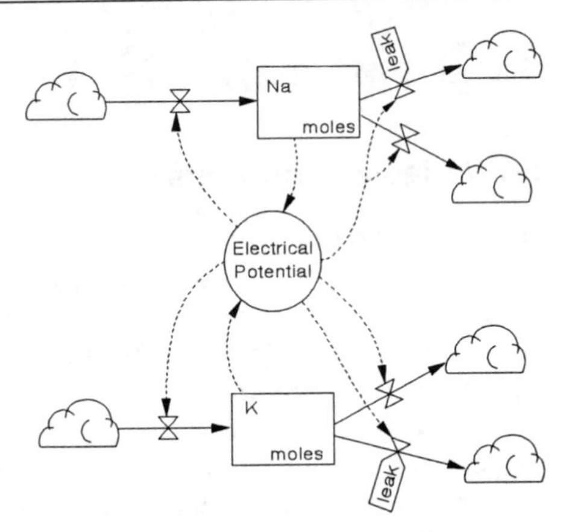

Figure 3.9: Forrester diagram when multiple flow variables are used. Unlabeled material transfers are assumed to be losses or gains caused by ion pumping.

An example of variables at different levels of organization is variables describing the size of individuals and population size. In models in which the growth rate of the population is influenced not only by the current numbers of individuals in the population, but also by the average body size (e.g., through the feeding rate), both quantities must be modeled. Obviously, these are two very different kinds of quantities and it is absurd to suppose that they can be related by a material transfer (solid arrow in a Forrester diagram). It makes no sense to say that average body size "flows" into numbers of individuals. Consequently, in a model, these two variables must be kept separate.

To illustrate this concept graphically, consider a very simple model of nerve cell activity. The activity level is measured as the electrical potential across the nerve cell membrane. This is determined by the relative concentration of K^+ and Na^+ on the inside. Ions of K and Na flow into the cell through ion-specific channels at rates that depend on the current electrical potential of the cell. Figure 3.9 shows one implementation of the integration of the dynamics of K and Na to determine electrical potential. Since K and Na are different quantities, they are not interchangeable and therefore must have different inputs, outputs, and level representation.

Care must be exercised when diagramming to recognize differences in units between state variables. Units that are superficially the same can in some circumstances be completely different. Often these differences are hidden by the mathematical equations. For example, if our interest is in the flow of carbon between components of a plant (e.g., leaves and roots) in a plant growth model, then an atom of carbon in the leaves can actually become incorporated into the roots. In contrast, suppose our interest is in a model of the population dynamics of a species of plant and its herbivore and the "flow" variable of interest is numbers of individuals in each population. It does not make sense to say that individual plants flow into individual

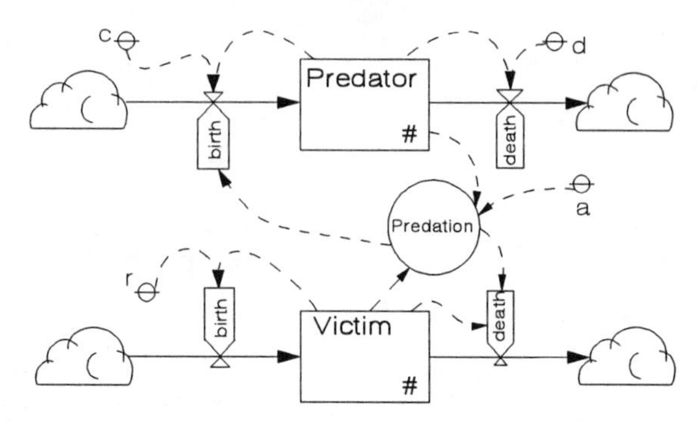

Figure 3.10: Simplified Forrester diagram for linked population models based on numbers of individuals.

consumers. The biomass of the plant in fact does become incorporated into the biomass of individual herbivores, but the numbers in the population are created by processes of birth and death. The basic concept here is one of "conserved" and "nonconserved" flow quantities. Grams of carbon is a conserved quantity; it is the mass of a physical object. And, except under unusual physical circumstances, an atom of carbon is never created or destroyed. Numbers of individuals are not conserved in the same way. Prior to birth the individual did not exist, although all of its atoms were present in other forms. At its death, the individual is destroyed, but its constituent atoms persist.

This distinction influences the way Forrester diagrams are drawn for some types of models. In predator–prey models, when numbers of individuals are modeled, the units are actually numbers of prey individuals and numbers of predator individuals. These units are as incompatible with each other as were the units in Fig. 3.9 and the diagram should use parallel models. Consequently, we should use a Forrester diagram similar to the simplified form shown in Fig. 3.10.

3.4 Errors in Forrester Diagrams

Below is a short list of some of the errors that can be made in drawing Forrester diagrams (see Fig. 3.11).

1. Using any symbols other than those defined in Fig. 3.2. For example, there is no symbol like a solid line with no arrowhead attached (Fig. 3.11a).
2. Failing to label all boxes, variables (auxiliary and driving), and parameters with names and units (where appropriate, Fig. 3.11a).
3. Showing sources or sinks influencing rates (Fig. 3.11b).
4. Showing rates influencing state variables (Fig. 3.11c).
5. Showing material flows (solid arrows) between objects other than state variables and sources and sinks (Fig. 3.11d).

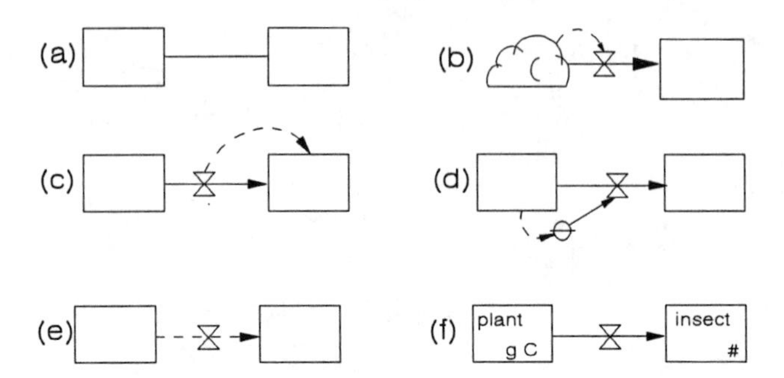

Figure 3.11: Examples of incorrect Forrester diagram fragments.

6. Showing an influence on a quantity that cannot change (e.g., a parameter, Fig. 3.11d).
7. Showing information flows between state variables (Fig. 3.11e).
8. Using incompatible units of flows or state variables (Fig. 3.11f).
9. Using state variables that are not in the model (objectives or equations) or not including state variables that are in the model.

3.5 Advantages and Disadvantages of Forrester Diagrams

Many modelers and theoreticians do not use Forrester diagrams and believe they only get in the way. There is an important element of truth in this view. The equations are the primary objects of interest. Their solutions, not the diagram, produce the output used to address the model objectives. Moreover, the diagrams are not always a compact representation of the model. As the number of state variables, parameters, and relations between objects become large, the size of the diagram increases. Complex diagrams can span several pages, in which case much of the heuristic value is lost.

There are, however, three situations in which Forrester diagrams are useful. First, in learning the rudiments of the modeling process, it is helpful to separate the trauma of mathematical equations from the conceptual issues of the nature of system objects, the characteristics of the material flows between them, and the controls on the dynamics by internal influences. A graphical language has this potential.

Second, many people who are not mathematicians and to whom a model must be explained react favorably to the graphical representation. For most variables and flows, there is a natural correspondence between a material flow and a physical or biological process (e.g., consumption in a foodweb), or between a state variable (boxes) and a compartment (e.g., population). These are concepts with which most people have some experience. As a consequence, understanding is more quickly attained, and constructive criticism (or, even, agreement) is more readily achieved. Moreover, although mathematics offers opportunities for an extremely compact representation of complex relationships, models attempting to achieve a high degree of precision or realism will often require complicated equations. The mathematical expressions for driving variables are often an example of this since they can represent seasonal effects on physical variables such as temperature. Forrester diagrams can reduce some of this complexity by subsuming the mathematical details in a simple symbol.

Finally, Forrester diagrams can be a valuable aid in organizing the computer simulation program. Each level effectively becomes a program module; the set of input and output arrows are components that increase and decrease the finite difference equations. The parameters are the data on which program module operates. Input information flows and parameters indicate arguments to the subroutine; output information flows indicate subroutine side effects (changed variables).

Clearly, there is a point at which diagram complexity obfuscates the basic structure of the model and frustrates attempts to effectively communicate. Just as we must provide objectives for models, we must also recognize our objectives in presenting a model in one form or another. The choice will depend on whether we are communicating with politicians, managers, mathematicians, computers, or our biologist colleagues.

3.6 Principles of Qualitative Formulation

> *The first rule of discovery is to have brains and good luck. The second rule of discovery is to sit tight and wait for a bright idea.*
>
> — Polya (1973)

Qualitative model formulation is one of the sub-problems in the modeling activity. We wish to discover the simplest description of a system that will satisfy the objectives. This section describes a few basic principles that apply to all attempts to formulate a qualitative compartment model using Forrester diagrams. Many of the principles will also apply to other modeling approaches. Based on the Forrester diagrams shown thus far, it should be clear that the purpose of the principles is to help you

- Identify the state variables (levels)
- Identify the flows among the state variables
- Identify the controls on the flow rates
- Identify the auxiliary and driving variables.

To accomplish the above, answer the following questions.

1. *What are the questions to be answered?* Write down all the questions for which the objective requires answers. If you cannot do this, then you do not understand the problem. For example, in the population doubling model, the question was: "When (at what time) will the population be twice its current value?"

2. *What quantities are needed to answer the questions?* In compartment models (and almost all others), objective questions are answered with specific numbers or series of numbers. Write down the required quantities and their units.

   In the population doubling problem, it is the "year" when the population has doubled. The size of the population at doubling is of minor concern in this problem (indeed, given the initial condition, it is trivial to compute).

3. *What equations will answer the questions?* Can you write an explicit dynamic equation (e.g., finite difference equation) whose value at some time will constitute an answer? In the population doubling problem, the answer is "no." We did not solve the problem by writing an equation describing the doubling time. We wrote an equation for population growth and *from this* determined doubling time. If the question had been, "What will the population size be in 1999?" then a dynamic equation would answer it.

   If you can, in principle, answer the question directly with a dynamic equation, then this is at least one of the state variables in the model and it becomes a level in a Forrester diagram. (You do not write the equation at this stage, but simply recognize that such an equation, when written, will answer the question.) If a dynamic equation will not immediately answer the question, then (a) you need an auxiliary equation to compute the answer from another variable, and (b) you need another quantity and state variable that will serve as input to the auxiliary equation. An information flow (dotted line) will connect these two objects. Figure 3.8 illustrates the concept in the relation between Harvest (g N) and Profit ($). The units of the state variable and the auxiliary variable will almost certainly be different, for otherwise a dynamic equation would have answered the question.

4. *What other primary flow quantities are needed?* From the objectives and prior knowledge or data, write down the quantities that will flow into and out of the state variables that contribute to the question. These flows determine the dynamics of the level. The flows will connect to additional levels by material-transfer arrows in the Forrester diagram. For descriptive purposes only, we will call these the *primary* state variables. In the simple population doubling problem, a single state variable suffices, so there are no others. In Fig. 3.8, a single state variable influences the primary quantity needed for the objectives (Profit). But the objectives refer to pesticide and fertilization effects, and we know (or presume) from prior information that the harvest dynamics will be influenced by the size of the crop in the field (Field Crop), and this will be influenced by insect consump-

tion (**Pests**). Prior knowledge also tells us that fertilizer is applied to the soil and is subsequently removed from a pool of N contained in the soil. Thus, we hypothesize that a sufficient model would be one that contained the state variables (levels) shown in Fig. 3.8 (i.e., **Soil Nutrient Pool**, **Field Crop**, **Pests**, and **Harvest**).

5. *Is an explicit spatial representation required?* Do the objectives refer to or require knowledge of events at different places? If so, then a transport model (Chapter 1) may be appropriate or the primary state variables should be replicated at each discrete spatial location. Typically, the state variables at the different spatial locations will be connected by material transfers (immigration or advection).

6. *What are the controls on the flow rates between the state variables?* The controls become influences or information transfers in Forrester diagrams. For each state variable, list the factors influencing the rates of flow into the level and influencing the rates of flow out of the level. In general, there will be four sources of influences: (1) parameters, (2) auxiliary variables whose inputs are from the primary state variables, (3) driving variables, and (4) inputs (possibly via auxiliary variables) from state variables other than the primary state variables. Type (1) is illustrated in Fig. 3.10 by the influence of parameter "c" on "birth rate." Type (2) is illustrated in Fig. 3.9 by the loop between "K," "Electrical Potential," and flow rate into "K." Type (3) is illustrated in Fig. 3.8 by the influence of "Fertilization" on the flow rate into "Soil Nutrient Pool." Type (4) occurs, for example, when the primary state variables are defined on one level of biological organization (e.g., population), but secondary state variables at another level of organization (e.g., individual body size) are required to implement hypothesized flow rate controls at the population level. For example, populations with large average body size consume resources faster than populations with small body sizes. If type (4) controls are present, then the secondary state variables must be implemented as levels in a parallel model (Fig. 3.9).

7. *Do you know any parameter names?* If the objectives or prior knowledge suggests important parameters, these should be included in the Forrester diagram. Most of these do not become known until explicit equations are suggested for flow rates and auxiliary variables.

3.7 Model Simplification

Thus far, we have emphasized the mechanics of qualitative model formulation. For a number of practical and esthetic reasons, we wish our models and explanations of biological phenomena to be as simple as possible. On the other hand, biological systems are complex, having many processes and variables that interact in complicated, non-linear ways. It is, therefore, natural when creating a model from a general objective statement, such as we used in our example of pesticide effects on farm profit, to create a model that is more complicated than needed or desirable. There is some evidence that models of intermediate complexity are best (Costanza and Sklar 1985). Being able to simplify a model is almost as important as the ability to formulate it in the first place. Think of it as editing the first draft of an essay. Moreover, in Chapter 2 we stressed the importance of evaluating alternative models in parallel. An excellent approach to creating a family of alternative models is to create a gradient from simple to complex. So, the process of model simplification and its converse, model elaboration, are valuable tools for hypothesis testing. Logan (1994) has formalized this philosophy in what he calls the *composite-modeling* approach. In this approach, one designs an initially large model that contains most of the relevant processes and relations. Afterwards, one reduces the large model into progressively simpler, mathematically more tractable versions that, although simple, maintain links and similarities with the more complete model. The end result is a family of models and tools each of which have uses and applications. Because model simplification is central to these ideas, we now present a few principles for simplifying models (see also Shannon 1975).

Eliminate State Variables Every state variable must have a dynamic equation (differential equation or finite difference equation) as well as parameters and initial conditions. There are two ways to reduce model complexity arising from state variables.

1. *Convert a state variable into a constant (e.g., a parameter) or an auxiliary variable.* For example, in Fig. 3.8 we represented Profit as being influenced by harvested crop nitrogen, whose dynamics were determined by the size of the field crop. However, given that alfalfa is harvested by mowing and collecting a fraction of the field crop, a simpler model would be one in which profit is determined from the current field crop and a parameter representing the simple fraction harvested. If we wished to retain the concept that harvesting occurs at fixed time intervals, we could replace the Harvest state variable with an auxiliary variable that is influenced by Season, Field Crop, and a parameter representing the fraction of the field crop harvested. Profit, then, would be determined by season and the harvestable fraction of field crop.

2. *Aggregate state variables.* In Fig. 3.8, we separated soil nitrogen and crop nitrogen to examine the potential interaction between the timing of applications of fertilizer and pesticide. If we would be willing to drop this aspect of the objectives, then we could lump plant and soil nutrients into a single state variable.

Make "Stronger" Assumptions Complexity also enters models in the form of the equations and functional relationships. For example, we compared the models of population growth with and without density effects on reproduction. The former is more complex than the latter. There are several approaches for simplifying functional relationships, and while we will explore the quantitative relationships in more depth in Chapter 4, we can list two possibilities here.

1. *Convert functions of state variables into constants.* Equation 3.2 hypothesizes that effective birth rate decreases with increasing density. If we assume that this function does not exist, then we have simply a constant (r) that describes birth rate (Eq. 3.1).
2. *Convert nonlinear relationships into linear relationships.* Equation 3.2 is a linear relationship between current population density and birth rate. It is not difficult to imagine a more complex relationship that is a curvilinear function. Thus, Eq. 3.2 is already a relatively simple model.

Remove Temporal Complexity Models with temporal variability have a layer of complexity that can be eliminated as follows.

1. *Convert random models into deterministic models.* As discussed briefly in Chapter 1 and in more detail in Chapter 10, random effects on dynamics can be achieved by allowing parameters to vary randomly in time. These types of models have more parameters than their deterministic counterparts and can produce significantly more complicated dynamics that require greater effort to analyze and understand. Removing randomness simplifies the model.
2. *Convert driving variables to constants.* Driving variables or other time-varying perturbations are another means of allowing parameters and processes to vary in time, due to causes not modeled by internal system dynamics. Removing these variables will simplify the model by reducing the number of parameters and amount of data used as well as simplifying dynamics. The simple population models we have discussed so far have no driving variables.

Remove Spatial Complexity As with time, removing spatial complexity is an important simplification tool. The usual method is to convert a model that explicitly models spatial events to one that ignores spatial differences. In Fig. 3.8, we already made this simplification, because we did not attempt to model spatial differences within our alfalfa field. If we had incorporated spatial effects, then (in one possibility) we would have had additional state variables. This would require, essentially, duplicating the four state variables shown for each of the spatial areas we wished to discriminate. For example, we might distinguish the effects of pesticides and fertilizers on the border of the field from those in the interior of the field. If so, then we would need state variables for Pests_Inside, Pests_Border, Field_Crop_Inside, Field_Crop_Border, and so on. Adding space to a model usually greatly increases its complexity, so assuming *spatial homogeneity* is a simplifying assumption.

3.8 Other Modeling Problems

In Chapter 1, we introduced four broad classes of models: compartment, transport, particle, and finite state. Forrester diagrams were designed for and are especially useful in describing compartment models. This modeling approach is an extremely powerful and general framework that has many applications in biology, from ecosystems to enzyme kinetics. It is most useful when the system can be decomposed into flows of material or energy among a finite, but possibly large, number of discrete "pools" or compartments. It can also be used when we are interested in quantities that superficially do not "flow," for example, blood or water pressure in animal and plant physiological systems. By linking many compartments together in complicated ways, compartment models can address complex interconnection networks (e.g., foodwebs of many species). Compartment models can also incorporate elaborate control relationships between variables (e.g., the relationship between fertilization schedules and profit). Nevertheless, the remaining three model classes are conceptualizations of systems for which this approach is not optimal or useful.

3.8.1 Transport Models

Of the remaining three classes of models, transport models are closest to compartment models. In transport models, we have a substance [energy (heat) or a quantity of matter] that flows from spatial point to point. A simple example is the flow of a pollutant along a stretch of river after it is emitted from a point source (e.g., a sewage outfall). A central concept shared with compartment models is a quantity that flows, but a major difference is that there is no clear concept of a finite number of compartments in which the substance resides. Instead, there are, in the continuous formulation, infinitely many points along the river at which some quantity of the substance exists. When we model spatial flows across space in this way, we are using an *Eulerian* frame of reference: the origin of the spatial coordinate system is fixed and the substance moves over this coordinate system.

There are many forces that could influence the flow of the pollutant, but the following simplified view uses two that will illustrate the qualitative model formulation. Advection moves the substance with a physical flow of water from point to point (river flow). Diffusion moves a substance in any direction according to the concentration of the substance around each point. Consider an infinitely short segment of the river along its x direction

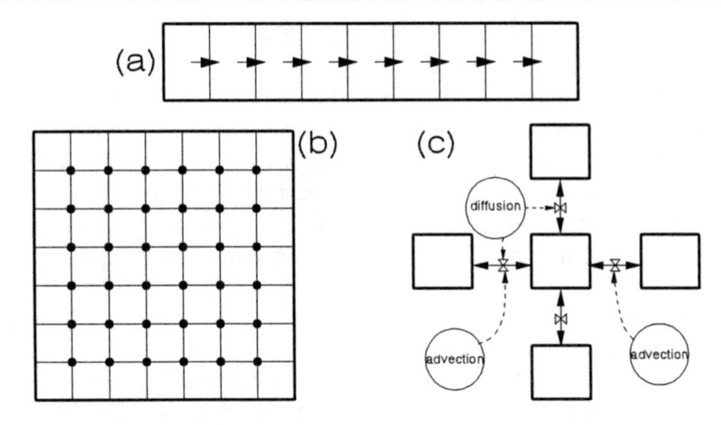

Figure 3.12: (a) Flow between imaginary compartments in a continuous one-dimensional system. (b) Discrete grid system used in two-dimensional transport models. (b) A close-up of five grid points showing the similarity to compartment models.

($\Delta x \to 0$). Figure 3.12a illustrates water and pollutant flows between these infinitely thin segments of river. Since we have rate functions dependent on two variables (space and time), we use partial differential equations based on partial derivatives. For functions of two or more variables [e.g., $f(x, t)$, where x is a spatial dimension and t is time], $\partial f / \partial t$ is the partial derivative of f with respect to t when the spatial variable is held constant. Similarly, $\partial f / \partial x$ is the derivative of f with respect to x when t is fixed. Using this notation, we can write a conceptual rate equation for each segment as:

$$\begin{aligned} \frac{\partial p(x,t)}{\partial t} &= \begin{pmatrix} \text{Advection} \\ \text{In} \end{pmatrix} - \begin{pmatrix} \text{Advection} \\ \text{Out} \end{pmatrix} + \\ &\quad \begin{pmatrix} \text{Diffusion} \\ \text{In} \end{pmatrix} - \begin{pmatrix} \text{Diffusion} \\ \text{Out} \end{pmatrix} + \\ &\quad \begin{pmatrix} \text{Pollutant} \\ \text{Creation} \end{pmatrix} - \begin{pmatrix} \text{Pollutant} \\ \text{Destruction} \end{pmatrix}, \end{aligned}$$

where $p(x, t)$ represents the concentration of the pollutant in the water at a point x in space and t in time. Because of the continuous nature of space in this conceptualization, compartment models do not do well here. [There may, of course, be compartments within the river (e.g., fish tissue) wherein the pollutant is stored which we may wish to model and for which compartment submodels will be appropriate.]

However, it happens that many of these models require numerical computations to obtain a solution. This typically requires that we discretize space by imagining it composed of many very closely spaced grid points at which we have obtained a numerical solution and know the pollutant concentration. Figure 3.12b illustrates this for a two-dimensional transport model where we assume the advective flow is unidirectional from left to right and diffusive flow can occur in both directions.

By discretizing space, we have introduced the element that previously distinguished the transport model from the compartment model: a finite number of storage compartments. Figure 3.12c shows a simplified Forrester diagram that illustrates how a compartment model framework could describe the system at one grid point. However, even though we can, after spatial discretization, force the system into the compartment model mode, this does not mean that a Forrester diagram is a felicitous description of the modeled system. It illustrates the forces and processes at a point, but it would be foolish to attempt to represent the spatial scale of Fig. 3.12b with a series of drawings like Fig. 3.12c iterated at each grid point. Since all discrete points are identical, no new information about the structure of the model is revealed by Forrester diagrams at different points.

A second kind of transport model uses a much coarser spatial resolution than that implied by the discretized continuous system above. In ecosystem models, we are often interested in flows of energy or material through a complex foodweb. The foodweb and other processes affecting dynamics, however, are frequently different in space. For example, an ecosystem model of a lake would describe nutrient flow from the physical compartments to plants to herbivores and up through several levels of fish species. Such a model might describe several species at each of these trophic levels, each having complex equations describing nutrient uptake. However, the set of species inhabiting the edges of lakes (littoral zone) differs from those in the open water habitat (pelagic zone), and nutrient inputs from the land obviously will enter the littoral zone. A modeling approach to this framework is to divide the lake ecosystem into two spatial compartments and to divide each of these into the trophic compartments of the biotic part of the system. When such a coarse level of spatial resolution is used, the compartment modeling approach is applicable and a Forrester diagram could be used by separating each biotic compartment in each spatial compartment.

In summary, a compartment model paradigm, in general, and the Forrester diagram approach, in particular, are not always appropriate. This is particularly true when the system is modeled as spatially continuous with small spatial resolution. Nevertheless, at least in early model formulation stages, the compartment model concept can be useful for transport models.

3.8.2 Particle Models

Particle models describe systems in which the variables are physical objects (e.g., billiard balls, or individual organisms) that change in some way according to dynamic equations. This is called the *Lagrangian* frame of reference, as opposed to the *Eulerian* approach of transport models. In general, there can be any finite number of these objects. The objects are characterized as having *essential properties* that are appropriate to the system being modeled and that change according to the dynamic equations.

Most often, especially in physics, the equations define how objects move through space (e.g., planets in a gravitational force field). In this case, the essential properties of objects are their physical position in a coordinate systems [e.g., (x, y, z) in a three-dimensional Cartesian space]. But biological (and physical) models can use a generalization of this framework to include not only spatial position, but other essential properties (e.g., physical properties: mass, momentum, velocity; biological properties: biomass, water content, hunger level). Recently, considerable interest has developed in this class of models in ecology using the name "individual-based modeling" (Huston et al. 1988; DeAngelis and Gross 1992) and human population sciences using the name "micropopulation modeling" (Dyke and MacCluer 1973; Ackerman et al. 1993).

Particle-based models that alter physical position do not fit the compartment model paradigm well, although it is possible. Figure 3.13 shows the physical system and a Forrester diagram for a single prey individual and a single predator individual moving in a 2D space that possesses a refuge for the prey. The state of the prey and predator is defined by their position in space [i.e., their (x, y) coordinates]. It is meaningless to speak of a substance flowing into or out of the "x" or "y" "levels" of the prey or predator, so here the arrow pointing into the position level indicates a small *increase* in the position (e.g., $\Delta x > 0$) and an arrow pointing to the cloud indicates a small *decrease* in the position (e.g., $\Delta x < 0$).

In addition to the artificiality of interpreting position change as a "flow," the compartment model paradigm fails for the same reasons as the discretized transport model. Typically, particle models simulate hundreds or thousands of objects. For complete accuracy, the diagram should be iterated for each of these objects just as it should have been iterated at each spatial point in the discrete transport model. This would add little new information and, in the case of Fig. 3.13, would require a huge number of dotted information transfer lines to indicate the effects of distances between many individuals. So, as with the transport model, Forrester diagrams can be useful for initial model formulation and detailing a subset of the objects and interactions. But it is not useful to describe all of the objects this way.

3.8.3 Finite State Models

Of the four classes of models, finite state models are the furthest from compartment models. As described in Chapter 1, finite state models have no explicit representation of a quantity that flows among pools. In the formulation of the model, we articulate the important states *a priori* and these are the only possibilities allowed. A useful qualitative tool is the state transition graph, which serves a role analogous to that of the Forrester diagram of a compartment model. Each node represents a state and an arrow between nodes represents possible alteration of the system from the state at the end of the arrow to the state at its terminus. Simple finite state models

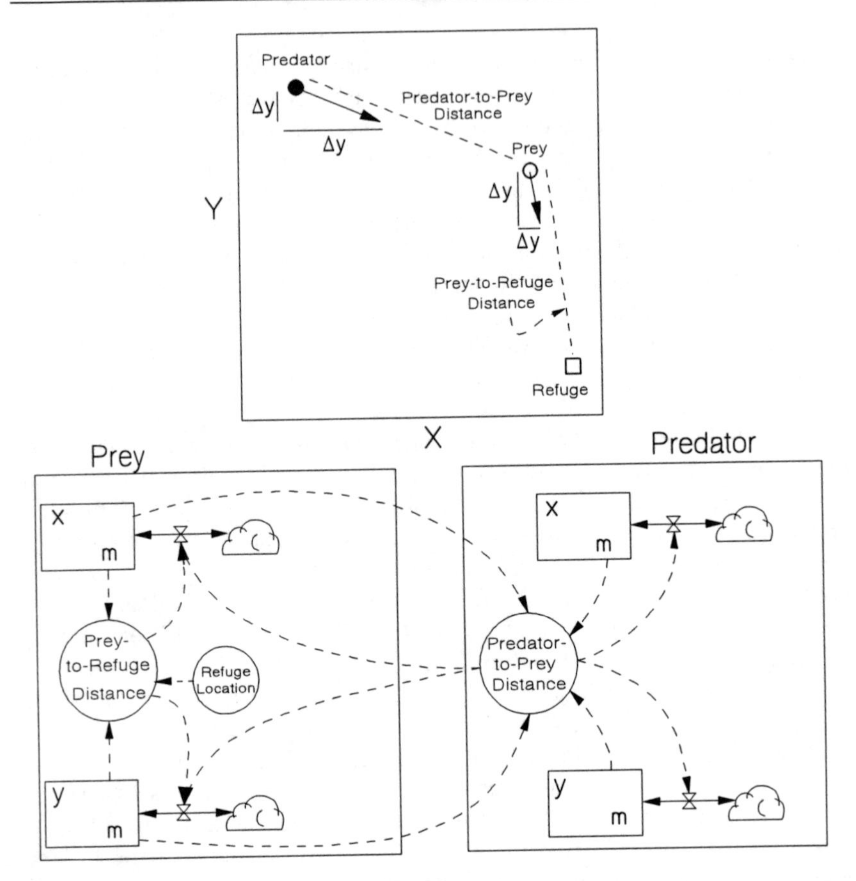

Figure 3.13: Diagram of physical system and Forrester diagram for a particle movement model showing a single predator chasing a prey. The Forrester diagram attempts to represent change in position (Δx, Δy) as a flow to a sink (decrease Δx) or to a level (increase Δx).

(e.g., Markov processes) are stochastic where the arrow is the probability of transition from one state to another; only the current state and the probabilities can affect the outcome. Figure 3.14 shows the transition graph and one stochastic realization for the finite state weather model (Chapter 1). Weather can take one of three states: **G**ood, **I**ntermediate, and **B**ad. A simulation of weather using the transitions probabilities shown on the arrows (Fig. 3.14a) produces a sequence of the three states (Fig. 3.14b). More complex models are possible where, for example, the state of previous time steps can affect the transition probabilities, or other events and conditions in the system can affect the probabilities. These models can be written as finite difference equations with appropriate discretization of the states. Similarly, the model can also be represented as a Forrester diagram (Fig. 3.14c), but it is a clumsy approximation of the transition graph and

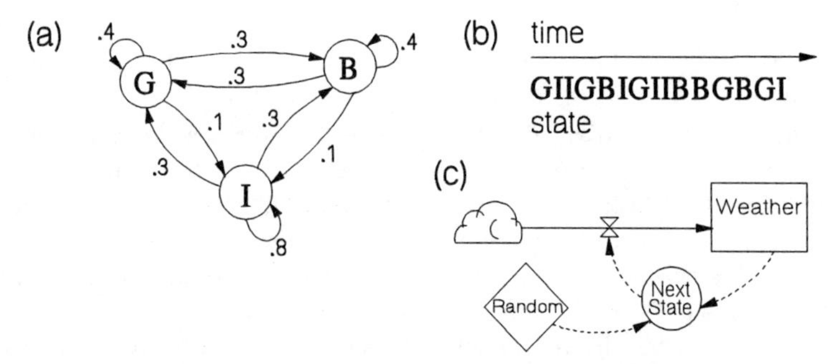

Figure 3.14: A finite state weather model represented as a state transition graph (a), where the numbers represent the probabilities of the transitions denoted by arrows. (b) One stochastic realization of the graph showing the resulting dynamics of states. (c) A Forrester diagram of the model.

the implied flow does not correspond to a physical flow.

3.9 Exercises

1. Assume a substance enters and exits the cell only by passive diffusion. The rate at which passive diffusion transports a substance across a membrane is directly proportional to the difference between the external and internal concentrations. Draw the Forrester diagram for a model in which the ambient concentration is a constant using one state variable, one auxiliary variable, and one rate equation.
2. Consider a substance ("A") that diffuses as above but also is transformed into another substance ("B"). The rate of transformation depends on both the quantities of A and B. Both A and B leave the cell by passive diffusion. Draw a Forrester diagram.
3. Simplify the model represented in Fig. 3.8.
4. Elaborate the model in Fig. 3.8 to include the use of a biological control agent to reduce insect pests on alfalfa. Assume the control agent is a wasp that lays eggs on pest larvae.
5. The classical Lotka–Volterra predator–prey model is:

$$\text{Prey:} \quad V_{t+1} = V_t + rV_t - aV_tP_t$$

$$\text{Predator:} \quad P_{t+1} = P_t + abV_tP_t - dP_t$$

   Assuming the units are a conserved quantity (e.g., g C), draw the Forrester diagram. The parameters are defined as: r = prey per capita rate of increase, a = rate of consumption of prey by predator, b = conversion of prey consumed to new predators, and d = predator death rate.
6. Discuss the relation between Levins' concept of model structure based on generality, precision, and realism and each of the strategies for

model simplification. Which strategies generate which type of model structure?

7. Draw a Forrester diagram of a model that describes the dynamics of the vertical position of an aquatic algae cell based on the following description of flotation in prokaryotic aquatic plankton. Blue-green algae use *gas vacuoles* to manipulate their position in the water column. A single gas vacuole consists of closely packed cylinders each of which is enclosed in a pseudo-membrane of pure protein. The vacuoles are continually produced at a relatively constant rate. The vacuoles collapse when their external pressure exceeds a critical threshold. Their gaseous contents are in equilibrium with the surrounding water. The position of the algal cell is regulated by the number of vacuoles. At high light intensities, cytoplasmic turgor pressure (external to vacuoles) increases beyond the critical threshold for vacuole collapse. This both increases the density of the cell medium and causes the cell to sink. Turgor pressure increases because the light stimulates the uptake of K^+ ions and by-products of photosynthesis (e.g., sugars). At low light levels, the turgor pressure is reduced, the gas vacuoles increase in number, and the cell is more buoyant.

8. Draw a Forrester diagram for the dynamics of blood glucose concentration based on the following simple description of the mammalian blood sugar regulation system. Ingestion of glucose raises stomach levels of glucose, which in turn raises blood glucose levels. This causes the pancreas to secrete insulin, which causes increased transport of glucose into the interior of cells. There it is either used as a source of respiratory energy or is stored. In the liver, glucose is stored as glycogen, which is a form that can be easily released to the bloodstream if blood glucose levels fall below a threshold. The liver acts as buffer to maintain blood glucose levels within acceptable limits between bouts of ingestion.

Chapter 4

Quantitative Model Formulation

4.1 From Qualitative to Quantitative

ONE way to understand a complex, mathematical model is to stare at it until it is obvious. This advice can be less than helpful if you do not know what you are looking for. The approach we follow here exploits the fact that biological models are composed of a relatively few, recurring algebraic constructs. Once these patterns are assimilated, building and reading models becomes a matter of knowing when to use the appropriate component.

We cannot begin, however, until we have a qualitative model for a system that specifies the objects; their basic, qualitative interrelationships; and the underlying hypotheses. The next step is to translate these ideas into mathematical equations. One of the major strengths of Forrester diagrams is the relative ease with which the equations can be generated from the diagram. We can now state a few elements of the method to introduce the material that follows.

The boxes of Forrester diagrams represent the objects of interest: the variables whose dynamic quantities we wish to determine over time. For each of these, we must supply a *state (dynamic) equation* that relates the value of the variable at the next point in the future with the current value and all of the inputs to and outputs from the variable's box. Inputs represent absolute rates of gain, and outputs represent absolute rates of loss. Each of the rates are, in general, calculated by complex, nonlinear equations that combine the flow relations and control relations among system components. The rate equations will therefore involve the *parameters*, *auxiliary equations*, and *driving variables* as specified by the Forrester diagram. Summing all of the rate equations for a given state variable yields the net rate of change for that variable at the current point in time. After incrementing time, this calculation is repeated using the state variable values from the previous iteration until the necessary number of solutions is obtained. In the remainder of this chapter, we will provide some general rules for the

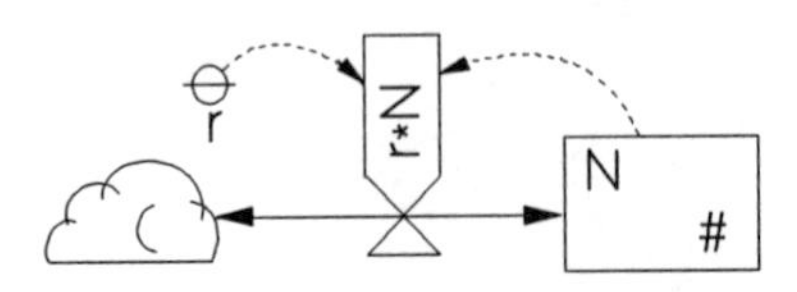

Figure 4.1: Forrester diagram for density-independent population growth.

specification of the rate equations. While I will use specific examples to illustrate the general principles, the equations will vary among disciplines (e.g., enzyme kinetics *vs* ecosystem dynamics). Additional examples are contained in *Part II: Applications.*

4.2 Finite Difference Equations and Differential Equations

4.2.1 Finite Difference Equations

Previously, we have used what I called *finite difference equations* or *recursive finite difference equations.* These have the general form:

$$N_{t+1} = f(N_t). \tag{4.1}$$

The function $f()$ can be arbitrarily complicated, incorporating nonlinear equations (e.g., state variables raised to powers), and positive and negative terms. For some $f()$, we can isolate N_t as a separate element on the right-hand side:

$$N_{t+1} = N_t + f(\text{state variables, parameters}, t). \tag{4.2}$$

Other $f()$ have nonlinear terms that prevent us from writing Eq. 4.1 as 4.2. For a special form of $f()$ in Eq. 4.2, the equation can be simplified and solved analytically, without computer simulation. We do this now to illustrate why these equations are termed "recursive."

Suppose $f() = rN$, which is the classical ecological model for density-independent population growth. This has the Forrester diagram shown in Fig. 4.1 and the following difference equation:

$$N_{t+1} = N_t + rN_t. \tag{4.3}$$

Notice that the figure and the equation match up in a nice way. The label in the box is the state variable that is being projected in time. The parameter r and variable N_t both influence the total rate of change (Eq. 4.3, second term on right-hand side), as indicated by the information flows in the Forrester diagram. The only item missing from the equation is the cloud, but this is precisely what the cloud means: a source or sink that is not modeled.

This equation projects one discrete time step into the future. For additional times, we repeat the process by substituting the left-hand side into the appropriate locations in the right-hand side.

This procedure is a solution to our problem to determine the future values of N. It is possible, however, to also solve the basic equation (Eq. 4.3) analytically, without having to compute intermediate times, by exploiting the recursive nature of the equations. By repeatedly (i.e., recursively) substituting previously computed values of N_{i-1} we have:

$$\begin{aligned}
N_1 &= N_0 + rN_0 = N_0(1+r) \\
N_2 &= N_1 + rN_1 = N_1(1+r) = N_0(1+r)(1+r) = N_0(1+r)^2 \\
N_3 &= N_2 + rN_2 = N_0(1+r)^3 \qquad (4.4) \\
&\vdots \\
N_{t+1} &= N_t(1+r) = N_0(1+r)^{t+1}
\end{aligned}$$

The terminus of the sequence in Eq. 4.4 is the classical analytical solution to the density-independent growth model in discrete time. Not all recursive equations of the general form of Eq. 4.2 can be reduced to the form of Eq. 4.3. Moreover, many of the equations used in population ecology do not have analytical solutions; so, this technique is not generally useful. For other analytical solution techniques, see a mathematics text in difference equations such as Grossman and Turner (1974).

When we use difference equations, we must be clear as to the assumptions we are making about the underlying biology. Recursive finite difference equations assume time is discrete. Indeed, time, in one sense, does not appear in the equations. We have only an arbitrary *index* which here we have symbolized by t and interpreted as *time*. This implies that no events or processes occur between increments of time. Although it is true that we can interpret these time steps to be physical time units as small as we wish (e.g., year, day, second, etc.), the conceptualization is still one of discrete increments. Many biological systems match this situation to a satisfactory degree. An example is the life cycle of an insect that breeds synchronously in the fall, after which all adults die, and the eggs or larvae overwinter to become adults in the spring. Birth and death in this case defines the discrete nature of time. Other systems cannot easily be represented this way, for example, the continuous, unsynchronized reproduction of humans.

In short, when we use finite difference equations we are asserting that time and biological processes are discontinuous and that the equations are exact representations. In the next section, we discuss the case when time is assumed to be continuous, but we discretize time with small time steps to approximate the true situation.

4.2.2 Differential Equations

Differential equations are the continuous time version of finite difference equations written in the form of Eq. 4.2. They have analogous analytical solutions, and as we will see later by discretizing time, their true solutions can be approximated to arbitrary exactness with numerical (computer) methods. But first we will review a bit of basic calculus to better see that the use and solution of differential equations is not a huge step beyond the mathematics we may have learned earlier in our careers (or so the author fervently hopes).

Aside on Derivatives and Integrals

Consider a function such as $y = x^2 + C$, shown as one of the curves in Fig. 4.2a. The derivative of the function at a point x^* is related to the slope at x^*. "Slope" has the usual meaning: "change in y (Δy) divided by change in x (Δx)." Of course, we can numerically compute the slope only for finite values of Δy and Δx. Technically, if we want the slope at a point x^*, then there is no interval over x or y to use. But we recognize that if we take very small intervals around x^* and the corresponding y^*, then we will have a good approximation to the slope. The smaller the interval, the better the estimate of the slope, and if we let the intervals go to zero, we will converge to the derivative at the point. The derivative of a function y with respect to a single variable x is

$$\frac{dy}{dx} = \lim_{\Delta x \to 0} \frac{y_{x+\Delta x} - y_x}{\Delta x}, \tag{4.5}$$

where $\lim\limits_{\Delta x \to 0}$ means "let Δx go to 0" or "let the interval around x^* get arbitrarily small."

Figure 4.2 shows that the numerical value of the slope is different at different values of x^*. The derivative of a function tells us how the slope changes with different values of the independent variable. In this case, the derivative of $y = x^2 + C$ is

$$\frac{dy}{dx} = 2x.$$

This is plotted as the heavy line in Fig. 4.2b. Remember from elementary calculus that the original function $y = x^2 + C$ is the anti-derivative (the integral) of the derivative. For the purposes of the discussion to follow, we will describe two general approaches to obtaining the integral.

The first method treats the integral as a summation: the total area under the derivative curve (Fig. 4.2b) from 0 to 8 (in this case). We approximate the area using discrete increments of the x-axis ($\Delta x = 1$). From Fig. 4.2b, note that the total area of the discretized curve is the sum of the columns. Note also that, by definition, the height of each column is dy/dx. This fact gives us a simple recursive formula for summing the columns if they are indexed from left to right. Column $i + \Delta x$ is the sum of column i plus the derivative times the size of Δx:

$$y_{i+\Delta x} \doteq y_i + \underbrace{(2x_i)}_{\text{derivative}} \Delta x. \tag{4.6}$$

If we begin with $i = 0$ and $y_0 = 0$, then recursively applying Eq. 4.6 N times (using $x_0 = 0$, $x_1 = 1$, etc.) will yield the sum of N columns.

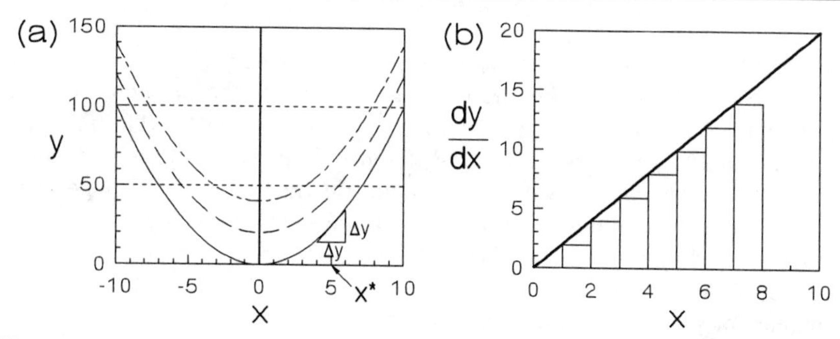

Figure 4.2: (a) The parabola $y = x^2 + C$, where C is an arbitrary constant. (b) The derivative of $y = x^2 + C$, $dy/dx = 2x$ (solid line) and a discretization of the derivative.

The expression $2x_i$ is the derivative whose integral we desire at point x_i. The formula will work for any derivative, if we substitute the appropriate equation for the derivative. The accuracy of the approximation increases as Δx decreases.

The second method to calculate the integral is to simply apply the rules of integration that we all memorized in elementary calculus and remember to this day. The simple derivatives of elementary calculus have a common property that makes this method easy to use. The derivatives have a right-hand side that does not involve the dependent variable. The parabola and its derivative is an example. Consequently, when we apply the rules of integration we are applying a technique known as separation of variables. Below, we apply it to the derivative of the parabola.

$$
\begin{aligned}
\frac{dy}{dx} &= 2x \\
dy &= 2x\,dx && \leftarrow\text{separate variables} \\
\int dy &= \int 2x\,dx \\
\int dy &= y + C_1 && \leftarrow\text{integrate left side} \\
\int 2x\,dx &= x^2 + C_2. && \leftarrow\text{integrate right side}
\end{aligned}
$$

Equating these integrals gives

$$y = x^2 + C.$$

This is trivial because the integral of the separated left-hand side does not involve the dependent variable. Differential equations relax this restriction, and their solutions are more difficult.

Integrating ODEs

An ordinary differential equation (ODE) is any equation involving a derivative of a dependent variable with respect to its independent variable. We are interested in the special case when the independent variable (x, in the above), is time. Unlike the easy derivatives in the previous section, the equation can contain the dependent variable explicitly. The previous section discussed a special case of differential equations. It is significant that ODEs allow the derivative to depend on the value of the dependent variable. This is fundamental to almost all physical and biological systems.

To connect the previous discussion with differential equations of interest to biologists, consider the continuous form of the familiar density-independent population model in ecology:

$$\frac{dN}{dt} = rN. \tag{4.7}$$

The independent variable (t) is time and we interpret the derivative as being a rate of change. In its basic form, this is similar to the derivative of the parabola: it has a derivative on the left-hand side and a function on the right-hand side. Unlike the earlier derivative, this function depends on the dependent variable (N) and not on the independent variable. The integral of $dy/dx = 2x$ gave us the parabola $y = x^2 + C$. This latter equation has a property important to us now: given any value of x, we can compute the value of y. In the current case, if we could find the integral of Eq. 4.7, then given any t, we could compute the value of N. In other words, if we have the integral, we can predict future values.

There are here, as before, two general strategies for finding the integral: apply the rules of integration, or approximate the area under a curve by summing. To show that this differential equation is a simple extension of the calculus we have already learned, we will employ both strategies. We begin with the use of integration rules.

Earlier, we separated the independent and dependent variables and integrated each part separately. In this simple differential equation, we can do the same.

$$\begin{aligned}
\frac{dN}{dt} &= rN \\
\frac{dN}{N} &= rdt && \leftarrow\text{separate variables} \\
\int \frac{1}{N}\, dN &= \ln N + C_1 && \leftarrow\text{integrate left side} \\
\int rdt &= r\int dt = rt + C_2 && \leftarrow\text{integrate right side} \\
\ln N &= rt + C_3 \\
N_t &= e^{rt+C_3} = e^{C_3}e^{rt} = N_0 e^{rt}.
\end{aligned}$$

After setting $t = 0$, we interpret the constant e^{C_3} to be the initial number of individuals in the population (N_0). The last equation in the above series is the *solution* of the differential equation.

Not all differential equations have the simple structure that allows their variables to be separated in this way. Some of these others can be solved with substitutions or other tricks. But if none of the tricks work, then we must use the summation technique to get the integral. It works the same as in the previous derivative, except we discretize t instead of x. This gives

$$N_{t+\Delta t} \doteq N_t + \underbrace{(rN_t)}_{\text{derivative}} \Delta t. \tag{4.8}$$

This equation is clearly similar to a FDE except that we have Δt equal to some number other than 1. Beyond this, however, is the fact that Eq. 4.8 is viewed to be an approximation to the true integral and the FDE was viewed to be an exact representation.

4.3 Quantitative Representation of Fundamental Processes

The previous section demonstrated that (a) the solutions of differential equations are not fundamentally different from the integrals of derivatives as we learned them in elementary calculus, and (b) the form of the numerical solutions can be similar to the discrete, finite difference equations we used to solve dynamic problems (e.g., the leaky bucket). In the future, we will stress the use of differential equations to represent biological models.

One of the main points to be made in this book is that the differential equations used in the various subdisciplines of biology are similar. The models are composed of algebraic components [e.g., (rN)] that recur in many different fields, sometimes in slightly different guises, but still representing fundamentally similar processes. In this section, we describe some mathematical formulations that occur frequently in biological models. Before proceeding, we will need a few basic rules pertaining to translating Forrester diagrams to equations. [In Section 4.4.3 we give a more complete list.] The first rule is that every level in a Forrester diagram is a state variable that requires a differential (or difference) equation. The left-hand side of the differential equation represents the rates of change as they are altered by the objects of the system. The right-hand side describes how these changes occur. The second rule is that, at a minimum, every material flow into and out of a state variable requires an explicit algebraic expression. The sum of these expressions associated with the inflow and outflow arrows is the right-hand side of the differential equation. Grouping all the inflows together and all the outflows together, a general differential equation for a single state variable is

$$\frac{dx}{dt} = \sum \text{inflows} - \sum \text{outflows}.$$

Although the expressions for the inflow and the outflow can be quite complex, take heart in the fact that they will all reduce to the above simple form. Therefore, our problem in quantitative model formulation is "simply" to find the appropriate set of expressions for the inflows and the outflows.

The third rule is that although biological systems are complex, many of them share a few basic processes that have similar mathematical expressions. When viewed across the many relevant hierarchical levels (biochemical, cellular, physiological, ecological), the diversity of living systems is, indeed, immense. It would seem there would be little similarity in the mathematical representations used by the different disciplines to model the variables and processes specific to their domains. This is true to a certain extent, but, nevertheless, there are recurrent mathematical forms that appear in many systems. In this section, we discuss these general forms both for their own value in all biological modeling as well as to illustrate the basic method of creating quantitative models. In later chapters, we discuss specific models and concepts and equations germane to different subdisciplines of biology.

The approach we take here is a *tool-kit* approach to model construction. We will identify a relatively small set of biological processes and their mathematical representations (the tools) and link these together according to the biological hypotheses to form the complete model. In the sections that follow, we present some of these basic processes and their corresponding mathematical implementation. The description proceeds from simple to more complex biological processes and relations.

4.3.1 Constant and Bulk Flow Rates

The simplest process of interest occurs when there is only material flow and no information transfer between a state variable and the inflow or outflow rates. The rate is, therefore, constant and determined by a parameter. An example of this type of flow is shown in Fig. 4.3, which illustrates the Forrester diagram, the differential equation, the relation of the rates to the affected state variable, and the resulting dynamics. The plot of the rates against the quantity of S is interesting for its contrast to later examples that illustrate feedback. For now, simply note that the hypothesis that the absolute rates are constant implies that the dynamic values of the state variable can have no effect on the rates.

The hypothesis that flows are independent of state variables can be extended to multiple compartments (Fig. 4.4). The model, in this case, is a system of three differential equations:

$$\begin{aligned} \frac{dS_1}{dt} &= F_{01} - F_{12} - F_{13} \\ \frac{dS_2}{dt} &= F_{12} + F_{32} - F_{24} \\ \frac{dS_3}{dt} &= F_{13} - F_{32} - F_{34}. \end{aligned} \tag{4.9}$$

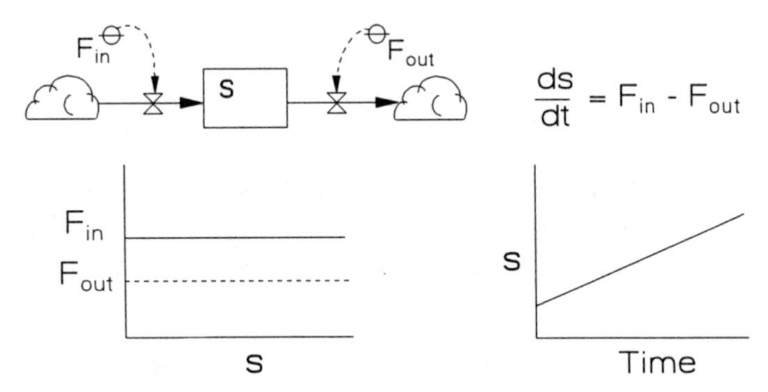

Figure 4.3: Constant rate of flow into a state variable.

Notice the pattern of the arrows and the right-hand side of each equation. Also note that for flows between two compartments an inflow arrow to one compartment (e.g., F_{32} into S_2) is an outflow arrow from another compartment (e.g., S_3). This relationship is reflected in the signs attached to the flows in Eq. 4.9. Finally, it should be obvious that, since each F_{ij} is a constant number, we could collapse the equations so that the right-hand side of each is a single number. These numbers will be positive or negative depending on the relative magnitudes of the F_{ij}. This simple model is frequently used in models of large complex systems (e.g., whole, terrestrial ecosystems) where it is difficult to perform experiments that reveal the internal system controls that influence the flows. There are very few dynamical systems that satisfy the basic assumption that rates are constant.

4.3.2 Dynamic Relative Rates

A more common model is one in which it is hypothesized that the rates are influenced by one or more state variables. A fragment of such a model is shown in Fig. 4.5, where A is the effect of S_1 on the rate and B is the effect of S_2. A and B can be simple or complicated algebraic expressions, but a common method of incorporating these effects into the differential equation is to multiply the auxiliary variable by the current quantity of the

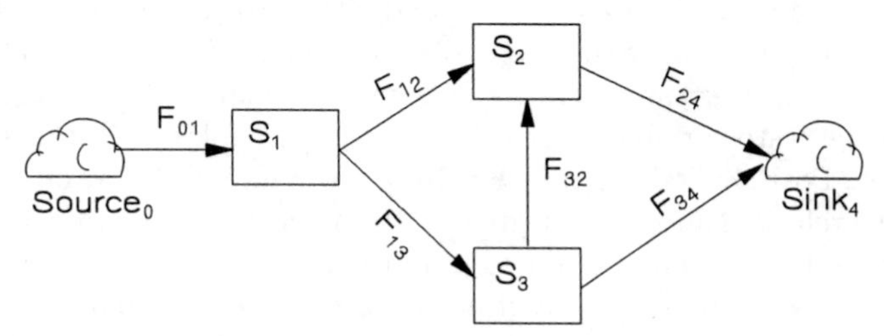

Figure 4.4: Modified Forrester diagram for constant rates of flow among three state variables.

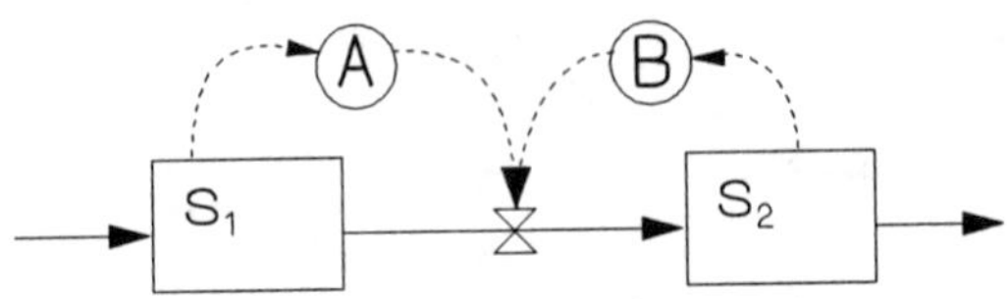

Figure 4.5: Simple information transfer illustrating the influences of state variables on rates.

state variable. For example, several different possibilities might be

$$\frac{dS_1}{dt} = \ldots - (A)S_1 - \ldots \tag{4.10}$$

$$\frac{dS_1}{dt} = \ldots - (B)S_2 - \ldots \tag{4.11}$$

$$\frac{dS_1}{dt} = \ldots - (A)S_1(B)S_2 - \ldots \tag{4.12}$$

The quantities A and B are *relative* or *per capita* rates. They are the contribution of one unit of the state variable to the flow. When multiplied by the current quantity of the state variable, we compute the *absolute* rate for that particular flow.

When influence B is absent (Eq. 4.10), we say the flow is *donor controlled*, since the "donating" variable (S_1) determines the rate. When A is absent (Eq. 4.11), the flow is *recipient controlled*. This jargon is not particularly enlightening since it is common for flow rates to be determined by both donor and recipient variables (Eq. 4.12). The key point, however, is that an extremely common mathematical form is the multiplication of the controlling variable (S_i) by the auxiliary variable that represents the mechanism by which the control occurs. This mechanism is frequently cast as a relative rate (Eq. 4.7). You will have come a long way when you are able to perceive this form in unfamiliar models.

4.3.3 Feedback

Feedback is pervasive in biological systems and is one of the fundamental processes that is contained in almost all interesting models. It refers to the relationship in which increases or decreases of the value of one or more *controlling variables* affect the *rate* at which a process occurs. The action on the rate can be *direct* or *indirect* and either *positive* or *negative*. The action is direct when only the single variable affected is involved. The value of the state variable influences its own rate of change. If the mechanism affecting the state variable involves other state variables, then the feedback is indirect. Positive and negative feedback are endpoints on a continuum of dynamical relationships. The degree to which a feedback relation is positive or negative depends on the function and parameters. Any given relation can be either strongly or weakly negative or positive. The balance between the two produces the possibility of sustained oscillations (i.e., dynamics that neither blow up nor return to an equilibrium).

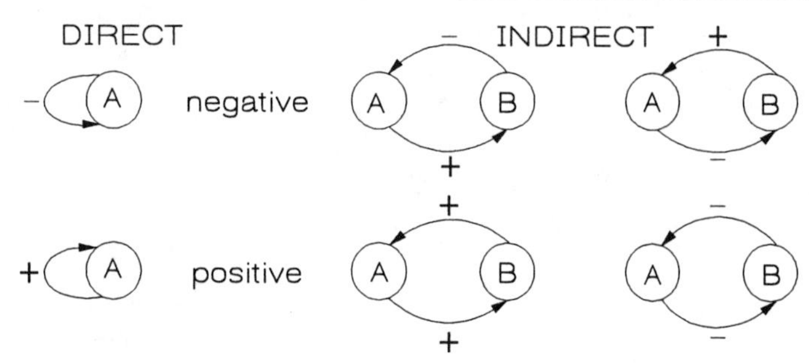

Figure 4.6: Qualitative analysis of direct and indirect effects of system influences producing either positive or negative feedback. The sign on each arc represents the effect of the influencing variable on the variable that terminates the arc.

The qualitative nature of these relationships is revealed by *loop analysis* (Levins 1974). Some very simple examples are shown in Fig. 4.6. The "+" or "–" symbols attached to the arrows indicate the direction of the effect on the future values of the state variable (i.e., positive or negative, respectively). The basic test of feedback direction on a state variable (e.g. A) is to determine whether A, if it is increased in quantity, will decrease or increase as determined by following the effects around a loop. For example, the upper left indirect loop in Fig. 4.6 is negative because an increase in A will increase B which will then decrease A.

Positive Feedback

The simplest form of positive feedback is direct, and occurs when the absolute rate of change of a state variable is an unbounded, increasing function of the state variable. (Recall that absolute rate of change is the rate associated with a flow into or out of a state variable.) In other words, the more there is of the state variable, the greater the positive rate of change of the state variable. The traditional example of this is unrestricted, or exponential population growth (Eq. 4.7), as shown in Fig. 4.7. However, positive feedback can also cause a variable to become more negative. A simple example is the spatial position of a frictionless ball that is confined to rolling in one dimension down a slope that falls away in the negative x direction. As the object moves further in the negative direction, the rate of increase in the negative direction increases. The value of the state variable (position along the x-axis) becomes more negative.

Any number of equations can produce this behavior and it can result from both direct and indirect causes. The critical feature is that the rate increases without bound.

Negative Feedback

Negative feedback is any feedback that is not positive. In other words, the rate of the process is bounded for positive values of the controlling variable. The rate of change does not increase to infinity as the variable

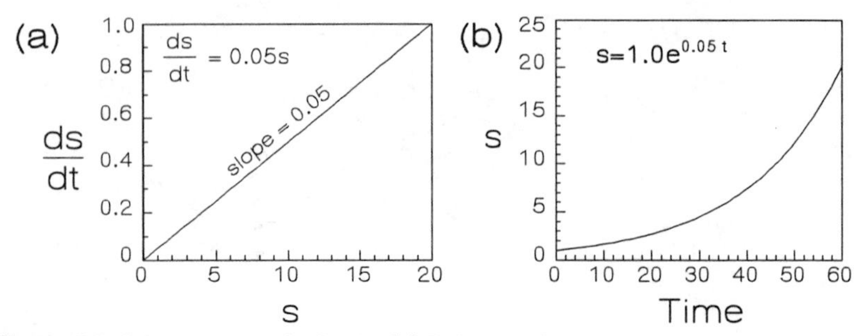

Figure 4.7: Direct positive feedback. (a) Relation of absolute rate of change in a state variable to the value of the variable and a differential equation that behaves in this way. (b) The resulting dynamics.

increases. There are three primary mathematical methods by which this condition can be implemented: feedback by *self-inhibition*, *limitation by extrinsic factors*, and *process saturation*.

Self-Inhibition When a system shows direct negative feedback based on per capita mechanisms, there is a negative relation between the value of the controlling variable and the per capita rate. The variable inhibits its own further growth. The most familiar example is density-dependent, *logistic* population growth. In this model, the relevant rates of change are

$$\frac{dN}{dt} = \left[r\left(1 - \frac{N}{K}\right)\right]N$$

$$= rN - \frac{r}{K}N^2 \tag{4.13}$$

$$\frac{dN}{dt}\frac{1}{N} = r - \frac{rN}{K}, \tag{4.14}$$

where r is the maximum per capita rate of change, and K is the carrying capacity of the population. Since this model has a single state variable, N is the controlling state variable. Equation 4.14 represents the per capita rate and is shown in Fig. 4.8a. Equation 4.13 represents the absolute rate of the process (population growth) and is plotted in Fig. 4.8b. These plots show that negative relations between per capita rates of change and the variable N, produce bounded rates of increase.

Extrinsic An extrinsic factor may limit a process. Consider a beaker of cold water that is warming up to ambient temperature. We note the following facts:

1. The water temperature is initially below ambient and does not surpass it.
2. The rate of temperature change is initially large and decreases over time.

These facts are consistent with the hypothesis that the rate of temperature rise is a function of the difference between the current temperature

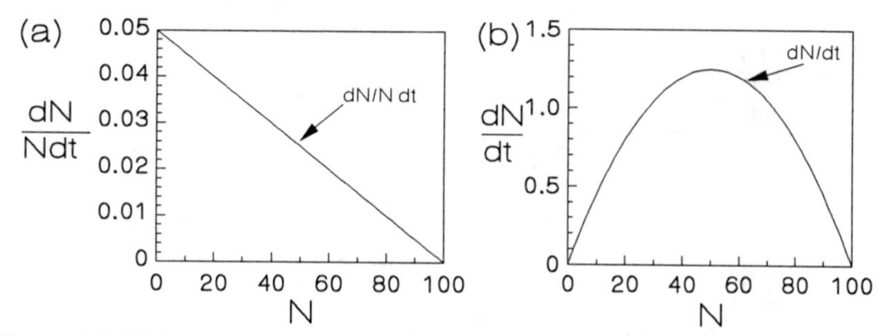

Figure 4.8: (a) Per capita rate of change in density-dependent model as a function of population size. (b) Absolute rate of change in density-dependent model as a function of population size.

and the ambient temperature. A simple model (Newton's Law of Cooling) is based on a linear equation

$$\frac{dT}{dt} = k(T_a - T), \tag{4.15}$$

where T_a is constant ambient temperature and k is a constant of proportionality that is determined by the physical characteristics of the fluid.

This model simply hypothesizes that the rate of warming is proportional to the difference (i.e., the *gradient*) between the container temperature and the ambient temperature. This differential equation has a solution whose time course looks like a hyperbola: T asymptotically approaches T_a, and the absolute rate of change goes to zero. Clearly, the derivative is bounded, and the bound is determined by the ambient temperature.

The basic concept here is that a rate of flow into or out of a state variable (T) is controlled by the difference between a quantity associated with T (e.g., the temperature of the container) and a *similar* quantity associated with the environment of T or another state variable. By an *extrinsic* factor, we mean any quantity "outside" of the state variable to which the differential equation applies. This other quantity may be in the nebulous "unmodeled" environment (e.g., ambient temperature) or it may be the current state or associated auxiliary variable of another, modeled state variable.

Extrinsic factors are particularly important when we model a flow of materials or energy over a physical distance. In the warming beaker example, this was exemplified by the flow of heat energy from the beaker to the environment. It is also applicable to diffusion of molecules across a barrier, where the relevant gradient is the difference in concentrations on both sides of the permeable barrier. In organ-level physiological models, substance concentration is modeled as being moved by bulk transport along with a carrying medium (e.g., O_2 in blood). The rate of flow of blood between organs (e.g., liver and kidney) is proportional to the difference in blood pressure at the two sites. In ecological systems, the migration of a population of animals between habitats (e.g., forest and grassland) may

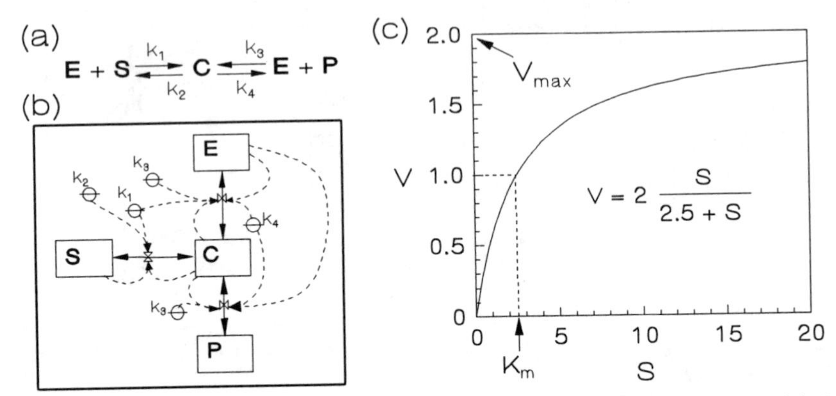

Figure 4.9: Michaelis–Menten saturation feedback control of chemical dynamics: (a) the chemical diagram, (b) the Forrester diagram, and (c) the rate of formation (V) of the product (P). E represents enzyme concentration, S is the concentration of the substrate, C is the complex formed by the chemical binding of E and S, and P is the product. k_i are the rates of conversion.

be modeled in analogy with diffusion, i.e., proportional to the difference in densities of animals at the two sites. All of the above examples use the differences between quantities to calculate the rates of flow.

In cases where a process is determined by several gradients, we must combine the effects in some way. For example, nerve cell voltage potential across the membrane is determined by the ionic gradients associated with Na, K, and Cl. A standard approach is to model the net potential as being proportional to the sum of the gradients of each ion (Deutsch and Deutsch 1992). A similar approach would be appropriate in models of animal dispersal among neighboring, discrete patches of habitat. The rate of flow from a given patch to any of its neighbors would be proportional to the sum of the differences between the pairs of patches.

Saturation Negative feedback frequently emerges in systems through an interaction between the quantity of the donor variable and the ability of the recipient to convert the donor substance. By analogy with chemical dynamics where this is common, negative feedback puts bounds on rates by *saturating* the recipient. Basically, this is nothing more than a bottleneck effect. Saturation is a case where the relation has elements of both positive and negative feedback: the rate neither decreases to 0 nor does it increase indefinitely. The overall dynamical effect is feedback intermediate between positive and negative which permits persistent oscillations to occur.

The Michaelis–Menten model of enzyme kinetics is an excellent example. This model describes the dynamics of the formation of a product (P), in which we may be interested for its own sake or because its concentration is an important component of a larger system (e.g., a step in the Krebs Cycle). Figure 4.9a shows a pictorial representation of the chemical reactions involved in the interaction between an enzyme (E) and a substrate

(S) that combine (C) to form the product. A plausible Forrester diagram is shown in Fig. 4.9b. The differential equations are

$$\frac{dE}{dt} = -k_1ES + k_2C + k_4C - k_3EP \quad (4.16)$$

$$\frac{dS}{dt} = -k_1ES + k_2C \quad (4.17)$$

$$\frac{dC}{dt} = k_1ES - k_2C - k_4C + k_3EP \quad (4.18)$$

$$\frac{dP}{dt} = k_4C - k_3EP, \quad (4.19)$$

where the k_i are rate constants.

Note the relation among the equations, the chemical diagram, and the Forrester diagram. This is a perfectly good model of the system, but usually the rates of formation and breakdown of C are very fast compared to the rates of formation of the product. Since we are primarily interested in P and not C, we want to simplify the model by eliminating the need to track C. We do this by assuming that (a) the experiments are performed when P_t is present only at negligible concentrations (i.e., initially absent) and (b) the rate of formation of C equals its breakdown rate. After suitable algebraic manipulation, the rate of P formation is described by the Michaelis–Menten equation

$$V = V_{max}\left[\frac{S}{K_m + S}\right], \quad (4.20)$$

where V is the rate of P formation. [See Rubinow (1975) or Murray (1989) for detailed derivations.]

Note that Eq. 4.20 describes an increasing, nonlinear curve (Fig. 4.9c). The independent axis is S and the expression in brackets is a curve that asymptotically approaches 1.0. This basic curve is scaled (*parameterized*) by two parameters: V_{max} (the maximum reaction velocity) scales the velocity to which the curve is asymptotic at large S; K_m scales how "fast" the curve rises toward the asymptote. The shape of the curve is scaled so that $V = 0.5V_{max}$ when $S = K_m$ and is, therefore, called the *half-saturation* constant. Low K_m describes a rapidly rising curve; large K_m describes a slowly rising curve.

This equation is significant for two reasons. First, the Michaelis–Menten equation defines a limit to the rate of the reaction (V_{max}). Properties of the enzyme (e.g., the time required to join with S, alter the substrate's molecular configuration, and disassociate from the complex leaving P) and quantities of E limit the rate of the reaction. Thus, the saturation of the enzyme has produced negative feedback. Second, we represented a control on a rate by a basic nonlinear relation $[S/(K_m + S)]$ multiplied by a constant (V_{max}). This is a very common strategy in quantitative model formulation: hypothesize a basic relation, then multiply it by a constant to scale it multiplicatively for a particular process.

Besides chemical reactions, this basic relation is also used to model the effects of the concentrations of dissolved nutrients on phytoplankton growth and the foraging rates of predators. In the latter case, the equation is rewritten using different parameter definitions. The new form is also based on the general equation for a hyperbolic relation: $y = \frac{Ax}{(B+x)}$ (see Section 4.5, *Useful Functions*). With suitable rearrangement, this is also the form for the *Holling disc equation* (Holling 1959) which relates the numbers of prey (y) consumed by a predator in a fixed period of time (e.g., 1 day or 1 experiment duration) to the density of the prey available. The typical parameterization is

$$y = aT_T \left[\frac{x}{1 + abx} \right], \tag{4.21}$$

where a is successful search rate (units: prey/time) times the probability of detection, T_T is the total time available for foraging (units: time), b is the handling time per prey (units: time/prey), and x is the concentration of prey. The Holling disc equation is one form for the Type 2 functional response of predators. Analogous to the rate of product formation in chemical reactions, the rate of prey consumption is saturated by properties of the predator (handling time and hours in the day available for foraging). Other such asymptotic functional forms are shown in Section 4.5, *Useful Functions*. But again, note the similar form for representing saturation feedback: a basic asymptotic relationship times a variable (i.e., aT_T) to scale the rate to the process.

Combined Feedback Interactions In some systems, saturation or positive feedback can combine with inhibition to produce more complicated relations between variables and rates. In this case, at low values of the variable the response is positive to the addition of a unit of the variable (e.g., during a saturation process). But at high levels of the variable, adding a unit of the variable produces a decrease in the rate. For example, at low light levels, the rate of photosynthesis of a leaf increases until saturation occurs; further increases in light cause a decrease in photosynthesis because of *photoinhibition*. An example from population ecology is the effect of population density on per capita births. At low densities, females have difficulty in finding mates; per capita births will increase as the number of males (and females) in the population increases. Eventually, however, birth rates will decline at high densities due to competition. This is known as the *Allee effect*.

Usually, this general phenomenon of combined feedback is produced by the action of two or more biological mechanisms (e.g., light saturation of photoreceptors and degradation of enzyme systems at high light intensities, or mate location and competition). Consequently, this situation is frequently modeled as the product of two separate factors. For example, photoinhibition can be modeled as follows (Steele 1962):

$$P = P_{max}\Big[\underbrace{(aI)}_{\text{increase}}\ \underbrace{(e^{1-aI})}_{\text{decrease}}\Big], \tag{4.22}$$

where P is the photosynthesis rate, P_{max} is the maximum photosynthetic rate, I is light intensity, and a is a shape parameter. Again, note the use of a relative rate (Eq. 4.22 in brackets) scaled by a third parameter, P_{max}.

4.3.4 Mass Action

A biological process that recurs in many models is *mass action*. The chemical dynamics just presented (Eqs. 4.16–4.19) used the concept extensively by modeling some rates as proportional to the product of the concentrations of two molecules. The Law of Mass Action states that the rate of a reaction is proportional to an integral power of the concentrations of all substances taking part in the reaction (Carson et al. 1983).

If P and Q are the concentrations of two substances and R is a rate of transformation of substance Q, then a general model of R is

$$R = aQ^{\alpha}P^{\beta},$$

where a is a constant of proportionality, and α and β are integer powers. The *order* of the reaction relative to P or Q is β and α, respectively. The order of the overall reaction is the sum of the powers. In zero-order reactions, the rate of change is a constant, independent of the dependent variable ($\alpha = \beta = 0$). In first-order reactions, the rate is proportional to the concentration of only one of the substances. Second-order reactions may be caused by an interaction of two substances ($\alpha = 1$, $\beta = 1$) or a second-order function of one substance (e.g., α or β equal to 2).

The values of the orders of the relations are often determined by the *stoichiometric* or weight relations of the compounds involved in the reaction. For example, suppose we have this chemical reaction:

$$\mathrm{A} + 2\mathrm{B} \underset{k_2}{\overset{k_1}{\rightleftarrows}} \mathrm{C}.$$

The corresponding differential equation using mass action for C is

$$\frac{dC}{dt} = k_1AB^2 - k_2C,$$

where B is raised to the power of 2 because two molecules of B are required.

The mechanistic hypothesis underlying this functional form is analogous to that of the probability of encounter among randomly moving particles. For example, in a reaction in which $\alpha = 1$ and $\beta = 1$, we hypothesize that a reaction will occur whenever two molecules of the two substances are brought together to the same place at the same time. Since we are dealing with the concentrations of the substances, this is similar to saying that the rate of the reaction is proportional to the probability that the two molecules will collide. Q and P are not true probabilities, of course, since they can have values greater than 1.0. In Eq. 4.17, both first- and second-order reactions were hypothesized.

While these relations are fundamental in chemical dynamics, they have also been applied in ecology. The classical Lotka–Volterra predator–prey equations are a good example:

$$\frac{dV}{dt} = rV - aVP \tag{4.23}$$

$$\frac{dP}{dt} = abVP - dP, \tag{4.24}$$

where the victim (V) grows in a density-independent fashion with rate r. Predators (P) die at a constant per capita rate d. The term aVP (Eq. 4.23) quantifies the rate at which prey (V) are consumed by predators (P), so a is the search rate. Predators convert the food consumed into new predators with an energetic efficiency b. Since we generally apply these equations to densities of prey and predators, we assume that the prey are removed according to the probability that individuals of the two species will coincide in time and space.

4.3.5 Conservation of Mass and Energy

The concept of conservation of mass is important to almost all biological disciplines. It plays a role in biochemical dynamics, nutrient and pollutant flows in ecosystems, and transport of material in space. The central idea is that material or energy that flows from one place to another is lost from the first and an equal amount is gained by the second place. If this is not the case, then there must exist one or more additional sinks for the outflow material. If the mass or energy is to be conserved, then all sources and sinks must be accounted for. We will discuss two situations in which the concept occurs. The first treats a system that has no spatial extent and the "places" for flow are biological compartments (e.g., a state variable constituted by a set of herbivores). The second assumes the system has spatial extent and part of the equation to conserve mass involves its movement from one geographical location to another (e.g., a pollutant moving along a river).

The biochemical system described by Eqs. 4.16–4.19 is a good example of a material flow that leaves one compartment and arrives, in equal amount, in another compartment. For example, compartment C loses mass to one sink at the rate $-k_2C$ (the minus sign indicates loss), and compartment E gains mass, through this pathway, at a positive rate of $+k_2C$.

Ecosystem Example

To gain more experience with the equations and to see the application of these ideas to another area of biology, we will examine a simple model

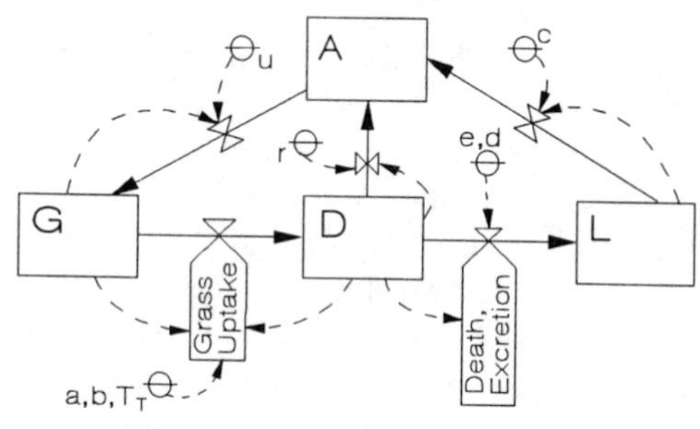

Figure 4.10: Carbon flow in a simple terrestrial ecosystem.

of carbon flowing through an ecosystem. Figure 4.10 shows the Forrester diagram for the system. A possible model that is consistent with Fig. 4.10 is the following set of equations.

$$\frac{dD}{dt} = \left(\frac{aT_TG}{1+abG}\right)D - rD - D(e+d) \tag{4.25}$$

$$\frac{dG}{dt} = uG - \left(\frac{aT_TG}{1+abG}\right)D \tag{4.26}$$

$$\frac{dL}{dt} = D(e+d) - cL \tag{4.27}$$

$$\frac{dA}{dt} = -uG + rD + cL. \tag{4.28}$$

The verbal definitions of the parameters are contained in Table 4.1. The details are given below.

In Eq. 4.25, we assume that deer (D) are limited by their resource (i.e., $G = 0$ implies no growth of D) and by restrictions on foraging behavior (e.g., foraging time, handling time). The Holling disc equation is used and describes the rate of consumption of g C by a single deer. We multiply this by the number of deer present to obtain the total amount of grass (G) removed by deer. We further assume that a fixed rate (proportion) of the carbon in D is respired away to the atmosphere: $-rD$. This is a simple, linear equation; it assumes that if the deer population gets very large, the amount of carbon respired also gets very large: it is not bounded by a saturation feedback. We also use linear relationships to describe the

Table 4.1: Parameter definitions for a carbon flow model.

Parameter	Definition
a	Deer successful search rate for grass
b	Deer handling time while eating grass
c	Rate of decomposition of feces and dead deer by bacteria
d	Rate of feces production by deer
e	Fraction of deer carbon becoming rotting corpses
r	Rate of deer production of gaseous carbon (respiration)
T_T	Total time for foraging
u	Rate of atmospheric carbon uptake by grass

loss of carbon from deer to a lumped compartment (L) of all decaying by-products of deer (dead carcasses, feces, urine, etc.). Here, we describe just two of them: the rate that deer die (e), and the rate of feces production (d). These are all the inputs and outputs to the deer compartment that we hypothesize as important.

Equation 4.26 shows only two flows: a single input and output. The input is the removal of CO_2 from the atmosphere ($+uG$). This expression assumes that grass (G) growth rate is not limited by atmospheric carbon. This flow is a recipient-controlled flow and it assumes that grass can consume as much CO_2 as necessary at a rate that is proportional to the amount of G present. It does not depend on the amount of A present, and this is an important biological assumption. The output from G is the expression for the Holling disc equation just as it appears in Eq. 4.26. This is an instance of conservation of mass: the amount that left G entered D.

The equation for decaying deer by-products (L, Eq. 4.27) also shows conservation of mass. These losses from D are the inputs to L. In addition, we assume that bacterial decomposition of these by-products (expressed as the amount of carbon entering the atmosphere) occurs at a rate that is proportional to the amount of decaying matter present ($-cL$). This assumes that there are no other variables (e.g., moisture or temperature) that control or influence this flow.

Finally, Eq. 4.28 assumes that the atmosphere is essentially a passive compartment whose rate of change is determined by the requirement to conserve carbon in the system. Grass removes as much carbon as needed ($-uG$), independent of the amount of carbon in A, and A is replenished by losses of gaseous CO_2 due to deer respiration ($+rD$) and bacterial decomposition ($+cL$). Once we have made what we hope are reasonable assumptions for the biological compartments, the equation for A simply contains the same flows but with reversed sign.

Spatial Flows

We next discuss the case where the flows are physical flows between spatially separate compartments. We have already introduced these ideas in Chapter 3. When the spatial resolution is such that only a few, large regions are modeled (such as broad areas in a lake), then the problem can be treated just as we treated the carbon flow problem. We write ordinary differential equations (analogous to Eqs. 4.25–4.28) for each spatial area with appropriate flows between the various spatial regions. The important distinction here is not the size of the region, but rather the extent to which the region is an isolated and discrete entity. In situations where we can not reasonably assume homogeneous regions (i.e., where there is a continuous gradation of the spatial structure), we must use a different conceptual framework.

In these cases, the framework we use is based on *partial differential*

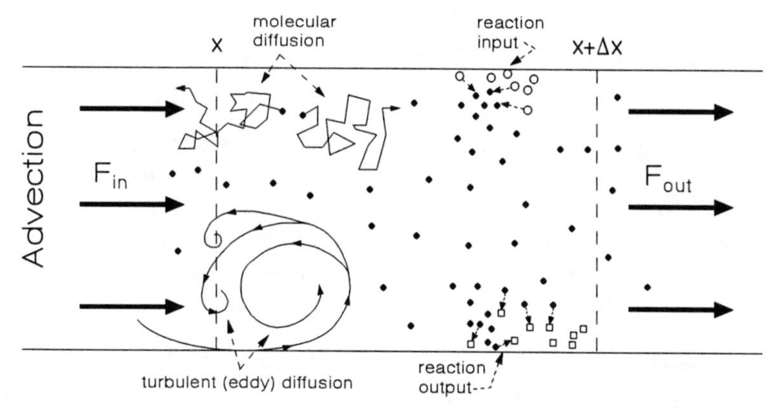

Figure 4.11: Flows and processes in one-dimensional fluid flow. Advection flow is from left to right. Solid dots represent particles of the substance of interest. The vertical dotted lines represent arbitrary, imaginary boundaries located at x and $x + \Delta x$.

equations (PDEs). These equations form a very important and difficult part of applied mathematics. Formulating and solving models using PDEs is not easy, and it is recommended that interacting with an applied mathematician will be helpful. Here, we only give some background and a brief introduction to some of the solution strategies. We emphasize fluid dynamics, especially flows of solutes (C) in water.

Envision a medium that flows in one dimension in which a solute (C) is dissolved. This might be a very simple model of a pollutant in a river. We wish to model the concentration of C at all points along the one dimension and over time. Thus, we now have two independent variables (time and space) over which the state variable (C) varies. Four fundamental processes affecting fluids and solutes recur in these models: (1) advection, (2) molecular diffusion, (3) turbulent diffusion, and (4) reaction. We discuss each in turn.

Figure 4.11 shows the basic physical flow system with the four components pictured. The continuous spatial dimension is arbitrarily divided into discrete segments bounded by x at the left and $x + \Delta x$ on the right. Fluid, containing the substance of interest at concentration C_{in}, enters the segment of interest at x with velocity F_{in}. While the molecules of the substance are in the segment, they may move randomly because of diffusion caused by thermal energy. The molecules may also be caught in eddies generated by turbulence. Molecules of the substance may be created or destroyed within the segment as a result of chemical or biological processes. Finally, molecules may be carried out of the segment along with the fluid, which leaves $x + \Delta x$ at velocity F_{out}.

In earlier examples, when we were concerned only with ordinary differential equations having a single independent variable, time, we thought of the system as moving forward through time in discrete steps (Δt) according to the currently computed NetChange in time:

$$y_{t+1} = y_t + \Delta t[\text{NetChange(t)}]. \tag{4.29}$$

In considering spatial changes, we use an analogous concept. First, assume the system is in temporal equilibrium in order to ignore changes in time for the moment. In a segment of the spatial dimension (Δx) we have an inflow (F_{in}) and an outflow (F_{out}). By conservation of mass and analogy with time, we have a finite difference equation based on discretized space:

$$C_{out} = C_{in} - \Delta x[\text{NetChange(x))}],$$

where C_{in} is the concentration at x and C_{out} is the concentration at $x+\Delta x$.

More conventionally, we write

$$C_x = C_{x+\Delta x} + \Delta x[\text{NetChange(x)}]. \tag{4.30}$$

NetChange(t) in Eq. 4.29 is the right-hand side of a differential equation (e.g., dy/dt). Analogously, NetChange(x) in Eq. 4.30 is also the right-hand side of a differential equation: dC/dx.

$$\lim_{\Delta x \to 0} \frac{C_x - C_{x+\Delta x}}{\Delta x} = -\frac{dC}{dx} \tag{4.31}$$

When we add time and require conservation of mass, we must insure that the temporal changes in C equal the spatial changes in C. Since C is being changed by processes both in time and space, we use the partial derivatives to represent the two modes

$$\frac{\partial C}{\partial t} = -\frac{\partial F}{\partial x}, \tag{4.32}$$

where F represents a complex function of several physical processes. This simply says that the rate of change of the concentration in a segment must equal the inflow minus the outflow. To see this, imagine a stream of fluid having a cross-sectional area of A and flowing in one dimension from left to right. The velocity of fluid coming into a segment of length Δx will be F_x, and the velocity out of the segment will be

$$F_{out} = F_{x+\Delta x} = F_x + \frac{\partial F}{\partial x}\Delta x.$$

The change in mass M of the solute in the segment over a time interval Δt is

$$\Delta M = A\Big[\underbrace{F_x}_{F_{in}} - \underbrace{\left(F_x + \frac{\partial F}{\partial x}\Delta x\right)}_{F_{out}}\Big]\Delta t.$$

This is a statement of the principle of conservation of mass. Dividing both sides by $A\Delta x$ converts mass to concentration (C_x). Dividing by Δt and taking limits gives Eq. 4.32. This basic equation will change slightly when we add reaction processes below. But for now we will keep this one and expand it with equations for advection and diffusion by writing expressions for F.

Advection Advection is the flow of media and the solute from point to point. If the velocity is a constant U over a small spatial interval, then the flux of C is simply

$$F = UC,$$

so

$$\begin{aligned} \frac{\partial C}{\partial t} &= -\frac{\partial F}{\partial x} \\ &= -\frac{\partial (UC)}{\partial x}. \end{aligned} \tag{4.33}$$

Diffusion Molecular diffusion is the movement of mass due to random motion of individual molecules. Figure 4.11 shows two hypothetical paths. Based on Fick's Laws (Berg 1983), the flux F through a plane is proportional to the spatial gradient of the concentration over a small Δx. Or, after letting $\Delta x \to 0$

$$F = D\frac{\partial C}{\partial x}. \tag{4.34}$$

For diffusion alone and substituting Eq. 4.34 into Eq. 4.32, the conservation equation is

$$\begin{aligned} \frac{\partial C}{\partial t} &= -\frac{\partial F}{\partial x} \\ &= \frac{\partial D \frac{\partial C}{\partial x}}{\partial x} \\ &= D\frac{\partial^2 C}{\partial x^2}, \end{aligned}$$

where D is a constant called *diffusivity* and is assumed here to be constant over x.

Putting advection and diffusion together to find changes in C, we have

$$\frac{\partial C}{\partial t} = D\frac{\partial^2 C}{\partial x^2} - \frac{\partial (UC)}{\partial x}. \tag{4.35}$$

This is an extremely common form for biological PDEs called a *conservation equation.* You will see it often, especially in spatial chemical dynamics and morphological development (Edelstein-Keshet 1988, Murray 1989). We develop a model of insect movement in Chapter 15 that uses an equation of this form.

The second manifestation of diffusion is turbulent diffusion, which is too hard for us to describe here. Turbulent diffusion is hard because it is scale dependent: the fluxes due to turbulence depend on the size of Δx one chooses. The larger the Δx, the larger the eddies and fluxes involved (Fig. 4.11). Simulation of turbulence is an active research topic in theoretical physics and involves some very subtle programming and physical details that we cannot address here. Consequently, we will sweep this big problem under the rug by assuming that our time scale is long enough that the average effect of turbulent diffusion can be treated as a component of the advection term (U in Eq. 4.33). Smaller scale phenomena will be lumped in empirical measurement of molecular diffusivity.

Reactions Reaction processes are any processes other than advection and diffusion that change the concentration of a solute inside the spatial interval Δx. These may be chemical interactions (e.g., the substance going in or out of solution), or biological uptake and excretion (e.g., the uptake of nitrogen by plants). These processes are treated mathematically as an ordinary differential equation. For example, suppose nitrogen is removed from solution by plants (P) according to a Michaelis–Menten relation and excreted by fish (T) in proportion to the amount of fish present. In addition, advection and molecular diffusion occurs. Then the conservation equation is

$$\frac{\partial N}{\partial t} = \underbrace{D\frac{\partial^2 N}{\partial x^2}}_{\text{diffusion}} - \underbrace{\frac{\partial (UN)}{\partial x}}_{\text{advection}} - \underbrace{\mu_{max}\frac{NP}{K_m + N}}_{\text{uptake}} + \underbrace{eT}_{\text{excretion}}. \tag{4.36}$$

In this equation, *uptake* and *excretion* are the two biological processes constituting the *reaction*.

In general, we must describe material transport in three spatial dimensions. For the processes described above, we add the spatial fluxes. For example, advection in three dimensions is

$$\frac{\partial N}{\partial t} = \frac{\partial U_x N}{\partial x} + \frac{\partial U_y N}{\partial y} + \frac{\partial U_z N}{\partial z}.$$

Obviously, we must have estimates for each of the average flux rates in the x, y, and z directions (i.e., the U_i above). Diffusion is treated similarly.

Like simple ODEs, some simple PDEs have analytical solutions that describe the value of the variable for any t and any x. Often, however, the equations are too complex for a complete analytical solution, and we must use numerical methods. This is a complex subject, but in Chapter 5 we will discuss one numerical method that has intuitive appeal, is simple to code, but is not particularly fast.

4.3.6 Multiple Controlling Factors

We have seen how negative feedback can arise because of a single limiting factor (either extrinsic or by saturation). Another important feature of interconnected systems (e.g., biochemical cycles, physiological systems, ecological foodwebs) is to have multiple factors controlling a rate. There are two different, common situations. First, the equation for the rate is a univariate function of a primary influencing variable (i.e., the x-axis, such as available light intensity), and one or more of the parameters of this

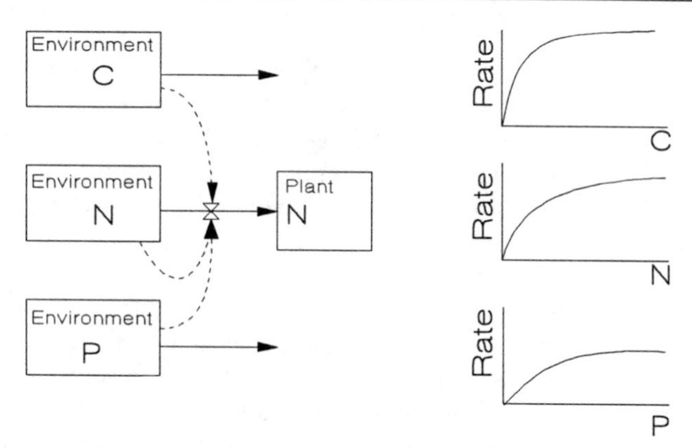

Figure 4.12: Plant growth in which three nutrients interact. On the right is shown Monod growth curves as determined by single-variable experiments that hold the other two nutrients constant.

equation is modeled as a function of a second controlling factor (e.g., g C). Second, the rate is the outcome of several interacting factors that combine to create a function having multiple independent variables.

An example of the first case is a simple model of net photosynthesis rate in plants when it is controlled by both light intensity (I) and carbon availability (C). The primary variable of the rate equation is I and we assume an asymptotic relationship analogous to the Michaelis–Menten relation

$$P = \frac{\alpha I P_{\text{max}}}{\alpha I + P_{\text{max}}}, \tag{4.37}$$

where P is the net photosynthesis rate, P_{max} is the maximum rate, and α is an empirically determined constant. The effect of carbon is to alter linearly the maximum rate

$$P_{\text{max}} = bC, \tag{4.38}$$

where b is the effect of carbon (C) on P_{max}. Substituting, the new equation for net photosynthesis is

$$P = \frac{\alpha b I C}{\alpha I + bC}. \tag{4.39}$$

The second case concerns multiple factors affecting a process that requires all of the factors. Consider the biological case of plant growth in the presence of three nutrients (carbon, nitrogen, and phosphorus). All resources are required for growth (i.e., one cannot be substituted for another). Since the resources have different units, we use parallel, or multiple models, but here we focus our discussion on the N component (Fig. 4.12). The rate of uptake of N is determined by the total growth of the plant, but this is affected by the supplies of the other two nutrients. If one of these is in very low supply, total growth will be small and N uptake will also be small, even though N is plentiful. The modeling problem is to portray mathematically this basic biological fact.

With a single controlling variable N, we could measure growth at experimentally controlled levels of N and fit an equation to the resulting responses. With two controlling variables (e.g., N and C), we could use the same procedure, but using a more complicated experimental design that varies N and C in combination. We could again fit an equation to this two-dimensional response surface and thereby predict growth from simulated values for N and C.

With three (or more) variables (e.g., N, C, and P) the cost of the experiments and the complexity of the equation needed to fit the results often becomes prohibitive. Instead, we seek an intermediate solution in which a series of single-variable experiments are performed (i.e., vary N alone, C alone, and P alone), each response is fit by an equation, and then the three equations are mathematically combined to incorporate the interactions between the variables. These interactions are not measured or exactly known, of course, but we hope that our clever tricks to combine the equations will accurately reflect the interactions. Below, we discuss four general methods to combine the controlling variables: *Liebig's Law of the Minimum, Multiplication, Arithmetic Averaging, Mean Resistance.* We also introduce a fifth candidate specifically designed for combining Michaelis–Menten relations.

To begin, consider the simple case with just a single limiting resource (N). Nutrient uptake across cell walls is mediated by ATP and enzymes, so we use Michaelis-Menten kinetics to relate biomass increase to nutrient concentration. When applied to growth rates, we have the Monod equation

$$\frac{dB_N}{dt} = \underbrace{\overline{\mu}_N \left(\frac{N}{K_{m_N} + N} \right)}_{\mu} B_{N_t},$$

where $\overline{\mu}_N$ is the maximum rate of incorporation of N into plant material per g N of plant material (i.e., a relative or per capita rate). The product of $\overline{\mu}_N$ and the expression in parentheses is μ (the actual relative rate). Now we turn to the situation where multiple factors affect μ.

Liebig's Law of the Minimum

If we assume that a process (biomass growth) proceeds at the rate of the slowest sub-process (uptake of individual nutrients), then we use

$$\mu = \overline{\mu} \left\{ \min \left[\left(\frac{C}{K_{m_C} + C} \right), \left(\frac{N}{K_{m_N} + N} \right), \left(\frac{P}{K_{m_P} + P} \right) \right] \right\},$$

where $\overline{\mu}$ is an average maximum growth rate and min[ ] is a function that returns the smallest of the three numbers. This is Liebig's Law of the Minimum, and it assumes that the limiting effects are independent.

Multiplicative Rates

Alternatively, we could assume that the limiting processes interact. This means that as the growth declines because of limitation due to one nutrient, the ability to grow at the current concentrations of the other nutrients also declines. One method for combining concentrations of the nutrients to implement this hypothesis is to multiply the concentrations:

$$\mu = \overline{\mu}\left[\left(\frac{C}{K_{m_C} + C}\right) \cdot \left(\frac{N}{K_{m_N} + N}\right) \cdot \left(\frac{P}{K_{m_P} + P}\right)\right].$$

Since the expressions in parentheses are all less than 1.0, as we increase the number of limiting nutrients, the growth rate decreases dramatically. Empirically, this form sometimes predicts slower growth rates than observed.

Arithmetic Average Rate

The arithmetic average of the limiting effects is

$$\mu = \overline{\mu}\frac{1}{3}\left[\left(\frac{C}{K_{m_C} + C}\right) + \left(\frac{N}{K_{m_N} + N}\right) + \left(\frac{P}{K_{m_P} + P}\right)\right].$$

This expression has the advantage that it models an interaction between the limiting nutrients, but does not allow the overall growth rate to have extremely low values. Its disadvantage is that the largest value will greatly influence the overall average. This approach may predict an unrealistically high growth rate.

Mean Resistance (Harmonic Mean)

The fourth method analogizes the effect of multiple limitation to the flow of current through an electrical circuit that has resistors in parallel. To illustrate this for our plant growth model, we define an auxiliary variable, *substrate effect*, as the fraction of the maximum growth rate possible:

$$S_{eff} = \frac{S}{B + S}, \tag{4.40}$$

where S represents the concentrations of the limiting nutrients (e.g., C, N, P). So, we have a C_{eff}, a N_{eff}, and a P_{eff}. The *integrated effect* (I_{eff}) is computed from

$$\frac{1}{I_{eff}} = \left(\frac{1}{C_{eff}} + \frac{1}{N_{eff}} + \frac{1}{P_{eff}}\right),$$

or, in general

$$\frac{1}{I_{eff}} = \left(\sum_{i=1}^{n} \frac{1}{S_{eff,i}}\right).$$

If $1/I_{eff}$ is multiplied by $1/n$, we have the harmonic mean. It has the advantages of the arithmetic mean, but emphasizes most the smallest growth rate (i.e., the *most* limiting of the nutrients).

Additive Rates

O'Neill et al. (1989) extensively compared eight families of methods for combining Michaelis–Menten relations. Although this is a specialized function, since it so ubiquitous a relation between substrate and process, especially in ecological and biochemical models, it is germane to a large number of models. They developed a theory of combining two processes based on arrival times of "molecules" necessary for a "reaction" to occur. One is not restricted to chemical reactions here; their results apply to arrival times of prey and predators as well. They developed a new method called the *additive* method (translated to the notation above):

$$PI_{\mathit{eff}} = P\frac{CN}{k_2N + CN + k_1C},$$

where N and C are the concentrations of two substrates, and k_i are constants to be estimated. This function is similar to the inverse of the Mean Resistance formula given above (i.e., I_{eff}, above), when only two substrates are used.

O'Neill et al. (1989) compared the ability of the eight methods to fit 11 different data sets. Overall, in their opinion, the additive and mean resistance models performed best and virtually identically in terms of accuracy to the data. The methods differed in the value of one of the parameters fitted. However, for the data sets on which the Law of the Minimum produced meaningful values, it often had the overall best fit. Unfortunately, there were data sets in which it failed altogether to provide biologically interpretable values. This property disqualified it in the eyes of O'Neill et al. (1989). They concluded that the additive method had an edge over mean resistance because the former reduced exactly to the Michaelis-Menten equation when only one substrate was present. An advantage of mean resistance is that it can apply to functional forms other than Michaelis–Menten.

Summary of Multiple Controls

To summarize this discussion, we can make the following recommendations. Either replace a constant with a function of the secondary controlling variables (case 1), or use a form of competing factors (case 2). In the latter case, *mean resistance* and the *Law of the Minimum* seem to be the most reasonable forms to use, but this can depend on the system. If the individual functional forms are Michaelis–Menten, then consider using the *additive* method. Regardless of which method is used, the basic principle is this:

1. Define a maximum rate for each nutrient or limiting variable and a function (e.g., Monod: Eq. 4.40) that alters the rate at different levels of the variable.

2. Combine the rates of all of the limiting processes into a single RATE using one of the above methods.
3. Incorporate this rate into the final differential equation:

$$\frac{dB}{dt} = \text{RATE} \cdot B_t + \ldots \tag{4.41}$$

4.3.7 Discontinuous Functions

All of the equations we have discussed so far to describe dynamics and auxiliary variables have been continuous; there were no sharp jumps in the value on the dependent variable with small changes in the independent variable. We can argue whether any phenomena at the space and time scales of biological systems (i.e., non-quantum mechanical systems) can be truly discontinuous. Some would say that examining sufficiently small steps on the independent variable would reveal a continuous, albeit extremely steep, change in the dependent variable. In any case, for reasons of simplicity and convenience if nothing else, we often choose to represent the phenomena as discontinuous. A hypothetical example is

$$R = \begin{cases} 2x & \text{if } 0 \leq x < 0.5 \\ 1.0 & \text{if } 0.5 \leq x < 1.0 \\ 1.0 - bx & \text{if } 1.0 \leq x. \end{cases}$$

where R is some quantity used in a differential or difference equation. This example describes a function that (1) increases linearly from 0.0 to 1.0 as x goes from 0.0 to 0.5, (2) is exactly 1.0 for x from 0.5 to 1.0, and (3) decreases linearly for x greater than 1.0. This kind of equation is used when biological morphology interacts with continuous functions. For example, water transpiration from a leaf is determined by the opening of the stomata on the leaf surface. The amount that the stomata are opened is determined by an interaction between the pressure of the guard cells and that of the surrounding epidermal tissue. It is possible to choose reasonable parameters of this interaction such that at sufficiently high epidermal pressure, the calculated stomatal aperture would be less than zero. It is nonsensical to speak of negative aperture opening, so we would use a discontinuous function such as the following:

$$a = \begin{cases} b_g P_g - b_e P_e & \text{if } b_g P_g > b_e P_e \\ 0.0 & \text{otherwise,} \end{cases}$$

where P_g is guard cell pressure, P_e is epidermal cell pressure, and b_g and b_e are proportionality constants. While perfectly legal, this kind of equation can make mathematical analysis difficult. Computers, however, have no difficulty with this type of equation. It is coded as an `if-then-else` construct.

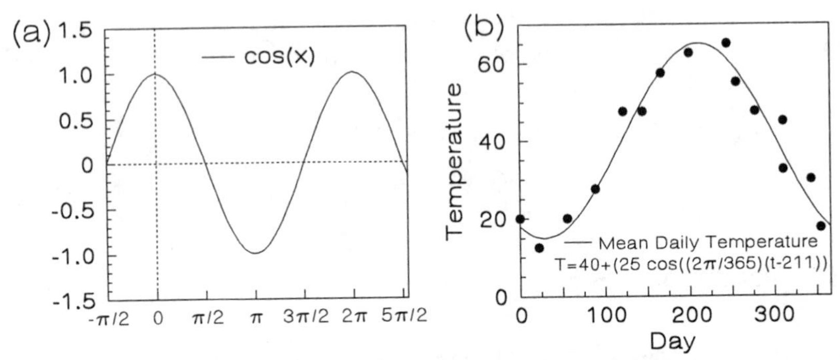

Figure 4.13: (a) An untranslated cosine function. (b) General cosine function with parameters fitting a hypothetical time series of seasonal temperature values.

4.3.8 Time and Driving Variables

Time can be an explicit component of differential equations simply by appearing directly in an equation that varies with time (e.g., "season" in Fig. 3.8). These equations typically describe driving variables. As an illustration, a cosine function is a reasonable function to fit to the yearly cycle of temperature in the northern latitudes. To fit a cosine function to a time series of temperature values, we translate the function vertically and horizontally and adjust its frequency until it matches the oscillations of the data.

Figure 4.13a shows a simple cosine function that completes one cycle in 2π radians and oscillates between ± 1. Real driving variable data (e.g., temperature) are not constrained to these values, so we must use the general equation for a cosine function that permits us to vary these properties

$$y = M + A\cos(\omega(t - t_o)), \tag{4.42}$$

where M is the mean value of the function (e.g., temperature). A is the amplitude of the peak above the mean. $(t - t_0)$ shifts the peak by t_0 physical units. ω is the angular frequency per physical unit; it scales the frequency of oscillations of the function to the physical frequency. Angular frequency has units of radians per physical unit, e.g., radians/time, where time is the period of one cycle in physical units (e.g., 365 days, 24 hours, etc.). We need to choose these four variables appropriately to fit a cosine function to the data.

As an example, suppose a time series of mean daily air temperatures has a mean of 40° F, an amplitude of 25° F, a period of 365 days, and the position of the first peak is on July 30 or Julian day 211 (Fig. 4.13b). This temperature time series is modeled as:

$$T = 40 + 25\cos\left(\frac{2\pi}{365}(t - 211)\right).$$

A second approach to incorporate time in functions used in computer simulations is a look-up table. This method uses the actual data during

A second approach to incorporate time in functions used in computer simulations is a look-up table. This method uses the actual data during a simulation and does not attempt to fit a function. A look-up table of daily temperatures requires two sets of numbers. One set is the Julian days 1...365. The second set is the temperature on that day. The look-up method is computer code that finds the temperature that corresponds to a given day.

4.4 Using the Toolbox of Biological Processes

There are three simple rules for creating a model. Unfortunately, nobody knows what they are. — JWH and W. Somerset Maughan

We have identified and described some mathematical formulations for eight basic biological processes that occur frequently in models: (1) constant rates, (2) relative rates, (3) feedback, (4) mass action, 5) conservation of mass, (6) limitation by multiple controls, (7) discontinuous functions, and (8) time dependence. These are the basic tools in our toolbox for reading and constructing models. These eight do not describe all processes, and within each there are many mathematical variants we have not discussed. Nevertheless, an approach to successfully reading and constructing quantitative models is to combine these basic formulations in ways that represent the biological hypotheses. This is a skill that is achieved only with practice and attention to published models of similar systems. However, we can provide some simple verification and simplification techniques as well as list a few rules of thumb that will aid you in thinking about the equations.

4.4.1 Checking Units

The physical units of the derivative must match the units of the equation on the right-hand side. This will check for two types of errors: (a) inappropriate expressions (e.g., dividing when you should subtract) and (b) bad logic that requires parameter values with incorrect units. The procedure is simply to replace every variable and parameter with its units and to cancel units until no further reduction is possible. If the final expressions of the units of the two sides of the equation are not equal, there is an error.

For example, consider the logistic equation (Eq. 4.13). The units on the left are `numbers/time`. The units of K are `numbers`, and r are `1/time`. So the units are

$$\frac{\texttt{numbers}}{\texttt{time}} = \frac{1}{\texttt{time}}\texttt{numbers}\left(\texttt{unitless} - \frac{\texttt{numbers}}{\texttt{numbers}}\right)$$
$$= \frac{\texttt{numbers}}{\texttt{time}}.$$

4.4.2 Conversion to Dimensionless Format

Often in deriving differential equations, the resulting expressions will contain many parameters that occur in combinations. It is possible to reduce the number of parameters by converting the differential equation to a dimensionless form, thereby creating new variables and parameters, but also eliminating many old variables and parameters. The net gain is fewer parameters. We achieve this by writing each state variable and the time variable as the product of two components: one with units denoted as $\bar{x}$ and one without units denoted as $\breve{x}$. For example, the numbers in a population will be written $N = \bar{N}\breve{N}$. The objective, then, is to manipulate the equation to replace all parameters and variables with dimensionless quantities. Applying this procedure to the familiar logistic equation gives

$$\begin{aligned}
\frac{dN}{dt} &= rN\left(1-\frac{N}{K}\right) && \\
\frac{d(\bar{N}\breve{N})}{d\bar{t}\breve{t}} &= r\bar{N}\breve{N}\left(1-\frac{\bar{N}\breve{N}}{K}\right) && \leftarrow\text{create unitless variables} \\
\frac{d(\bar{N}\breve{N})}{d\breve{t}} &= \bar{t}r\bar{N}\breve{N}\left(1-\frac{\bar{N}\breve{N}}{K}\right) && \leftarrow\text{multiply by } \bar{t} \\
\frac{d(\breve{N})}{d\breve{t}} &= \bar{t}r\breve{N}\left(1-\frac{\bar{N}\breve{N}}{K}\right) && \leftarrow\text{divide by } \bar{N} \\
\frac{d(\breve{N})}{d\breve{t}} &= a\breve{N}(1-\breve{N}), && \leftarrow\text{dimensionless form} \qquad (4.43)
\end{aligned}$$

where the new quantities are $a = \bar{t}r$ (unitless parameter) and $\bar{N} = K$. We have reduced the number of parameters from 2 to 1, and we have essentially scaled time by $1/r$ and population size by K. (We can recover the original quantity N as $\breve{N}K$.) The mathematical clarity and savings in parameters can be even greater when we apply this technique to models with several state variables.

4.4.3 Conservation Principle

If a model uses a conserved quantity (e.g., g C) all of whose sources and sinks are accounted for, then a state variable can be eliminated from the system of equations. Suppose a fixed amount K of carbon flows among three state variables (x_i), each described by an ODE. Since $K = x_1 + x_2 + x_3$, and K is a constant, we can rewrite any one of the x_i in terms of the other state variables and total C: $x_3 = K - x_1 - x_2$. x_3 effectively becomes an auxiliary variable and we can substitute $K - x_1 - x_2$ anywhere x_3 is used.

4.4.4 Rules of Thumb

In addition to the above approaches which help us understand and verify the correctness of the equations, there are several maxims of model formulation that can be generally applied.

1. *Know the question.* Study and understand the objectives, model question, hypotheses, and available data. These give hints to answers of the basic questions to address in model formulation: *How is feedback present in the system?* Negative feedback implies that the rate is a declining function of a state variable. *Are the flow variables conserved?* If yes, then all pathways must be expressed in the state equation and flows between compartments will be expressed in both state equations (gains in one, losses in the other). *Do multiple factors control the process?* If so, then we must write state equations that incorporate all the factors.
2. *Understand the objects.* Every state variable (level or box in a Forrester diagram) must have an explicit ODE or FDE. Auxiliary and driving variables are not described with differential or difference equations.
3. *Reconcile the diagram with the rate equation.* Out-bound material flow arrows are subtractions from the rate equation; in-bound flows are additions.
4. *Check the units.* The units of every state equation (ODE or FDE) will be identical on the left and right sides of the equality.
5. *Extrapolate the functions.* The rate equations must make sense for all legitimate values of their parameters and variables. Check that the function produces valid biological quantities (e.g., yields only positive concentrations) by examining extreme values (e.g., 0 and ∞) of the independent variables of the rate equations.
6. *Simplify the model.* All things being equal, simple models are better than complex models, but understand when and why it is not always desirable to simplify. If it is possible, try these techniques:
   - Reduce the equations to dimensionless variables.
   - Aggregate state variables.
   - Exploit conservation principles.
   - Use linear functions initially.
   - Use descriptive, phenomenological representations before detailed, mechanistic processes. When objectives or model failure require it, increase the level of details.
   - Assume homogeneous space.

4.5 Useful Functions

$\pi =$ *Yes. I need a drink, alcoholic, of course, after the heavy sessions involving quantum mechanics.* —Miller (1981)

Many of the biological processes can be represented by a variety of equations (e.g., hyperbolic saturation as either Michaelis–Menten or Holling disc equation). Some are nearly identical in shape, but use different parameters. Choosing among these, unless there are theoretical reasons, is largely a matter of taste and the appropriateness of the normal interpretation of the parameters. For example, the half-saturation constant in Michaelis–Menten can be applied to either enzyme kinetics or animal foraging. However, the handling time parameter in the Holling disc foraging equation may not be a natural concept in enzyme kinetics.

Figure 4.14 lists the equations and demonstrates the shapes of common nonlinear functions. In all curves and equations shown, y is the dependent variable and x is the independent variable. The plots do not show the behavior of the function for all x values. Beware of potentially undesirable y values for some values of x. For example, a straight line with a negative slope will have negative values if x is allowed to be sufficiently large. To avoid this, you must truncate (using a discontinuous function) the function to restrict y to desirable values. Most of the equations can be generalized by translating the curve along either the x-axis or the y-axis. To translate along the x-axis, add or subtract a value from the variable x. (This is illustrated in a few cases below.) To translate along the y-axis, add or subtract a value from the variable y (i.e., subtract or add from the left-hand side of the equations). Some equations range from 0 to 1.0; their shape can be complemented by subtracting the value from 1.0.

1. Linear:

$$y = Ax + b$$

   If A is negative, then the y-axis intercept (b) and the slope (A) define the line, but note that the x-axis intercept may also have a biological interpretation (e.g., K in the density-dependent per capita function for growth rate).

2. Exponential: Shown in Fig. 4.14A, the equation is

$$y = k_1 e^{k_2 x}.$$

   Parameter k_1 scales the y-axis intercept; k_2 determines the shape: large values produce steep curves.

3. Power: Shown in Fig. 4.14B, the equation is

$$y = k_1 + k_2 x^{k_3}.$$

   As with the exponential function, k_1 scales the y-axis intercept, and k_2 determines the steepness of the slope. If $k_2 < 0$, the curve decreases. If $k_2 > 1$, the curve is concave upward with an increasing slope. If $0 < k_2 < 1$, the curve is convex upward with a decreasing slope. k_3 scales the height of the curve. This function is frequently used to represent allometric growth relationships.

4. Saturation: Hyperbolic and Exponential: Shown in Fig. 4.14C, the

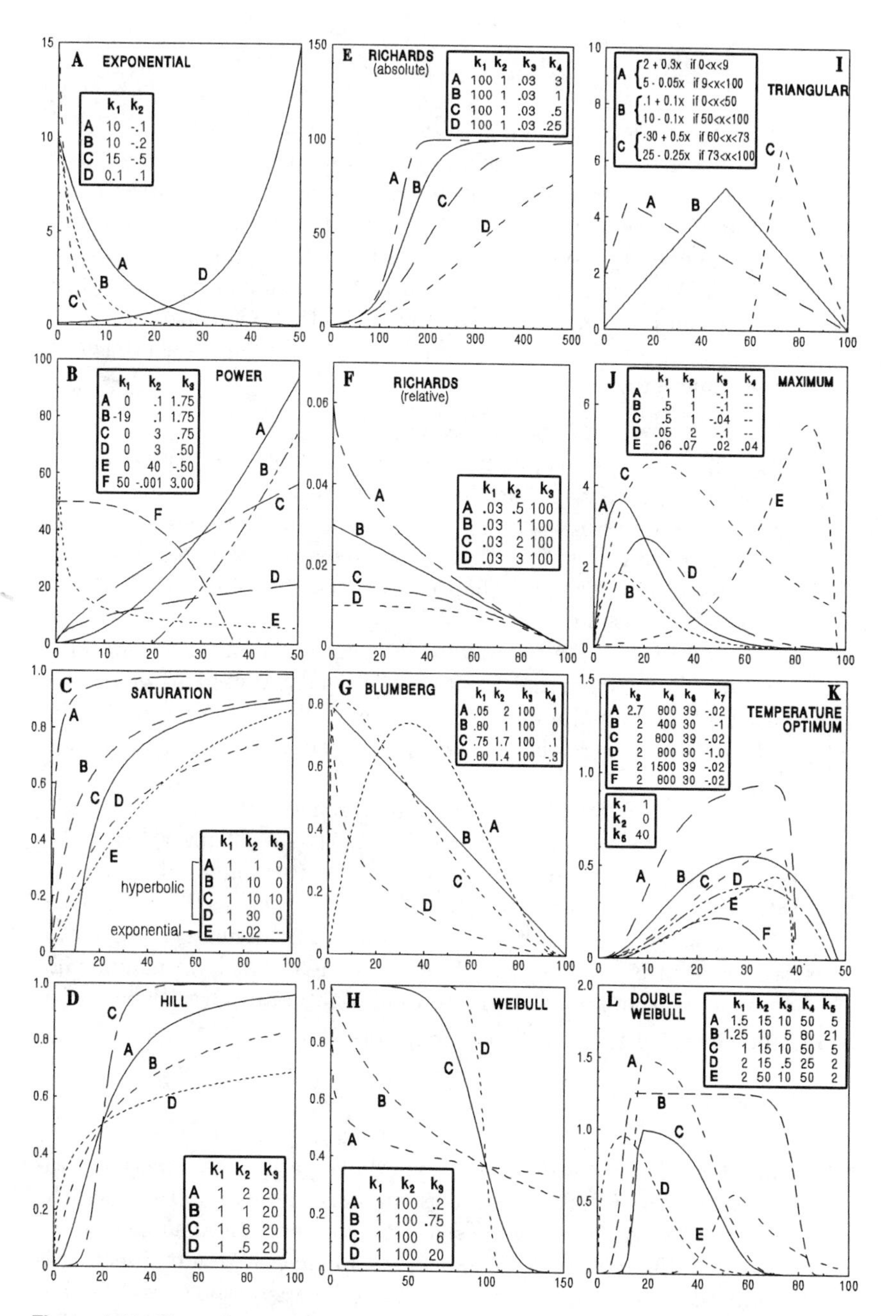

Figure 4.14: Plots of common nonlinear functions at different parameter values. Refer to the text for the meaning of the parameters.

equation for hyperbolic saturation is

$$y = k_1 \left(\frac{(x - k_3)}{k_2 + (x - k_3)} \right).$$

Parameter k_3 determines a threshold on the x-axis below which the function has a negative value. This is useful when the function is used to model microbial growth to describe a threshold nutrient concentration below which no growth occurs. This is a case when a truncation is necessary. When $k_3 = 0$, this function produces the classical Michaelis–Menten equation. k_1 scales the maximum value to which the function is asymptotic. k_2 is the half-saturation constant.

Also shown in Fig. 4.14C is one example of the exponential saturation function

$$y = k_1(1 - e^{k_2 x}),$$

where k_1 scales the maximum value and k_2 determines the steepness of the curve (large values produce steep curves). When k_2 is negative, the curve approaches k_1 from below. When $k_2 > 0$ and $x > 0$, the function declines from 0. Notice that the exponential and hyperbolic functions produce similar shapes, but that the slope of the latter increases more rapidly at low x values.

Both functions can be used for foraging functions or chemical dynamics. The exponential function is frequently used to model the growth of individual animals. Note that neither function has an inflection point (where the slope changes from accelerating to decelerating).

Another function that superficially resembles the saturation functions is $\tanh(x) = (e^x - e^{-x})/(e^x + e^{-x})$ (see Chapter 12). This function has the property that when $x > 0$, $\tanh(x)$ rises asymptotically to 1.0 and when $x < 0$, $\tanh(x)$ decreases asymptotically to -1.0. Therefore, it is useful for functions whose domain can take positive or negative values. This function is used widely in mammalian physiology as an empirical description of laboratory relationships.

5. Hill: Shown in Fig. 4.14D, the equation is

$$y = k_1 \left(\frac{x^{k_2}}{k_3^{k_2} + x^{k_2}} \right).$$

This is a generalization of the hyperbolic saturation function (Rubinow and Segel 1991). k_1 scales the maximum value to which the function is asymptotic; k_2 is a shape parameter; k_3 is analogous to the half-saturation constant. If $k_2 = 1$, the Michaelis–Menten function is produced. If $k_2 < 1$, a steeper version of the hyperbolic results. For $k_2 > 1$, an "S-shaped" function results. With $k_2 = 2$, this function can be used for Type 3 functional responses of predators. For other integer values of $k_2 \geq 1$, the Hill function is used extensively in enzyme kinetics for systems in which there exist several reactive sites on the enzyme (e.g., cooperative dimers; see Rubinow 1975).

6. Richards: The standard Richards equation (Richards 1959) is a generalization of the logistic growth equation (Fig. 4.14E)

$$y = \frac{k_1}{\left(1 + \left(\frac{k_1}{k_2} - 1\right) e^{-k_3 k_4 x}\right)^{1/k_4}},$$

where x is normally interpreted as time, k_1 is the maximum to which the function is asymptotic, k_2 is the value at $x = 0$, k_3 describes the steepness of the curve, and k_4 scales the location of the inflection point along the x-axis. The logistic curve is obtained when $k_4 = 1$.

   The Richards equation, as written above, describes the absolute values of a process (e.g., population size). A relative rate version exists that, when applied to population growth, describes the per capita rate of change of the population. The relative curve is shown in Fig. 4.14F and has equation:

$$y = \frac{k_1}{k_2} x \left[1 - \left(\frac{x}{k_3}\right)^{k_2}\right],$$

where k_1 scales the process on the vertical axis and k_3 corresponds to the maximum value (e.g., population size). $k_2 = 1$ gives the classical logistic relative rate of a linear decrease in the rate as x increases. $k_2 < 1$ gives a concave curve that shows a rapid decline at small x; $k_2 > 1$ produces a convex curve and has a slow decline at small x, but a rapid decline at large x. Note that k_3 is the intercept of the x-axis.

7. Blumberg: Blumberg's equation (Blumberg 1968, Buis 1991), also known as the *hyperlogistic*, generalizes the Richards relative-rate equation by adding a fourth parameter. The curve for the relative rate is shown in Fig. 4.14G and its equation is

$$y = k_1 x^{k_4} \left[1 - \left(\frac{x}{k_3}\right)^{k_2}\right],$$

where k_1 scales the curve on the y-axis, k_3 is the maximum value, and k_2 and k_4 are shape parameters. Be aware that when $k_2 > 1$ and $k_4 < 1$, the function is 0 when $x = 0$. This is illustrated in Fig. 4.14G (curve D) where the relevant curve decreases sharply to 0 when $x < 1$. Used as a relative rate, this function is useful in a wide range of models.

8. Complemented Weibull: Shown in Fig. 4.14H, the equation is

$$y = k_1 \exp\left(-\left[\frac{x}{k_2}\right]^{k_3}\right),$$

where k_1 scales the maximum value, k_2 controls the point along the x-axis at which the function is approximately 0, and k_3 is a shape parameter that specifies whether the function is concave or convex. It is a very powerful function that is useful in many situations including the probability of surviving from one age to another. It is related to the Richards equation. When $k_1 = 1$, the function ranges from 1 to 0. Consequently, a common form is the Weibull cumulative distribution function: $1-y$, which produces a positive relation between the x and y. This form behaves very much like the Hill equation (Fig. 4.14D); it has been generalized by Bradley and Price (1992).

9. Triangular: Linear functions can be combined to represent processes with maxima. Their use require truncation using discontinuous functions. The general formula is

$$y = \begin{cases} k_1 + k_2 x & \text{if } x < k_3 \\ k_4 - k_5 x & \text{if } x > k_3. \end{cases}$$

Three examples are shown in Fig. 4.14I.

10. Maxima: Shown in Fig. 4.14J, the equation is

$$y = k_1 x^{k_2} e^{k_3 x}.$$

This produces a maximum by using the product of two functions: one increasing, the other decreasing with increasing x. To produce a function with a maximum, we must have $k_3 < 0$. For most purposes, using $k_2 = 1$ fits a wide range of phenomena. Its primary attraction is its simplicity, but it cannot produce curves skewed toward large x. To skew curves to the right, use

$$y = k_1 e^{k_2 x} \left(1 - k_3 e^{k_4 x}\right),$$

as shown in Fig. 4.14J, curve E.

11. Temperature Optimum: Many biological processes have a maximum that is skewed toward large values of x. Logan (1988) described the relation of temperature on a process as

$$y = \frac{k_1 (x-k_2)^{k_3}}{k_4^{k_3} + (x-k_2)^{k_3}} - \exp\left(k_7 - \left(\frac{k_5-(x-k_2)}{k_5-k_6}\right)\right)$$

(Fig. 4.14K). The first expression on the right-hand side is similar to the Hill equation. k_1 scales the overall curve on the y-axis, k_2 is the lower temperature at which the process is 0, k_3 is a shape parameter for the rising part of the curve, k_4 is roughly analogous to the half-saturation constant, k_5 is the maximum temperature at which the process is positive, and k_6 is the temperature at which the value of the process is maximal. Note that there are complex interactions among the parameters in this complicated function and that choices can be made such that some actual quantities (e.g., largest temperature for positive values) do not match the corresponding parameter definitions.

12. Double Weibull: This function is the product of the Weibull distribution and its complement. It is shown in Fig. 4.14L and has the form

$$y = k_1 \left(1 - e^{-(x/k_2)^{k_3}}\right) e^{-(x/k_4)^{k_5}}$$

The parameters have the same meaning as described above for the Weibull function. This is one of the most flexible functions used in biological modeling.

13. Trigonometric: Extremely complex series of data over either time or space can be represented by the sum of general sine and cosine functions by choosing different values for mean, amplitude, phase, and angular frequency:

$$y = \sum_{i=1}^{N} M_i + A_i \cos(\omega_i (x - x_{0_i})).$$

14. Cubic Splines: Another method for modeling complex data series is to fit adjacent subsets of the data (e.g., sets of four datum points) to separate polynomial equations:

$$y = k_0 + k_1 x + k_2 x^2 + \ldots + k_n x^n.$$

Cubic splines is such a method that uses a third order polynomial for each subset of the data and smoothly joins the separate cubic equations together. This method is used widely in microcomputer graphics applications and is being more frequently used in dynamic simulation (Jørgensen 1986, Coleman and Gay 1990). While good fits to data are possible, this method uses a relatively large number of parameters that do not have empirical meaning.

15. Polynomials: Sums of integer powers of the independent variable can produce complex forms:

$$y = a_0 + a_1 x + a_2 x^2 + \ldots + a_n x^n$$

Even more complex forms are possible using *rational* functions of the form:

$$y = \frac{a_0 + a_1 x + a_2 x^2 + \ldots + a_n x^n}{1 + b_0 + b_1 x + b_2 x^2 + \ldots + b_n x^m}$$

4.6 Examples

Below are four examples to illustrate the procedures of quantitative model formulation. The difficult problem is to go from a verbal or diagrammatic statement of the system (which may include data or functional forms for some processes) to the equations.

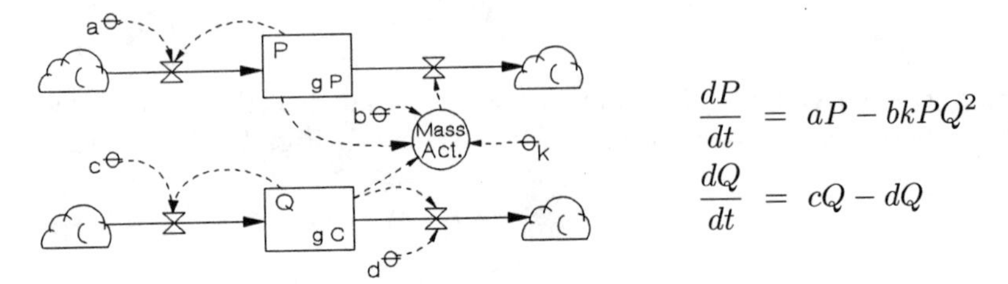

$$\frac{dP}{dt} = aP - bkPQ^2$$
$$\frac{dQ}{dt} = cQ - dQ$$

Figure 4.15: Flows with different units.

4.6.1 Flows with Different Units

This is a hypothetical example that does not apply to any particular biological system. Suppose we are modeling the dynamics of a compartment of phosphorus (P) and a compartment of carbon (Q). Phosphorus increases by a constant fraction at each time step and decreases as a third-order mass action effect between P and the square of Q. Carbon increases by a constant fraction each time step and decreases by a constant fraction each time-step. The conversion of C to P is a constant ratio (k). Normally, the expression bk would be represented as a single parameter. Figure 4.15 shows the Forrester diagram and equations.

Since the state variables have different units, we must use a parallel model, with information flows between state variables and rates to indicate the interactions. The problem states that both variables increase by a constant fraction of their values. This implies a relative or per capita rate that does not change with the value of the state variable (i.e., it is not density-dependent). The equation is the product of a constant (the fraction) and the variable. The loss from P depends on the value of Q, and we use an auxiliary variable to represent that relation. The loss from Q is another constant fraction equation.

4.6.2 Driving Variable

Suppose a state variable has three inputs; two are constant rates and one is a fixed per capita rate. There are three outputs; one is a constant rate, one a fixed per capita rate, and one is a hyperbolic function of temperature that varies with time. Figure 4.16 shows the Forrester diagram and equations.

"Constant rate" implies a rate that is simply constant and does not involve any state variables. The absence of an information flow from a state variable to the rate illustrates this assumption. We could have used any one of several functions to represent the hyperbolic relation noted in the problem. However, the implication of this relationship is that temperature is the independent variable, which occurs in the exponent of e as shown in the equation.

4.6.3 Riding a Bike

This example illustrates feedback control when there is not an obvious physical unit that flows between compartments. The problem is to describe the dynamics of the front wheel of a two-wheeled bicycle when it is driven (a) with hands in the normal position (left hand on the left handlebar, right hand on the right handlebar) and (b) with hands reversed.

When people learn to ride a bicycle using the normal hand position, they have learned how to implement a negative feedback control system. We will hypothesize that when the front wheel deviates from a fixed direction (assumed to be 0 degrees) toward the left, we put greater pressure on the left hand than on the right hand and thereby cause the wheel to move to the right. We do the opposite if the wheel deviates to the right. So, the problem and our hypothesis calls for a model that describes the dynamics of the wheel position and the pressure applied to each hand.

Figure 4.17 shows a Forrester diagram and equations when the hands are in the normal position. It is assumed that a deviation of the wheel to the left is a negative deviation and that to the right is positive. r and l are the pressure applied to the right and left hands, respectively. D is the deviation of the wheel from the desired orientation of 0 degrees. a, b, and c are positive constants of proportionality.

If the hands are reversed, it is not clear how the brain is confused, but there is no doubt that it is difficult to keep the bicycle upright. Apparently, if the wheel deviates to the right, the eye–brain system tells the body to increase pressure on the right hand regardless of its position (i.e., not the hand on the right handle bar). With hands reversed, this is a positive feedback system because deviations to the right are accentuated by increased pressure on the left handlebar (via pressure on the right hand). We can model this by multiplying dD/dt by -1.

4.6.4 Brewing Beer

In its simplest form, brewing beer involves putting sugar and yeast together in a vessel so that alcohol is produced as a by-product of the metabolism of sugar by yeast. Actual beer fermentation is much more complicated than this, but this will serve as an initial conceptual model. Two important

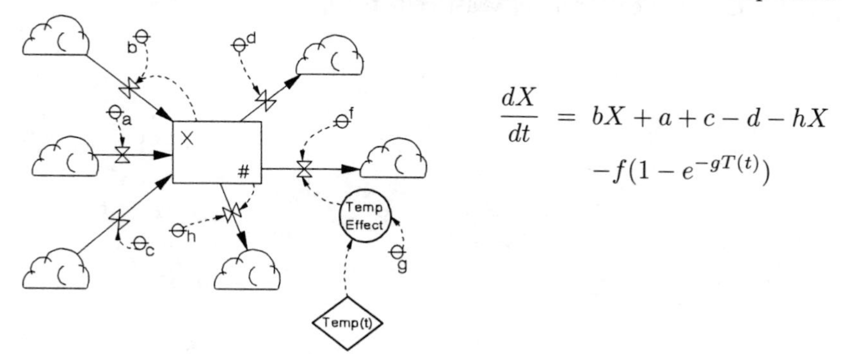

Figure 4.16: Driving variable and multiple input and outputs.

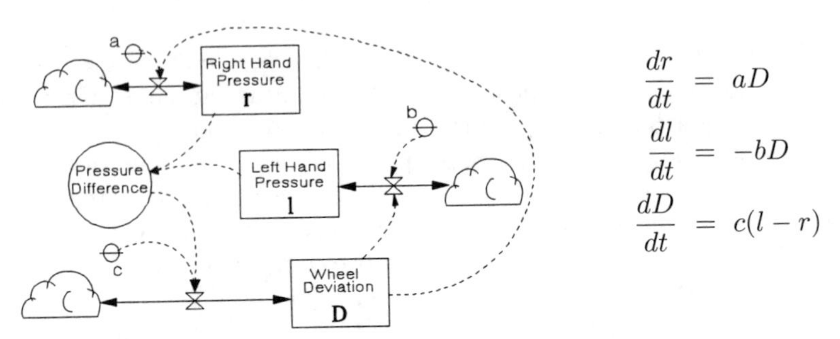

$$\frac{dr}{dt} = aD$$
$$\frac{dl}{dt} = -bD$$
$$\frac{dD}{dt} = c(l - r)$$

Figure 4.17: Feedback control for riding a bicycle.

facts associated with this situation are: (1) there is only a finite amount of sugar at the beginning and it is depleted over time, and (2) excessive alcohol will kill yeast cells.

To model this, we analogize the relation of yeast and sugar to a predator–prey or an enzyme–substrate interaction. We can also think of the effects of alcohol molecules on yeast cells in a similar way: alcohol "preys" on yeast. For the purposes of this example, we assume that we measure yeast in terms of cell counts, sugar in mg-sugar/liter, and alcohol as mg-alcohol/liter. Therefore, to account for incommensurate units, the Forrester diagram (Fig. 4.18) shows parallel models. To keep the mathematics simple, we assume that the rates of sugar consumption and yeast mortality due to alcohol follow mass action laws. We also assume that the rate of alcohol production is proportional to the rate of sugar consumption.

In Fig. 4.18, S is sugar content in mg/liter, A is alcohol content in mg/liter, and Y is yeast cells per liter. The auxiliary variable *S:Y Mass Action* is the equation aSY. Since this expression occurs three times in the model, assigning it to an auxiliary variable simplifies model presentation. The parameters are defined as: a = rate of sugar breakdown, b = fraction of sugar breakdown that yields alcohol, f = fraction of sugar breakdown that yields CO_2, c = rate of yeast cell formation per unit breakdown of sugar, and d = death rate of yeast cells per unit of alcohol.

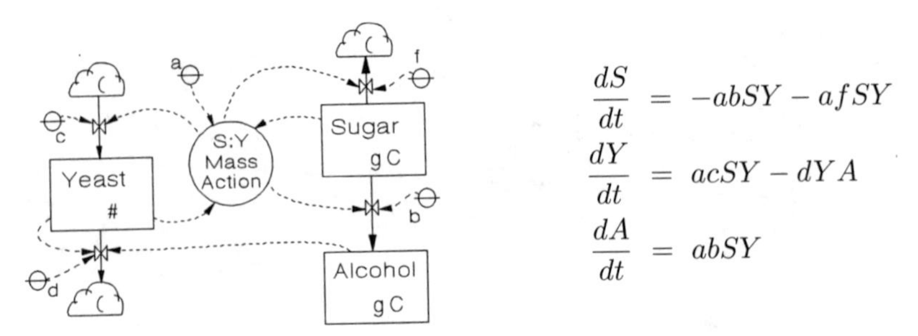

$$\frac{dS}{dt} = -abSY - afSY$$
$$\frac{dY}{dt} = acSY - dYA$$
$$\frac{dA}{dt} = abSY$$

Figure 4.18: Alcohol production by yeast in beer fermentation.

4.7 Exercises

1. In Eq. 4.20, show that when $S = K_m$, the velocity of the reaction is $0.5V_{max}$.
2. Rearrange Eq. 4.21 into the standard form of a hyperbolic relation and verify that the units are correct.
3. Draw the Forrester diagram and sketch the dynamics of a system containing a single state variable with a constant input rate (Fig. 4.3).
4. A simple foodweb (1 prey, 2 predators) is modeled in units of grams of carbon. A prey species (x) grows according to density-dependent growth in the absence of either predator. The consumption of x per unit of predator 1 (y) is a saturation feedback function and the consumption of x by predator 2 (z) is a fixed fraction of x. The per capita death rate of y is constant. The per capita death rate of z is a negative exponential. Draw a Forrester diagram and write differential equations with the above hypotheses.
5. Add a time-varying temperature effect to the decomposition component of the ecosystem carbon flow model (Eq. 4.27).
6. Foraging Saturation Based on Gut Capacity. Derive a saturation (hyperbolic) expression relating predator feeding rate and prey density when the predator is limited both by prey density and gut capacity. Assume the prey are randomly distributed in two dimensions at density S, the predator forages with velocity v (meters/hour), for h hours per day, and has a detection radius of r meters, and a gut capacity of $J_{\max}$ (m^3). Further assume that for a given S, the predator evacuates its gut at a fixed proportion of the current gut content.
7. Formulate the bike riding model as a finite difference equation and simulate its dynamics using code modified from that given in Chapter 5 for the predator-prey model.
8. An alternative model for riding a bike is

$$\frac{dr}{dt} = \begin{cases} -ar & \text{if } D < 0 \\ 0.0 & \text{otherwise} \end{cases}$$

$$\frac{dl}{dt} = \begin{cases} bl & \text{if } D > 0 \\ 0.0 & \text{otherwise} \end{cases}$$

$$\frac{dD}{dt} = c(l - r).$$

   This model also causes the deviation (D) to stay near 0. What other dynamical behavior does it have that suggests that it is a poor alternative? Simulate this model and compare with the original model.
9. The simple bike riding model may not capture basic biological and psychological mechanisms. Specifically, will humans react the same way to large deviations as to small deviations? What does the model

assume? Make a simple x–y plot that depicts the model assumption and a more realistic alternative. Write a new model (possibly using the functional forms in Section 4.5) that incorporates the new hypotheses. Does the new model produce more realistic dynamics?

10. Create another bike riding model that assumes that the nervous system operates at a very, very much faster time scale than does the dynamics of the bike (which has greater mass than nerves and muscles, has momentum, and slow speed, etc). In other words, assume that the dynamics of pressure changes is irrelevant. Run simulations of this new model and compare its ability to control the bike to that of the original model.
11. Modify and simulate the beer equations so that yeast growth uses the equation for Temperature Optimum shown in Fig. 4.14, curve A. Let temperature oscillate around 20° C with an amplitude of 8°and a period of 1 day with a peak at 12:00 noon. Choose other parameters so that sugar is exhausted in about 5 days.
12. Modify the beer equations and Forrester diagram so that a conserved quantity (e.g., g C) flows among the three compartments.
13. What are the first 15 digits of π?

Chapter 5

Simulation Paradigms

5.1 Computer Simulation

INSIGHT and illumination are what we really want from a mathematical model. But when we write equations so complex that they cannot be solved with paper and pencil, we must enter the murky and sometimes dissatisfying world of numerical approximation and computer simulation. In these nether regions, while we grope for pattern among the numbers, a healthy sense of humor is essential, for much of the programming for computer simulation is boring, repetitive, and an impediment to the use of our bright ideas to understand systems and address real-world problems. Clarity, grace, and style are our coding goals, but along the way we face an abundance of arcane input/output minutiae, complex argument passing, and a mass of numbers whose meaning may be obscure at best and undiscoverable at worst. To appreciate this and the alternatives, we first consider the structure of a typical program to simulate finite difference equations.

Figure 5.1 is a flow chart of the basic steps required for all continuous system simulation models we will discuss. The first thing we must do is read data required both to control the simulation run and for the particular model being simulated. Simulation control data include maximum iterations to simulate, printing and plotting intervals, file names for input and output data, and so on. The model-specific data include initial conditions and parameter values. After reading data, we initialize any variables or arrays that must contain specific, known data (e.g., the variable for the current time or iteration of the simulation).

The core of the simulation program is the loop that iteratively solves the dynamic equations. Inside the loop, for each state variable in turn, we calculate its new value based on its current value and the current values of other state variables if needed. We do not overwrite the current value with the new value immediately because the FDEs of other state variables (which we have not calculated yet) may need the current value of the

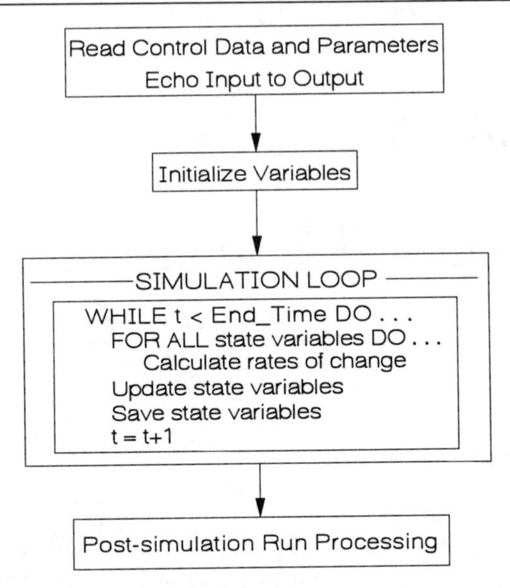

Figure 5.1: Flow diagram for a simple simulation model.

state variable we are currently computing. After the FDEs for all state variables have been computed, we update the current values and save them in arrays for later processing. The final step is to increment time, the variable that controls the simulation loop. When the time variable has been incremented past its maximum value, the simulation run is completed and we can perform post-simulation processing. This is usually printing, on-screen graphical plotting, and statistical analysis (e.g., calculating time averages, finding maxima, etc.).

To provide a sense of the simulation process in a programming language, we give an example using C. If you are unfamiliar with C, simply read the comments and glance at the code to get a feel for the details that are necessary when we implement the basic structure of Fig. 5.1. The following code simulates the classical Lotka–Volterra predator–prey model (see Eqs. 4.23 – 4.24 and Chapter 4):

$$V_{t+1} = V_t + rV_t - bV_tP_t$$

$$P_{t+1} = P_t + bcV_tP_t - dP_t.$$

```
#include <stdio.h>
#define MAXPRN 500        /* maximum number of results to save */

main()
{
      /* simulation control */
  int t,              /* current time */
      maxt            /* maximum time */
      prndelt,        /* time interval to print */
      prntime;        /* next time to print */
```

```
    /* state variables */
double V,         /* current victim numbers */
       newV,      /* new victim numbers */
       P,         /* current predator numbers */
       newP;      /* new predator numbers */

    /* parameters */
double r,         /* prey per capita growth rate */
        b,        /* attack rate */
        c,        /* prey-to-predator conversion factor */
        d;        /* predator death rate */

    /* results arrays */
int     toprint;            /* next array index to store */
int     tarray[MAXPRN];     /* time values */
double Varray[MAXPRN],      /* victim values */
        Parray[MAXPRN];     /* predator values */

    /* loop locals */
int    i;

    /* read simulation control parameters */
printf("Enter max. time and print interval (eg, 200 2): ");
scanf("%d%d",&maxt,&prndelt);
printf("Enter prey initial num and growth rate (eg, 50 0.02): ");
scanf("%lf%lf",&V,&r);
printf("Enter pred initial numbers, attack rate, conversion, \n");
printf("      and death rate (eg, 4 0.002 .1 0.2): ");
scanf("%lf%lf%lf%lf",&P,&b,&c,&d);
    /* Echo back the input */
printf("\n\nInput:\n\tPrey: Init=%lf  growth=%lf\n",P,r);
printf("\tPred: Init=%lf  attack=%lf  convers=%lf  death=%lf\n",
        P,b,c,d);

    /* Initialize */
tarray[0]=t;              /* save initial conditions */
Varray[0]=V;
Parray[0]=P;
t=1;
toprint=1;
prntime=t+prndelt;        /* next time to print */

    /* Top of Simulation Loop */
while ( t<maxt ) {
  newV=V + r*V - b*V*P;   /* FDE for prey */
  newP=P + b*c*P - d*P;   /* FDE for predator */
      /* update */
  V=newV;
  P=newP;
  if(prntime <= t) {      /* time to save results?     */
    tarray[toprint]=t;    /* save current time,        */
    Varray[toprint]=V;    /*      current victims,     */
    Parray[toprint]=P;    /*  and current predators    */
    prntime += prndelt;   /* increment next print time */
    toprint++;
  }
  t++;
} /* simulation run complete */
  /* Post-Simulation */
  /* -- just print results */
printf("%10s\t%10s\t%10s\n","Time","Prey","Predator");
```

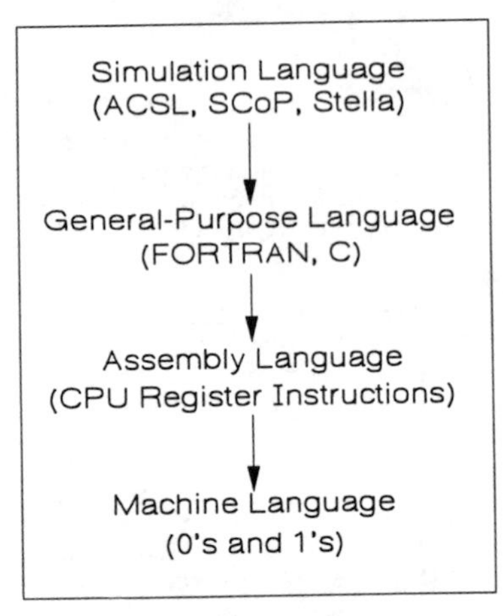

Figure 5.2: A hierarchy of computer languages.

```
  for(i=0;i<toprint;i++) {
    printf("%10.2d\t%10.2lf\t%10.2lf\n",
             tarray[i],Varray[i],Parray[i]);
  }
}
```

In the above code, notice the relative number of lines that are actually used to calculate new state variable lines (4) compared to the number devoted to input/output (12). This is fairly typical of programming.

5.2 Hierarchical Levels in Languages

One way to reduce the amount of programming required is to teach the computer how to do it without human intervention. Computer scientists have not yet designed programs that write programs that solve general, arbitrary problems (but see Section 19.5). In specific problem areas (such as simulation models), we can simplify programming by using high-level constructs that a machine can translate into appropriate low-level code.

Figure 5.2 illustrates the hierarchical nature of computer languages. The most general languages are at the bottom, and the most specific languages are at the top. Machine language is the actual set of bits (0s and 1s) that are fed into the CPU (Central Processing Unit) to compute (e.g., to read and write to memory locations, add numbers, etc.). Assembly language is a set of mnemonic instructions in English that code for machine language binary forms. Assembly language instructions manipulate the CPU and its components directly, but use simple English words so that humans do not have to remember sequences of bits.

General-purpose languages are those with which most of us are familiar: `BASIC`, `FORTRAN`, C. Among these languages there is a great range of facilities for implementing algebraic operations and data manipulation, but they all restrict the programmer's ability to directly access CPU components.

As a consequence, a single line of a general-purpose language represents a great many lines of assembly or machine language code. This means that while using a general-purpose language, the programmer cannot explicitly control and perform certain operations: the language compiler decides how best to implement a procedure at the level of CPU components. Simulation languages further restrict the programmer's ability to manipulate the CPU. For example, whereas general-purpose languages allow the programmer to reserve and manipulate computer memory in the form of arrays, many simulation languages do not permit the programmer to do this. Such details are handled by the simulation language compiler automatically; the compiler prohibits the programmer from intervening.

Below is a simple example from a hypothetical simulation language to solve the population growth equation. Uppercase keywords signify functional blocks.

```
PARAMETER
r = 0.05   K = 200

INITIALIZE
N = 4      t = 1994

EQUATION
dydt  = N*r*(1-N/K)

PLOT
N vs t
```

Contrast this with the C program listed above. While the above example is not a real language, it illustrates the small amount of information that changes from one model to the next. All we really need to specify are the equations, the parameters and initial conditions, and how to view the results. All the rest of it (e.g., reading data, calling functions, passing arguments, producing output) is ancillary coding details from which the modeler can be unburdened.

To remove programmer involvement from such coding details may seem to be an undesirable loss of control. However, for every instance that a programmer loses freedoms of "micro-control," he or she gains a freedom to visualize and solve the problem without concern for messy details. Sometimes we need to muck around in these messy details. Often, computation time can be reduced by programmer fine-tuning the code for a particular problem to squeeze maximum performance from the computer. At other times, computing speed is not the most important criterion. Simulation languages and the modeling facilities provided by them are usually written in a generic way that sacrifices speed for ease of use. This is appropriate whenever the time to program the computer is more important than the time it takes for the computer to calculate the answer. As deadlines for projects approach (e.g., the end of the school term), programmer (i.e., stu-

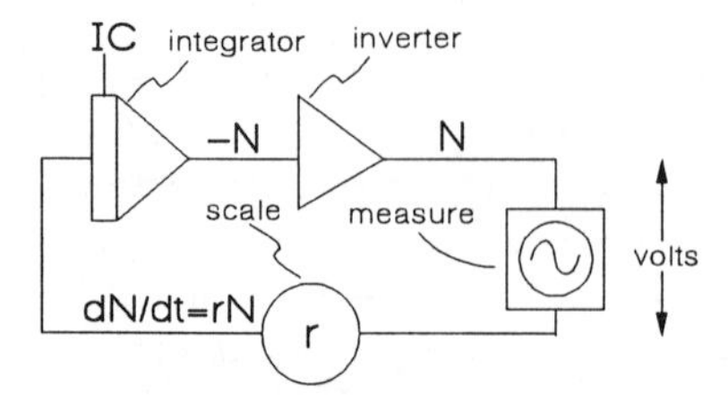

Figure 5.3: Diagram of an electrical circuit acting as an analog computer that integrates $dN/dt = rN$. The current value of N is the voltage potential between the top branch of the circuit and electrical ground, as displayed on the oscilloscope. Scaling is achieved with a variable resistor ("pot"). The integrator shown outputs the inverse of the input and requires an inverter. IC = initial condition.

dent) time becomes very valuable (to the student). It is times like those when we would give a lot to avoid worrying about the subtleties of C.

In the sections that follow, we review some aspects of analog and digital computer simulation languages. We also examine four different approaches to digital simulation environments to gain a sense of the interfaces available.

5.3 Simulation Languages

The computing scientist's main challenge is not to get confused by the complexities of his own making. — Dijkstra (1988)

5.3.1 Electronic Analog Computers

One of the oldest simulation languages is the *block diagrams* used to create models on electronic analog computers. These are computing devices that function by using the electrical properties of circuits (e.g., voltage, current flow) to represent numbers. Voltage is analogous to a quantity, as distance along a slide rule is analogous to a quantity.

Electronic circuits that can perform basic mathematical functions relevant to simulation modeling are manufactured in discrete, interchangeable modules. For example, an operational amplifier can multiply an input voltage by a constant or integrate the input over time. If voltage is used as the physical embodiment of a quantity, then an amplifier can integrate an equation whose current value is represented by volts. By interconnecting these components on a *patch board*, it is possible to create an electrical circuit in which the dynamics of the voltage potentials, at key points in the circuit, represent solutions to the differential equations. Figure 5.3 is an example that integrates the density-independent population growth equation. The voltage measured after the inverter and displayed on the oscilloscope is the solution: the electrical analog of population numbers. By connecting this output back to the input of the integrator, a new output based on the old output is generated. A different model would require that a new set of components be connected on the patch board.

The legacy of analog computers is the concept of modular components that perform basic operations and that are interconnected to complete a dynamic model. The electrical modules and their symbols (Fig. 5.3) became the basis first of an analog simulation language and then a digital simulation language after the modules were implemented in software. This form of representation is still widely used by physiological modelers (e.g., Coleman and Gay 1990).

5.3.2 Aids to Digital Simulation

In the digital world, there are three classes of programming aids to simulation: (1) libraries of functions that can be called from a general-purpose programming language, (2) special-purpose languages similar to the brief example given above, and (3) simulation *environments* that provide a language as well as a menu or graphical user interface that allows easy viewing of simulation results, editing of parameter values, and statistical analysis.

The above tools simplify modeling by providing code that performs the common features of most modeling problems. These include: (1) solve differential equations, (2) generate and use random numbers, (3) format and receive input data, (4) display of results graphically, (5) statistically analyze results, (6) allocate computer memory to store variables, (7) edit parameter values and perform repeated runs, and (8) perform error checking and recovery. The use of these tools is not without costs, and the advantages and disadvantages of general-purpose *vs* simulation languages are listed in Table 5.1.

From the very simple beginnings of electronic circuit block diagrams, computational scientists have developed many examples of all three of the above simulation aids. Describing these could be a book in its own right, but a few of the historically important languages should be mentioned. CSMP (Continuous System Modeling Program) is a block-based system that emulates the approach of analog computers. DYNAMO (Dynamic Models) was developed and used by Jay Forrester (of Forrester diagram fame) for continuous systems. ACSL™ (Advanced Continuous Simulation Language, Mitchell and Gauthier Associates, Inc.) is a modern version of earlier simulation languages used widely in the aerospace industry. Many of these and other simulation languages also run on personal computers.

We will illustrate the nature of simulation languages by discussing in more detail four radically different paradigms. SCoP™ (Simulation Resources, Inc.) is a text-based system for continuous systems that executes on a variety of platforms including MS-DOS® (Microsoft Corp.) compatible computers. Stella® (High Performance Systems, Inc.) is a graphical-based system for continuous system simulation on Macintosh® (Apple Computer, Inc.), and Microsoft Windows® (Microsoft Corp.) compatible computers. Spreadsheets are general-purpose computing environments that can be adapted for simulations. Finally, Mathcad® (MathSoft, Inc.) is an

Table 5.1: Advantages and disadvantages of simulation languages and general purpose languages.

General-Purpose Languages	
Advantages	**Disadvantages**
1. Minimum number of restrictions on output format	1. Longer programming time per model
2. Models portable to many machines with a standard compiler	2. More debugging
3. Flexibility in model structure	3. Complex I/O programming

Special-Purpose Languages	
Advantages	**Disadvantages**
1. Less programming time per model	1. Limited flexibility in output formatting
2. Better error checking	2. Less flexibility of model structure
3. Flexibility to include user-defined routines	3. Limited ability to optimize model running speed
4. Easy display of output	4. Decreased portability to other computers
5. Automatically controls management of storage needed by simulation runs	

example of computer-aided mathematics software that is capable of performing a broad class of mathematical activities, including solving ODEs.

5.4 SCoP

SCoP (Simulation Control Program, Kootsey et al. 1986) was developed by researchers at Duke University with funds from the National Institute for Health. It was primarily designed for simulation of physiological and biomedical systems, but it is a general package that can handle any set of first- and second-order differential equations as well as a restricted set of partial differential equations. It can also solve algebraic equations representing equilibrium conditions. It is similar to other simulation packages (e.g., ACSL), so the concepts learned for SCoP will carry over to these. Because it was developed with public funds, it is available for nominal cost. The system supports MS-DOS compatible VGA graphics with simple but adequate plotting functions. Hardcopy output of graphical results is restricted to screen dumps, but ASCII files of results can be written and exported to other plotting packages.

SCoP is a preprocessor that reads a program written in the SCoP language and translates it into a C program that is a series of C function calls. A C compiler is then invoked that compiles the program and links in the extensive set of SCoP library functions. The SCoP library contains code for basic simulation functions: generating the menu system that the modeler uses for simulation control, numerically integrating the equations, altering parameter values and initial conditions, graphically viewing the output, performing post-simulation analyses, and saving results to files.

A simple example will illustrate the savings in coding time. Consider the model of yeast dynamics developed in Chapter 4. The equations to solve are

$$\frac{dS}{dt} = -abSY - afSY = -aSY$$

$$\frac{dA}{dt} = abSY$$

$$\frac{dY}{dt} = acSY - dYA,$$

where S is sugar content in mg/liter, A is alcohol content in mg/liter, and Y is yeast cells per liter. The parameters are defined as: a is the rate of sugar catalysis, b is the fraction of sugar catalysis that yields alcohol, f is the fraction of sugar catalysis that goes to CO_2, c is the rate of yeast cell formation per unit of sugar catalysis, and d is the death rate of yeast cells per unit of alcohol. We assume that sugar catalyzes to either alcohol or CO_2.

The SCoP code for this model is

```
TITLE Yeast Alcohol Model
: Model of anaerobic fermentation in yeast
: in which the biochemical transformation of sugar to alcohol
: is coupled with yeast population dynamics.

CONSTANT
{
  a = 0.00002          : sugar breakdown rate per unit of yeast
  b = 0.66667          : fraction of sugar catalysis transformed
                       :to alcohol
  c = 0.03000          : yeast growth rate per sugar breakdown
  d = 0.00000035       : yeast death rate per unit of alcohol
}

INDEPENDENT
{
  time FROM 0 TO 400 WITH 200             :independent variable
}

PLOT yeast, sug, alco VS time             :visualization control

STATE                                     :state variables (levels)
{                                         : with plotting ranges and
  yeast     FROM 0 TO 750 START 100.0     : initial conditions
  sug       FROM 0 TO 500 START 52000.0
  alco      FROM 0 TO 500 START 0.0
}

DERIVATIVE MoreBeer                       :differential equations
{
  sug'  =  -a*sug*yeast
```

```
    alco' =  a*b*sug*yeast
    yeast'=  a*c*yeast*sug - d*yeast*alco
}

EQUATION                                    :integration technique
{
    SOLVE MoreBeer
}
```

As illustrated above, SCoP decomposes the model into blocks identified by reserved words. The order of the blocks is not important. `CONSTANT` identifies the parameters in the model (four, in this case). Braces are used as block identifiers to permit multiple entries. `INDEPENDENT` defines the name of the independent variable (`time`). Within the `INDEPENDENT` block the modeler can define the range of time over which a solution is desired (`0 to 400`) and the total number of solutions that are printed (graphed) in the entire interval (`WITH 200`). `PLOT` defines the default curve to plot during a simulation run (which may be altered during a simulation session). In the above example, all three state variables are plotted together on a single screen. `STATE` identifies the state variables, the allowable plotting range (`FROM x TO y`), and the initial conditions (`START x`). The basic model equations are defined in the `DERIVATIVE` block which can include an optional name (`MoreBeer`). The apostrophe (`'`) stands for time derivative. Finally, the commands to instruct SCoP to solve the equations are in the `EQUATION` block with the `SOLVE` instruction. In the case of differential equation models, this instruction uses one of eight numerical integration methods that may be specified in the model using a command syntax such as: `SOLVE` *derivative_name* `METHOD` *method_name*. Among the methods available are Euler, fourth-order Runge–Kutta, time variable versions of these, two advanced predictor–corrector methods, and a method for stiff equations. These are discussed in Chapter 6.

After successful compilation, a normal MS-DOS executable file is produced. When the program is executed, a menu system is displayed that controls the simulation session. The `SCoP Main Menu` provides options either to run the model with current parameter values and initial conditions, invoke an editor to alter those values, alter the default datafile of parameter values or continue an interrupted simulation, or alter the variables plotted. Following completion of the run, another menu permits the user to calculate some statistics, re-plot the results, define conditions for interrupting subsequent simulations, or call the user-defined *terminal* procedure.

The results of one run of the yeast model are shown in Fig. 5.4. In addition to the simple ODE model above, SCoP can solve arrays of ODEs similar to the method of lines (Chapter 6) and a restricted set of PDEs. It can solve time-lagged and unlagged finite difference equations. SCoP models can call user-defined SCoP functions (using the SCoP block struc-

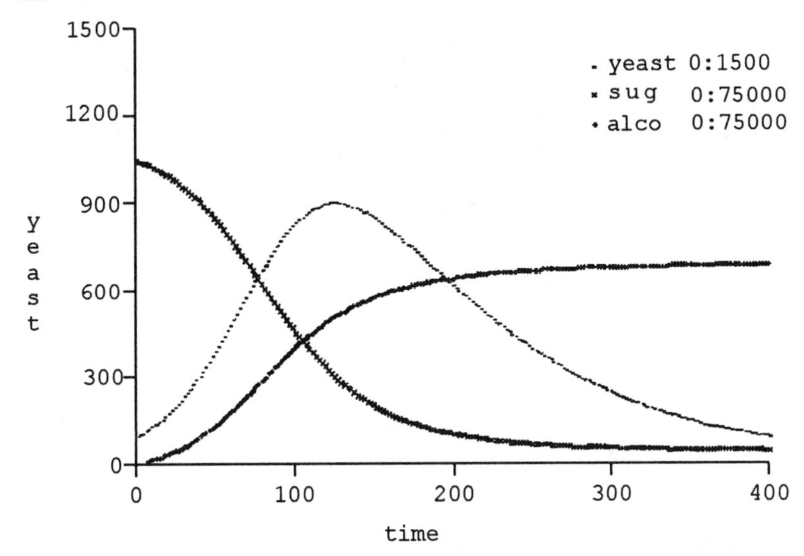

Figure 5.4: SCoP results of the yeast model (screen dump to a laser printer). Parameter values used: $a = 0.00002$, $b = 0.66667$, $c = 0.03$, $d = 3.5 \times 10^{-7}$.

ture `FUNCTION`) as well as user-supplied unadulterated `C` code using the `VERBATIM` block. There are many built-in functions for random number generation, mathematical functions, and various kinds of forcing functions (i.e., driving variables: step, ramp, sine, etc.). There are also facilities to simplify repetitive alterations of parameter values for parameter sensitivity analyses. A separate program (`SCoPFit`™) finds a least-square fit for parameter values using iterative simulations compared to a dataset.

5.5 Stella

Stella II (High Performance Systems, Inc. 1992), is a graphical approach to aid the user in defining and solving dynamical equations. It was originally developed for Apple Macintosh, but a version for Microsoft Windows 3.1 was released in 1994. The graphical language by which models are specified uses a modified Forrester diagram. Essentially, Stella II takes as input a Forrester diagram and equations and converts these into a simulation model. Models are constructed from a series of menus and *views* that are manipulated in standard Macintosh ways (pull-down menus, double clicking, etc.).

In Stella II, a model can be viewed in two different forms: as a Forrester diagram and as a set of equations. The Forrester diagram views of the model are constructed within Stella II using drawing tools to create and position levels (boxes, state variables), flows between boxes and clouds (sources and sinks), and auxiliary variables (Fig. 5.5). There is a strong similarity between Stella II's desktop and the patch board of an analog computer (Fig. 5.3). The symbols represent different concepts, but the use

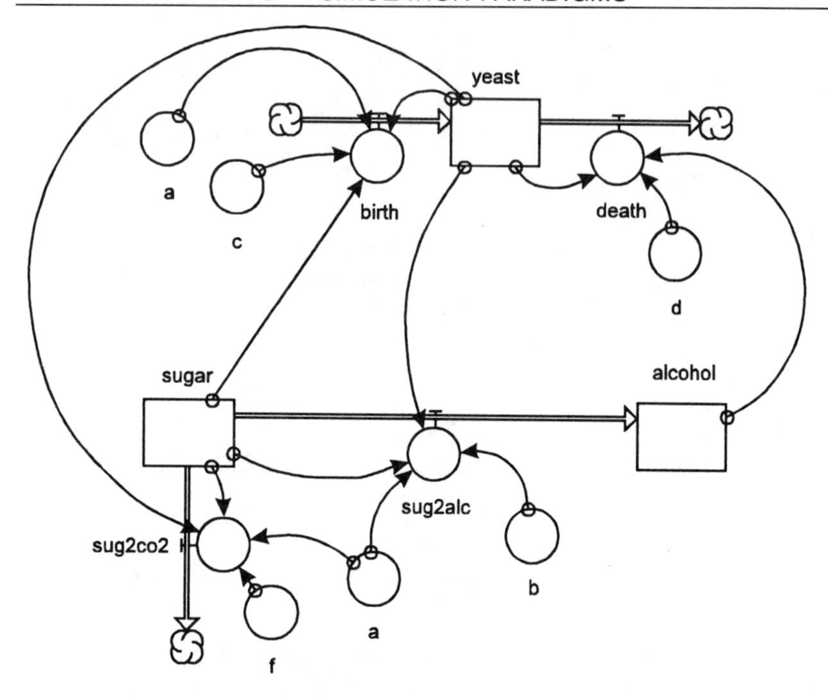

Figure 5.5: Stella II primary diagram view of the yeast model.

of a graphical/physical representation of the model is the same.

Once the Forrester diagram has been specified, the equations view prompts the user for algebraic equations that define the differential equations needed by the diagram. Figure 5.6 shows the equation window for the completed yeast model. Double clicking on any of the components [rate equations (INFLOWS, OUTFLOWS), parameters (circles), initial conditions (rectangles)] brings up a dialog box for editing values or equations. Components that have not been defined are flagged with a question mark.

Because virtually any algebraic expression can be used in the differential equation, Stella II can do little more to help the modeler than provide a template to "fill in the blanks." However, since the diagram defines the state variables, parameters, and auxiliary equations, a considerable amount of error checking can be done by Stella II. The user is prevented from using variables in equations that would contradict the arrangement of the arrows in the diagram. Stella II also complains if not all of the information transfers defined in the diagram are used in the equations. Numerical values of the parameters and initial conditions are specified in this menu as well (Fig. 5.6). A large set of built-in mathematical and random number generation functions are available. Stella II cannot do partial differential equations, but a Stella II object called a Conveyer can approximate a discretization of a simple one-dimensional flow. Stella II can do discrete event simulation as well as solve time-lagged and unlagged finite difference

```
□alcohol(t) = alcohol(t - dt) + (sug2alc) * dt
   INIT alcohol = 0
   INFLOWS:
         sug2alc = a*b*sugar*yeast
□sugar(t) = sugar(t - dt) + (- sug2alc - sug2co2) * dt
   INIT sugar = 52000
   OUTFLOWS:
         sug2alc = a*b*sugar*yeast
         sug2co2 = a*f*sugar*yeast
□yeast(t) = yeast(t - dt) + (birth - death) * dt
   INIT yeast = 100
   INFLOWS:
         birth = a*c*sugar*yeast
   OUTFLOWS:
         death = d*yeast*alcohol
O a = .00002
O b = .6667
O c = .03
O d = .00000035
O f = 0
```

Figure 5.6: Stella II equation view for inputting and editing the differential equations and parameters for the yeast model.

equations.

Models are executed through the `Run` menu, where simulation controls (i.e., time duration, time step, integration method) are also selected. The dynamics of all variables defined (e.g., state variables and rates of change) may be displayed as x–y plots or tables. The characteristics of the output (i.e., type, variables, ranges, etc.) are specified by pull-down menus accessible by double clicking on the axes or tabular column. All views of the model and output can be printed. Figure 5.7 shows the graphical output for the yeast model.

5.6 Spreadsheets

A spreadsheet can be viewed as a large matrix of functions. Each element is addressed by a letter designating a column and a number that references the row (e.g., `D121`). Elements can contain constants or complex equations. Although spreadsheets were not originally designed for simulation, they can be used as simple simulation languages for finite difference and differential equations. One approach to modeling in spreadsheets defines one column as time and another set of columns as the state variables. The elements of the state variable columns are equations that use the immediately previous row for those columns of the interacting state variables.

Figure 5.8 illustrates a simple predator–prey model. Time is in the column labeled `time`, the prey and predator values are stored in the columns

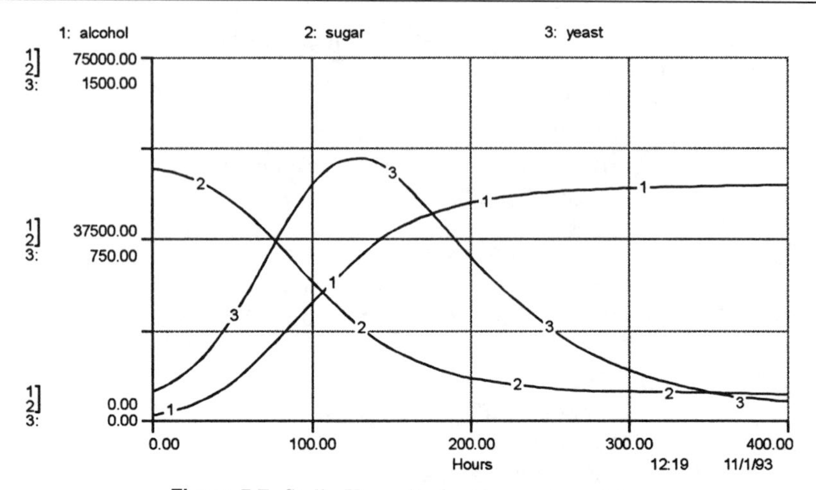

Figure 5.7: Stella II results for the yeast model.

indicated. Initial conditions are put in B1 and C1.The equations for the prey for the next time step are given for element B2, and similarly for the predator for C2. By performing a *relative copy* of these equations down the respective columns, the new values of the state variables will be calculated using values in the row immediately above. (In a relative copy, the form of the equation is copied, but references to cells in the equation are altered to maintain the same relative position as the original version.) Most spreadsheets can plot the columns (e.g., B and C *vs* A) as well as print the columns as ASCII files for access by other software packages. Stent and McCallum (1995) give a more sophisticated method using *self-referencing formulae.* Modern spreadsheets are considerably more powerful than the first versions of the early 1970s that ran in 32K of memory on 8-bit microcomputers. Sophisticated subroutines called *macros* are able to integrate differential equations using advanced techniques such as Runge–Kutta. The distinction between specialized simulation software and general-purpose accounting tools such as spreadsheets is becoming blurred.

5.7 Computer-Aided Mathematics

In the last few years, another class of software has been developed that we can call *Computer-Aided Mathematics* (CAM). This software originated in the early programs that performed basic symbolic algebraic manipulations. But, CAM has grown far beyond these attempts, and now there are several packages available that offering, in one environment, symbolic manipulations, numerical solutions, differential and difference equations, matrix manipulations, programming, graphical plotting, and word-processor quality report generation. Two of the many examples of this software are Mathcad and Mathematica® (Wolfram Research, Inc.). Both of these programs are available on a variety of operating systems and computer platforms in-

Columns

Rows	A time	B prey	C pred	D
1	1	10	5	
2	2			
3	3			
4	4			
5	5			

B2=B1+B1*0.05-0.1*B1*C1

C2=C1+0.1*0.5*B1*C1-0.03*C1

Figure 5.8: A spreadsheet implementation of the Lotka–Volterra predator–prey model. Columns are indicated as letters; new state variable values (current row) are computed using values of row above and equations as shown.

cluding Microsoft Windows, Apple Macintosh, and UNIX® (AT&T). Here, we illustrate the approach by implementing the yeast model in Mathcad (Fig. 5.9).

Mathcad is an entirely graphically oriented, what-you-see-is-what-you-get environment (WYSIWYG). A Mathcad project is contained on *worksheets* that combine text, equations, and graphical elements. Text can be simple one-line comments or multiple paragraphs that span several printed pages. This facility makes it possible to combine "live" mathematics with extensive textual discussion and documentation. Equations are typeset by Mathcad as they are typed; they appear on the screen and on paper as equations appear in mathematics books and journals. For example, square roots appear as radicals, quotients are graphically depicted as numerators above denominators separated by a horizontal line of the appropriate length, powers are superscripts, and so on. Special keystrokes or menu selections create the graphical forms of the equations. Many standard and specialized mathematical functions are available (e.g., ODE integration, trigonometry functions, Bessel functions, probability distributions, etc.). A wide variety of 2D and 3D plots and histograms are possible.

Figure 5.9 is the Mathcad worksheet that simulates the Sugar–Alcohol–Yeast model. The bold sans-serif font is text used for documentation and explanation; model components are shown in Times Roman font. Variables are created as they are typed on the worksheet and assigned values with the $:=$ operator. The state variables are elements of a vector $\mathbf{x}$ and defined when they are assigned the initial conditions. The derivative equations are stored in another vector ($\mathbf{D}$). The simplest explanation of the relation between these two vectors is to show how Mathcad updates the state variable vector using Euler integration:

$$\mathbf{x}_{t+1} = \mathbf{x}_t + \mathbf{D}\Delta t.$$

The role of $\mathbf{D}$ is the same if more sophisticated numerical methods are used, but the equation would be more complex.

Sets of expressions are evaluated in order from left to right and from

MoreBeer: Sugar-Alcohol-Yeast Model

$$x := \begin{bmatrix} 52000.0 \\ 0.0 \\ 100.0 \end{bmatrix}$$

// initial conditions as a vector
// x0=sugar, x1=alcohol, x2=yeast

// parameters . . .

$a := .00002$ $b := 0.6667$ $c := .03$ $d := .00000035$

// D( ) is the system of ODEs (derivatives) as a vector . . .

$$D(t,x) := \begin{bmatrix} -a \cdot x_0 \cdot x_2 \\ a \cdot b \cdot x_0 \cdot x_2 \\ a \cdot c \cdot x_0 \cdot x_2 - d \cdot x_1 \cdot x_2 \end{bmatrix}$$

// rkfixed( ) solves the ODEs
// z is a vector with the results
// notice the free form of the statement
// positions

$z := \text{rkfixed}(x,0,500,500,D)$

$n := 0..400$

$z_{n,3} := z_{n,3} \cdot 50$ **// scale yeast for plots**

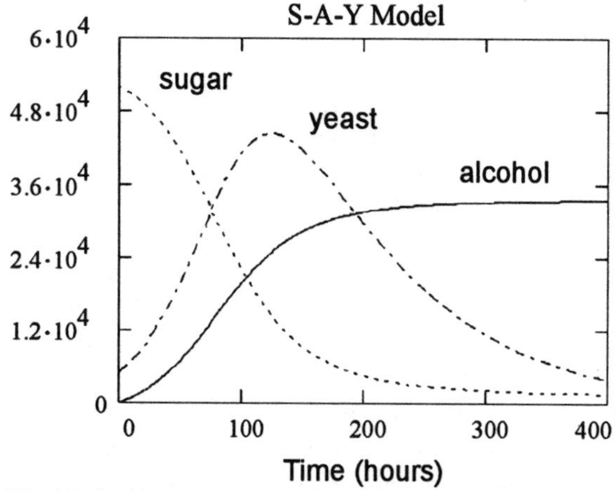

Figure 5.9: The Mathcad worksheet for the yeast model. A model is defined by a vector of state variables (x) and a vector of derivatives for each state variable (D). Runge–Kutta integration [`rkfixed( )`] solves the three ODEs.

top to bottom. As soon as the invocation of the integration function

$$\mathbf{z} := \text{rkfixed}(\mathbf{x}, t_0, t_{\max}, N, \mathbf{D})$$

is encountered, the system of equations $\mathbf{D}$ are solved from time t_0 to $t_{\max}$ using initial conditions in $\mathbf{x}$. N results are stored in the output vector $\mathbf{z}$. Sets of simulations with different parameter values are performed simply by changing the parameters and reinvoking the integration routine to store new results in a different vector. These may be listed and plotted together.

5.8 Summary

SCoP supports a wide variety of model types including ordinary and partial differential equations and linear and nonlinear algebraic (static) equations. The quality of graphical hardcopy output is low, but tabular output can be saved. Simulations execute quickly, since SCoP creates compiled programs. There is a good selection of different integration techniques. The language is simple and relatively easy to learn.

Stella II solves ODEs using only a few integration methods and cannot do partial differential equations. It can do discrete event models. The size of the model is ultimately limited by the size of the Macintosh desktop onto which the diagram is made, but new features make this effectively very large. The amount of abstract thinking that must be performed in model development has been reduced to nearly the smallest amount possible. The quality of graphical output is very high. Model solution is relatively slow since Stella II does not compile the model directly into machine language. Therefore, a Stella II model is not a stand-alone program (as is a SCoP model) and requires the complete Stella II application or a simplified run-time version to execute. It is basically an interpreted language, but it does this relatively quickly.

In recent years, spreadsheets have become very powerful and flexible. Several excellent spreadsheets packages are available on all computing platforms. A model implemented as a spreadsheet is essentially an interpreted program, so it will execute relatively slowly. Graphical output from spreadsheet applications has become very flexible and of high quality. Because spreadsheets were not designed for simulation models, their use in this context is somewhat awkward. However, since many users will already own a spreadsheet program, the monetary and learning time cost is reduced. To use some spreadsheet packages for simulation modeling will require that the user either write programs or buy specialized modules.

Mathcad is the most flexible and extensible of the four. It employs a WYSIWYG interface that presents the model to the user using familiar mathematical notation. It has an extensive library of mathematical functions; it can perform both numerical and symbolic (algebraic) solution to a wide variety of mathematical problems. The current version has good support for ODEs and graphics. Other CAM applications are also available on a variety of computer platforms.

Chapter 6

Numerical Techniques

6.1 Mistakes Computers Make

SOME people think computers make mistakes whenever their behavior departs from human expectations. In this sense, their mistakes can be disturbingly frequent, especially when we program in C. Often, the correct solution is to alter our expectations, but this does not always work because inherent hardware limitations can prevent computers from being correct. In this chapter, we discuss what these limitations are and how to work around them.

Recall that we interpret a finite difference equation as an exact representation of the biological system. Therefore, the numerical solution is also exact and not an approximation. Differential equations are different and their numerical solutions are only approximate and are error prone. In the remainder of this chapter we examine various problems, considerations, and techniques related to the numerical solutions of differential equations. We will emphasize solutions to ordinary differential equations: those that do not describe spatial processes. However, we will also describe one method for solving partial differential equations by converting them to a set of ordinary differential equations. We begin with a general discussion of errors in numerical techniques, but to understand and appreciate these, we must realize how different kinds of numbers are represented and stored in computers.

6.1.1 Representations of Numbers

For our purposes, a *bit* is the logical representation of the electrical state of a computer component called a logic gate. A bit cannot be decomposed into a set of lower-level states or machine components. All other data types (e.g., integers, real numbers, etc.) are defined in terms of bits. In most scientific programming, we are interested in three data types: characters, integers, and real numbers. All data types must be stored using a finite number of bits, and this fact produces the opportunity for error.

In most programming languages, a *character* is a set of eight bits, also know as a *byte*. Bit 0 is called the *least significant bit*, and bit 7 is called the *most significant bit*. Characters are distinguished by the patterns of 0s and 1s in the eight positions. Since each of the eight positions can be in one of two states (0, 1), a byte or character can represent $2^8 = 256$ different numbers (0–255). Depending on the context, the value of a character can be interpreted as a number (an 8-bit integer) or as a printable character. If it is interpreted as a character, then a code is required to convert the bit pattern into alphanumeric symbols (e.g., "A"). The most common code is the ASCII (American Standard Code for Information Interchange) code.

Most programming languages also define an *integer* data type. The number of bits used for integers depends on the hardware to which the programming language compiler is targeted. Personal computers use either 16 or 32 bits for integers; minicomputers and supercomputers use 64 bits. A 16-bit integer can represent $2^{16} = 65536$ different numbers.

Basic integer arithmetic operations such as addition and multiplication use standard binary arithmetic rules. For example, `1+1=0`, and carry a 1 to the next higher position. Since there are only 16 bits, a problem occurs when we attempt to describe a number larger than 65535. To see this, consider a simpler, hypothetical case where we use only three bits to represent integers. Such a number might be: `001 + 101 = 110` (in decimal: $1 + 5 = 6$) Since only a finite number of bits can be reserved to hold the result of an arithmetic operation, it is possible for *overflow* to occur (e.g., `111 + 1 = ???`). A compiler can resolve this dilemma by *wrap around* (result equals `000`), or *truncation* (result equals `111`). In either case, we cannot represent numbers larger or smaller than those that can be represented in the number of bits reserved for the data type.

Similar problems occur in *floating point* numbers. A floating point number is a real number (i.e., not an integer) represented in such a way that the decimal point can float so that a fixed number of significant digits is always represented, no matter how large or small the absolute value of the number. This is simply the scientific notation using powers of base 10 (e.g., 1.234×10^{-2}). A floating point number is composed of a mantissa (e.g., 1.234) and an exponent (e.g.,–2), either one of which may be positive or negative. Both of the components must be represented as a bit pattern. Consequently, not all decimal numbers can be represented. The number of bits used to represent the exponent determines the size of the number that can be represented. The number of bits used for the mantissa represents the precision (number of significant digits) of the number. The standard method of coding is the IEEE Standard 754. A *single-precision* floating point number (i.e., `float` in `C`) is one that uses a total of 32 bits (1 for the mantissa sign, 23 for the mantissa, and 8 for the exponent). (The exponent does not have an explicit sign bit; the upper half of the possible range is assumed to be positive, the lower half assumed to be negative.) A *double-precision* number (`double` or `long float` in `C`) uses a total of 64 bits (52 for the mantissa, 11 for the exponent, plus the sign bit).

Since a mantissa and an exponent are simply a series of bits like integers, operations on these floating point components have the same possibility of overflow. If the exponent is negative and the operation on the exponent causes an overflow in the exponent bit pattern, the condition is called *floating point underflow*, since the operation attempted to create a number smaller than that which could be represented. If the exponent is positive and the exponent bit pattern becomes too large, then the floating point number *overflows*.

6.1.2 Round-Off, Truncation, and Propagation Errors

Errors arise in numerical calculations because of the limited computer memory available to store floating point numbers and the nature of the algorithms. Storage limitations produce overflow or underflow and these become *round-off* errors. Floating point storage round-off occurs because the number of significant digits in floating point numbers are limited by the number of bits in the mantissa. This error occurs most frequently when we add a very small number to a large number. For example, suppose we wish to add $1 \times 10^{-2} + 100 \times 10^{3}$. To accomplish this we first *align the exponent* by rewriting the smaller number so that it has the same exponent as the larger number. This is 0.00001×10^{3}, so the number has been changed from using one significant digit to five significant digits. In most computers, this is a minor increase in digits. However, if the smaller number is many times smaller than the larger (e.g., $10^{-10} + 10^{10}$), then we can come to the point where aligning the exponents will require more bits in the mantissa than are available. Since we cannot use more bits than defined for the data type, the mantissa of the smaller number will contain only the smallest number representable, not progressively smaller numbers as actually needed. In other words, we have rounded off the smaller number, and it has become larger than intended.

Two other kinds of errors occur depending on the operations used in the algorithm. These errors occur regardless of the storage constraints. Numerical algorithms often have to calculate the value of an unknown function. An important mathematical tool for representing an unknown function with some arbitrarily close approximation is to use an infinite series. *Truncation errors* occur because the algorithm approximates a function as an infinite series truncated after the first n terms. These kinds of approximations occur in many algorithms, but the value of n is specified by programmer/analyst so the error is easily controlled. Nevertheless, it may be costly in computer time to reduce the error. *Propagation errors* are errors made at every stage of an iterative algorithm (such as solutions to differential equations) and that accumulate over the entire solution.

In an iterative procedure, these sources produce two types of error: local error (at every solution step) and global error (deviation from the true solution). Local error due to truncation can be estimated by increasing the number of terms used in the approximation (e.g., the solution step size Δt) and calculating the relative change (or improvement) in the answer. Global error usually cannot be estimated since in general we do not know the true solution.

6.2 Numerical Integration

In Chapter 4, we noted that a differential equation and its solution are different manifestations of the same model. The former portrays the functional dependencies of the rates of change; the latter form gives the values over time. The integral is the anti-derivative, and it is possible to go back and forth between the two forms. This concept is central to understanding the approximations used to obtain numerical solutions to equations that cannot be solved analytically. A *slope field* is a concept that unites the two forms.

6.2.1 Slope Fields

We restrict attention to ordinary differential equations in which we have simple derivatives with respect to time. The solutions of these types of equations can be plotted in a two-dimensional space in which the y-axis is the dependent variable and the x-axis is time (t). One point in this space [i.e., a (y, t) pair] satisfies the solution equation. Furthermore, taking the derivative of the solution function (generally unknown) at a point on the time axis will give the numerical values of the original differential equation for the particular (y, t) pair. If we calculate the derivative at many of these pairs, we will produce a field of slopes (i.e., the slope field). There are multiple slopes at each t because each different initial condition produces its own trajectory of slopes. Figure 6.1 shows the slope field for one differential equation.

Also shown in Fig. 6.1 are the true solutions for this equation (solid lines). Usually we do not know the true solution, but we can compute the slope field from the differential equation. The problem in numerical approximation of the true solution is to find the subset of slopes in the slope field that corresponds to the true solution. The subset of particular interest is the sequence of slopes that begins at the known initial condition. There are an infinite number of true solutions (one for each initial condition) and, therefore, there are infinitely many incorrect sequences. Our problem is to stay as close as possible to the correct sequence that lies on the solution curve. Below, we discuss two different methods.

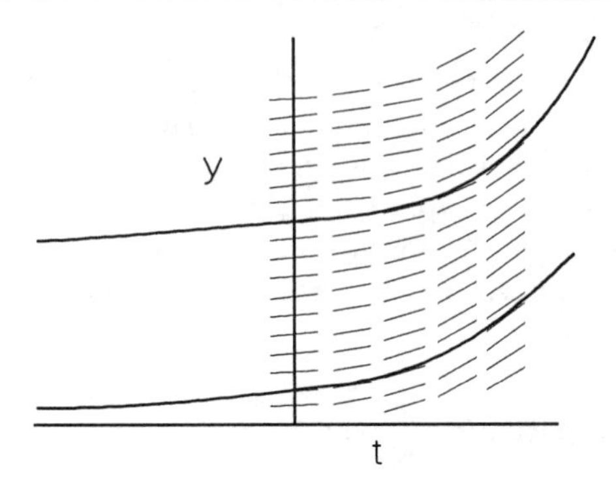

Figure 6.1: Slope field and two true solutions of a differential equation.

6.2.2 Euler's Method

All the methods to solve the differential equation are similar to the simulation models discussed thus far. Given that we are starting at a solution point (the initial condition), the strategy is to move from the initially correct slope in the slope field to the next correct slope, from there to the next correct slope, and so on.

The Euler method is the simplest, most straightforward approximation. This formula was derived in Chapter 4:

$$y_{t+\Delta t} \doteq y_t + \Delta t \cdot f(\ldots). \tag{6.1}$$

Figure 6.2 shows the relation between the correct solution and the Euler approximation. The solid line is the true solution; the straight-line segments are the approximations. The dotted lines show why the approximations of this function underestimate the true solution. The slope at $t = 0$ is exactly correct since the solution at that time is simply the initial condition. Since the true slope is continuously increasing (in this function), but our approximation over Δt is not, the approximation is too small. The approximation continues to get worse (error propagation), because the new slope at $t + \Delta t$ uses the approximated value of y, not the true y at that time. This yields a slope calculation from the differential equation below

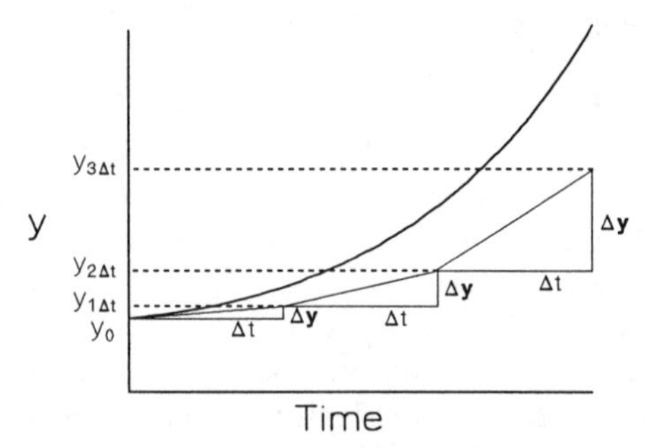

Figure 6.2: A series of Euler approximations (straight lines) to a true solution (curved line) over Δt solution intervals.

that of the true solution at $t + \Delta t$.

Typically, we must solve several differential equations simultaneously and these equations are a system in the sense that their derivatives are functions of the other state variables. As emphasized earlier in the computer code for finite difference equations in Chapter 5, we must use care that we do not alter the current values of state variables inside a time step. The analogous equation is

$$\begin{aligned} \frac{dV}{dt} &= rV - bVP \\ \frac{dP}{dt} &= bcVP - dP. \end{aligned} \tag{6.2}$$

In the Euler method, these continuous equations are replaced by the approximations:

$$\begin{aligned} V_{t+\Delta t} &\doteq V_t + [rV_t - bV_tP_t]\Delta t \\ P_{t+\Delta t} &\doteq P_t + [bcV_tP_t - dP_t]\Delta t. \end{aligned} \tag{6.3}$$

Because both derivatives in Eq. 6.2 depend on the current values of both state variables, the expressions in brackets on the right-hand sides of Eq. 6.3 must be computed before variables are updated so that the order of the equations does not influence the calculations. Hence, we should always first calculate the rates (derivatives), then update the states.

6.2.3 Runge–Kutta Basics

The primary advantage of the Euler method is its simplicity. But it has many disadvantages; the foremost among them is that it is inefficient: very small Δt and many iterations are required to obtain acceptable accuracy. Acceptable accuracy is a relative term, of course, and depends on the objectives of the model. Nevertheless, there are many other, better methods. As a general method with wide applicability, the Runge–Kutta (RK) method has many advantages. It is easy to code; its numerical behavior is less sensitive to the size of Δt than the Euler method. In addition, it is remarkably efficient: a large Δt provides accurate solutions.

In contrast to the Euler method, which uses a single evaluation of the derivative to extrapolate into the future, the Runge–Kutta method uses several estimates of the slope of the function. In this regard the Runge–Kutta is actually a family of algorithms in which the members are distinguished by the number of slope calculations performed. Here we will describe in detail the so-called *second-order* Runge–Kutta and briefly the more useful *fourth-order* method.

The basic steps in the method are listed below and illustrated in Fig. 6.3.

1. Make a first ($i = 1$) tentative step from t to $t + \Delta t$:
   (a) Evaluate $dy/dt = f(y_t)$ at t.

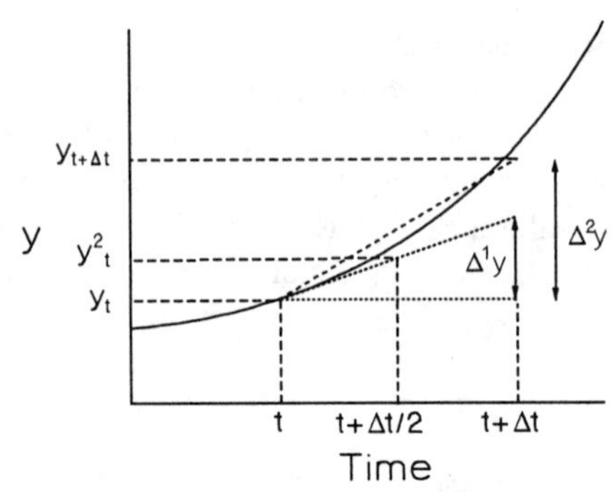

Figure 6.3: Second-order Runge–Kutta integration. $\Delta^2 y$ is the second estimate of the rate of change based on the midpoint value y_t^2; $y_{t+\Delta t}$ is obtained by weighting $\Delta^2 y$ and adding to y_t.

   (b) Calculate $\Delta^1 y_t = f(y_t)\Delta t$ and $y_t^1 = y_t + \Delta^1 y_t / n_1$ using the Euler method.

   (c) Set $i = 2$.

2. Choose n_i, the number of intervals to divide Δt for the ith pass through the algorithm.

3. Calculate the ith tentative change from t to $t + \Delta t$:

   (a) Calculate the ith intermediate y from the previous step: $y_t^i = y_t + \Delta^{i-1} y_t / n_i$.

   (b) Calculate the ith tentative change: $\Delta^i y_t = f(y_t^i)\Delta t$.

4. If the RK order is greater than 2, repeat step 2 above, using the most recent tentative y_t^i at which to evaluate the derivative and to calculate a new (tentative) $\Delta^i y_t$.

5. Combine all estimates of $\Delta^i y_t$ in a single $\Delta^* y_t$ using weighting factors w_i: $\Delta^* y_t = \sum w_i \Delta^i y_t$.

6. Compute the updated variable: $y_{t+\Delta t} = y_t + \Delta^* y_t$.

In second-order RK (RK-2), step 2 is done once; in third-order RK-3, it is done twice; in fourth-order RK-4, it is done three times, and so on. Thus, RK-2 has two Δys to combine and RK-4 has four. The last two steps combine the estimates of $\Delta^i y_t$ and obtain the final new $y_{t+\Delta t}$. The different Runge–Kutta orders use different values of n_i and weights (w_i) for combining the intermediate derivatives. The values to use for Euler, second-order, and fourth-order Runge–Kutta are listed in Table 6.1

The calculations for one time step of the RK-2 on the equation $dy/dt = ay$, with $y(0) = 10.0, a = 0.5, \Delta t = 1.0$ are:

$$\Delta^1 y_t = (0.5)(10)(1.0) = 5.0$$

$$y_t^1 = 10 + 5.0 = 15.0 \qquad \longleftarrow \text{Euler (not used)}$$

$$y_t^2 = 10 + 5.0/2 = 12.5$$

Table 6.1: Time steps and weights for three numerical integration algorithms. n_i and w_i are the number of intervals and combination weights used in the ith tentative step.

	n_1	n_2	n_3	n_4	w_1	w_2	w_3	w_4
Euler	1	—	—	—	1	—	—	—
RK-2	1	2	—	—	0	1	—	—
RK-4	1	2	2	1	1/6	1/3	1/3	1/6

$$\begin{aligned}
\Delta^2 y_t &= (0.5)(12.5)(1.0) = 6.25 \\
\Delta^* y_t &= 5.0 * 0 + 6.25 * 1 = 6.25 \quad \longleftarrow \text{ Weighted } \Delta^i y \\
y_{t+\Delta t} &= 10.0 + 6.25 = 16.25
\end{aligned}$$

Compare this estimate with the true solution: $y_{1.0} = 16.4872$.

Table 6.2 compares the accuracy of the Euler method with second- and fourth-order Runge–Kutta and the true solution. This illustrates that (1) all methods become less accurate over time, (2) the Euler method becomes more accurate as Δt decreases, (3) the Euler method is less accurate than the Runge–Kutta method even when the methods use the same number of derivative calculations [e.g., Euler ($\Delta t = 0.5$) *versus* RK-2, and Euler ($\Delta t = 0.25$) *vs* RK-4], and (4) RK-4 is remarkably accurate for this simple ODE.

6.3 Numerical Instability and Stiff Equations

Some numerical methods applied to specific equations may produce answers in which errors due to round-off interact with algorithm truncation to produce large errors that increase as the solution unfolds. Such methods are "instable" and are obviously undesirable. One may envision instability arising because the solution jumps around in the slope field, possibly alternating on either side of the true solution with increasing deviation as the solution unfolds. In most cases, decreasing the step size will reduce the rate of increase of these errors. Desirable methods are those that reduce the errors more effectively at large step sizes. RK is generally more effective for many more problems than Euler, but RK fails for certain equations.

A prime example of these are *stiff* equations. Stiffness can arise when the equations use several, very different time scales. Different time scales in equations often cause the solution algorithm to add very large numbers to very small numbers. This is a situation that produces large round-off and

Table 6.2: Comparison of Runge–Kutta and Euler methods solving $dy/dt = ay$, $a = 0.5$, $\Delta t = 1.0, 0.5, 0.25$.

Time	Euler $\Delta t = 1.0$	Euler $\Delta t = 0.5$	Euler $\Delta t = 0.25$	RK-2 $\Delta t = 1.0$	RK-4 $\Delta t = 1.0$	True
0.0	10.0	10.0	10.0	10.0	10.0	10.0
1.0	15.0	15.625	16.0181	16.2500	16.4844	16.4872
2.0	22.5	24.400	25.6579	25.3900	27.1735	27.1828

truncation errors. Some examples of systems whose differential equations may be stiff are:

1. Algal Nutrient Uptake and Cellular Division: Nutrient uptake is a rapid process that occurs over microseconds; cell division requires several hours (Abbott 1990).
2. Photosynthesis and Enzymatic Reactions: Oscillating light levels will produce a rapid change in enzyme kinetic parameters but a relatively slow change in photosynthesis at the leaf level (Gross 1982).
3. Rotating Rocket Orbiting Earth: The rocket rotation is fast compared to the orbiting time (Rice 1983).
4. Refinery Control: Chemical reactions occur rapidly compared to the temperature response of the large vats (Rice 1983).

Additional examples from the physical sciences can be found in Brackbill and Cohen (1985). There are two broad approaches to solving this problem of multiple time scales. The first method is most applicable to computer simulation in which we create submodels that correspond to the subsystems having different time scales. For example, we could build a model of nutrient uptake and a separate model of cell division. Integrating the dynamics of the submodels is a problem. The usual approach is to build a simulation program that has a global clock controlling all processes. At fixed, large intervals of the clock, a subroutine to update the slow time scale submodel is executed. At smaller intervals, the subroutine for the fast time scale submodel is executed. Effectively, this approach assumes that between the large intervals, the slow process does not occur. However, as exemplified by the cell division problem, the two processes depend on each other. Since the fast submodel generates many values between executions of the slow submodel, the modeler must decide which value(s) will be used to influence the slow process. Should it be the average value, the final value before execution of the slow submodel, the mid-interval value, or the integral of all values over the time interval? While there are, as indicated, problems arising from this approach, it has the benefit of forcing the modeler to propose specific hypotheses for each of the subsystems. In essence, this approach forces us to explain the origin of the time scales by modeling the subsystems explicitly.

The second approach comes from physics and does not attempt to identify and model specific subprocesses that account for the existence of the time scales. An example of this is a rotating rocket that orbits the earth. From a physical perspective, the complex motion is a result of continuous forces acting on the rocket: angular momentum, gravity, and so on. Rather than modeling these as separate subsystems, a numerical approach is to find a better method of integrating the equations. The problem of stiff equations in this context arises simply because the parameters in the system of ODEs vary over several magnitudes. Press et al. (1992) give a concrete example. Suppose we have the following differential equations:

$$\frac{du}{dt} = 998u + 1998v$$
$$\frac{dv}{dt} = -999u - 1999v.$$

Without going into details, these equations produce solutions for u and v that are the sum of negative exponentials, one of which is e^{-1000t}. This term requires a very small Δt to accurately approximate the solution (too large a Δt will miss the dynamics caused by this term by "stepping over" the changes). There are two possible solutions: (1) decrease the step size appropriate to the fastest time scale, and (2) use a different numerical method. Solution (1) is inefficient, but for many biological simulations this is not an important issue, especially as desktop computers become faster. Option (2) is feasible since many good algorithms are available (e.g., implicit methods), but one must choose the proper method for the problem at hand, and the methods are inevitably more complex and difficult to program than RK or Euler. The programming problem is not critical as libraries of numerical functions in all common languages become available (see Press et al. 1992).

In conclusion, time scales and stiff equations are a potential problem because biological dynamics occur over many different time scales. It is advisable, when studying equations with which one does not have much previous experience, to monitor the net rates of changes of each state variable. The relative net rates of change should stay within reasonable bounds. As a very crude check, if $(1/x_i)(dx_i/dt) > 0.2$ in any time step, then you should consider reducing the time step or using methods developed for stiff equations. Rice (1983) reviews the available software packages for dealing with such equations and Press et al. (1992) provide more references and C code. Below, we discuss one approach that can easily be incorporated into simulation models.

6.4 Integrating ODEs with Variable Time Steps

Using small time steps to deal with stiff or nearly stiff equations can be inefficient because small steps are not always needed. At times, all state variables are changing slowly and large time steps are appropriate and desirable. A way to accommodate this situation is to allow the time steps for integration to vary according to the most rapidly changing variable. Coleman and Gay (1990) advocate this for physiological systems using Euler integration. Given the dramatic efficiency of RK-4, a better solution is to allow RK time steps to be variable (Press et al. 1992). In this section, we describe how to do this.

The simplest approach to optimizing the time step for any integration method is to calculate, at every iteration, the estimate for the next value using the current time step and an estimate using a smaller time step. If these differ by an unacceptable amount, then the truncation error is too great and a smaller step size is needed. This test is repeated as many times as necessary within the current time step until the error criterion is satisfied. Of course, the penalty for choosing a smaller but more accurate time step is that we must perform additional calculations of the derivative.

For the Euler, the calculations are

$$\begin{aligned}
y_{t+\Delta t} &= y_t + \Delta t f(y_t, t) && \longleftarrow \text{full step} \\
y^*_{t+\frac{\Delta t}{2}} &= y_t + \frac{\Delta t}{2} f(y_t, t) && \longleftarrow \text{midpoint value} \\
y^*_{t+\Delta t} &= y^*_{t+\frac{\Delta t}{2}} + \frac{\Delta t}{2} f(y^*_{t+\frac{\Delta t}{2}}, t + \frac{\Delta t}{2}) && \longleftarrow \text{two half steps}
\end{aligned}$$

The absolute error estimate is

$$E_{\Delta t} = y_{t+\Delta t} - y^*_{t+\Delta t} \tag{6.4}$$

and the error relative to the current magnitude of the state variable is

$$e_{\Delta t} = \frac{E_{\Delta t}}{y_t}. \tag{6.5}$$

Instead of one derivative calculation, the above scheme requires two. While this formula is useful, we can take it one step further. Given this calculated $e_{\Delta t}$, we can calculate another $\Delta' t$ which is the time step needed to exactly produce the target or desired error. This permits us to both reduce the time step when the error is too large and increase it when the error is smaller than needed. To do this, we need to compute the largest step possible that does not produce error larger than desired. For reasons we will not go into here, the error estimates are proportional to $(\Delta t)^2$. But we use this fact to note that if $e_{\Delta t} \propto (\Delta t)^2$, then there is a target error proportional to some other time step: $e'_{\Delta t} \propto (\Delta' t)^2$. Using these two proportionalities, we have

$$\Delta' t = \Delta t \left(\frac{e'_{\Delta t}}{e_{\Delta t}} \right)^{1/2}. \tag{6.6}$$

This approach also applies to RK-4, but each of the four steps must be performed for both the full time step and the two half time steps. As with the Euler method, we must also apply the two calculations to each state variable at each stage. Therefore, in the step-doubling method for RK-4, we must calculate the derivative 11 times, as compared to 4 for the nonvariable method. Rather than discuss this approach further, we briefly mention the Runge–Kutta–Fehlberg (RFK) method which is an alternative that is described in detail in Press et al. (1992).

The RKF method also uses an estimate of the truncation error to determine the best time step. This method is a fifth-order RK method that requires six calculations of the derivatives. When these calculations are recombined in a different way, they produce a fourth-order estimate of the new $y_{t+\Delta t}$. The difference in the fourth-order and fifth-order estimates is the error, and this, once known, is used in the same manner as above to determine the best time step. The major feature of this algorithm is that it gives an error estimate using only six evaluations of the derivatives, rather than the 11 needed for the time step varying method described above. We will not discuss the details here, since Press et al. (1992) do an admirable job, and, conceptually, it differs from RK-4 only in the procedure for combining the trial solutions.

6.5 PDEs and the Method of Lines

Whereas the RKF method is a good, general method for ordinary differential equations, partial differential equations are more difficult and, if optimal performance is necessary, require more specialized numerical methods. We will not attempt a discussion of these in this introductory text, but only illustrate one solution method that reduces the problem to solving a large number of ordinary differential equations.

6.5.1 Discretization

In a spatially explicit system distributed over continuous physical space, the dynamical processes described by the differential equations operate at all points in the space (except perhaps at the boundary of the space). Obviously, these processes will also operate at some finite subset of points in the space. To obtain an approximate, numerical solution to the continuous equations, we discretize continuous space into a large, but finite, number of grid points. Since the dynamical processes operate at each point, we must translate continuous mathematical representations (e.g., second-order partial derivatives to represent diffusion) into finite differences. This is analogous to the problem of solving ODEs at a finite number of time values.

Imagine a one-dimensional spatial axis represented as a line with nodes at fixed intervals. The nodes are points where we will obtain solutions. Each node is given an index number, and we will focus on one of these nodes i. To the left of i is $i-1$; to the right of i is $i+1$. The first process that we translate is advection. A common approximation for advection is

$$U_x \left(\frac{\partial y}{\partial x} \right)_i \approx U_x \left(\frac{y_{i+1} - y_{i-1}}{2\Delta x} \right)_i, \tag{6.7}$$

where Δx is the finite space interval in physical units, y_i is the quantity of the state variable at node i, and we assume that the flux rate (i.e., velocity) in the x direction (U_x) is independent of position (i). This expression simply states that the change at node i is the inflow (from $i+1$) minus the outflow (to $i-1$). Of course, the direction of flow could be in the opposite direction, but this is accounted for by the sign of the coefficient.

Likewise, a reasonable approximation for the second-order diffusion process is

$$D\left(\frac{\partial^2 y}{\partial x^2}\right)_i \approx D\left(\frac{(y_{i+1}-y_i)+(y_{i-1}-y_i)}{\Delta x^2}\right)_i$$
$$\approx D\left(\frac{y_{i+1}-2y_i+y_{i-1}}{\Delta x^2}\right)_i . \tag{6.8}$$

As the first equation above indicates, Eq. 6.8 is simply the sum of the gradients on either side of node i. This is the basic diffusion concept we developed in Chapter 4.

In typical mass transport models (Chapter 4), the processes that move mass (or energy and momentum) are additive in two or three dimensions. This means that the above discretizations can be rewritten for other dimensions by changing the spatial index (e.g., x to y). Mass transport models also have a term describing the rate of change of the variable (i.e., $\partial y/\partial t$). This term can also be discretized with a finite difference scheme so that all dimensions (space and time) are discrete.

The above method of discretization is called *central differencing* because the scheme is centered around the node currently being evaluated (i in Eqs. 6.7 and 6.8). Once the PDEs have been discretized, they must be solved. There are two broad families of methods (Kahaner et al. 1992). If time and space are both discretized, the classical finite difference or finite element methods based on solving a set of algebraic equations are used (see Press et al. 1992 for methods). If time is not discretized, but space is, we use the *method of lines.* Since this builds on our previous discussions, we present this method here as one that is generally useful and understandable.

6.5.2 Method of Lines and ODEs

Consider the flow of a contaminant in a river (p) with advection, molecular diffusion, and bioaccumulation in biotic components (b). A plausible model might be

$$\frac{\partial p}{\partial t} = -U_x\frac{\partial p}{\partial x} + D\frac{\partial^2 p}{\partial x^2} - kb\left(1-\frac{b}{B}\right) \tag{6.9}$$

$$\frac{\partial b}{\partial t} = kb\left(1-\frac{b}{B}\right) - q_x\frac{\partial b}{\partial x}, \tag{6.10}$$

where the velocity in the x direction is U_x, D is diffusivity, and contaminant uptake by biota (k) decreases as the amount of the biota (b) increases to a maximum biomass level (B). This relationship resembles the logistic equation of population growth. The differential equation for the biota describes the uptake and bioaccumulation as well as an advective flow away from the node at the rate q_x. A physical, chemical, or biological process such as chemical uptake is called a *reaction.* Consequently, equations such as Eqs. 6.9–6.10 are called *reaction-diffusion* equations.

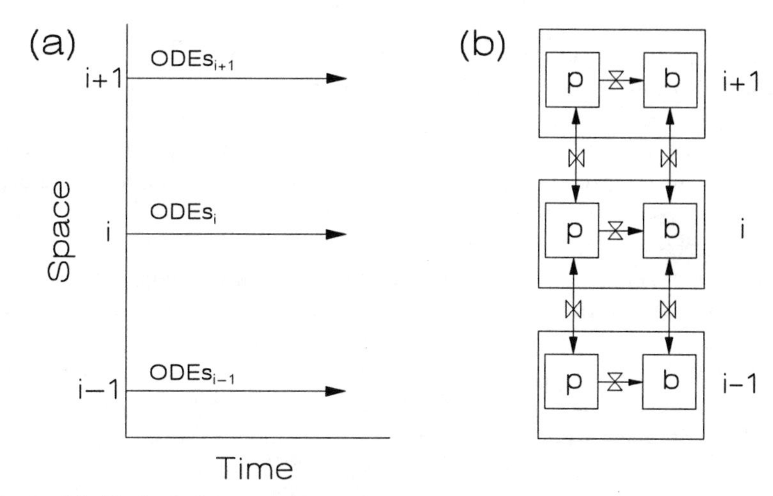

Figure 6.4: Method of lines representation for a one-dimensional advection-diffusion-reaction model of contaminant flow and bioaccumulation. (a) Three discrete spatial lines moving forward in continuous time. (b) Forrester diagram of ODEs solved at each space node.

In the method of lines, we discretize space but not time. Figure 6.4 shows this relationship and a simplified Forrester diagram for three of the nodes. Note that we basically replace each node with a set of compartments (p and b) that interact with each other and the relevant compartments at neighboring spatial nodes.

Using i to index the nodes, the ODEs that must be solved at each node are

$$\frac{dp_i}{dt} = -U\frac{p_{i+1} - p_{i-1}}{2\Delta x} + D\frac{p_{i+1} - 2p_i + p_{i-1}}{\Delta x^2} - kb\left(1 - \frac{b}{B}\right) \quad (6.11)$$

$$\frac{db_i}{dt} = kb\left(1 - \frac{b}{B}\right) - q\frac{b_{i+1} - b_{i-1}}{2\Delta x}. \quad (6.12)$$

All of the dp_i/dt and db_i/dt must be solved simultaneously using an ODE method such as RKF. The spatial scale (Δx) must be chosen to adequately represent the rates of mass movement. Thus, the time step and the spatial grid size are interrelated. If the grid size is too large, we may not correctly represent the dynamics at any node. If the grid size is too small, we will perform unnecessary calculations. This is an important issue with the method of lines, because the number of calculations can become large. For example, if the stretch of river to be modeled is 1000 m, and we wish to describe changes every meter, then, in the above model, we must solve 2000 ODEs at each time step. If the problem is two-dimensional, then the number of equations to solve increases with the square of the number of nodes along one linear dimension. To double the spatial area modeled or halve the gird resolution requires four times as many equations to solve.

In three dimensions, the number of equations increases as the cube of the number of nodes. For a three-dimensional grid 100 m on a side at a resolution of 1 m between nodes, the number of equations to be solved is 10^6 nodes times the number of ODEs per node. To decrease the spatial resolution to 10 cm requires 10^9 nodes. Even if it takes only one microsecond to compute all the ODEs associated with a single node, 1 minute of simulated time will require over 15 minutes of computer time. (No wonder we can't predict the weather!) More specialized and sophisticated methods for solving PDEs can improve this considerably, but the basic problem remains. Be prepared for long runs if your model is spatially explicit and requires high resolution. This is not a hypothetical problem; one spatially explicit model of a wetland ecosystem solves 19,832 equations with a time step of 1 week to simulate a period of 22 years (Maxwell and Costanza 1993).

One final detail is unresolved. Equations 6.11–6.12 will work well for grid nodes that are on the interior of the space being simulated. We must treat the boundary nodes differently because they do not have all the neighbors required by the equations. Two basic approaches are commonly used: (1) force the values of the boundary nodes to specific values (which may vary in time), (2) set the fluxes into or out of the boundary nodes to some specific magnitude (which may also vary in time). Whatever the condition chosen, in the method of lines, special equations are solved that apply to the boundary points.

6.6 Exercises

1. Graph all of the slopes (Δy^i) used in the fourth-order Runge-Kutta method.
2. Consider a finite difference equation of the form $N_t = R^t N_0$. By how much is the final N_t altered if R is halved and t is doubled? Write a general equation for the relationship for any proportional change in R and t.
3. Scale mortality measured in years to a weekly time step so that 52 iterations of the weekly basis produces the same yearly mortality rate as the original.
4. Create a table analogous to Table 6.2 using finite difference equations. In other words, let $a = 0.5$ and solve for two time steps, then let $a = 0.25$ and solve for four time steps.
5. The Torricelli model (Chapter 1) produces negative volumes when Δt is only moderately large. How small must Δt be to prevent this in the Euler method? How small in the RK method?
6. Solve the Torricelli model using a variable time step Euler method. Plot the step size over time.

Chapter 7

Parameter Estimation

7.1 The Problem

The universe does not seem to have been designed by an information retrieval specialist. — Anderson (1974)

EVERY model that is used to make quantitative predictions contains parameters whose values must be specified. Even very simple models can easily contain a dozen parameters needing estimation: the Lotka–Volterra predation model with only two equations and simple, linear relations has four parameters. It is to be hoped that all of the parameters can be estimated in principle (i.e., have operational definitions), but even if this is true, performing the necessary experiments to estimate these values is often difficult in practice.

The following example illustrates the concept. Suppose we wish to model the population dynamics of a single population of an animal in which reproduction is limited at high densities. Basic ecological considerations lead us to perform a series of laboratory experiments in which we control the population density, run the experiment long enough to allow most females to produce offspring, then calculate the average number of offspring each female produced. We assume we are careful in our procedures and design to ensure that the number of adult females does not change significantly during the experiment.

From these experiments, we obtain a set of paired numbers and a graphical (functional) relation (Fig. 7.1). We wish to use this functional relation as the basis for our population dynamics model, so we must translate it into an equation. Using functions from Section 4.5, we might choose the power function: $y = k_1 + k_2 x^{k_3}$, where y is the offspring per female and x is the number of females. This equation has three parameters whose values must be determined. This is the parameter estimation problem.

In general, the basic problem is that given a functional form with a dependent variable and one or more independent variables, and given data such that the observed dependent variable can be plotted against the observed independent variable(s), we wish to know the estimates of the pa-

Females	Offspring
1	5
10	4.9
25	4.5
45	4.0
60	3.0
70	1.5
75	0.5

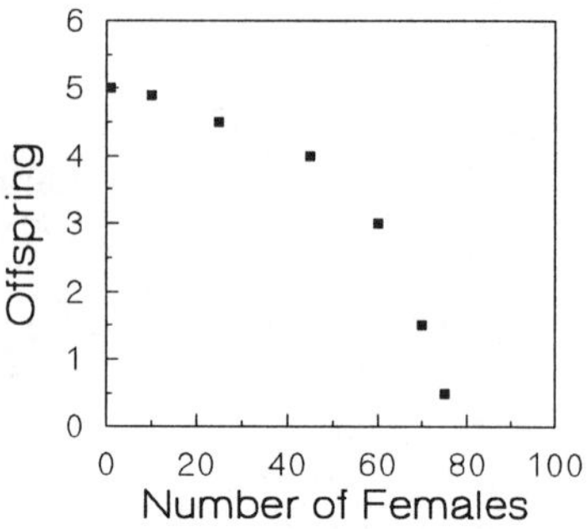

Figure 7.1: Experimental results for number of offspring per female at different densities of females.

rameters of our function that provide the best fit to the data. There are several difficult words in that statement, particularly "estimate" and 'best." A good introduction to these topics is in Richter and Söndgerath (1990).

In the following discussion, we assume that we have a mathematical function to fit (e.g., $y = ax + b$), a set of parameters used in the function (e.g., a, b), and a set of observations [i.e., a matched set of xs (independent observations) and ys (dependent observations)]. We wish to find the parameter values that provide the "best" fit to a particular data set. That is, for functions (y) of a single independent variable, x, we have data pairs of the form (x_i, y_{ij}), where we may have more than one y observation at a given x value. The statistical model we use is

$$y_{ij} = f(x_i, p_k) + \epsilon_i, \tag{7.1}$$

where p_k are k parameters for which we wish the best estimates and ϵ_i is the error associated with the ith value of the independent variable. But this depends on what we mean by "best," that is, how we will measure ϵ. The standard definition of best is the least-squared difference, which attempts to minimize the error term:

$$\min \sum_i \epsilon^2 = \min \left(\sum_i (y_{ij} - f(x_i, p_k))^2 \right). \tag{7.2}$$

This criterion has many nice features (e.g., unbiased, identical to maximum likelihood estimator for some conditions). We will emphasize this approach in the following sections. This method does not, however, tell us which function to use. If we wish only to obtain a good fit with a function that passes through as many points as possible, then a cubic spline fit would be a good choice (Chapter 4). Usually, however, we wish to use functions with few parameters or to use a particular function, one perhaps that was derived from first principles. In this case, we can use some of the techniques described below.

7.2 Simple Linear Regression

One of the simplest functions we can attempt to fit to data is the linear function ($y = mx + b$), where m is the slope and b is the intercept. Simple linear regression, which involves only a single independent variable, should be familiar to the reader from introductory statistics. However, using regression to estimate model parameters often requires careful thought about the structure of the data and the model being fit. By being clever, one can often obtain the estimates from data which on the surface may appear to be nonlinear.

7.2.1 Static Applications

The easiest case to which linear regression applies is a simple experiment with a single independent variable. models may be fitted to linear equations from in which a single independent variable is controlled a single effect. This is a classical application of linear regression in which the slope and intercept are the parameters of interest. For example, we might perform a feeding experiment in which the density of prey is controlled (varied) and feeding rate (numbers eaten in a trial period) observed. Assuming the data were approximately linear, we could model this as $f = mp$ (f is feeding rate, p is prey density) and estimate m using linear regression. This approach to parameter estimation is covered in many introductory statistics books.

7.2.2 Dynamic Applications

The models and systems discussed here have all been dynamic. Data taken from dynamic sequences of observations can often be used directly for parameter estimation by linear regression. For example, the density-independent model

$$\frac{dN}{dt} = rN$$

is itself a linear equation with the slope equal to r. Therefore, to estimate r we have only to make observations of a population growing according to the equation at discrete times. From these data, we can calculate absolute population change ($\Delta N/\Delta t$) and regress these values against the corresponding N_t. So, although this is not an experiment in the classical sense, we can use dynamic data in linear regression to obtain the parameter r.

It is sometimes necessary to perform simple transformations on these data to obtain estimates for more complex models. For example, the density-dependent model is

$$\frac{dN}{dt} = rN\left(1 - \frac{N}{K}\right).$$

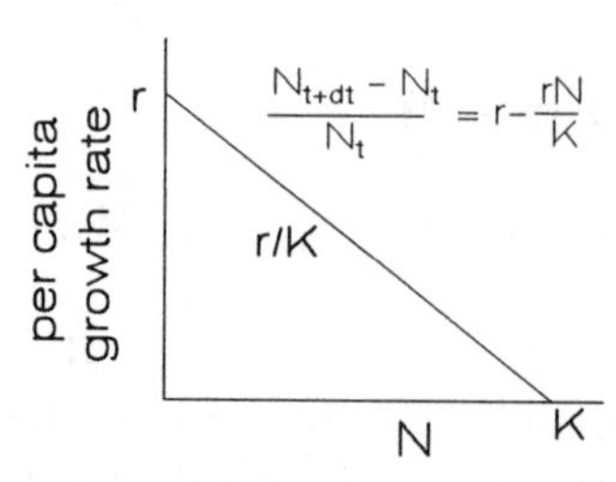

Figure 7.2: Relation of per capita growth rate to the parameters r and K in the density-dependent model.

We note that this is a nonlinear equation (it has an N^2 term), and, therefore, we cannot obtain estimates from simple linear regression of the absolute population change against N. However, by dividing both sides by N we produce a new dependent variable $1/N \cdot dN/dt$ that is a linear function of N and has a negative slope (Fig. 7.2). Both of these examples illustrate that careful reflection on the structure of the equations and the types of information that can be obtained from observations is necessary to effectively estimate parameters. A problem with these applications is that the independent variable (N) is not usually known exactly. Ricker (1973) described one method for treating this problem.

7.2.3 Linear Regression on Transformed Equations

Regardless of the source of the data for regression (i.e., from static experiments or dynamic data), often the relations are nonlinear. In these cases, we can transform the equation to a linear form. This is commonly taught in introductory statistics courses and need not be emphasized here. We give only a few examples to make the point and then give some cautions on the use of this technique when better methods are available.

Division by a Variable This method was shown above when we created the per capita growth rate by dividing both sides of the differential equation by N.

Logarithms Power functions are expressions in which the parameter to be estimated is part of the power of a constant or independent variable. These equations can be made linear by a log transform. For example,

$$\begin{aligned} y &= Ax^b \\ \log y &= \log A + bx. \end{aligned} \tag{7.3}$$

This transform creates a new variable ($\log y$); by regressing this against x we can estimate A as the anti-log of the intercept. The slope is b.

Inverses Hyperbolic functions can be linearized by inverting the function. A famous example is the Michaelis–Menten relation for enzyme kinetics:

$$y = \frac{Ax}{B + x}$$

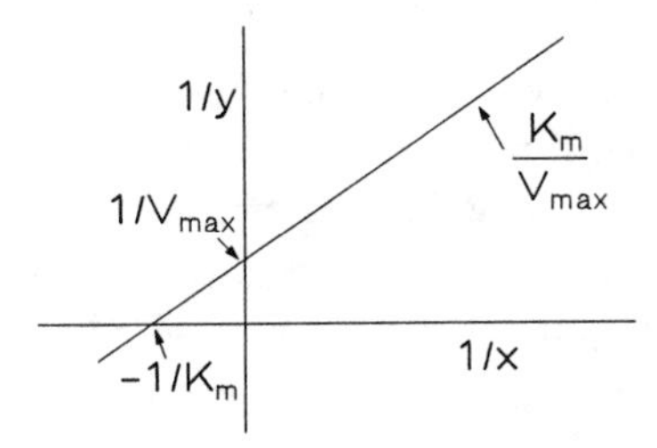

Figure 7.3: Lineweaver–Burk plot to obtain the Michaelis–Menten parameters.

$$\frac{1}{y} = \frac{B}{A}\frac{1}{x} + \frac{1}{A}. \tag{7.4}$$

This relationship and Fig. 7.3 are known as the *Lineweaver–Burk plot.* The maximum rate of the reaction (V_{max} in Fig. 7.3 and A in Eq. 7.4) is the inverse of the intercept with the y-axis. The half-saturation constant (K_m in Fig. 7.3 and B in Eq. 7.4) is the slope multiplied by V_{max}. This equation is still commonly used in biochemical and physiological studies.

7.2.4 Problems with Transformations

All things considered, use of linear regression for parameter estimation of nonlinear equations is a poor method. There are several reasons for this.

1. More advanced and better methods are commonly available in easy-to-use desktop computer statistical packages.
2. Linear regression can estimate only two parameters. Many nonlinear equations use more than two parameters; using linear regression requires that other methods be used to estimate the remaining parameters. For example, the sigmoid curve and its linear transformation is

$$y = \frac{A}{1 + Be^{nx}}$$

$$\ln\left(\frac{A}{y} - 1\right) = \ln(B) + nx. \tag{7.5}$$

   There are three parameters (A, B, and n), and one of these must be assumed in order to estimate the other two.
3. Inversion transformations can produce clustering of the resulting transformed data; this can produce spurious statistical correlations between the variables.

   Consider a set of values evenly placed every 0.5 units between 0.5 and 3.0. The inverse transformation converts this to the sequence: 2.0, 1.0, 0.67, 0.5, 0.4, and 0.33. Most of the numbers are clustered near 0 and there is now an isolated point at 2.0. In extreme cases, this condition can produce isolated groups of datum points that can incorrectly inflate the degree of association between the dependent and independent variables.

4. Inverse transformations turn small numbers into large numbers. Often, the measurement of small quantities has large relative errors. These errors will be magnified after transformation.
5. Since the log of a number less than or equal to 0.0 is undefined, logarithms can require that data be discarded.
6. One rationale for transforming data is to cause the errors between data and the functions to better fit the assumptions of linear regression. This does not always occur and depends on the data and transforming function (Seber and Wild 1989).
7. In parameter estimation for modeling purposes, we almost always are interested in the parameter values stated in their original (untransformed) units. This requires that we "detransform" the numbers (e.g., take the anti-log of the intercept). Sometimes this detransformation will produce biased results (Seber and Wild 1989).

7.3 Nonlinear Equations Linear in the Parameters

There are powerful analytical techniques for estimating parameters in a special class of nonlinear functions. The class is characterized by being linear in the parameters. This means that although the equation is nonlinear with respect to the independent variable (i.e., x), the parameter (a) is not involved in a nonlinear expression. The polynomial equation $y = ax^2$ is linear with respect to a. Some examples of equations that are nonlinear in the parameters are

$$\begin{aligned} y &= \frac{ax}{b+x} \\ y &= a\exp(bx) \qquad (7.6) \\ y &= ax^b. \end{aligned}$$

If the equations are linear in the parameters, we can use several analytical techniques (*nonlinear* or *polynomial* regression). If they are nonlinear, we must use iterative techniques. Below we discuss the polynomial regression and in a later section a few of the iterative techniques.

7.3.1 Multiple Linear Regression

If the equation can be represented as a sum of terms, each of which is linear in the parameters (such as a polynomial equation), then multiple linear regression can be used to estimate the parameters. For example, if the equation is

$$y = a + bx + cx^3,$$

we notice that if we consider x^3 to be a separate variable (call it w, for example), then the equation is linear, and any of several software packages that can perform multiple linear regression will estimate c.

7.3.2 Polynomial Regression

A more general approach is to use nonlinear least-squares regression. I will describe this technique for the special case of a polynomial, but it will work with any equation that is linear in the parameters. The discussion below develops the theory only to the point of estimating the parameters.

The general model for the relation of an observed dependent variable to a function evaluated at various observed independent variable points is

$$y_{ij} = f(x_i, p_k) - \epsilon_i, \tag{7.7}$$

where y_{ij} are multiple observations of the dependent variable at the x_i observations, and ϵ_i is the error between the predicted [$f()$] and observed values of the dependent variable. The x_i are assumed to be known exactly.

To implement the least-squares criterion, we wish to choose the p_k in order to minimize the sum of squared errors (ϵ_i in Eq. 7.7) over all the x_i observations. That is, we want the p_k such that

$$\min \sum_i \epsilon^2 = \min \sum_i \left(f(x_i, p_k) - y_{ij}\right)^2. \tag{7.8}$$

We illustrate the method for the particular function

$$f(x_i, p_k) = A + Bx_i + Cx_i^2,$$

where the problem is to find A, B, and C that satisfy our minimization criterion. So, we have (dropping the j subscript on the multiple y_i observations)

$$\begin{aligned} \epsilon_i &= (A + Bx_i + Cx_i^2) - y_i \\ g = \sum_i \epsilon_i^2 &= \sum_i \left((A + Bx_i + Cx_i^2) - y_i\right)^2. \end{aligned} \tag{7.9}$$

After expanding,

$$\begin{aligned} g &= \sum_i [\,(A^2 + 2ABx_i + 2ACx_i^2 - 2Ay_i) + (B^2x_i^2 + 2BCx_i^3 - 2Bx_iy_i) \\ &+ (C^2x_i^4 - 2Cx_i^2y_i + y_i^2)]\,. \end{aligned} \tag{7.10}$$

Recall from calculus that the minima and maxima of functions relative to a variable can be found by setting the derivative of the function to 0. We wish to minimize g with respect to three "variables" (A, B, and C) simultaneously. To do this, we form three derivatives: $\partial g/\partial A$, $\partial g/\partial B$, and $\partial g/\partial C$. This yields

$$\begin{aligned} \frac{\partial g}{\partial A} &= \sum_i (2A + 2Bx_i + 2Cx_i^2 - 2y_i) \\ &= 2A\sum_i 1 + 2B\sum_i x_i + 2C\sum_i x_i^2 - 2\sum_i y_i \end{aligned}$$

$$\frac{\partial g}{\partial B} = 2A\sum_i x_i + 2B\sum_i x_i^2 + 2C\sum_i x_i^3 - 2\sum_i x_i y_i$$

$$\frac{\partial g}{\partial C} = 2A\sum_i x_i^2 + 2B\sum_i x_i^3 + 2C\sum_i x_i^4 - 2\sum_i x_i^2 y_i \tag{7.11}$$

The error function g will be minimized at those A, B, C that cause each of the above partial derivatives to be 0. Therefore, we set the partials to zero to get three equations in three unknowns:

$$An + B\sum_i x_i + C\sum_i x_i^2 = \sum_i y_i \qquad (\partial g/\partial A = 0)$$

$$A\sum_i x_i + B\sum_i x_i^2 + C\sum_i x_i^3 = \sum_i x_i y_i \qquad (\partial g/\partial B = 0)$$

$$A\sum_i x_i^2 + B\sum_i x_i^3 + C\sum_i x_i^4 = \sum_i x_i^2 y_i. \qquad (\partial g/\partial C = 0)$$

Equations such as these can be easily solved once they are re-written in matrix notation:

$$\begin{pmatrix} \sum_i y_i \\ \sum_i x_i y_i \\ \sum_i x_i^2 y_i \end{pmatrix} = \begin{pmatrix} n & \sum_i x_i & \sum_i x_i^2 \\ \sum_i x_i & \sum_i x_i^2 & \sum_i x_i^3 \\ \sum_i x_i^2 & \sum_i x_i^3 & \sum_i x_i^4 \end{pmatrix} \begin{pmatrix} A \\ B \\ C \end{pmatrix}$$

$$\mathbf{D} = \mathbf{S} \qquad \mathbf{P}.$$

$\mathbf{P}$ is the vector of unknown parameters whose values will be known if we can isolate $\mathbf{P}$ on one side of the equation ("solve" for the elements of $\mathbf{P}$). Using matrix operations, we do this by premultiplying both sides of the equation by the inverse of $\mathbf{S}$ (denoted $\mathbf{S}^{-1}$)

$$\mathbf{S}^{-1}\mathbf{D} = \mathbf{S}^{-1}\mathbf{S}\mathbf{P} = \mathbf{I}\mathbf{P} = \mathbf{P}, \tag{7.12}$$

where $\mathbf{I}$ is the identity matrix (1 along the main diagonal and 0 everywhere else). So, voilá: $\mathbf{D}$ and $\mathbf{S}$ are both known from data, and the only remaining problem is to determine $\mathbf{S}^{-1}$. This we can do by hand for small matrices or by using a general-purpose statistics package.

7.4 Equations with Nonlinear Parameters

There are some equations that are not linear in the parameters and that cannot or should not be transformed. Iterative methods must be used to estimate their parameters. We discuss two different methods: gradient and simplex. But we set the stage with the following geometric picture of the problem.

We again use the least-squares as the error function (Eq. 7.8) to minimize. This function depends on both the fitting function (f) and the data (y_i). For fixed f and observed y_i, the error function takes a different value

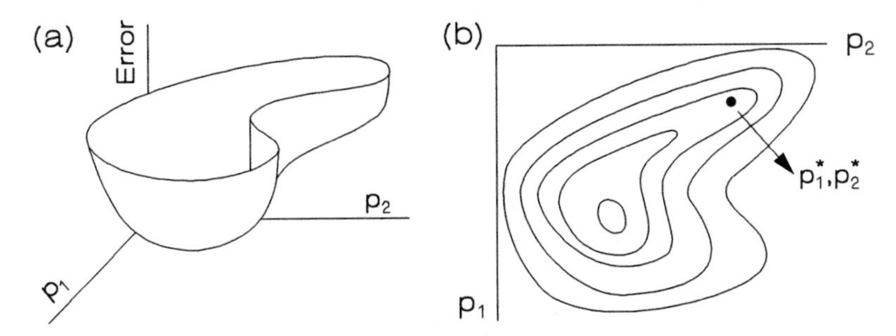

Figure 7.4: Error surface (a) and contour plot (b) for hypothetical fitting function in parameter space. Point p_1^*, p_2^* is a particular set of parameters.

for each combination of parameters. This produces a surface in parameter space as shown in Fig. 7.4.

The general problem in parameter estimation is to find the minimum point (i.e., the combination of parameters that correspond to minimum error). Iterative methods start at some arbitrary point in the space [(p_1^*, p_2^*) in Fig. 7.4b)] and move from a parameter combination corresponding to large error to a combination with small error. That is, these algorithms move down the slope of the surface stopping only when the current parameter set is sufficiently close to the minimum. The problem is to create an algorithm that does this efficiently.

7.4.1 Gradient Methods

Gradient iterative methods calculate the slope of the surface at the current set of parameters and base the direction to change parameters on the direction of greatest change in the error surface. This can be a powerful method, but since the shape of the error surface is not known, the derivative must be numerically approximated. This involves approximating the surface with a truncated Taylor series, which can be computationally expensive. Although there are many methods and variants, three are of fundamental importance (Sorenson 1980).

Gauss This method truncates the Taylor series at the first-order terms. (In actuality, Gauss' method does not expand the error surface function, but a function related to it.) In other words, it approximates the surface at the current solution point to a flat surface (e.g., plane). This method requires that a matrix composed of first-order derivatives be computed and inverted. An explicit step-size parameter in the algorithm controls the error associated with the linearization.

Newton–Raphson This method is similar to Gauss' method, but approximates the surface to a quadratic function by truncating the Taylor series after the second-order terms. This requires that a complex matrix of first- and second-order derivatives be computed and inverted. It has an explicit step-size parameter.

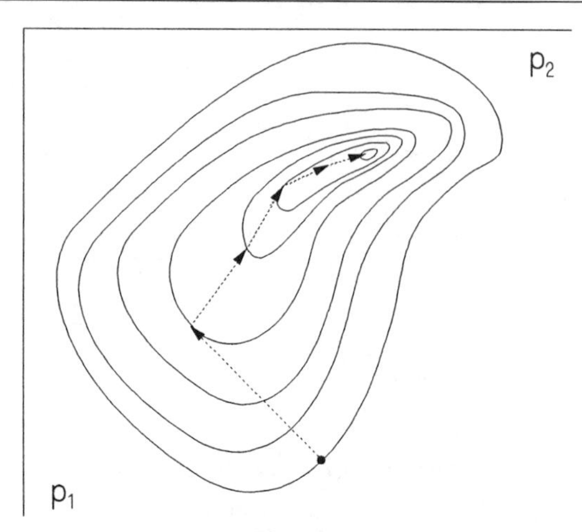

Figure 7.5: The Powell–Zangwill method of direct search. Contour lines represent error isoclines. Arrows are movements through the parameter space toward the minimum.

Steepest Descent This is a simplification of the Newton–Raphson method. It eliminates the second-order derivatives and the matrix inversion, but retains the step-size parameter.

7.4.2 Direct Methods

Because of the computational cost of numerically approximating derivatives and performing matrix inversion, *direct* methods are an attractive alternative. They do not require derivatives and choose the direction for the next move by directly evaluating the error surface in the neighborhood of the current point. For example, a naive algorithm might be to create a square grid of points around a current point in parameter space [e.g., point (p_1^*, p_2^*) in Fig. 7.4b] and test each vertex of the grid. The vertex with the smallest error is chosen to be the next current point. This is inefficient since we do not need to test all eight points (assuming a two-dimensional parameter space), nor do we need to be restricted to fixed distances between points. More efficient methods are described below.

Powell-Zangwill

This direct method does not have a fixed step size, but permits the structure of the error surface to determine how far to go in a given direction. Refer to Fig. 7.5 for the following heuristic description of the method.

The algorithm cycles through n iterations until a stopping criterion is satisfied. The first step is to move in an initial direction until the slope *in that direction* is 0 (tangent to a contour line). Next, cycle through the remaining parameter directions (only one other in the simple case of Fig. 7.5) to get a second intermediate point. Update the parameter choice by moving in the search direction until the slope is 0. Calculate a new search direction

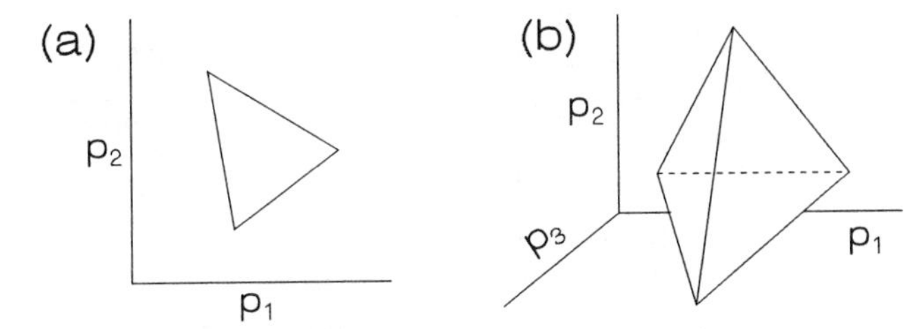

Figure 7.6: A simplex in (a) two- and (b) three-dimensional parameter space.

and new parameter directions (not necessarily perpendicular to the search direction) and repeat. The process stops when the improvement of the estimate between successive iterations falls below a specified threshold.

This method hinges on the choice of new search directions and parameter directions. These choices are based on mathematical structures (conjugate vectors) that are intimately related to the algebra of minimizing functions. This is a very efficient method that converges for most error functions in a few iterations. However, in each iteration the method must determine, for each dimension, the minimum of the function along a line in the search direction. Brent (1973), Sorenson (1980), and Press et al. (1992) give more details. The point of illustrating this method is to emphasize two characteristics of these derivative-free direct methods: (1) variable step sizes that are large when the parameters are far from the minimum and (2) choice of search directions that depend on the error surface.

Simplex

A graphically more appealing method that does not require the minimization of the function along a line is the *simplex method*. The simplex method (Nelder and Mead 1965; Caceci and Cacheris 1984) should not be confused with a method of the same name used in the optimization of linear equations (linear programming). This parameter estimation method is based on moving a geometric object (simplex) through parameter space until the object encloses the best estimate.

A simplex is a geometric figure with one vertex more than the dimensions of the space in which it is embedded. For example, if the space has two dimensions (Fig. 7.6a), then the simplex has three vertices (i.e., a triangle). If the space is three-dimensional, the simplex is a tetrahedron (Fig. 7.6b). The vertices of the figure correspond to points in parameter space so that each vertex is a combination of parameters that may satisfy our stopping criterion for the approach to the true parameters. The simplex method is an algorithm that alters the location of the simplex in parameter space so that when the stopping criterion is satisfied, the "best" values of the parameters are contained within the edges of the simplex.

An overview of the process is as follows. In a space of $n-1$ parameters, the simplex algorithm begins with n known starting points; these are the

Table 7.1: Fundamental operations on a simplex (see Fig. 7.7).

Reflection	Extend a line d units long from W to the midpoint of the B–O edge and d units beyond. The end of the line $2d$ units long is the trial vertex (W′).
Expansion	If W′ is an improvement, continue the extension of the line another d units in the same direction to W″.
Contraction	If reflection shows no improvement, extend a line $d/2$ units long from W to the midpoint of the B–O edge. Create a new vertex (W′) at this point.
Shrinkage	If none of the above, create two new vertices, one at the midpoint of the B–O edge and the other at the midpoint of the B–W edge.

vertices of the first simplex. Each vertex corresponds to a parameter set for the function. At each of these vertices, we calculate the error. Typically, the error is the square of the difference between the function and all of the datum points, but it could be another criterion. Of the n vertices, one will be best in the sense that its error will be smallest (vertex **B**), one will have the next smallest error (vertex **O**), and one will be the worst with the largest error (vertex **W**). Using these results, we transform the simplex into one that is closer to a point that minimizes the error function using four fundamental operations (Table 7.1).

These operations are designed so that the magnitude of the transformation is dynamic during the search. When the current solution is far away from the minimum, we wish the algorithm to take big steps (make large transformations). When it is close to the minimum we want the algorithm to take small steps. Further, when the slope of the error surface is shallow, the algorithm takes big steps; the converse occurs when the slope is steep. We illustrate the approach for functions with two unknown parameters (i.e., the parameter space is two-dimensional). Refer to Fig. 7.7 for notation.

Figure 7.8 shows an example of the movement of a simplex. The curved lines are contour lines representing the error function. The initial three guesses for the two parameters are in the upper right corner; the minimum error is in the center of the figure. From the initial simplex (vertices **012**), we reflect and then expand to simplex **123**. From this, we again reflect and expand to get **134**. We then reflect **134** to simplex **345**, but expanding in the direction of vertex **5** makes the estimate worse, so we stop searching in this direction. We then reflect **345** to get **456**, but expansion fails. We reflect and expand **456** to **467**, then reflect only to get **478**. Reflection, but no expansion, gets us **789**, then **7 9 10**. Reflection of the latter simplex fails, so we contract to **7 10 11**. This process continues until the differences in the error of the three estimates (one estimate at each vertex) is less than a threshold. The final parameter estimate is the average of the parameter

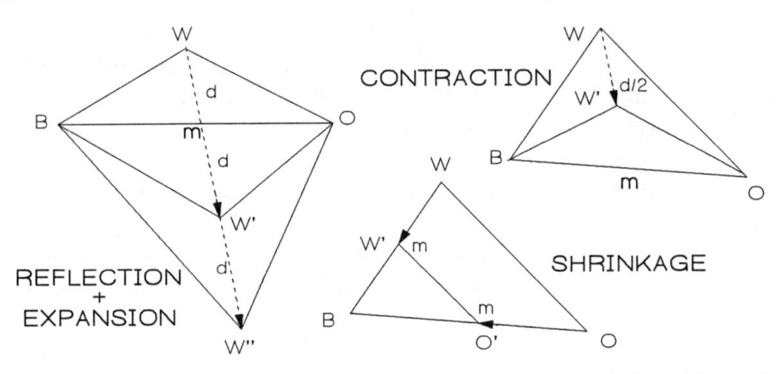

Figure 7.7: The four operations on the vertices of a two-dimensional simplex. W, O, B = worst, intermediate, and best vertex; m = midpoint of an edge. See Table 7.1 for other definitions.

sets at each of the vertices.

Marsili-Libelli (1992) generalized the simplex method to incorporate dynamically varying amounts of expansion and contraction. Of the two direct methods discussed, Powell–Zangwill is apparently the most efficient (Press et al. 1992), but the simplex method can be extremely robust (i.e., it rarely blows up by searching in the wrong direction).

7.5 Calibration to Dynamic Data

Above we were concerned with data sets in which the independent variable was not time. These data are typical of situations in which we can find functional relations between variables (e.g., between per capita growth rate and population size). Another approach to fitting parameters in a dynamic

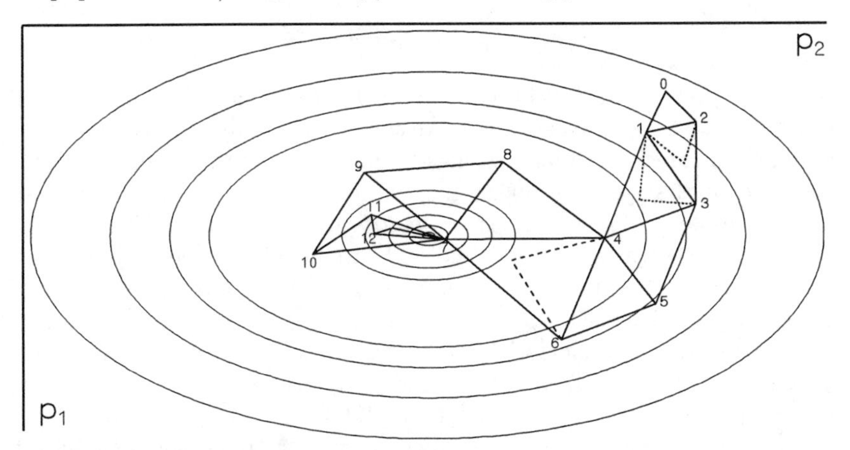

Figure 7.8: An example of simplex convergence on a minimum error function. The axes are the parameters of an equation. The ellipses are the contours of the error between the function and a fixed data set (high values at edges). The triangles are the simplexes as they move over the surface from simplex "012" to converge on the minimum in the center of the contours.

model is to find a set of parameters that minimize the sum of errors between the dynamic model output (e.g., numbers *vs* time) and similar observed dynamic trajectories over the entire time period simulated. There are two cases to consider: (1) the function to fit is an analytical solution to a differential equation and (2) the function to fit is the results of a simulation model.

The first case requires no new concepts. For example, a dynamic model based on density-dependent growth is sufficiently simple that we can solve the differential equation

$$\frac{dN}{dt} = rN\left(1 - \frac{N}{K}\right)$$

$$N(t) = \frac{K}{1 + e^{\beta - rt}},$$

where r is maximum per capita growth rate, K is carrying capacity, and β is related to the starting population size $[N(0)]$. We can estimate all three parameters by fitting the function $N(t)$ to experimental data consisting of population size over time. Obviously, $N(t)$ is nonlinear in the parameters so we must use one of the techniques for nonlinear regression (gradient or direct methods).

If the model consists of a set of interrelated linear differential equations, then general analytical solutions to the dynamics can be stated. For example, if the model is

$$\frac{dx}{dt} = ax + by$$

$$\frac{dy}{dt} = cx + dy,$$

then the dynamics $[x(t)$ and $y(t)]$ are be written as a sum of exponentials (see Section 9.3.2). In other words, we can find an analytical solution whose parameters can be estimated using the methods described above. In this special case of systems of linear differential equations, the parameter estimation problem is known as *system identification.* Spriet and Vansteenkiste (1982) give a lengthy review of methods applicable to linear systems and some simple nonlinear systems. Carson et al. (1983) apply these methods to models of physiological systems. Several monographs give a compendium of nonlinear dynamic models with analytical solutions commonly used in biological modeling as well as the estimation equations necessary (e.g., Seber and Wild 1989, Richter and Söndgerath 1990).

A more complicated case arises when we wish to use dynamic data to estimate parameters in a differential model that we cannot solve analytically. From an estimation perspective, this problem is no different than other applications. We wish to compare data to a function $f(x, t)$ that is the numerical solution of the differential equations. We do not know $f(x, t)$

until we simulate the model. So, this estimation problem is complicated by the fact that the model must be run with the current parameters over the entire time period in order to calculate the total error. A new set of parameters requires another run to determine the error. Consequently, a large number of runs may be required to converge on the best parameters. This dynamic aspect to the error function complicates the calculation of derivatives needed by some methods. Therefore, the direct methods are effective on this problem. The SCoP simulation package (see Chapter 5) contains a set of routines for calibrating models using the Powell–Zangwill direct method on dynamic data. Recently, Marsili-Libelli (1992) applied the simplex method to this problem. Since the discrepancy between model output and observations is dynamic, this approach to calibration can incorporate decisions to permit large errors at certain times (e.g., early in the simulation) and to achieve very small errors at other times. Whether this is something to consider depends on the objectives of the model. Accurate prediction of the final state of a system may be more important than prediction of the model trajectories by which it occurred.

7.5.1 Evolutionary Techniques

Parameter estimation is an optimization problem, and radically new approaches have been introduced recently. These methods are based on analogies with the evolution of biological systems, since one naive view of the evolutionary process is that it will produce organisms that are optimized to their environment by having maximum *biological fitness*. Many biologists would disagree with this caricature of evolution, but the analogy has been extremely productive in computer science. The new methods are members of a loose family of algorithms called *evolutionary computation*. The basic idea applied to parameter estimation is that the parameter space is searched by a large set of "organisms" that are defined by their position in the space. Their fitness is the value of the error function at that point in parameter space. Organisms with low fitness (large error) are discarded. Surviving organisms mate and produce slightly different offspring by combining the positions of the two parents to form a new location in parameter space. This process is repeated until organisms do not show further improvement. These methods are proving to be very effective on error surfaces that are complex with many hills and valleys. We discuss these methods more fully in Chapter 19.

7.6 Parameter Estimation Cautions

7.6.1 All Methods

1. Beware of transformations. Nonlinear regression or iterative methods are preferred.

2. Examine your data for obvious outliers. You may need to filter the data (e.g., compute running averages) or apply some other method for eliminating extreme data points.
3. Beware of *extrapolating* beyond your data. Brown (1990) shows a fifth-order rational function (i.e., a quotient of polynomials) that fits one cycle of a periodic function with $r^2 > 0.99$, that goes to positive and negative infinity outside this range. (This is quite unlike the sine function being fit, of course.) Some situations in some methods can also make *interpolating* between datum points dangerous. A quotient of two fifth-degree polynomials fits a data set with multiple observations at each x value with $r^2 = 0.973$. The curve, however, is not continuous between sets of observations so that the function predicts correctly if given the original x values, but not if given any others between these. Rational functions should not be used for data sets with multiple observations.
4. Beware of using a simple statistical index (e.g., r^2) to determine the function to use. An equation with sufficiently large numbers of parameters can be fit to match every little jog in a noisy data set with high r^2, but may fail to reveal a simpler representation.
5. Use a graphics package to view your data and fitted curve. Be suspicious of any obvious departures. In general, use common sense and remember why we fit parameters in models: we wish to obtain a *simple* and *general* description of the observations. Simplicity in the form of equations with small numbers of parameters is usually preferable to complicated equations with a good fit to a particular dataset. The equation is the object of interest, not the r^2. (The model objectives may influence this; models that must achieve accurate predictions may require particular, specific functions.)

7.6.2 Problems with Iterative Methods

1. Nonevolutionary, iterative methods find only local minima. Use several starting points to search for the global minimum. Initial guesses can be obtained from previous knowledge or linear regression on transformed data. You should also restart the search at a random point near the local minima when it is located. This will help verify that numerical conditions (e.g., round-off) have not caused the algorithm to stop prematurely. The newer methods using evolutionary programming appear to be better at finding the global minima (or maxima).
2. Methods requiring derivatives can be slow and sensitive to the "roughness" of the error surface. Steep gradients and sudden reversals can cause numerical approximation of derivatives to go astray. Methods such as simplex that do not use derivatives are less sensitive to this.

Test the results with several step sizes.

3. Most iterative methods do not give exact r^2 values. Approximate values can be obtained by boot-strapping or by fitting a polynomial to the error function after a good fit is found. Bootstrapping is a computational method in which statistics are calculated based on randomly chosen subsets of the original data. In parameter estimation, a series of subsets is chosen, an estimate obtained for each, and the mean and variance of the estimates calculated from these. If a polynomial is fit to the error function, it is wise to verify that the conditions over which the methods are known to be valid hold in your application. An important condition is the curvature of the surface; see Seber and Wild (1989) and Ratkowsky (1983).
4. If the error surface around the minimum is flat, then convergence to the stopping criterion may be slow. Most iterative methods use two stopping criteria: one based on the relative change in the residuals and the other a ceiling on the number of iterations performed. After the algorithm has stopped, verify that sufficient iterations were allowed to ensure that the first criterion (not number of iterations) was used to stop the search.

7.7 Exercises

1. The equations for the parameters of a simple linear regression are:

$$Intercept: \; A = \hat{y} - B\hat{x}$$

$$Slope: \; B = \frac{\sum xy - (\sum y \sum x)/n}{\sum x^2 - (\sum x)^2/n},$$

where $\hat{x}$ and $\hat{y}$ are the means of the independent and dependent variables, respectively.

Using logic analogous to the derivation of equations for polynomial regression, derive these equations starting with $y = ax + b$. (In so doing, you will prove that standard linear regression does, indeed, minimize the sum of squared error.)

2. (Advanced: requires matrix algebra.) Below are data that approximate a parabola. Calculate the matrices $\mathbf{S}$ and $\mathbf{S}^{-1}$, and the vector of parameters $\mathbf{P}$.

x:	1.0	5.0	70.0	85.0
y:	0.00115	0.00560	0.04850	0.0485
x:	95.0	120.0	160.0	179.0
y:	0.05	0.0434	0.0202	0.00110

3. A friend of yours claims to have developed a direct method of parameter estimation that uses just N iterations, where N is the number

of parameters. Her method was to pick a starting point, find the minimum along the first parameter axis, use this point as the starting point to find the minimum along the second parameter axis, and repeat this for each of the parameters. The process stops when all N parameters have been minimized in this way. Draw an error contour plot that illustrates that this method can find a minimum. Draw another error contour plot that causes the method to fail.

4. Below are data from Gause (1935) for density-dependent population growth of *Paramecium*. Assume $N(0) = 2$. Estimate r and K using
   (a) linear regression on the transformed solution (N *versus* t),
   (b) the simplex method on the untransformed solution,
   (c) linear regression on per capita growth rates,
   (d) polynomial regression on absolute growth rates,
   (e) the simplex method on absolute growth rates, and

   Discuss the differences among the methods and determine which is best.

Day:	0	2	3	4	5	6
Pop:	2	17	29	39	63	185
Day:	7	8	9	10	11	12
Pop:	258	267	392	510	570	650

5. (Advanced) Choose an integral function that is tabulated in a handbook of mathematical and physical functions. Two good examples are the complementary error function [erfc( )] and the gamma function. Since these are integrals, an integration routine is needed to numerically evaluate them (hence, the tables). An alternative is to fit the integrals with a suitably complex function such as a high-order polynomial or rational function. Do this for your function using nonlinear regression and the simplex method. Try several polynomial orders and tabulate the errors. Try several error functions (e.g., least-squares, absolute value of difference, chi-square, minimum of the maximum deviation)

Chapter 8

Model Validation

8.1 Insight and Illumination

MODELING, like computing, should produce insight, not merely numbers (Hamming 1962). Up to this point, we have stressed numbers and methods for generating them. Now we discuss tools that help evaluate the meaning of the numbers. We will focus on three general areas.

- *Validity:* Validation concerns the degree of our faith in the quality of the model with respect to the external world. Below we discuss statistical methods and problems in evaluating model adequacy and usefulness.
- *Uncertainty:* Ignorance and uncertainty occur at many points in the modeling process: in the equations, the parameters, and in the definition of the system itself. We will discuss tools for evaluating the contribution of this uncertainty to model output.
- *Behavior:* The change of state variables over time is the lowest level of system understanding. To grasp fundamental interactions, we need to visualize the covariation between coupled variables, and identify system conditions in which the dynamics of the variables are qualitatively similar.

In this chapter, we discuss validation and model quality. The following chapter covers uncertainty analysis, especially sensitivity analysis, and behavior with emphasis on stability analysis.

8.2 Validation: When Models Go Bad

When we consider model validation, we are interested in the quality of the model. This is a more difficult problem that one might suppose. Indeed, there is significant disagreement over the word to use. Most authors agree that model quality is not truth or veracity (Caswell 1976). In line with this, we previously used *verification* to mean establishing the correctness of an algorithm or computer code. Therefore, the system scientists who use the word *validation* use it to mean model quality with respect to the objectives of the modeling project (Shannon 1975, Sargent 1984). More recently, however, several authors have argued for using *corroboration* or *confirmation* for validation (Reckhow and Chapra 1983a; Swartzman and Kaluzny 1987). They favor this usage because (1) they feel that "valid model" refers to "correct model" and does not permit degrees of quality, and (2) there is a precedent set by certain philosophers of science for "corroborate" and "confirm." [For my part, in light of the rather small number of well-tested models in biology and the generally low rigor of the tests, I think the adjective *plausible* more accurately reflects the nature of tested biological models and the skeptical attitude we should adopt (Carson et al. 1983). To a more cynical observer, the dictionary definition of "specious" might also come to mind.]

In any event, two points emerge from all the discussions and definitions: (1) model quality, if it is quantifiable at all, is a continuous variable and perfection is probably not achievable, and (2) the process of model evaluation is unending. In the following, I do not take sides in the semantic debate, but acquiesce to the weight of common opinion and use "validate."

There are many components to quality and these depend on the uses to which the model will be put. Earlier, we discussed three main uses: control, understanding, and prediction. These provide important criteria for quality, but a more complete list is:

- usefulness for system control or management
- understanding or insight provided
- accuracy of predictions
- simplicity or elegance
- generality (number of systems subsumed by the model)
- robustness (insensitivity to assumptions)
- low cost of running or constructing the model.

All of these concepts are, to varying degrees, legitimate components of quality; none are mutually exclusive. The model objectives will determine the weighting to be given to the different components. Generality, simplicity, increasing understanding, and qualitative correctness of model behavior are concepts that are more relevant to purely theoretical studies, where the quantitative behavior of the real world is relatively unimportant. Usefulness, accuracy, and cost are more important to applied problems such as control and management. Here, we will emphasize accuracy of predictions.

Ideally, we would like to treat our dynamic mathematical models and our data in the same way we treat a statistical null hypothesis and the data. We would like to perform an objective, rigorous hypothesis test in which we can ascribe a definite quantity of faith (i.e., the probability level) that the model is correct. Before describing the very serious difficulties that may prevent our achieving this goal, it is useful to recognize the logical bases of validation.

8.2.1 The Logic of Falsifying Complex Simulation Models

The validity of an argument does not guarantee the truth of its conclusion.
— Copi (1954)

An Aristotelian syllogism is a sequence of logical steps that in totality is true regardless of the truth or falsity of the component steps. The basis of the modern concept of scientific falsification (Popper 1968) is a syllogism called *modus tollens*:

Form:	Example:	
$A \Rightarrow B$	*if Spock is human, then he will act illogically.*	(8.1)
$\neg B$	*Spock does not act illogically.*	
$\neg A$	Therefore: *Spock is not human.*	

where $\neg$ means "NOT" or logical negation.

In applications of this argument in science, "A" is the general hypothesis (law) and "B" is the implication or prediction that follows from the law in a particular instance. Popper proposed this as the basic logical construct for the hypothetico-deductive method. He distinguished this logically correct argument from the fallacy that he claimed underlies the approach of the logical positivists (Nagel 1961). The fallacy is that of *affirming the consequent*:

Form:	Example:
$A \Rightarrow B$	*if Frodo loses the ring, then he will be ill.*
B	*Frodo is ill.*
A	Therefore: *Frodo has lost the ring.*

Although the above is, indeed, a logical fallacy (not a syllogism), it summarizes the central problem of the *confirmationist* philosophy. Even though one observes many instances of the major premise (Frodo losing the ring and becoming ill), this neither establishes it as a law nor permits one to infer the conditional based solely on the observation of the prediction.

Modus tollens is difficult to implement in mathematical models because the law ("A" in Eq. 8.1) is actually a conjunction of a large number of separate assumptions. For example, in a mathematical model there are several equations that constitute a conglomeration of hypotheses and generalizations; there are also parameters and initial conditions that must be specified. So in reality the argument form is

$$\begin{array}{l} (a_1 \wedge a_2 \wedge a_3 \ldots \wedge a_n) \Rightarrow B \\ \underline{\neg B \qquad\qquad} \\ \neg(a_1 \wedge a_2 \wedge a_3 \ldots \wedge a_n), \end{array}$$

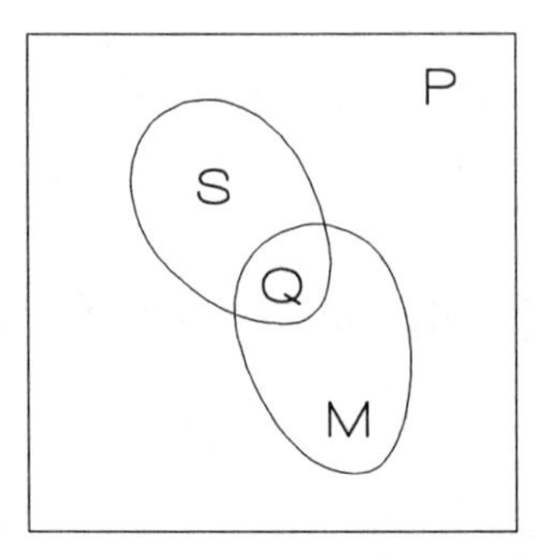

Figure 8.1: Relations of sets of observations on the system (S) and model (M) for validation. Q is the set of correct predictions. (From Mankin et al. 1977, Fig. 1.© 1977 Simulation Councils, Inc. Reprinted with permission Simulation Councils, Inc., publisher.)

where $\wedge$ means "AND." The last line above is a negation of a conjunction and is defined as $\neg a_1 \vee \neg a_2 \vee \ldots \vee a_n$ (i.e., "not a_1 OR not $a_2 \ldots$ OR not a_n"). In general, we do not know which of the a_is is false. This problem has prompted some to assert that mathematical (simulation) models cannot be used as a tool of the hypothetico-deductive method (Romesburg 1981). The situation is not completely hopeless. We can perform independent experiments to estimate parameters, perform parameter sensitivity analysis to evaluate their effects on model response, or create and investigate alternative models. Although the philosophical and logical problems are real, we will not discuss them further here, but rather proceed to discuss practical problems associated with testing models.

8.2.2 The Geometry of Validation

Truth is the intersection of independent lies. — Levins (1970)

Mankin et al. (1977) provide a useful conceptual framework that encompasses different validation problems and situations. They considered validation in terms of the relation of sets of measurements that can be made on systems and models (Fig. 8.1). **P** is the set of all possible observations on the class of systems studied (e.g., ecosystems). **S** is the set of all observations actually made on the study system. **M** is the set of model outputs, and **Q** is the intersection of **M** and **S** (i.e., the overlap of data and model predictions). Also imagine, since we advocate the use of alternative models, that there may be several $\mathbf{M_i}$s, each with different $\mathbf{Q_i}$ that may themselves overlap.

There are several qualitative relations between these sets that help us understand different validation situations and ways that models can fail (Fig. 8.2). If **Q** is empty (Fig. 8.2a), there is no intersection between model and observation, and the model is *useless.* If **Q** is nonempty, we say the model is *useful.* At the other extreme (Fig. 8.2c), the model may predict all of the system observations and make no predictions that are not observed

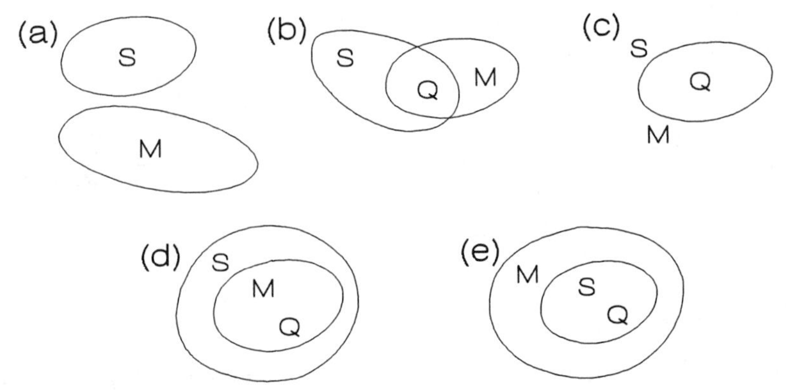

Figure 8.2: Relations of model predictions and system observations.

(i.e., **M** and **S** are exactly the same set: only in your dreams). The more typical situation is intermediate (Fig. 8.2b): the model predicts a subset of the observations and makes some predictions that are not observed. Two other special cases can be imagined: (1) the model never makes a mistake (Fig. 8.2d), but is incomplete; and (2) the model is complete, but makes mistakes (Fig. 8.2e).

Mankin et al. (1977) also suggested that *model reliability* is the ratio of the size of **Q** to the size of **M**. *Model adequacy* is the ratio of the size of **Q** to the size of **S**. For example, in Fig. 8.2d the model is relatively inadequate (although it makes no incorrect predictions), but reliable. In Fig. 8.2e, the model is relatively unreliable, but very adequate (it predicts all of the observations). Certainly, there are problems in defining a measure of the sizes of the sets, but this conceptualization emphasizes that many and varied comparisons, both quantitative and qualitative, can be made between data and predictions. We must investigate both reliability and adequacy. Most published validation exercises focus on the size of **Q** or, at best, on model adequacy. Below, we will stress quantitative comparisons and model adequacy, but the broader picture (Fig. 8.2) should be kept in mind. To address model reliability, the model must be tested in imaginative ways. For example, it should be tested against (1) different systems [e.g., different organisms, or habitats (aquatic *vs* terrestrial)]; (2) different geographical areas; or (3) using different parameter values and environmental driving variables and perturbations.

8.2.3 Variables and Levels for Validation

While the logic of validation may be clear enough, in practice it is not obvious exactly what comparisons should be made. Usually, the model objectives will dictate which quantities should be compared between model and data. The most common are the dynamics of the state variables or *derived measures* in the form of Forrester auxiliary variables. The latter may be (1) functions of individual state variables [e.g., a state variable scaled

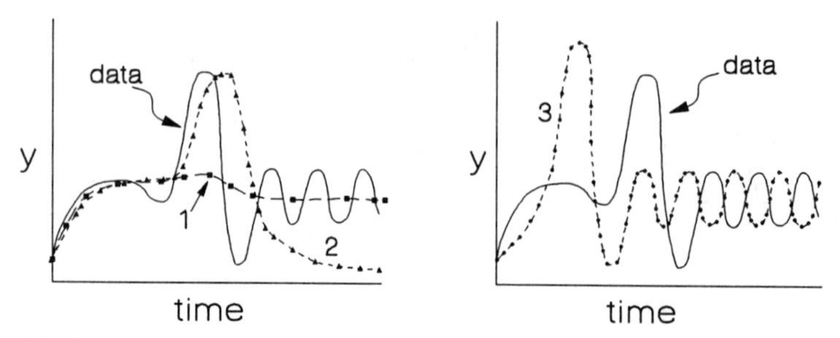

Figure 8.3: Comparisons between data (solid line) and the predictions of three hypothetical models.

to other units (concentration computed from an absolute quantity)], (2) the time or spatial averages or frequency distributions of a state variable, (3) the maximum of a state variable, or (4) the time that a state variable achieves a particular value (e.g., its maximum). We can also use auxiliary variables that are computed from two or more state variables (e.g., species diversity in foodweb models), or ratios of state variables (e.g., root/shoot ratios).

In addition to choices of variables, there are levels of comparisons. At one extreme are theoretical models whose object is understanding with only vague reference to qualitative similarity between model predictions and common knowledge about the system. The other extreme is rigorous statistical testing of model predictions with replicated field or laboratory experiments. The intermediate ground is broad and involves a wide range of techniques and problems.

To illustrate this point, consider Fig. 8.3, which shows three comparisons of model output (broken lines) and data (solid line). Each model output fails in different but important ways. Model 1 generally captures the long-term trends of the data by passing through the mean of the cycles at the end of the time series. It misses, however, the strong peak in the middle of the data. Model 2 hits the peak, but misses the cycles. Finally, model 3 has both the peak and the cycles, but the size of the peak and the timing are wrong.

Are any of these models satisfactory, and, if so, which is the best? Ask three different modelers these questions, and you will get three different answers (especially if the models are their own creations). Unfortunately, there are no definitive answers to the questions. It depends not only on the objectives, but also on what one thinks the *defining pattern* of the data to be. Is it the peak, the cycles, or the long-term trends? There are rational arguments for all of these features. Familiarity with the system can help in these cases, but there is danger that an expert's preconceived notions and pet hypotheses may influence which patterns are emphasized. Because of this, we seek objective, statistical criteria to compare models and data. This satisfies our urge to be rigorous, but we should not lose sight of the fact that models can have large statistical errors, but still capture the essence of the data (e.g., model 3 in Fig. 8.3). By doing this, they maintain their utility even if they fail statistical validation.

8.2.4 Conditions for Validation

In dynamic models, validation is usually concerned with the comparison of two time series: observations and model output. These comparisons have four attributes that will influence the kind of validation that is possible: data independence, number of system responses, number of time points, and degree of replication. Below, we discuss some of the issues and methods that are appropriate depending on the attributes.

Data Independence

An essential condition that must be met in any rigorous comparison of data and predictions is *data independence.* The data used for model validation must be separate from and independent of any data used to formulate model hypotheses and estimate parameters. This condition motivated the revision of the standard view of the modeling process to include multiple working hypotheses and models tested in parallel (Chapter 2). When independent data are difficult to obtain, we must be careful to avoid a circular comparison of model output with data used at some point in model formulation as part of validation. If the comparison data are not independent of those used to construct the model, then we are only doing calibration and not true validation. There are, therefore, no validation techniques we can apply.

Single and Multiple Responses

In almost every system and model, we can measure or compute a number of different quantities that could be compared. For example, in all but the simplest systems, there is more than one state variable. Each of these can be measured or computed, and, therefore, each of these is a response that we can use to evaluate model predictions. Our validation test procedure must decide how many and which of all possible responses will be used. If we choose to validate using more than one response, then we must decide if we will compare system and model for each response separately or produce a synthetic validation that incorporates all responses simultaneously. If we analyze the responses separately, then we have the problem of deciding overall model quality if model predictions are acceptable for some, but not others. Multivariate statistical techniques (discussed below) can perform comparisons simultaneously. If we do not use these methods, then we can either report each individual comparison separately and make a subjective evaluation, or we can combine errors of all responses in an index (Shannon 1975). This latter approach, although it is quantitative, has only the aura of objectivity, because typically there will be no statistical test to determine if the index is large or small. So, if rigorous statistical evaluation of overall model quality for many response variables is our goal, we should use multivariate techniques.

Single and Multiple Comparison Points

We can choose to validate the model either at a single point in time or at several points in time over the series. If we choose to evaluate the model at a particular point in time, then we must have a criterion for determining what the point will be (e.g., at the end of the growing season, or when a particular condition has occurred). If only a single time is used, then the problem of statistical bias due to serial correlation in the time series does not arise. If multiple time points are used, then we must use care in applying standard statistical tests.

Unreplicated Systems and Models

Model validation using statistical tests requires some form of variability in either model predictions or observations. In real systems, variability is usually produced from replicated observations. It can be produced in stochastic models from repeated runs that differ in the sequence of random numbers used to generate the modeled randomness (Chapter 10). Variability without replication can be found in some kinds of regression. Here we discuss the case when there is no replication, but there may be variation.

> *A proof is an argument that convinces someone who knows the subject.*
> — Davis and Hersh (1981)

Turing Tests If there is variability neither in the model nor in the data, and we wish to compare model and system time series, then classical statistical inference is not possible. Consequently, we are restricted to a qualitative assessment that the model behavior is "reasonable." Often this assessment is done informally by presenting the reader with a plot of dynamic model output and system measurement on the same graph. A more formal method is the *Turing test.*

Alan Turing was a British mathematician instrumental in the design of early British computers and interested in theoretical biology and artificial intelligence. He proposed to validate computer models simulating human verbal behavior by putting one human (the interrogator) in a room with a computer terminal connected to two other rooms containing a human test subject and a computer, respectively. The interrogator asks questions of both the computer and the human to determine which room contains the computer. If the computer's program is successful, its verbal responses will fool the interrogator, and he will fail to guess the location of the machine. Thus, a computer model passes a Turing test if it fools the expert. Or, to put it in a semiquantitative way: *A good model is one that fools 80% of the experts 80% of the time.*

This approach can be used for biological models by asking experts to distinguish similarly prepared figures or reports of genuine and simulated system dynamics. The format of the simulated output must be similar to the norm for the genuine system. In most cases, this could be x–y plots of time traces of key variables (e.g., net plant productivity during a growing season). Other systems may require specialized documents.

Schruben (1980) used this approach to evaluate a model of the flow of patients among a set of operating rooms. The model was validated by testing the ability of the facility director to discriminate between simulated and genuine reports of room use. On the first test, the director easily distinguished the simulated and actual reports. (In part, this was due to the fact that the modelers forgot to remove from the computer output excessive significant digits in reporting minutes of room use!) On the second try, the director's suggestions on model hypotheses were incorporated, but she was still able to identify the simulated reports. The third model incorporated more suggestions by the director, and eventually she failed to discriminate the two sets of reports.

Overall, it is difficult to interpret this type of test. One can apply rigorous statistical analyses (e.g., the *kappa* statistic of agreement, Fleiss 1973), but in the above example there is a disturbing repetitive loop between model structure and test. Moreover, as Schruben (1980) admitted, the expert became better at noticing small differences between genuine and simulated reports, so that achieving a high quality model became more difficult with each additional test. Too much of this sort of thing would discourage even the Red Queen of Wonderland.

Observed vs Predicted Regression Even without replication, linear regression is sometimes used to test that a model is statistically indistinguishable from the data. While there are situations when this approach is legitimate, after describing the method, we discuss some problems.

Consider the case when the deterministic model output and unreplicated system trajectory are paired such that we can associate a prediction for every time t at which we have an observation. The linear regression approach regresses the observations (y axis) onto the predictions (x axis). If the model were perfect, all of the points would fall on the 1:1 (45°) line, and the regression slope would be 1.0 and its intercept 0.0. The correct approach is to test for these two values *simultaneously* (unlike the tests in standard statistics texts). Dent and Blackie (1979) provide the required formula as an F statistic

$$F = \frac{(n-2)\left(na^2 + 2n\overline{X}a(b-1) + \sum X_i^2(b-1)^2\right)}{2nS^2}, \qquad (8.2)$$

where a is the estimated intercept, b is the estimated slope, X_i are the individual model predictions, $\overline{X}$ is the mean of the model output, and n is the number of system–model pairs. S is the standard error of the estimate (i.e., the residual mean squared error) and is computed as

$$S = \frac{\sum \left(Y_i - \hat{Y}_i\right)^2}{n-2},$$

where

$$\hat{Y}_i = \overline{Y} + b(X_i - \overline{X}),$$

where Y_i are the individual system observations and $\overline{Y}$ is the mean system value.

Many of the standard linear regression computer packages will compute S, $\sum X_i^2$, and $\overline{X}$, so it is an easy task to compute Eq. 8.2. Some packages will calculate Eq. 8.2 directly. This statistic follows the F distribution with 2 and $n-2$ degrees of freedom. If the original model has merit, we will fail to reject the null hypothesis that slope is 1.0 and intercept is the 0.0. Consequently, small values of F mean our model is a good fit.

In addition to testing the parameters, an overall test for lack-of-fit can be made (Zar 1984). As its name suggests, this statistic measures the degree that the model does not fit the observations. The model is corroborated if we do not reject the null hypothesis. A disadvantage of this approach is that poorly replicated studies are likely to have large sample variances, while studies with many replicates will more likely have small variances. We are more likely to reject the lack-of-fit null hypothesis in the latter case than in the former, and therefore models are more likely to be corroborated in poor observational studies than in carefully designed studies with good replication.

The scatter plot associated with this method is a powerful visual tool to bolster belief in the model. For this reason, it is common for modelers not to use regression, but to perform a correlation analysis (Zar 1984) between model output and the observations. As in regression, the r^2 measures the strength of the straight-line relation between model and data. While statistical analyses exist to test $\rho = 0$ (no correlation), there is no specific non-zero values of ρ against which to test. For example, there are no reason to test for $\rho > 0.6$, unless this value was an element of the model objectives.

A shortcoming of either regression or correlation is that the temporal aspects of the deviations between data and model are lost in the scatter plot, but this can be made explicit with plots of the deviations over time. More serious problems occur when the method is applied to situations in which the assumptions of linear regression are not satisfied (Mayer et al. 1994). Those assumptions that are especially important are: (1) the X_i must be known exactly, (2) the variance of the errors must be constant for all values of X_i, and (3) the Y_i are independent. Although we are always uncertain that a model and its parameters are correct, assumption (1) is normally satisfied, given a *particular* deterministic model, with *particular* parameter values. However, we must recognize that the statistical inference applies only to that complete set of conditions; specifically, we cannot extrapolate the inference to the same model using different parameter values. If the X_i are not assumed to be exact, then the regression procedure is more complicated and problematical (Ricker 1973).

Assumption (2) is probably not true because (a) we often have greater errors in measuring small numbers than large numbers, and (b) if the dynamics are monotonically increasing, then differences between the data and the model may diverge over time (as the X_i grow). However, linear regression is relatively robust to violations of (2), although it should always be verified.

Assumption (3) is particularly important because linear regression is sensitive to it and it is often difficult to determine when it is violated. It will be violated when observations are made repeatedly over time on the same experimental unit (e.g., growth of an individual organism or dynamics of a variable measured at a particular location in a lake). The interpretation becomes fuzzy when a random subset of objects is repeatedly selected. For example, if the weight of fish in an experimental enclosure is repeatedly measured on a random selection of fish in the enclosure, has the assumption of independence been violated? What about the case of a small random sample of fish from a lake with a very large population of fish?

Indices In addition to determining if the model deviates from ideal by testing the slope and intercept of a regression, a variety of indices have been developed as diagnostic tools to assess the nature of the deviations. One important approach is due to Theil (1961), who defined an *inequality coefficient* as

$$U = \frac{\sqrt{\frac{1}{n}\sum(X_i - Y_i)^2}}{\sqrt{\frac{1}{n}\sum X_i^2} + \sqrt{\frac{1}{n}\sum Y_i^2}}, \tag{8.3}$$

where X_i, Y_i are the model output and observations at the *i*th time point, respectively, and n is the number of paired points. U varies between 0 and 1. U can be decomposed into three components associated with (1) differences between the model and system means (i.e., a nonzero intercept or *bias error*): U_M, (2) differences between the variance of model output and the variance of observations (i.e., slope-not-unity error): U_S, and (3) the deviation of the correlation of model and observation values from 1.0 (i.e., random error): U_C. The relationships are

$$U^2 = U_M^2 + U_S^2 + U_C^2, \tag{8.4}$$

where the equations for these three components are respectively

$$U_M = \frac{\overline{X} - \overline{Y}}{D} \tag{8.5}$$

$$U_S = \frac{(S_X - rS_Y)^2}{D} \tag{8.6}$$

$$U_C = \frac{(1-r^2)S_Y^2}{D} \tag{8.7}$$

$$D = \frac{\sum(X_i - Y_i)^2}{n}, \tag{8.8}$$

where r is the correlation of X and Y, and S_Y and S_X are the standard deviations of the X and Y variables.

Dividing the right-hand side of Eq. 8.4 by U^2 normalizes the three components so that each represents the proportion of total error due to its respective cause: bias, slope, or random. This is useful for diagnosing why a model has failed. Rice and Cochran (1984) used this approach to identify the bias error (U_M) as the most important component of error in a fish bioenergetic model.

A number of additional indices and calculations can be made to further quantify model error (Power 1993, Mayer and Butler 1993). To a certain extent, these indices can be thought of as measures of model adequacy (Mankin et al. 1977). Typical of these is *bias*

$$\text{bias} = \frac{\sum(X_i - Y_i)}{n},$$

where X_i is the ith model prediction, Y_i is the ith system observation, and n is the number of comparison points. Many variants of this basic idea are possible: (a) normalize it by the standard deviation of all model predictions, (b) convert it to a percent, and (c) use the absolute value or squares of the deviations. With a few exceptions, these do not have inferential capabilities, but can be used to measure the degree of departure of model output from observations. Halfon (1989), however, used boot-strapping (a statistical randomized re-sampling technique) to compute the probability that calculated validation statistics were within acceptable limits.

Goodness-of-Fit Tests For sufficiently long time series, a frequency distribution of the observed and predicted values can be created by counting the occurrences of discrete intervals of response values (i.e., frequency distributions of the state variable values). If the model is accurate, the two distributions should be indistinguishable. χ^2 and the log-ratio G tests are two common approaches (Sokal and Rohlf 1981). This test, like regression, collapses the dynamics. Statistical independence is also a problem.

Replicated Systems or Models

Replication in the system observations means that we have multiple, independent observations at points in time. Model replication means we have a stochastic model that has been run several times or a deterministic model that is run several times with randomly selected parameter values. Naturally, it is possible for both the data and the model to be variable. Whether we can legitimately use this variability to test statistically for differences between model output and the data depends on whether we are comparing a single value or a time series of values. An excellent tool for visualizing model or data variability, especially stochastic model output, is the *box plot.* This is a graphical representation of a set of numbers in which the sample size, mean, median, the range, and other measures are all represented. A thorough description of this technique in stochastic modeling is given by Reckhow and Chapra (1983b).

Single Value If only a single value (e.g., the maximum of a state variable) is being tested, then standard statistical testing can be done (e.g., t-tests or ANOVA). If variability exists in only one component (e.g., the data) then we use a single-sample t-test (H_0: $\mu_M = \mu_D$). This compares the mean of the replicated data with the single number of the unreplicated number (model prediction). If both model and data are variable, the standard, two-sample t-test is used. Standard statistics texts give the appropriate formulae for one- and two-sample t-tests.

Time Series As with unreplicated situations, time series introduce autocorrelations. Certainly, model values are correlated over time, since we use previous states to calculate current states, according to the equations. Measured values in real systems also tend to be correlated. These correlations can violate basic assumptions of standard statistical analyses so that extreme care must be exercised in their applications. When the model is stochastic, the standard linear regression approach described above is inappropriate since the X_i (model output) are not known exactly. There are methods for analyzing this case, but their use is not so straightforward as classical linear regression. More appropriate techniques use single-factor repeated measures analyses and split-plot designs (Mayer and Butler 1993) or the multivariate profile analysis (Steinhorst 1979; Balci and Sargent 1982).

Single-factor repeated measures and *split-plot* designs are types of analysis of variance (ANOVA, see Winer 1971). Single-factor repeated measures designs use a single set of treatments applied sequentially to all of a single group of individuals (e.g., a sequence of drugs applied to patients). A split-plot design applied to repeated measures situations generalizes this approach to include multiple factors so that not all individuals receive all treatments (e.g., drugs partitioned by chemical properties). A split-plot design partitions the error among a main effect (e.g., system or location) and subdivides or splits each of these error components into effects associated with the treatments. Both approaches assume that the correlation of responses among treatments is known and is constant over time. This is usually not true, and caution and additional tests of statistical assumptions are needed if this approach is used. Because of this problem plus the fact that the method is discussed in standard texts (e.g., Winer 1971, Chapters 4 and 7), we will not give further details.

Profile analysis is a multivariate method that, as the name suggests, tests the hypothesis that the trajectories of data and model output are parallel. There are three major advantages to this method over other possible approaches. First, the approach does not make assumptions about the nature of the variance or covariance relationships of the variables, so it is a more general approach to repeated measures problems. Second, the approach also permits us to examine the relation of the data and the model for several output variables (i.e., several state variables) simultaneously. Finally, the statistical power of the test can be computed.

The null hypothesis tested is that the difference between model and data is 0.0 for each and all time values of comparison. This technique calculates Hotelling's T^2 statistic, for which probability tables are available. See Timm (1975) for an introduction, Steinhorst (1979) for an application to ecosystem models, and Balci and Sargent (1982) for a queuing system example.

Here, we only illustrate the method with a numerical example; for more generality and theoretical details the reader is referred to the references cited. First, some terms and assumptions are necessary. We assume that we have k time points at which we measure each of q biological responses. We also have k model predictions for the q model variables (usually state variables). So, we have a total of qk values to compare. Each system is replicated n times; a replicate might be a controlled experimental field plot, one of several sampling stations in a lake, and so on.

The null hypothesis is

$$H_0 : \tilde{d}(1) - \tilde{m}(1) = \tilde{d}(2) - \tilde{m}(2) = \cdots \tilde{d}(k) - \tilde{m}(k) = 0,$$

for all system responses measured. $\tilde{d}(i)$ is a vector of observations of all response variables at time i, and $\tilde{m}(i)$ is the model output of all response variables at time i. For example, suppose for concreteness that the first response is phytoplankton biomass (P, μg chlorophyll a/liter) and the second is zooplankton biomass (Z, μg weight/liter). We have samples from six independent systems (e.g., lakes) or locations (e.g., stations or transects within a lake) that constitute the replicates made at three different times. Thus, in this example $q = 2$, $k = 3$, and $n = 6$. The model is deterministic, so all samples are compared to the same model output. Some hypothetical data are shown in Table 8.1.

From the data in Table 8.1, we subtract the model prediction from each entry (Table 8.2) to create prediction *deviations*. We call the table entries for phytoplankton (P) deviations δ_{Pjk}, where j indexes the sample number and k indexes the time of the sample. Zooplankton (Z) deviations are δ_{Zjk}. Next, we subtract the data-model deviation at one time from the deviation at the next time $\Delta_{Pjk'} = \delta_{Pjk} - \delta_{Pj(k+1)}$, and $\Delta_{Zjk'} = \delta_{Zjk} -$

Table 8.1: Hypothetical data and model response for six replicates and three points in time for phytoplankton (μg chl-a/liter) and zooplankton biomass (μg/liter). Columns are time (1,2,3); rows are the replicates and the model prediction.

	Phytoplankton			Zooplankton		
Sample	1	2	3	1	2	3
1	2.5	4.0	1.0	10	50	20
2	2.0	3.9	1.3	15	60	18
3	2.3	3.8	0.9	12	55	22
4	1.9	4.1	1.2	9	48	19
5	1.5	3.2	0.7	18	60	18
6	2.2	3.8	1.1	16	64	21
Model	2.1	3.8	1.0	13	56	20

$\delta_{Zj(k+1)}$. These values will be the data on which we will perform the test for parallelism, since parallel lines will have equal slopes.

Finally, we arrange these time differences in a matrix (Table 8.3), so that the columns represent all of the replicate time differences (in temporal order) for all of the responses being tested. Thus, columns are arranged in groups first by response variables (e.g., P or Z) and then by time differences within response variable (e.g., response at time 1 minus response at time 2 and response at time 2 minus response at time 3). For example, column 1, row 1 will be $\Delta_{P11'} = \delta_{Pj1} - \delta_{Pj2}$, which is the deviation of the model prediction of phytoplankton from the data (δ) at time 1 minus the same deviation at time 2 for sample (replicate) 1. Column 1, row 2 is the same quantity computed for the second sample ($\Delta_{P21'}$), and so on for the remaining rows ($\Delta_{Pj1'}$). Column 3, row 1 ($\Delta_{Z11'}$) is the difference between time 1 and 2 using the prediction deviation for zooplankton biomass for sample 1. Using this convention on our example, the 2D matrix has six rows which are the replicates and four columns [two responses (P and Z) and two time differences (time 1 minus time 2, time 2 minus time 3)].

This is a one-sample multivariate test of the equality of means, and so is a generalization of the one-sample univariate test based on Student's t. The test in the general case is based on Hotelling's T^2 for which the general

Table 8.2: Deviations of data and model response for six replicates and three points in time for phytoplankton and zooplankton biomass. Columns are time, rows are replicates.

	Phytoplankton			Zooplankton		
Sample	1	2	3	1	2	3
1	0.4	0.2	0.0	-3	-6	0
2	-0.1	0.1	0.3	2	4	-2
3	0.2	0.0	-0.1	-1	-1	2
4	-0.2	0.3	0.2	-4	-8	-1
5	-0.6	-0.6	-0.3	5	4	-2
6	0.1	0.0	0.1	3	8	1

Table 8.3: Time differences of model-data deviations for 6 replicates of phytoplankton and zooplankton responses. Columns are differences, rows are replicates. The dot in the column label (e.g., $\Delta_{P.1'}$) denotes all of the replicates in a given column. Column means are shown in the last row.

Sample	$\Delta_{P.1'}$ $\delta_{P\cdot 1} - \delta_{P\cdot 2}$	$\Delta_{P.2'}$ $\delta_{P\cdot 2} - \delta_{P\cdot 3}$	$\Delta_{Z.1'}$ $\delta_{Z\cdot 1} - \delta_{Z\cdot 2}$	$\Delta_{Z.2'}$ $\delta_{Z\cdot 2} - \delta_{Z\cdot 3}$
1	0.2	0.2	3	-6
2	-0.2	-0.2	-2	6
3	0.2	0.1	0	-3
4	-0.5	0.1	4	-7
5	-0.0	-0.3	1	6
6	0.1	-0.1	-5	7
Means →	-0.03	-0.03	0.17	0.50

formula for data of this type is (Timm 1975):

$$T^2 = (n)(\mathbf{Y} - \mathbf{Y}_0)'\mathbf{S}^{-1}(\mathbf{Y} - \mathbf{Y}_0),$$

where n is the number of replicates; $\mathbf{Y}-\mathbf{Y}_0$ is a column vector of the average differences between observed ($\mathbf{Y}$) and expected ($\mathbf{Y}_0$) means; $(\mathbf{Y} - \mathbf{Y}_0)'$ is the transpose of $(\mathbf{Y} - \mathbf{Y}_0)$, and so is a row vector of the average differences, and $\mathbf{S}^{-1}$ is the inverse of the *variance–covariance* matrix (or, simply, the covariance matrix) for the test variables (columns in Table 8.3). $\mathbf{S}^{-1}$ has size $q(k-1) \times q(k-1)$. The variance–covariance matrix is a square matrix whose diagonal is the variance (of the samples) of a given response and time difference (e.g., $\Delta_{P.1'}$). Thus, each diagonal element is the sum of the squared deviations of replicates from the mean [i.e., $\sum(x_i - \overline{x})^2$] divided by $n-1$. The off-diagonal elements are the covariances. The covariances are the sum of the deviations of replicates of variable x from the mean of variable x times the deviations of replicates of variable y from the mean of variable y. Symbolically, the covariance is: $\sum[(x_i - \overline{x})(y_i - \overline{y})]/(n-1)$. The covariance between two variables is closely related to the degree of correlation of the variables. See Searle (1982) for a formal definition.

In this case, we are using the deviation of the model from the data; thus, the expected mean is 0, so Hotelling's T^2 for model validation is (Steinhorst 1979):

$$T^2 = (n)\mathbf{Y}'\mathbf{S}^{-1}\mathbf{Y}, \tag{8.9}$$

where $\mathbf{Y}$ are data-model deviations and $\mathbf{Y}'$ is the last row in Table 8.3.

The variance–covariance matrix computed from Table 8.3 is

$$\mathbf{S} = \begin{bmatrix} 0.0747 & 0.0067 & -0.2933 & 0.2600 \\ 0.0067 & 0.0387 & 0.3267 & -1.1600 \\ -0.2933 & 0.3267 & 10.9667 & -17.5000 \\ 0.2600 & -1.1600 & -17.5000 & 42.7000 \end{bmatrix}.$$

Provided that sufficient replicates are available, the inverse of $\mathbf{S}$ can be obtained from standard software packages as

$$T^2 = 6\left[[-0.03,-0.03,0.17,0.50]\begin{bmatrix} 30.44 & -147.43 & -4.28 & -5.94 \\ -147.43 & 1469.80 & 50.36 & 61.47 \\ -4.28 & 50.36 & 2.03 & 2.22 \\ -5.94 & 61.47 & 2.22 & 2.64 \end{bmatrix}\begin{bmatrix} -0.03 \\ -0.03 \\ 0.17 \\ 0.50 \end{bmatrix}\right]$$
$$= 0.0463.$$

To determine the significance level of T^2, we use a table (Timm 1975) of Upper Percentage Points of Hotelling's T^2 for $T^\alpha(p,\nu)$, where p is $q(k-1)$ [i.e., $2(3-1) = 4$], α is the probability level for the test, and ν is $n-1$ (i.e., 5). The values for our case corresponding to $\alpha = 0.01$, 0.05, and 0.10 are

$$T^{0.01}(4,5) = 992.494$$
$$T^{0.05}(4,5) = 192.468$$
$$T^{0.10}(4,5) = 92.434$$

Therefore, since $0.0463 < 992.494$, we cannot reject the null hypothesis that the profile of data minus model predictions is zero at $P = 0.01$. Thus, this test result validates (or confirms) the model.

If T^2 tables are not available, there is an alternative test procedure. If the null hypothesis is true, then T^2 is distributed as the F distribution, so H_0 is accepted at the α significance level if

$$\frac{n-p}{p(n-1)}T^2 \leq F^\alpha(p, n-p),$$

where $p = q(k-1)$ in our notation.

The identity of the T^2 and F distributions allows us to calculate the power of the test. Statistical power is $1-\beta$, where β is the probability of a Type II error: *failure to reject a false null hypothesis.* In practice, we calculate the power of a test when we fail to reject H_0. This is always important to do, but especially so in model validation because failure to reject H_0 means that we accept the model as a good fit. It could happen that we fail to reject H_0 because of small sample size: having few replicates often means large variability and small values of the test statistic (e.g., T^2). Power measures our ability to detect the observed difference between the model and our particular set of data, given the sample size n. Zar (1984) gives an introduction to power analysis and some basic calculations; Cohen (1988) gives a larger compendium of calculations.

For the one-sample profile test, power is determined for curves developed for the noncentral F distribution (the distribution to use when H_0 is false) as follows (Timm 1975). Calculate

$$\phi = \sqrt{\delta/[q(k-1)+1]},$$

where δ equals the estimated T^2. In our example,

$$\phi = \sqrt{0.043/5} = 0.464.$$

Using ϕ, choose the desired α for a Type I error (e.g., 0.01), determine the degrees of freedom for the F distribution numerator ($\nu_1 = p$) and denominator ($\nu_2 = n - p - 1$), and read the power from the curves (Timm 1975). In our case, $\alpha = 0.01$, $\nu_1 = 4$, $\nu_2 = 6 - 4 - 1 = 1$, and power is less than 0.10. This may seem like, and indeed is, an absurdly low number, but remember what is being calculated. We are calculating our ability to detect a very small difference between the data and the model, given our sample size. It is common knowledge in statistical design that in order to detect small differences we need large sample sizes. The low power of this test states that given that the model is apparently very accurate, we will need large sample sizes to falsify it. It is a quirk of these types of inferential statistics applied to model validation that (1) low sample sizes (and high variance) will tend to result in corroborated models, and (2) very accurate models will usually have low power. [Recently, the *post hoc* use of power analysis as described above has been questioned (Burnham 1995, Steinhorst *pers. commun.*). The main criticism is that since the estimate is based on the original data, it is biased and amounts to little more than a restatement of the test results.]

The approach described above works for any number of response variables and time intervals, and makes no assumptions concerning the structure of the variance–covariance matrix. T^2 is easy to compute using software that can manipulate matrices. A major disadvantage is that it requires moderately large numbers of replicates. To apply the method, we must have n replicates such that $n > q(k-1)$, where q is the number of system response variables and k is the number of times at which comparisons are made. This amount of replication is required in order to estimate the elements of the covariance matrix. If the model can predict these values, then an approach to validation related to profile analysis is possible with far fewer replicates (Feldman et al. 1984).

A second approach to time-series validation is to treat the model and the data as two time series and to measure the correlation between them using *cross-correlation* techniques. Qualitatively, this procedure attempts to quantify the correlation between two autocorrelated time series for a given *lag*. The lag accounts for the autocorrelation. This is a well-studied problem and Steinhorst (1979) summarizes the basic formulae to test the hypothesis that for a given lag interval there is zero correlation between the two time series. This method has the reputation of requiring large data sets. This requirement may limit its application in the ecological and environmental disciplines, but may not be a problem in biochemical and physiological systems.

A third approach to comparing time series is commonly published, but is not a rigorous test. One can simply plot model output and the data on the same graph and count the number of times the model output (or mean model output) falls within the data's 95% confidence intervals. These intervals are

$$\overline{X} \pm \left[t_{(0.05,n-1)}\right] s_{\overline{X}},$$

where $\overline{X}$ is the observed mean, $t_{(0.05,n-1)}$ is the theoretical Student's t distribution value for $\alpha = 0.05$ and $n-1$ degrees of freedom (n = number of observations), and $s_{\overline{X}}$ is the standard deviation of the sample.

One can further state an objective rule for judging model quality such as: "A model will be valid if model output falls within data 95% confidence limits for 80% of the model–data comparisons." Using the hypothetical data and model responses of Table 8.1, the 95% confidence intervals for the data are

Phytoplankton	Zooplankton
$t_1 : 1.17 - 2.97$	$t_1 : 4.18 - 22.48$
$t_2 : 2.99 - 5.97$	$t_2 : 40.04 - 72.30$
$t_3 : 0.48 - 1.59$	$t_3 : 15.47 - 23.87.$

From these values, we see that all of the model predictions fall within the 95% confidence intervals, and we would conclude that we have validated the model. See Van Henten (1994) for a real validation of a plant growth model using this technique.

8.2.5 Model Discrimination

If there is something wrong with every alternative, one tends to try a succession of wrong things in the hope that one of them will turn out, which it never does. — Boulding (1972)

The previous approaches assessed the degree to which a particular model deviated from a data set. This is important, but it does not address the method of multiple working hypotheses that we advocated in Chapter 2. We are, in principle, interested in the absolute difference between the observed system and model predictions, but as we have seen above, this is often difficult to achieve in practice. An alternative approach is to content ourselves with deciding among a set of models based on their *relative* adequacy. The process of discriminating between alternatives is basically a decision problem. Most decisions (e.g., Should I finish reading this book?, Should I change professions?) involve evaluating the probabilities of a set of events (e.g., the probability that I will get a raise, or that I will be happy, etc.). As we will see, calculating probabilities is central to model discrimination.

Model discrimination is fundamental to all statistical inference, so the problem is quite general, although we will discuss only a specific application. There are two broad types of model discrimination: *parametric* and *structural*. In parametric model discrimination, the form of the model is fixed (e.g., a straight line), the parameters are unknown, and the object is to find the optimal parameter set. We covered this problem when we discussed parameter estimation, and so we will not address it here. Structural model discrimination is more closely allied with model validation. There are two major approaches: *likelihood functions* and *Bayesian inference.* In the following, I have used extensively Reilly (1970), Blau and Neely (1975), Reckhow and Chapra (1983a), Carpenter (1990), and Reckhow (1990).

Likelihood Functions

As motivation, consider linear regression. The problem is to find the best set of parameters that minimizes the sum over all datum points of the square of the vertical distance between the model line and the data. Some parameter values will produce large sums, others will produce smaller sums. Likelihood functions are a similar idea.

The likelihood of a sample is the probability that the sample would be drawn from a specified probability distribution with known parameters (e.g., the mean and variance of the distribution). A likelihood function that calculates the likelihood of a sample is a mathematical function that results from applying a probability distribution to a particular sample in which one or more of the distribution parameters are allowed to vary as the function's independent variable. The dependent variable of the likelihood function is the *a posteriori* probability of the sample given the underlying probability distribution (Meyer 1975; Borowski and Borwein 1991).

Is the Die True? To see the utility and application of this concept, consider the problem of determining if a die (as in one half of a pair of dice) is true (Reilly 1970). A reasonable approach to this problem is to roll the die n times and observe the number (x) of occurrences of a particular face. If the observations deviate significantly from that expected from a true die, then we can conclude that the die in question is not true. For example, suppose we roll the die five times and observe two occurrences of the number 3. How likely is this outcome if the die is true? The underlying probability distribution for this kind of problem is the binomial distribution

$$b(x; n, \theta) = \frac{n!}{x!(n-x)!}\theta^x(1-\theta)^{n-x}.$$

This formula allows us to compute the probability that a particular event will occur, if we specify the unknowns. In the die problem, θ is a parameter of the distribution and is the probability that a given face will appear; in a true die, $\theta = 1/6$. n is the number of trials (five rolls of the die) and x is the observed occurrences of a face (two). We consider x and n to be data that are specific to a particular test. Inserting the data and parameters for an assumed true die, we find: $b(2; 5, 0.1667) = 0.16075$. In the problem, however, we do not know the true θ, so we form the likelihood function

$$L(\theta) = \frac{5!}{2!(3)!}\theta^2(1-\theta)^3$$

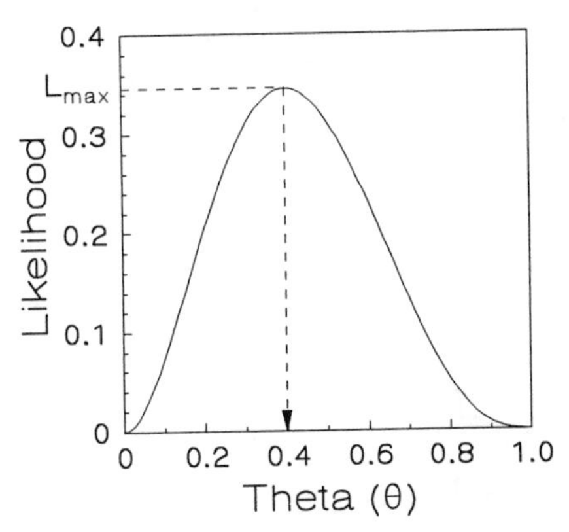

Figure 8.4: The likelihood function for the binomial probability distribution with $n = 5$ and $x = 2$. The maximum likelihood estimator is the θ associated with maximum of the function.

that pertains to the data of this particular experiment.

The graph of this function is shown in Fig. 8.4. From this we see that the probability of a face appearing that is associated with the maximum likelihood of the sample is 0.4, not 0.1667, which we would expect if the die were true. So this discrepancy between expected and most likely θ suggests that the die is not true. We quantify the amount of discrepancy by forming the *likelihood ratio* (R): $L(\theta_{0.4})/L(\theta_{0.17})$. In this case, the ratio is 2.15. So we say that the observed sample is 2.15 times as likely if $\theta = 0.4$ than if $\theta = 0.167$. We would, however, expect two 3s from a true die due to random chance, so is the discrepancy large enough to reject the hypothesis that the die is true? Since we have but a single estimate of the most likely θ (i.e., 0.4), we cannot say anything rigorously quantitative. A rule of thumb (Reilly 1970) states that if R is greater than 10, then we have a real difference. Under certain conditions, the *log of the likelihood ratio* ($\log R$) is distributed approximately as a χ^2 distribution, so that a probability can be associated with an observed R to assess if it is large enough to be due to factors other than chance (Sokal and Rohlf 1981). Dennis and Taper (1994) give further examples and a cogent introduction to the problem of ascertaining the ratio value at which to reject models.

Empirical Model Likelihood The above example is fine, if one manufactures dice, but it is not very useful in model discrimination. A better example is to choose among three structurally different models relative to a data set. Suppose we have three models and a data set as shown in Fig. 8.5. To compute the maximum likelihood for all three models, we need (1) parameters to maximize, (2) some data, and (3) a probability distribution that depends on the model parameters. Figure 8.5 shows the parameters as the a_i and the data we need. We also need a probability distribution that will compute the probability of observing the data, given

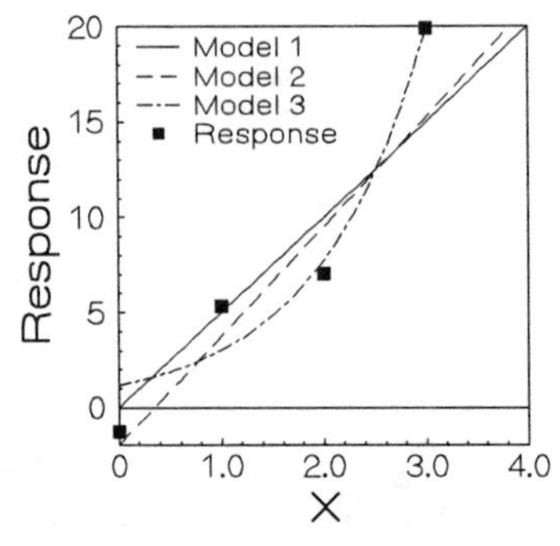

$$\text{Model 1:} \quad y = a_0 x$$

$$\text{Model 2:} \quad y = a_0 + a_1 x$$

$$\text{Model 3:} \quad y = a_0 e^{a_1 x}$$

Figure 8.5: Three models fit to hypothetical data as a basis for discriminating among them.

a model. (This is what the binomial distribution did for us in the die problem.)

For an intuitive grasp of the appropriate distribution, recall that in regression we think of each observed y value as being equal to a function plus an error term:

$$y_i = f(x_i, \theta_j) + \epsilon_{ij}, \tag{8.10}$$

where f is the function and ϵ_{ij} is the error associated with the ith x–y data pair and θ_j is a set of parameters. Regression chooses the θ_j parameters of f to make the error as small as possible. The error term, therefore, is related to the probability of observing a particular y_i, given $f(x_i, \theta_j)$. If, for a particular function and parameter set, the error is large, then the probability of observing y_i will be small, and vice versa. But in regression, as in Fig. 8.5, there are several y_i. They are incorporated into the computation of the probability of observing the total error around all of the y_i by multiplying the probabilities for individual datum points (the joint probability distribution). For example, if p_0 is the probability of observing y_0 [given $f(x_0, \theta_j)$], and p_1 is the probability of observing y_1 given the same function and parameters, then $p_0 p_1$ is the probability of observing both y_i given the function and parameters. The probability of the total error is just what we mean by the probability of observing the y_i. This, then, is the probability distribution we need for the likelihood of all the y. So, a *general likelihood function* is

$$L(\text{model,error}) = \prod_i^N p_i \tag{8.11}$$

(i.e., the product of the probabilities of obtaining each independent observation). To produce a particular likelihood function, we need an expression for p_i as a function of the error term in Eq. 8.10. We use one of the assumptions of linear regression: the errors are normally distributed and independent. The probability density function (pdf) for a single-variate normal distribution is

$$n(x;\mu,\sigma) = \frac{1}{\sqrt{2\pi\sigma^2}} \exp\left(-\left(\frac{(x-\mu)^2}{2\sigma^2}\right)\right), \tag{8.12}$$

where x is the independent variable, μ is the mean, and σ^2 is the variance. The latter two variables are the parameters of the distribution; x is the data. In our case of fitting a particular datum y_i to a model (Eq. 8.10), y_i is x and $f(x_i, \theta_j)$ is μ in Eq. 8.12. For a particular x and model as in Eq. 8.10, the difference between the observed y and the predicted y is a number drawn from $n(y_i; f(x_i, \theta_j), \sigma)$ (Eq. 8.12). Therefore, it is the probability of observing that particular y_i, given the model. The likelihood function for all datum points (all y_i), assuming model j, is

$$L_j(\theta) = \prod \exp\left(-\frac{(y_i - f(x_i,\theta_j))^2}{2\sigma^2}\right) \tag{8.13}$$

$$= \exp\left(-\frac{\sum_i^N (y_i - f(x_i,\theta_j))^2}{2\sigma^2}\right), \tag{8.14}$$

where N is the sample size. We drop powers of the scaling constant $(1/\sqrt{2\pi\sigma^2})$, since it does not alter the location of the maximum. This equation has the following important properties. (1) $(y_i - f(x_i, \theta_j))^2$ is the least-squared error between data and model. (2) Models and parameter sets (θ_j) that have large errors (poor fits) have small likelihood values. (3) There is a single maximum, which corresponds to the maximum likelihood. (4) The set of θ_j associated with the maximum is the best set of model parameters for model i. These θ_j are the *maximum likelihood estimators* of the parameters.

Based on the above, the maximum likelihoods for the three models in Fig. 8.5 and the ratio of likelihoods to the best model is (Reilly 1970)

Model	L_{max}	L_3/L_i
1	0.05	4000
2	1.0	202.2
3	202.2	1.0

We can summarize the above by saying that Model 3 is 4000 times more likely than Model 1 and 202.2 times more likely than Model 2. According to our rule of thumb on the sizes of likelihood ratios, this is a significant difference and Model 3 is better than the alternatives.

Mechanistic Model Discrimination A final application computes L_{max} for seven differential equation models of the dynamics of a radioactively labeled pesticide in an aquatic microcosm (Blau and Neely 1975; Carpenter 1990). An aquatic laboratory microcosm containing Water, Soil, Plants, and Fish was perturbed with ^{14}C-labeled Dursban to determine how much of the pesticide accumulated in the above microcosm components over time. Seven

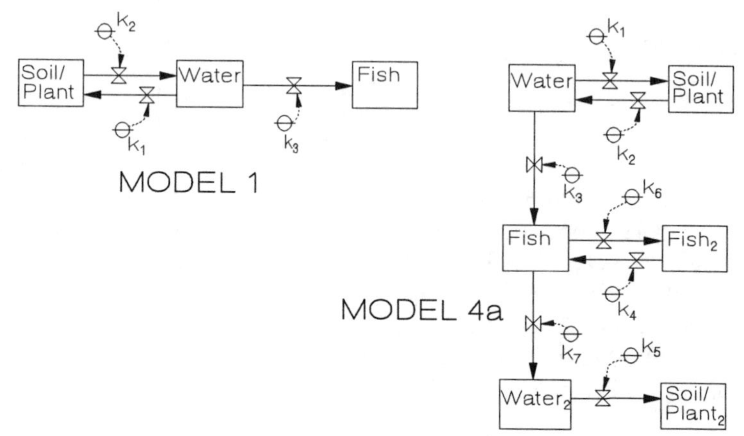

Figure 8.6: Two of the seven models of Dursban movement in an aquatic microcosm (Blau and Neely 1975). In model 4a, Fish$_2$, Water$_2$, and Soil/Plant$_2$ represent additional storage compartments for ^{14}C.

models based on linear differential equations were formulated as predictive tools. The relative merits of each as measured by maximum likelihood was assessed to discriminate among them. The models varied according to the number and relations of ecosystem components that each incorporated. The models differed from each other according to the number of flows and compartments in a series from simple to more complex. Two of these are illustrated in Fig. 8.6.

Blau and Neely (1975) fit each model to a single time series of results to obtain the best fit parameters (k_i). They applied Eq. 8.14 to each model and obtained the maximum likelihoods (relative to model parameters) in Table 8.4. They concluded that Model 4a was the best of the seven. However, a potential problem here is that the models have unequal numbers of parameters. Consequently, it is to be expected that the more complex models will fit better. In the likelihood approach, there is no rigorously quantitative method to accommodate such differences between models. Some authors (Reilly 1970) suggest increasing, in some nonobjective way, the threshold value of likelihood ratios needed for significant differences among models. Spriet and Vansteenkiste (1982) contains a lengthy discussion of the use of information measures that attempt to evaluate models based on both minimum error and parsimony of structure.

Bayesian Inference

> *Complete objectivity about one's own work is a little much to expect from a human being, even a scientist, but it is not too much to expect from one's colleagues.*
> — Efron (1986)

The likelihood method quantitatively ranks the adequacy of a set of competing models by their ability to fit the data, but it does not actually compute the probability that the models are correct. One method of calculating

Table 8.4: Maximum likelihood values and ratios for the seven models of Dursban movement. P is the Bayesian probability that the model is true.

Model	$L_{max,j}$	$L_{max,4a}/L_{max,j}$	$P(M_i \mid \mathbf{Y})$
1	1.44×10^{-389}	3.98×10^{385}	$< 10^{-8}$
2a	6.95×10^{-193}	8.24×10^{187}	$< 10^{-8}$
2b	4.16×10^{-62}	1.37×10^{57}	$< 10^{-8}$
3a	8.37×10^{-16}	6.84×10^{10}	$< 10^{-8}$
3b	8.95×10^{-16}	6.40×10^{10}	$< 10^{-8}$
4a	5.73×10^{-5}	1.00	0.97
4b	1.79×10^{-6}	32.0	0.03

this probability uses *Bayes Theorem.* This area of statistics is complicated and controversial; consequently, we will be able to provide only a heuristic introduction to its applications and strengths and weaknesses. Before defining this theorem, we relate the problem to more classical approaches.

Bayesian inference and statistical analysis based on Bayes Theorem provide an important alternative to the classical or *frequentist* statistics familiar to most biologists (e.g., t-tests, ANOVA, regression). In a nutshell, classical statistics use data to calculate a sample estimate of a test statistic (e.g., Hotelling's T^2). This estimate is compared with a frequency distribution of hypothetical samples of the same size (i.e., the probability tables for Hotelling's T^2). Thus, we compute the probability of observing a particular value of the test statistic, *given that the null hypothesis is true.* Advocates of the Bayesian approach argue that this is not the central focus of scientific questions. They claim that scientists are primarily interested in the *probability that the null hypothesis is true* (Reckhow 1990). Bayesian statistics were developed to address this question.

Bayesian statistics are based on different methods of calculating probabilities, and, in particular, include estimates of the truth of the null hypothesis *prior* to the test being made. Thus, they permit the inclusion of prior knowledge (e.g., data from other similar systems, historical data, etc.) in the test of the hypothesis for the given data set. Bayesian inference is still controversial among statisticians, but it is being applied to unreplicated data sets and to comparison of competing, alternative simulation models (Carpenter 1990).

The basis for this approach to inference is Bayes Theorem which in the present context is a recipe for calculating the probability that model i is true, given the observed data and a finite set of alternative models. The Bayesian probability is

$$P(M_i \mid \mathbf{Y}) = \frac{P(M_i)P(\mathbf{Y} \mid M_i)}{\sum_j^N P(M_j)P(\mathbf{Y} \mid M_j)}, \tag{8.15}$$

where N is the number of alternative models, $P(M_i)$ is the *prior* probability that model i is true, and $P(\mathbf{Y} \mid M_i)$ is the probability of observing $\mathbf{Y}$

values given that M_i is true (i.e., the *likelihood* of $\mathbf{Y}$). The denominator is a scaling factor that normalizes the likelihood of a particular model to the total likelihood of all the models.

There are two problems in computing Eq. 8.15: (1) specifying the prior probabilities and (2) computing the likelihood of observing the data, given a particular model. The solution to (1) is easy to state, but difficult to implement. The prior probability is simply our belief in model i before we collect the validation data. But this begs the question of how we quantify this belief. Some say we may use any evidence we have at hand: expert opinion, studies reported in the scientific literature, previous experiments, etc. When the prior probabilities are quantified from previous experience, they provide a solution to the major problem with the classical view of the modeling process (Chapter 2). Bayesian probabilities generated in earlier passes through the process can be used as the prior probabilities in later passes. Other users of Bayesian inference, however, recommend not using any previous experience. They suggest assigning the prior of each model an equal probability: $1/N$, where N is the number of models. The problem of the priors is the source of much of the controversy surrounding the use of Bayesian inference. It raises the issue of the role of subjective judgment in statistical inference.

The solution to (2) is difficult to describe, but the usual solution results in a relatively easy computation. The probability of observing a particular data set, given a model, is related to the error associated with fitting the model to the data. We saw how to do this in calculating the likelihood ratios of the three hypothetical empirical models (Fig. 8.5). So, the likelihood functions computed using the optimal fit of parameters to the data can be used as the $P(\mathbf{Y} \mid M_i)$ in Bayes Theorem.

This analysis has been applied extensively by Reckhow and Chapra (1983a) and Reckhow (1990) to a variety of management models. Carpenter (1990) has recently performed Bayesian analysis on the seven competing models for pesticide transport developed by Blau and Neely (1975). Since Carpenter chose not to incorporate other information about the prior probabilities of the seven models, he assigned each to have $P(M_i) = 0.1667$ (i.e., all equally likely). Using a normal distribution of errors for the $P(\mathbf{Y} \mid M)$, he calculated the probabilities that each model was true (Table 8.4, column 4). The posterior probabilities of five models were essentially zero. Another model had a probability of 0.03 of being true, while the remaining model's probability was 0.97. Thus, one model was clearly superior, given that all models were equally probable to be correct before the test was made.

8.2.6 Meta-Models

A recent alternative to classical model validation is the construction and validation of *meta-models* (Kleijnen and van Groenendaal 1992). This should not be confused with *meta-analysis*, which is the statistical analysis of the statistical analyses reported by other researchers. A meta-model is a nonlinear regression model of the output of a dynamic model. We like to think that the original dynamic model provides an understanding of the system, but too often, complex simulation models provide complex and confusing results that are themselves difficult to understand. Mathematical concepts such as nullclines and stability, which are described in Chapter 9, are one approach to understanding a model. Reducing complex model output to relatively simple regressions between model variables is another. The method developed by Kleijnen is as follows. (1) Use a series of original model runs to generate a data set. (2) Identify a set of potentially interesting relationships (*meta-relationships*, e.g., the relation of phytoplankton biomass to zooplankton biomass). Then, fit linear or nonlinear functions to the model data set. (3) Validate the meta-model by running the original model a second set of times with different input values (e.g., different driving temperatures). If valid, the meta-model should correctly predict the quantitative meta-relationships of the new runs. A valid meta-model will characterize the important dynamic relationships that are produced by the mechanistic relationships used in the original model. The meta-model can then be further validated against empirical data.

8.3 Précis on Validation

The relation of model validation and model discrimination has yet to be firmly established. They share important statistical similarities, and combined with carefully designed independent experiments, they have potential to address the logical problems associated with the use of complex simulation models in the hypothetico-deductive method of science (Romesburg 1981). Nevertheless, they represent different philosophies toward model evaluation. Likelihood ratios and Bayesian posteriori probabilities are, by themselves, not hypothesis tests. Using time series data and model output as the basis of likelihood functions is questionable because of the potential violation of the independence assumptions. On the other hand, model validation, as discussed here, has emphasized hypothesis tests for individual models without concern for the universe of alternative models. The issue of subjectivity in all aspects of model evaluation, whether it comes from model choice or prior probabilities, will continue to be hotly debated for many years. In practice, both approaches have emphasized model adequacy (Mankin et al. 1977), and neither approach has explored sufficiently the role of parsimony in model choice, despite the fact that techniques exist (Spriet and Vansteenkiste 1982, Costanza and Sklar 1985). Incorporating model complexity into our ultimate assessments of model performance is one approach to measuring model reliability (Mankin et al. 1977).

In the end, we are left with the evocative imagery of Swartzman (1980), who analogized modeling with shooting arrows through a mist toward a target that lies behind a brick wall. The archer becomes "distracted from the target by the shimmering colors of the mists." But, once loosed toward the target, the final resting place of the arrows cannot be ascertained because of the mists and wall. And, as if this were not enough, recalling the blind men and the elephant: "somewhere, way off behind the target, is the real system."

8.4 Exercises

1. Read Romesburg (1981) and discuss his claim that simulation models cannot be used in the hypothetico-deductive method.
2. If five rolls of a die produced five 1s, is the die true? Why?
3. Write the equations for the Dursban models illustrated in Fig. 8.6.
4. If we applied Hotelling's T^2 to a univariate sample, what elementary statistic would we compute? [Hint: Write the square of the formula for Student's t in a form that is similar to Hotelling's T^2.]
5. Examine published models from your discipline and rank them by the rigor and completeness of their validation efforts. Has your field, as a whole, produced well-validated models?

Chapter 9

Model Analysis: Uncertainty and Behavior

9.1 Analyzing Model Responses

VALIDATION is model analysis concerned with evaluating model quality relative to the real world using comparisons with empirical data. We now turn our attention to analyzing model performance by actively manipulating various components of the model. We will discuss two types of manipulations of the model structure. The first type manipulates the equations and parameters of the model to ascertain the extent and effect of modeler uncertainty. The second manipulation alters the values of state variables to determine if the system will return to the premanipulation levels.

9.2 Uncertainty Analysis

Among the many sources of uncertainty in biological modeling are (O'Neill and Gardner 1979):

- *Biological hypotheses and mathematical formulation.* We may be ignorant of the correct biological processes involved.
- *Parameter values.* We may be ignorant of the mean and variance of the population from which our parameter estimates are drawn.
- *Natural variation.* The system may have components that must be treated as stochastic (e.g., temperature). We will, therefore, be able to make only probabilistic predictions.

The first source is the most difficult to correct; this kind of uncertainty implies that we are ignorant of the underlying biology. There is little we can do about this other than to learn more, design better experiments, and be more clever in our mathematical formulation. Alternative models are one approach to formally investigating structure effects (Chapter 2). The effect on our predictions of our uncertainty in parameter values (the second source) can be investigated using a combination of *parameter sensitivity analysis* and *error analysis*. To address the problem of natural variation (the third source), we use *stochastic* models: models with output influenced by random effects on model variables or parameters (Chapter 10).

9.2.1 Parameter Sensitivity

The typical interpretation of a parameter estimate is the mean or expected value from a distribution. Parameter sensitivity analysis involves analyzing differences in model response to small differences in parameter values. I interpret parameter sensitivity to be addressing the question: "What are the dynamical effects of modeler uncertainty about the true mean value of the parameters?" (This interpretation differs from that of other authors, e.g., Swartzman and Kaluzny 1987.) Strictly speaking, we can ask this question of several model components: parameters, initial conditions, or driving variables. Typically, however, the analysis is applied to the parameters of the difference or differential equations.

Uses of Sensitivity Analysis

There are four major uses of parameter sensitivity analysis.

Validation Two different interpretations of sensitivity results pertain to our general judgments of model quality. First, we have an intuitive belief that most real systems will not respond violently to small changes in the values of the operating parameters or variables. That is, if we throw a pebble onto the quadrangle lawn, we do not expect to see mass hysteria, hurricanes, species extinctions, or the eruption of clouds of vile gases. If our model were to behave in this way after a similarly small change in parameter values, it would be evidence that we had not used correct mathematical formulations. Second, if we are relatively unconfident of the accuracy with which we have estimated a particular parameter and if the model is sensitive to a small change in that parameter, then we should not be confident in the accuracy of the model output. Alternatively, if the model is not sensitive to a change in the parameter, then we may conclude that our lack of confidence in the accuracy of the parameter estimate should not influence our faith in the model.

Research Design As we will see below, model response will be sensitive to some parameters and not to others. The sensitive parameters are those to which we should devote the greatest research effort so as to obtain the best estimates, given budget and time constraints.

An alternative interpretation, however, is not that one needs more imprecise parameter estimation, but rather greater precision (mechanistic detail) in the model formulation. We should place greater effort on formulating models with different biological processes or finer resolution in the state variables (e.g., additional compartments in physiological models or age-structure in population models, etc.).

System Control Managing a system requires that we can control the system. To control a system means that by altering parameters and variables we can produce desirable output. If varying a parameter does not alter system output (i.e., the system is insensitive to the parameter), then that parameter is not useful for control. Therefore, sensitivity analysis can be used to identify which parameters have potential as controllers.

Theory Often the model objective is to investigate a theoretical concept (e.g., conditions for system stability). The response of model output to different parameters may become the central question. For example, as we will discuss in *Part II*, complex dynamics in difference equations can emerge as a critical parameter is increased. At small values of the parameter, we might have steady-state dynamics; as the parameter is increased the dynamics may change to oscillations, until, as the parameter is increased more, the dynamics can become extremely complex to the point of being *chaotic*. Interesting theoretical questions are to determine which equations can show this behavior and which parameters are responsible for it.

Sensitivity Variables

Model sensitivity can be assessed by examining the responses of model state variables, quantities calculated by the model, or quantities that can be calculated from model output. Commonly used quantities are: the state variables at one or more fixed times, time averages of state variables, extreme values (e.g., maximum or minimum) of state variables over a run, and times within a run at which significant events (e.g., extreme values) occur. Simple combinations of state variables are also used, for example, sums, ratios. Which quantity to use should be obvious from the model objectives or question being asked.

Methods

When we perform sensitivity analysis we want to answer two questions: (1) How variable is the response? and (2) What are the ranges of model responses to the parameter changes? While we will treat these questions differently, they both share a common geometric interpretation as illustrated in Fig. 9.1.

The vertical axis is some measure of model response and may be presented in the units of variables calculated by the model (to answer question 2) or may be in sensitivity units (to address question 1). The other axes are the parameters manipulated in the sensitivity analysis. Regardless of

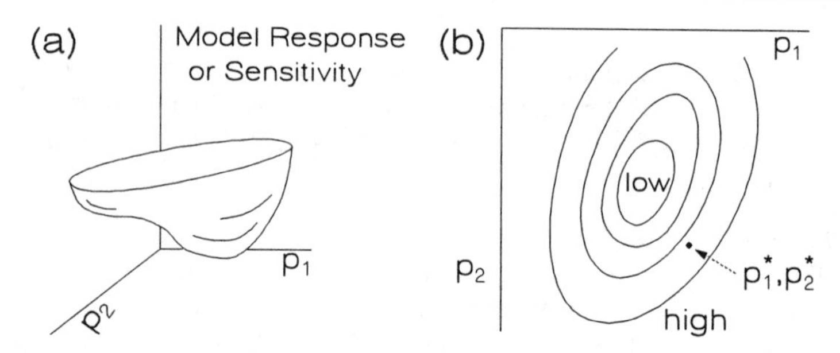

Figure 9.1: Model response to two parameters displayed as a surface (a) and contour lines (b). The point (p_1^*, p_2^*) is a particular set of parameters that shows moderately high sensitivity.

the interpretation of the dependent axis in Fig. 9.1, we do not know what this surface looks like. Sensitivity analyses provide us some clues.

To determine the nature of the surface, sensitivity analysis involves performing numerical experiments in which parameters are systematically changed and resultant model response analyzed. Normally, a set of parameters are identified as being the reference or *nominal* values. Usually, these are chosen because they are the arithmetic means of estimation experiments or are "typical" values from the literature or common knowledge. Ideally, we would like to examine a large region of parameter space, but in reality there are practical limits. If we wish to examine a large area (volume) of the parameters, then each parameter must be run at several values. The number of values per parameter used depends on the desired resolution of the surface. Moreover, the number of runs required increases in proportion to the number of values per parameter raised to the power of the number of parameters. (This assumes we wish to examine all combinations of parameter values.) For example, if we wish to examine four levels for each parameter and use all combinations of levels for six parameters, we would need $4^6 = 4096$ runs. In practice, we do not attempt all possible runs.

In light of these practical limits, there are two major strategies for varying the parameters relative to the nominal values (Fig. 9.2). First, we can vary only a single parameter at a time (Fig. 9.2a). The number of runs required is greatly reduced since we do not do all combinations. Thus, we can examine long transects across the space. The disadvantage is that we ignore interactions between parameters: model response when p_1 and p_2 are simultaneously increased by 20% may be much greater than the response when p_1 or p_2 is increased by the same amount separately. In nonlinear equations, these interactions may be important to our ultimate use of sensitivity analysis. The second strategy (Fig. 9.2b) recognizes this fact and explicitly performs analysis using combinations of parameters. This approach avoids huge numbers of runs by restricting the range of values per parameter and by restricting the set of combinations.

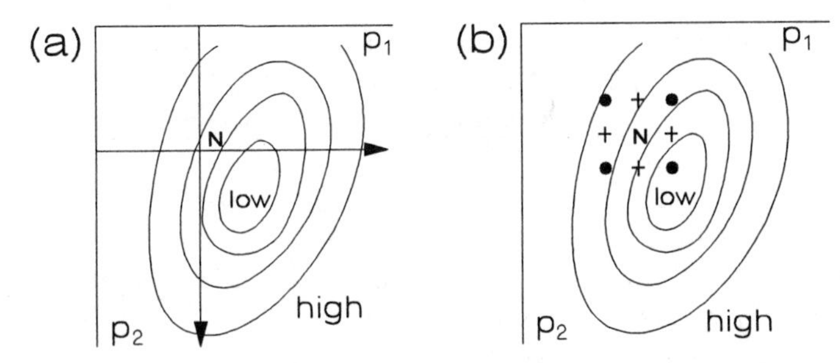

Figure 9.2: Two strategies for parameter sensitivity analyses. (a) Vary single parameters over a large domain and (b) vary multiple parameters over a small range. Contour lines are isoclines of model sensitivity. "N" represents nominal or best parameter values.

Single Parameter Sensitivity We characterize the sensitivity of a model with a simple index S that compares the change in model output relative to model response for a nominal set of parameters. In words, S is the ratio of the standardized change in model response (output) to the standardized change in parameter values (input)

$$S = \frac{\dfrac{R_a - R_n}{R_n}}{\dfrac{P_a - P_n}{P_n}}, \tag{9.1}$$

where R_a and R_n are model responses for altered and nominal parameters, respectively, and P_a and P_n are the altered and nominal parameters, respectively.

The question of which parameters and the degree of alteration to study is dependent on the objectives and the purposes of the sensitivity analysis. There are two strategies for determining the amount by which parameters are altered. In the *uniform* approach, all parameters are altered by the same percentage of the nominal values. Often, this is $\pm 10\%$, but values ranging from 2% and 20% are also used. The *variable* approach weights the altered interval by the variance of the parameter estimates, if this is known. This produces a more complex analysis since parameters will be altered by different amounts. It may, however, give a more accurate portrayal of real parameter variability.

As an example, suppose we are interested in the sensitivity of the density-independent growth equation at time $t = 10$ and $N_0 = 2.0$ and that the nominal parameter set is $r = 0.1$ and is altered by 20%. After running the model with both sets of parameters, we construct Table 9.1.

This table indicates that the model responds in the same direction as the parameter changes: positive changes produce increased output and negative changes produce decreased output. Numerically, the response is not linear: we do not observe a 20% change in the output. Also, parameter increases have a slightly greater effect on output than do identical

Table 9.1: Sensitivity of density-independent growth to r.

Parm	Nominal input	Nominal output	Altered input	Altered output	S
r^+	0.1	5.4	0.12	6.6	1.15
r^-	0.1	5.4	0.08	4.5	0.88

decreases in the parameter: parameter increases increased output by 15% and parameter decreases decreased output by 12%.

Normally, we are interested in more than a single parameter, so this table would have additional entries. Also, we are typically interested in more than one response variable, so sensitivities for these must also be computed. A useful technique for comparing these separate sensitivity analyses for different variables in the same model is the rank order of parameters from large to small sensitivity (Bartell et al. 1988). In addition, since we are performing only a single-parameter sensitivity analysis, we could examine a greater range in parameters (Fig. 9.2). A graphical presentation showing actual model response (not sensitivity, Fig. 9.3) can be more informative than Table 9.1.

Multiple Parameter Sensitivity Equation 9.1 works well for single-parameter changes, but it has problems when more than one parameter is altered from its nominal value. The numerator of Eq. 9.1 does not change, but we must replace the denominator by a distance measure that works in multiple dimensions. A reasonable choice would be the Euclidean distance:

$$d = \sqrt{(p_1 - p_1')^2 + (p_2 - p_2')^2},$$

where $p_1 (p_2)$ are the nominal values and $p_1' (p_2')$ are the altered values. Since d is always positive, we lose the ability to distinguish between positive and negative parameter changes in Eq. 9.1.

An alternative approach is based on a fractional factorial design (Shannon 1975; Steinhorst 1979; Swartzman and Kaluzny 1987). This approach treats parameter sensitivity analysis as if it were an experimental design for a statistical analysis of empirical data (ANOVA). The primary sensitivity index is not S, but the F statistic that is computed for analyses of

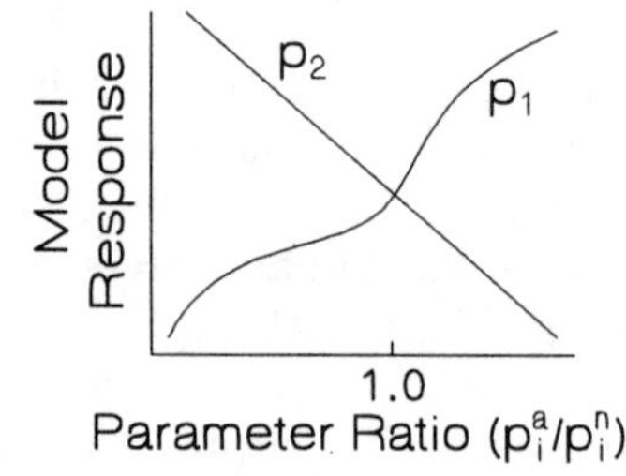

Figure 9.3: Model responses to relative changes in single parameters. The abscissa is the ratio of the altered parameter value (p_i^a) to the nominal parameter value (p_i^n).

variance. This is used only as a convenient index, and not as a variable for formal hypothesis testing, as it is in true ANOVA. Here, I only briefly sketch the approach and refer the reader to the literature.

A full factorial design is one in which experiments are performed for all possible combinations of levels and variables. We must be careful about what we mean by levels and variables in the context of parameter sensitivity. We are interested in the effects of increasing and decreasing parameters, so we treat the alterations of the parameters from the nominal values as the levels. Thus, with two parameters (variables) we would need four runs (experiments) corresponding to the circles in Fig. 9.2b. We do not need to perform the runs using parameters denoted by +, because we can calculate these knowing the responses at the corners. (We assume the surface is flat around the point "N".) We can also determine, from these experiments, interactions between the variables (e.g., response to p_1 is high at low p_2 and low when p_2 is high).

Thus, with this approach, we can gain much information based on relatively few experiments (simulation runs). Nevertheless, with many parameters it can require a large number of runs. For example, if there were three parameters (e.g., a, b, c) we would need $2^3 = 8$ runs. We can, however, distinguish the three main effects (due to the effects of a, b, and c), the three two-way interactions ($a \times b$, $a \times c$, and $b \times c$) and the single three-way interaction (a×b×c). For example, suppose we are investigating the effects of O_2, temperature, and relative humidity on plant photosynthesis. Suppose, further, that we wish to test O_2 at three levels, temperature at three levels, and relative humidity at two levels. The complete, full factorial design is a $3 \times 3 \times 2$ matrix of experiments. There would be 18 different experiments. This design permits us to test for significant heterogeneity among all of the main effects, all of the pairwise interactions, and the three-way interaction. The price we pay for this fine resolution is the number of experiments. We can eliminate some of the experiments, if we are willing to confound some of the effects with others. For example, we may be willing to assume that the three-way interaction is not significant. If so, we can perform a fractional factorial design in which we do not perform all of experiments, but we must be willing to make some assumptions about the statistical importance of some effects or interactions.

Steinhorst (1979) and Swartzman and Kaluzny (1987) suggest we use the same logic in order to reduce the number of sensitivity runs. In a large plankton (phytoplankton and zooplankton) simulation model, Swartzman and Kaluzny (1987) were interested in the sensitivity of five parameters varied at three levels. A full factorial would have required $2^5 = 64$ runs. Instead, they purposefully confounded main effects with high-order interactions. This allowed them to distinguish main effects and all pairwise interactions using just 16 runs, but they confounded three-way and four-way interactions with the main effects. They accepted this on the assump-

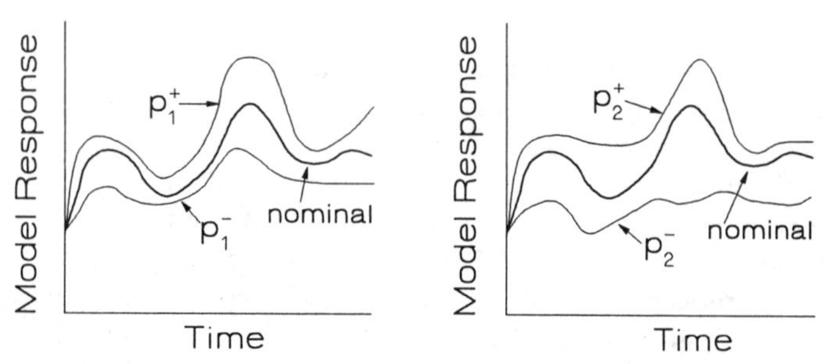

Figure 9.4: Dynamic sensitivity effects for combinations of parameters. Heavy lines are nominal model outputs, light lines represent model responses with altered parameters. p_i^+ and p_i^- are parameters that are increased and decreased (respectively) from the nominal parameters.

tion that these high-order interactions were unlikely to be important. The interested reader is referred to the original work for more details.

Dynamics of Sensitivity Regardless of the methods used to alter parameter values, it is important to remember that they produce dynamic changes in model responses. It is, therefore, useful to display the altered model behavior over time (Fig. 9.4). Tomovic (1963) described an analytical approach to dynamic sensitivity in which new differential equations for sensitivity of state variables to parameters are defined and solved in conjunction with the usual state variable equations. Such a graph can easily become too complicated to communicate important results, but one can limit the display to those combinations of parameters that produce large sensitivity. One can also graph the dynamics of the sensitivity index (e.g., Eq. 9.1) rather than the actual model response.

9.2.2 Error Analysis

Error analysis is similar to sensitivity analysis (many authors treat them as synonymous), but we will distinguish them here. Whereas sensitivity analysis is concerned with the effects of model response to small changes in the *mean* parameter values, I interpret error analysis to be concerned with changes of model response due to the *variance* of the parameter values. Before discussing practical methods for error analysis in simulation models, we must address the concept of *error propagation.*

Suppose we have a simple function $z = xy$ or $z = x/y$ in which there is uncertainty or error around the values of x and y. What is the resulting error around z? Before giving the answer, consider some general cases. In Fig. 9.5a, the errors around the mean $\overline{x}$ are *amplified.* That is, there is greater error around $\overline{z}$ than around $\overline{x}$. Error *compensation* (error reduction) is also possible (e.g., at $\overline{x}$ in Fig. 9.5b).

These examples illustrate that the propagation of error through a function evaluated at a point in the domain depends on the function and the evaluation point. A simple and elegant theory exists for calculating the errors around functions (Meyer 1975). The approach is based on the Tay-

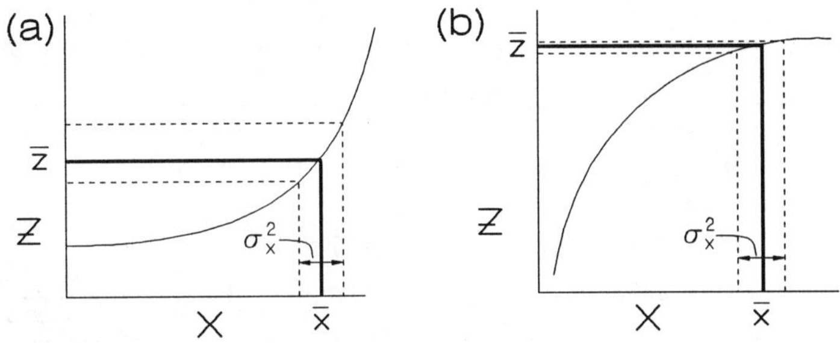

Figure 9.5: Error propagation in simple functions when there is error (σ^2) around the mean of the independent variable ($\overline{X}$). Depending on the function and the mean of the independent variable (X), the error may be *amplified* (a) or *compensated* (b).

lor series expansion of a function in the neighborhood of a point. The Taylor series is an infinite sum whose terms are progressively higher orders of derivatives of the function evaluated at the point. The Taylor series expansion of a single-valued function of x is

$$\begin{aligned} f(x) &= f(a) + \frac{\partial f(x)}{\partial x}(x-a) + \frac{\partial^2 f(x)}{\partial x^2}\frac{(x-a)^2}{2!} \\ &+ \frac{\partial^3 f(x)}{\partial x^3}\frac{(x-a)^3}{3!} + \cdots, \end{aligned} \tag{9.2}$$

where x lies within a small interval of a, and the partial derivatives are evaluated at the point a. There is also a multivariable form of the Taylor series for functions of more than one independent variables (see below).

It is impossible in practical calculations, of course, to use an infinite number of terms; so, the series is invariably truncated to relatively low orders of derivatives. The finite approximation of the series can be made exact by the inclusion of a remainder. These issues are typically dealt with in introductory calculus texts. In many applications, the series is truncated to include only the first-order derivatives and the remainder is ignored. For example, a function of three variables $f(x, y, z)$ has the first-order approximation:

$$\begin{aligned} f(x,y,z) &\approx f(a,b,c) + \frac{\partial f(x,y,x)}{\partial x}(x-a) + \frac{\partial f(x,y,z)}{\partial y}(y-b) \\ &+ \frac{\partial f(x,y,z)}{\partial z}(x-c) \end{aligned}$$

The first-order Taylor series is the approach we take for developing error propagation equations, the use of which we will call *analytical error analysis*. Suppose z is a function of two variables: $f(x, y)$ and we wish to approximate the variance of z given variance around x and y. The means of x and y are $\bar{x}$ and $\bar{y}$, respectively. By definition, $var(z) = (z - \bar{z})^2$, $\bar{z}$ is

Table 9.2: Variance formulae for simple functions with correlated and uncorrelated variables.

Function	Uncorrelated	Correlated
$z = x + y$	$\sigma_z^2 = \sigma_x^2 + \sigma_y^2$	$\sigma_z^2 = \sigma_x^2 + \sigma_y^2 + 2\sigma_{xy}$
$z = x - y$	$\sigma_z^2 = \sigma_x^2 - \sigma_y^2$	$\sigma_z^2 = \sigma_x^2 - \sigma_y^2 - 2\sigma_{xy}$
$z = xy$	$\sigma_z^2 = \bar{y}^2\sigma_x^2 + \bar{x}^2\sigma_y^2$	$\sigma_z^2 = \bar{y}^2\sigma_x^2 + \bar{x}^2\sigma_y^2 + 2\overline{xy}\sigma_{xy}$
$z = x/y$	$\sigma_z^2 = \bar{y}^{-2}\sigma_x^2 + \left(\frac{\bar{x}}{\bar{y}^2}\right)^2 \sigma_y^2$	$\sigma_z^2 = \bar{y}^{-2}\sigma_x^2 + \left(\frac{\bar{x}}{\bar{y}^2}\right)^2 \sigma_y^2 - 2\left(\frac{\bar{x}}{\bar{y}^3}\right)\sigma_{xy}$

the mean of z. The function is approximately

$$z \approx f(\bar{x}, \bar{y}) + \frac{\partial f}{\partial x}(x - \bar{x}) + \frac{\partial f}{\partial y}(y - \bar{y}),$$

where the partials are evaluated at the mean point $(\bar{x}, \bar{y})$. The expected value of $(x - \bar{x})$ and $(y - \bar{y})$ is 0. So, the expected value of the function is

$$\bar{z} \approx f(\bar{x}, \bar{y}).$$

The variance of z is the difference of z and $\bar{z}$, squared:

$$\begin{aligned}(z - \bar{z})^2 &= \left(f(\bar{x}, \bar{y}) + \frac{\partial f}{\partial x}(x - \bar{x}) + \frac{\partial f}{\partial y}(y - \bar{y}) - f(\bar{x}, \bar{y})\right)^2 \\ &= \left(\frac{\partial f}{\partial x}(x - \bar{x}) + \frac{\partial f}{\partial y}(y - \bar{y})\right) \\ &= \left(\frac{\partial f}{\partial x}\right)^2 (x - \bar{x})^2 + 2\frac{\partial f}{\partial x}\frac{\partial f}{\partial y}(x - \bar{x})(y - \bar{y}) + \left(\frac{\partial f}{\partial y}\right)^2 (y - \bar{y})^2.\end{aligned}$$

In general, for n variables

$$var(z) = (z - \bar{z})^2 \approx \sum_{j=1}^{n}\sum_{i=1}^{n} \frac{\partial f}{\partial x_i}\frac{\partial f}{\partial x_j}(x_i - \bar{x}_i)(x_j - \bar{x}_j) \qquad (9.3)$$

Note that the variance of x_i is $(x_i - \bar{x}_i)^2 = \sigma_i^2$ and that the *covariance* of x_i with x_j is $(x_i - \bar{x}_i)(x_j - \bar{x}_j) = \sigma_{ij}$ $(i \neq j)$. If the two independent variables are uncorrelated, then $\sigma_{ij} = 0$.

Using this general formula (Eq. 9.3), we can construct Table 9.2 when x and y are correlated and uncorrelated. This shows the variance in the dependent variable, given the variances and covariances in the independent variables. We apply the same logic in evaluating the errors of prediction when we are uncertain about one or more components of the predictive equation.

Analytical Error Analysis

Analytical error analysis uses the error propagation of functions, given that the model can be reduced to a single equation that predicts some quantity of interest. This is not possible for most biological models because the differential equations require numerical solution, but in some special cases these analytical solutions can be found and used with error analysis. A good example is by Reckhow (1979) who developed the following empirical model for phosphorus (P) loading in lakes given an input (L), mean depth (z), and mean residence time (τ, time required for 50% of lake volume to be removed)

$$P = \frac{L}{18z/(10+z) + 1.05\,(z/\tau)\,e^{0.012(z/\tau)}}.$$

By applying the first-order Taylor series, he calculated the variance of the prediction, given uncertainties in the parameters. Using this value, he calculated the total model error, S_T^2. This quantity permits us to make some very important and practical statements about the management of lake pollution. For instance, we can compute the 95% confidence interval around P. From this, we can compute the probability that a given target phosphorus level (e.g., minimum water quality standards) is acceptably close to our estimates of existing phosphorus levels. This allows us to couch pollution regulations and statements of violations in terms of probabilities.

To illustrate the role of the Taylor series approximation in performing this analysis, we will use a simpler problem. Suppose we wish to know the probability that a given population will go extinct. Certainly, if the population growth rate is negative, the population is doomed. But, one would think that if the environment is constant and the population is growing exponentially (unlimited resources), then the population has no chance of going extinct. Unfortunately, this is not the case because of *demographic stochasticity*. This is a form of random population growth that arises because populations are composed of individuals that have, in any given time interval, a certain probability of dying and of reproducing. These probabilities arise because of chance events conspiring to permit (or prohibit) individuals finding a mate, or to avoid (or not avoid) fatal interactions with predators. A small population with a positive growth rate can still go extinct if its individuals experience a sufficiently long string of bad luck in which no birth occurs and individuals die. The smaller the population, the more likely will extinction occur.

One simple model of the probability of exinction due to demographic stochasticity (Pielou 1977) is:

$$P = \left(\frac{d}{b}\right)^n, \tag{9.4}$$

where d is death rate, b is birth rate, and n is the initial population size. For example, if $d = 0.8$, $b = 0.9$, and $n = 10$, then $P = 0.31$. This very simple model assumes that the probability is not affected by density-dependent population growth nor by environmental stochasticity (e.g., catastrophic bad winters). For extinction models of the former situation, see Goodman (1987) and for the latter situation, see Mangel and Tier (1993, 1994). Nevertheless, this simple model permits us to address the important question: How certain are we that the calculated P is correct? Uncertainties of the true values of the parameters will propagate to create uncertainties of the predicted probability. We can apply Eq. 9.3 to Eq. 9.4 to estimate this prediction uncertainty. If the parameters are independent of each other,

$$\begin{aligned} var(P) &= \left(\frac{\bar{n}\bar{d}^{\bar{n}-1}}{\bar{b}^{\bar{n}}}\right)^2 \sigma_d^2 + \left(\frac{\bar{n}\bar{d}^{\bar{n}}\bar{b}^{\bar{n}-1}}{\bar{b}^{2\bar{n}}}\right)^2 \sigma_b^2 \\ &+ \left(\ln\left(\frac{\bar{d}}{\bar{b}}\right)\left(\frac{\bar{d}}{\bar{b}}\right)^{\bar{n}}\right)^2 \sigma_n^2, \end{aligned} \qquad (9.5)$$

where the three terms on the right-hand side are $\partial f/\partial d$, $\partial f/\partial b$, and $\partial f/\partial n$, respectively.

As illustration, suppose we have these values for means and standard deviations:

	d	b	n
mean	0.8	0.9	10
std. dev.	0.157	0.174	0.69

Then, the expected $P = 0.308$; the variance is $var(P) = 0.72$; and the standard deviation is 0.849. Assuming the error is normally distributed around the expected value, the 95% confidence intervals around the mean is

$$\mathrm{CI}_{\mathrm{lower}} = 0.308 - (1.96)(0.849) = -1.356$$

$$\mathrm{CI}_{\mathrm{upper}} = 0.308 + (1.96)(0.849) = 1.972$$

Obviously, the confidence intervals encompass the maximum and minimum values that P can have. With the uncertainties in the parameters indicated, we can not really say anything definitive about the expected probability of extinction of this population. All we can say is that the probability lies between 0 and 1.0, which is not terribly informative. Recall, however, that the formula used to calculate var(P) assumed that the parameters were uncorrelated. From Table 9.2 we note that the variances of quotients of correlated variables have the covariance subtracted. This effect could reduce the overall variance, if d and b are positively correlated.

Monte Carlo Error Analysis

Analytical approaches such as the above require a simple model in order to be performed: one that can be expanded by the Taylor series. Error analysis using Monte Carlo techniques (Chapter 10) can be applied to complex dynamic models and do not require extensive mathematical analysis. The method is to simulate repeatedly a system of equations using randomly selected parameter values. The output of each run is collected and statistically analyzed after all runs have been performed. The typical analysis is to display the frequency distributions of output (state) variables. Individual parameter values are selected from frequency distributions appropriate for each parameter; these may be theoretical distributions (e.g., the normal distribution) or empirical distributions obtained from replicated experiments.

We illustrate the method here using, not a dynamic simulation model, but the simple extinction model used above. For applications to dynamic ecological models, the reader should consult Gardner et al. (1980), O'Neill et al. (1980), Reckhow and Chapra (1983a), Bartell et al. (1986, 1988), or Summers et al. (1993).

Several practical problems arise when implementing a Monte Carlo analysis of error. Primary among these is the choice of the probability distribution from which to choose the parameters. If adequate data are available in the form of a distribution of values, then the empirical distribution can be used directly or the data can be fit to a theoretical distribution. If little data are available, analysis is more difficult. A variety of information might be available: the minima and maxima of the parameters, a statement of the mean and standard deviation, or estimates of the parameters of a probability distribution (e.g., dispersion and central tendency) that describe the distributions of the model parameters. Not all parameters in a model will be described with the same resolution. If the parameter distribution is unknown, another problem is to choose a distribution that is consistent with basic biological knowledge. In our example applied to the extinction model, all the parameters are positive, suggesting that a bounded distribution should be used. Further, we must calculate a ratio of parameters (d/b) which is restricted to be less than 1.0. So, not all combinations of values are appropriate, and this must be treated correctly in the simulation.

For the extinction model, the parameters were drawn from a log-normal distribution. The values listed in the above table describe the distributions. Parameter choices in which $d > b$ were rejected and new random parameters were drawn until $d < b$. The frequency and cumulative distributions are shown in Fig. 9.6. Note the discrepancy between the deterministic expectation and the mean of the simulated distribution. Also notice that the distribution is far from normal, contrary to the assumption of the above analytical error analysis. In Monte Carlo analysis, the 95% confidence intervals are determined directly from the cumulative distribution (Fig. 9.6) to be the range of values between 2.5% and 97.5% of the simulations. As a result, unlike the analytical analysis, the 95% confidence interval of the simulation lies between 0 and 1.0, but the interval is still very wide for these parameter values.

Figure 9.6b illustrates the effect of initial population size on the expected probability of extinction and the degree of uncertainty. Note that both the mean and uncertainty increases when population size is small.

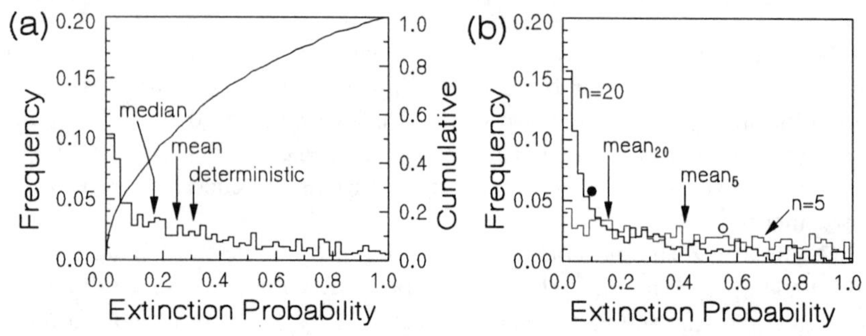

Figure 9.6: Distributions of extinction probabilities based on 1000 Monte Carlo replicates. (a) Frequency distribution and cumulative distribution of the probability of extinction for a population subject to demographic stochasticity and initial population size $n = 10$. Invalid parameter combinations were discarded. Monte Carlo descriptive statistics are indicated; the deterministic value is the probability computed using the mean parameter values. (b) The effect of changing initial population size: $n = 5, 20$. Arrows indicate Monte Carlo mean probabilities; filled and open circles are the deterministic means if $n = 20$ and $n = 5$, respectively.

These analyses are significant for all types of models because they force us to recognize the fallibility of deterministic models and to couch our predictions in terms of probabilities. From a philosophical perspective, error analysis is related to the Bayesian approach to validation (Section 8.2.5). By recognizing parameter uncertainty, we must address the issue of the probability distributions, and these, in effect, are one form of the prior probabilities needed in Bayesian analysis.

9.2.3 Aggregation Analysis

We mentioned earlier that model structure (i.e., the equations) was a source of uncertainty about which, basically, nothing could be done. That is not entirely true (O'Neill and Gardner 1979). One aspect of this problem is the number and nature of the state variables used in the model. We wish our models to be as simple as possible. One approach is to maintain a high degree of biological detail but curtail the extent of the system. For example, in ecosystem models, if our interest is the flow of energy, we are faced with a huge array of individual species that consume and process energy. We could reduce the complexity of the model by considering only one species (e.g., a species of tree). We might then be able to model energy flow through the individuals and population with great detail (e.g., differences between ages or sexes). But this sacrifices our ability to model interactions of the chosen species with other species in the system. So, another approach to simplification is to lump system variables together, for example, lump all trees together, as well as lump all herbivores and carnivores together. This strategy of lumping variables is known as "aggregation," and we want to estimate the errors that aggregation introduces into model output. Aggregating state variables is also one approach to scaling a model and is discussed in that context in Chapter 16.

In practice, all models are aggregated at some level of biological organization. Many models of human physiology at the level of the whole organism do not model individual cells. These are lumped into broad groups such as "tissues" or "organs." Similarly, models of cellular physiology do not model all biochemical pathways, but only those of interest. Other reactions are typically represented as loss or gain terms, that is, are aggregated into a single relationship. At the other extreme, ecosystem models do not represent each individual or even each species, but rather aggregate these into "functional" groups such as feeding guilds or trophic levels. Thus, an aggregated model is one in which the state variables of a more detailed model are lumped to form a subset.

Normally, in simplifying a model, we want the resulting dynamics of the simple model will be "similar", in some sense, to those of the original, complex model (Zeigler 1976). It is difficult, however, to adequately define the concept of *dynamic similarity* among structurally different models. Iwasa et al. (1987) developed an aggregation theory based on a restrictive definition. *Perfect aggregation*, by their definitions, is an aggregation that produces identical dynamics at each point of time considered. Obviously, the two models have different state variables, so cannot be directly compared. However, they assume a definite function that aggregates the values of the state variables of the detailed model to form quantities similar to those of the aggregated model. With this, we can solve or simulate the detailed model, apply the aggregation function to the results, and produce dynamics of the aggregated variables. We compare these dynamics with those produced directly by the aggregated model.

Using this concept, Iwasa et al. (1987) applied techniques from system control theory to a variety of ecological models to derive modeling conditions that must be satisfied in order for an aggregated model to reproduce the dynamics of the detailed model. While well-grounded mathematically, the results for many interesting ecological models are unfortunately restrictive. For example, suppose we have a density-dependent stage-structured model (see Chapter 13) in which there are n state variables that represent the numbers of individuals in different ages. To aggregate this model perfectly, we must combine variables. It is convenient to form a new model that uses two state variables that represent the juveniles and all of the remaining stages. Iwasa et al. (1987) proved that this aggregation was perfect if and only if fertility is proportional to body weight and net biomass increase is identical for each stage. Other relationships between these variables will not produce identical dynamics between the detailed and aggregated models. It is unlikely that these special relations will exactly occur in nature, and the concept of perfect aggregation will not be generally applicable. Nevertheless, it is mathematically rigorous and provides valuable bounds on the amount of error we can expect when aggregating models.

Additional progress has been made with a more relaxed attitude toward dynamic similarity. One such relaxation is that the equilibrium of the sum of the state variables of the detailed model should equal the equilibrium of the aggregated model. Applying this definition to linear models of two compartments, O'Neill and Rust (1979) showed that aggregated dynamics will be similar to detailed dynamics if the turnover rates of the two detailed state variables are equal. In particular, this latter condition will occur if the two output rates of the detailed compartments are equal. Cale et al. (1983) generalized this basic result to include any number of state variables in the detailed model and nonlinear growth terms exclusive of inputs and outputs.

While this set of models includes a large class of mass-balance models applicable to any level of biological organization, the final conclusion of these and other studies is that there will be few analytical tools to assess the amount of error that is made by our choice of state variables. This leaves us with Monte Carlo simulation of particular cases as the main tool to unravel errors that arise from lumping state variables. In a comprehensive study of 40 different models with varying arrangements of flows between compartments, Gardner et al. (1982) found that aggregation could produce errors of less than 10% even when turnover rates varied by more than three times. This suggests that within the set of ecological models considered, general patterns of dynamics are robust to errors in aggregation.

9.3 Analysis of Model Behavior

The model behavior that we have emphasized thus far is the dynamics that unfold from the initial conditions. These dynamics are often called the *transient* behavior. Many simple models, however, also have one or more *equilibria* in state space (points where rates of change are zero). It is useful to locate mathematically these points and explore their dependency on parameter values. In addition, it is interesting to know if the equilibrium dynamics will persist (i.e., will be *stable*) in the presence of small perturbations. We discuss both these topics in the next two sections.

9.3.1 Equilibria and Nullclines

A system of differential or difference equations is in equilibrium if the values of the state variables are not changing in time. Equilibrium analysis seeks to identify the values of all the equilibria. Here we do not distinguish, as do thermodynamicists and chemical engineers, between *steady state* and *equilibrium.*

Knowing the equilibria of a model is useful for several reasons. First, it characterizes the long-term behavior of the model by providing a set of algebraic equations that depend on the parameters and state variables. Second, knowing the location and number of equilibria for a model can help us interpret the transient dynamics that we observe from simulation. Third, the equilibria are the points at which we discuss the stability properties of the model (see below).

There are some difficulties and weaknesses of this analysis. First, we lose the dynamics that lead up to equilibria. Second, solving for equilibria in complex models may be difficult or impossible, except numerically. Third, there may be more than one equilibrium for any given model and if this number becomes large or dependent on many parameters, then our insight into system behavior is diminished. Last, not all models have simple equilibria as defined. Models with time-dependent driving variables (e.g., periodic changes in temperature) will likely not reach an equilibrium. Models with persistent cycles or complex, aperiodic behavior (e.g., chaos, Chapter 17) also do not reach constant dynamics.

Equilibrium analysis can be applied to both finite difference and differential equations. For simplicity, we discuss only the latter application. As a warm-up, consider the single state variable population model with density-dependent reproduction (the logistic equation):

$$\frac{dx}{dt} = rx(1 - x/K). \tag{9.6}$$

We wish to find the values of x at which the derivative is zero, which we will denote x^*. We proceed by setting the derivative to zero and solving for x^*

$$\begin{aligned} \frac{dx}{dt} = 0 &= rx^*(1 - x^*/K) && (9.7) \\ &= x^* - x^{*2}/K. && (9.8) \end{aligned}$$

This last equation shows that there are two equilibria which are the solutions to the second-order polynomial. There are several ways to determine the value of x^*. First, notice that Eq. 9.8 is a special case of a quadratic equation: $0 = C + Bx + Ax^2$ with $A = -1/K$, $B = 1$, and $C = 0$. Using the quadratic formula $[x_{1,2} = (-B \pm \sqrt{B^2 - 4AC})/2A]$ gives two roots: $x_1^* = 0$ and $x_2^* = K$ (the carrying capacity). This result accords nicely with the elementary textbooks. It says that if the population begins at 0 or K, it will remain at either of those two values forever. This analysis by itself does not assert that the long-term dynamics of the population will either be 0 when $x(0) = 0$ or K for any positive initial population size. For this, we need stability analysis (see below).

Equilibrium analysis is more interesting in cases with more than one state variable. The Lotka–Volterra predator–prey equations

$$\begin{aligned} \frac{dV}{dt} &= aV - bVP \\ \frac{dP}{dt} &= cbVP - dP \end{aligned}$$

are an easy example. Solving for the equilibria gives

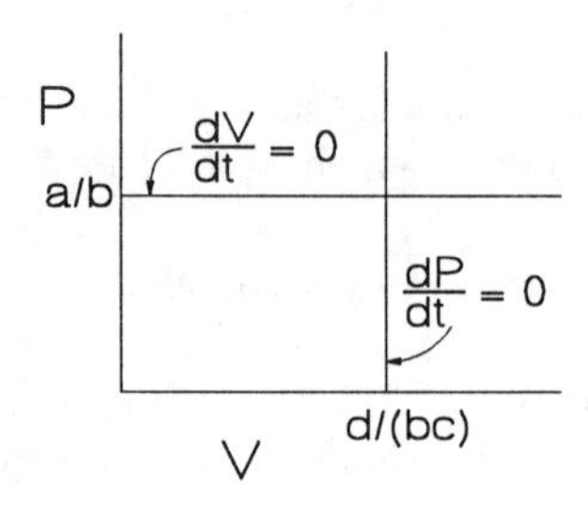

Figure 9.7: Lotka–Volterra predator–prey nullclines.

$$0 = aV^* - bV^*P^* \tag{9.9}$$

$$0 = cbV^*P^* - dP^*. \tag{9.10}$$

From Eqs. 9.9 and 9.10 we note that there are two equilibria: (1) both populations have zero values, and (2) both populations have nonzero values determined by the parameter values. Solving for V and P, we have

$$P^* = a/b \tag{9.11}$$

$$V^* = d/cb. \tag{9.12}$$

Notice that P^* and V^* depend only on parameter values: each is independent of its own value or of the other state variable. Note also that the relation between the value of the equilibria and the parameters are somewhat counterintuitive. For example, as b increases (the predators are more efficient at finding prey), the predator equilibrium numbers decrease. A bit of reflection should make it clear that as the predators become more efficient relative to the growth rate of the prey (a), they will drive down the prey population. This means there will be fewer prey to support the predator population, and its absolute growth rate ($cbVP$) will be reduced, producing lower predator numbers.

The equilibria tell us where in state space the system will not be changing, but they do not tell how the system will behave near the equilibria. We can learn more about this by plotting the *nullclines* (or *zero isoclines*) of the differential equations. The nullclines are the set of points in state space that satisfy the equilibria equations for *each* of the state variables. There is one nullcline for each state variable.

Since the two equilibria equations in the Lotka–Volterra equations are both constants, they are straight lines when plotted in state space (Fig. 9.7). $P^* = a/b$ is the nullcline equation for the victim population and $V^* = d/bc$ is the equation for the predator population. The entire system is at equilibrium where they intersect. If the system is started with initial Victim and Predator numbers corresponding to the intersection point, then no oscillations occur. If the system starts away from this point, the two state variables will oscillate around the equilibrium of the system. The amplitudes of the oscillations are larger when the system starts further from the equilibrium. (We will prove this in the next section.) Knowing the equilibria provides us with a tool for interpreting and guiding simulation exercises. For example, if we wish to examine system behavior in the vicinity of the equilibrium, then it is easy to calculate the appropriate initial conditions given the parameter values. Without this knowledge, we might spend a great deal of time doing short simulations searching for the equilibria associated with the parameters.

Although nullclines give important information about the locations of a model's equilibria, by themselves they say little about the behavior of the system near the nullclines and the equilibria. In particular, we would like to know whether a system near an equilibrium will move toward or further away from it when perturbed. Below, we discuss some tools that help us answer this question.

9.3.2 Stability

> *We must recall, however, that the existence of an unstable model for the solar system does not preclude the possibility that the Sun will rise every morning.* — Abraham and Marsden (1967)

Most systems of nonlinear differential equations cannot be solved analytically, but we can still obtain useful information by analyzing the behavior of the system in response to perturbations of the state variables. Such an analysis is called *stability analysis.*

Stability is a concept in dynamical systems that is subject to many interpretations (Innis 1975, Grimm et al. 1992). To some, it means no or little change in the rates of change (what we called equilibrium above). To others, it means persistent motion within a restricted region of state space. We will adopt the view common in most mathematical treatments. Intuitively, a system is stable following a perturbation of one or more of the state variables if the system returns to the specific *point* in state space or to a specific *orbit* (trajectory) in state space. The state space point of interest to stability analysis is invariably one of the equilibria, although, technically, it can be discussed relative to any point in the solution space.

In general, we are interested in the *global* response of the system to perturbations (i.e., where in state space the system will eventually be located). This is difficult for many nonlinear systems, and we usually are able to complete only a *local* (or *neighborhood*) analysis. Local stability analysis is a mathematical technique whereby, for a particular system of equations, a particular equilibrium is determined to be or not to be stable relative to *very small* perturbations. The analysis does not permit us to extrapolate to large perturbations. In particular, a system may be locally unstable but globally stable, but local analysis will not determine this. Before developing the techniques, we will briefly review possible dynamical responses to perturbations for a selected set of models.

A Menagerie of System Responses

Figure 9.8 illustrates eight possible responses to a perturbation. A single *linear* differential equation can only increase or decrease exponentially (Fig. 9.8a). The equation has an equilibrium only when the rate of increase is zero, and a perturbation will simply increase or decrease the state variable. Otherwise, if the rate of increase is nonzero, then a perturbation will produce continued growth or decline. If the equation is nonlinear, then the system may have a *stable equilibrium* (Fig. 9.8b) in which it returns to the equilibrium following perturbation. Conversely, the equilibrium of the nonlinear system may be *unstable* (Fig. 9.8c). *Multiple stable points* (or domains of attraction, Fig. 9.8d) are those in which small perturbations cause the system to return to the original equilibrium, but large perturbations may cause it to move far away, and become "trapped" in another domain of attraction. A stable equilibrium may show an oscillatory return (Fig. 9.8e). A nonlinear equation may show a *neutral limit cycle* (Fig. 9.8f): oscillations about an average that respond to a perturbation by simply moving further away from (perturbed away) or toward (perturbed toward) the average. A *stable limit cycle* (Fig. 9.8g) is a cycle in which the system returns to the cyclic trajectory following a perturbation. Finally, an *unstable limit cycle* (Fig. 9.8h) responds to a perturbation by moving far away from the cyclic trajectory.

Some of these behaviors will occur only when the system comprises more than one differential equation. In that case, the dynamics illustrated in Fig. 9.8 are best illustrated in the state space. Figure 9.9 shows some of the behaviors for systems with two state variables. Notice that in systems with saddle points (Fig. 9.9f), the "direction" of perturbation matters. This brief tour of system dynamics is not comprehensive; nonlinear systems can exhibit much stranger behavior than shown here, as discussed in Chapter 17.

Mathematical Analysis of Perturbations

We will proceed in two steps. First, we will examine the system behavior in the vicinity of a nullcline and illustrate how qualitative system dynamics will depend on parameter values and the position of the nullclines. Second, we will illustrate elementary mathematical analyses that quantitatively address stability. Since the mathematical analysis is based on linear equations, we will first develop the technique for linear models, then we will show how to convert a nonlinear model to a linear one so that our tools can be applied. If the reader finds the treatment presented here obscure, excellent introductions can be found in Vandermeer (1981) or Edelstein-Keshet (1988).

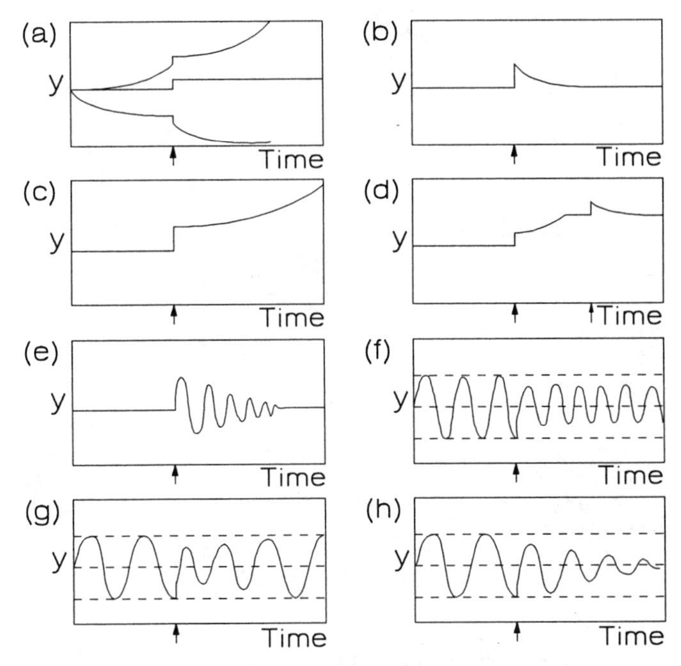

Figure 9.8: Possible responses of a single state variable to perturbations. Arrows indicate perturbations. (a) A single linear equation: neutrally stable if rate of increase is zero, unstable otherwise, (b) stable, nonlinear: system returns without oscillations, (c) unstable, nonlinear: system does not return, (d) two equilibria, nonlinear: system moves from unstable equilibrium to stable equilibrium, (e) stable, nonlinear: system returns with oscillations, (f) neutral, nonlinear limit cycle: system moves to another orbit, (g) stable, nonlinear limit cycle: system returns to original orbit, (h) unstable, nonlinear limit cycle: system no longer on a closed orbit.

Nullclines and Graphical Stability

Nullclines can give us a graphical and intuitive picture of stability analysis. Simple two-species competition is an easy example of this method. The basic differential equations and the resulting equilibrium equations are as follows:

$$\frac{dn_1}{dt} = r_1 n_1 \left(1.0 - \left(\frac{n_1 + \alpha n_2}{K_1}\right)\right) \tag{9.13}$$

$$\frac{dn_2}{dt} = r_2 n_2 \left(1.0 - \left(\frac{n_2 + \beta n_1}{K_2}\right)\right) \tag{9.14}$$

$$n_1^* = K_1 - \alpha n_2^* \tag{9.15}$$

$$n_2^* = K_2 - \beta n_1^*, \tag{9.16}$$

where i is the species index, r_i are the maximum intrinsic rates of increase, K_i are the capacities of the environment to support the species when growing alone, α is the effect of species 2 on species 1, andβ is the effect of species 1 on species 2.

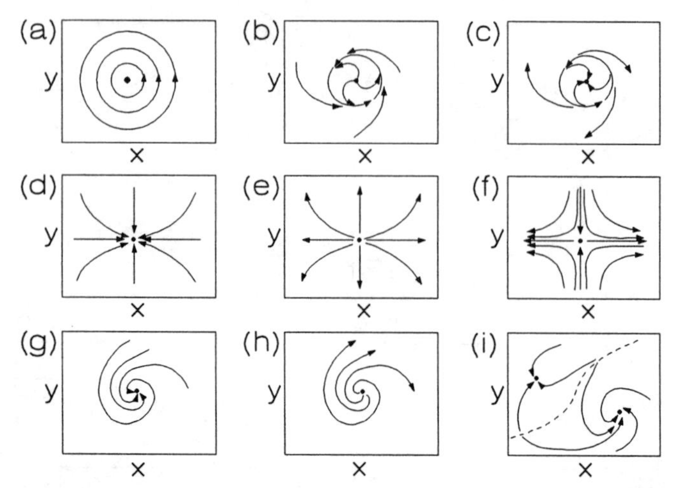

Figure 9.9: Possible responses of two state variables to perturbations. See Fig. 9.8. (a) Neutral limit cycle, (b) stable limit cycle, (c) unstable limit cycle, (d) asymptotically stable equilibrium, (e) unstable equilibrium, (f) saddle point, (g) stable equilibrium, (h) unstable equilibrium, and (i) multiple stable equilibria.

Notice that these equilibrium equations (Eq. 9.15–9.16) differ from those we have seen previously in that the right-hand sides depend on the equilibrium values of the state variables. We must think a bit about what these equations mean. Equations 9.15 and 9.16 are the set of points in state space at which n_1 and n_2 (respectively) are not changing. Each is a straight line in state space. Therefore, for all the points on the line, the associated state variable (species) is not changing, although the other variable may be changing. The point at which both lines intersect is the equilibrium for both species.

To understand the role of nullclines in stability analysis it is necessary to know the dynamics of points not on either line. We will develop the argument for n_1 only and leave the analysis of n_2 to the reader. Figure 9.10 shows the nullcline for n_1. At point A_1, the population is not changing in size. Point B_1 is directly below A_1 by an amount Δn_2. For clarity, let point A_1 be (n_{1A}, n_{2A}) and point B_1 be (n_{1B}, n_{2B}). So, the rate of change

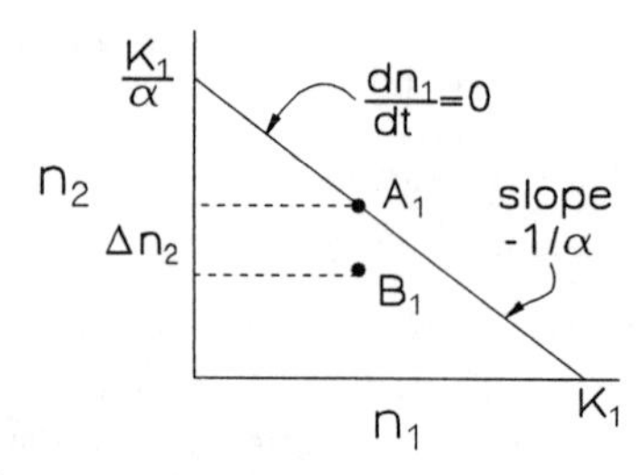

Figure 9.10: The n_1 isocline from the Gause competition equations.

of the population at A_1 is

$$\left(\frac{dn_1}{dt}\right)_{A_1} = r_1 n_{1A}\left(1.0 - \left(\frac{n_{1A} + \alpha n_{2A}}{K_1}\right)\right) = 0. \tag{9.17}$$

The rate of change of n_1 at B_1 is

$$\left(\frac{dn_1}{dt}\right)_{B_1} = r_1 n_{1A}\left(1.0 - \left(\frac{n_{1A} + \alpha(n_{2A} - \Delta n_2)}{K_1}\right)\right). \tag{9.18}$$

Rearranging Eq. 9.18 as

$$\begin{aligned}\left(\frac{dn_1}{dt}\right)_{B_1} &= r_1 n_{1A}\left(1.0 - \left(\frac{n_{1A} + \alpha n_{2A}}{K_1}\right) + \frac{\alpha \Delta n_2}{K_1}\right) \\ &= \left(\frac{dn_1}{dt}\right)_{A_1} + r_1 n_{1A}\frac{\alpha \Delta n_2}{K_1} \\ &= r_1 n_{1A}\frac{\alpha \Delta n_2}{K_1}.\end{aligned} \tag{9.19}$$

Thus,

$$\left(\frac{dn_1}{dt}\right)_{B_1} > \left[\left(\frac{dn_1}{dt}\right)_{A_1} = 0\right].$$

In other words, if population n_1 (or n_2) is below its nullcline, the population will increase. If they are above their nullclines, they will decrease. Note that the final equation of Eq. 9.19 is the rate of change of the population and that its magnitude depends on Δn_2: the larger the displacement from the isocline, the larger the rate of change.

Once we know how the system behaves in state space, we can qualitatively determine the stability of the equilibria. As Eqs. 9.15 and 9.16 indicate, the values of n_1^* and n_2^* depend on the parameters that determine the position of the nullcline. There are four possible orientations of the lines in space; these are illustrated in Fig. 9.11. The outcome of competition depends on the relationships of the parameters (i.e., which of the four cases holds) and on the initial numbers of the species. The arrows in Fig. 9.11 indicate direction and the approximate magnitude of subsequent time steps of change. Verify for yourself that the directions of the arrows are correctly drawn. This analysis shows that Case III is a stable equilibrium: after perturbations of the system away from the equilibrium value, the system returns. Case IV is an unstable equilibrium.

Linear Stability Analysis

Although visually compelling, the above analysis is only qualitative. A more rigorous approach is to examine the quantitative dynamics in the vicinity of the equilibria. For most nonlinear models, we cannot do this exactly, but we can for linear models. To address stability in nonlinear models, we must linearize the equations at the equilibria, then use the

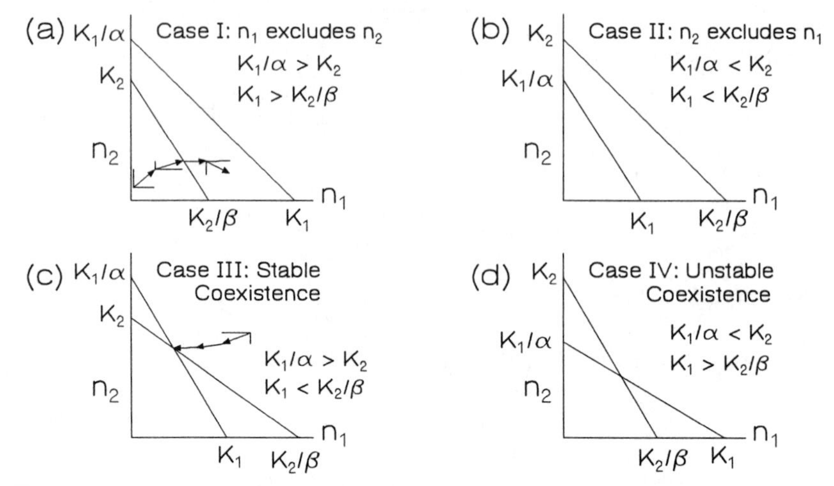

Figure 9.11: Four possible relationships between the two isoclines of the competition equations. Arrows indicate dynamics.

techniques that apply for linear equations. The linear approximation is valid only for small regions around the equilibrium.

To motivate the discussion, consider the standard linear model in population ecology, the density-independent growth equation:

$$\frac{dN}{dt} = rN \tag{9.20}$$

$$N_t = N_0 e^{rt}. \tag{9.21}$$

Equation 9.21 is the solution to the differential equation. From it, we can compute the future value of N for any t, once the initial condition and parameter are specified. We have discussed how the qualitative dynamics are controlled by the sign of r: $r < 0$ implies that the population decreases, $r > 0$ implies that the population increases. $r = 0$ is a special case where the population remains at its initial size (Fig. 9.8a).

To relate these facts to stability, suppose we have a population with $N = 0$ individuals (agreed: this is not terribly interesting from a biological point of view). Of course, this is an equilibrium point ($dN/dt = 0$). Now, suppose we perturb the equilibrium by adding one individual: Will the population return to the equilibrium or continue to move away? It depends on the value of r. If $r > 0$, the system will move away from the equilibrium and will be *unstable*. Otherwise, the system will be *stable*. If $r < 0$, the system will approach the equilibrium smoothly without oscillations. If $r = 0$, the system will not return to the equilibrium, nor will it move further away than the initial perturbation. This special case is called *neutral stability*. The important point is that the classification of systems as stable or not depends on the value of r, a single parameter that characterizes the overall dynamics.

In biological systems, we are rarely interested in a single state variable.

Characterizing the dynamics of linear models with two or more state variables is more complex, but conceptually identical to the logic just described for one state variable. We will find a solution to the differential equations and a quantity analogous to r from which we will determine stability.

A linear, two-state variable model is

$$\begin{aligned}\frac{dx}{dt} &= ax + by \\ \frac{dy}{dt} &= cx + dy.\end{aligned} \tag{9.22}$$

To get the solution, we use a technique that may appear suspect: we assume the answer is of a certain form. To biologists, this may seem as useful as rearranging the deck chairs on the Titanic. However, we can vindicate the approach if we show that differentiating the assumed solution gives us back the original differential equations. This is what we mean by a solution to a differential equation: if $f()$ is the integral, then $d[f()]/dx$ is the derivative. Proving this in our current special case is left as an exercise. In any case, we assume

$$x = Pe^{\lambda t} \qquad \text{and} \qquad y = Qe^{\lambda t}. \tag{9.23}$$

Notice the relation of λ in the above equation and r in Eq. 9.21. The behavior of the system depends on the value of λ, so we calculate it in the following way (Batschelet 1975).

Using Eq. 9.23, substitute the solution back into Eq. 9.22 and evaluate the derivative

$$\begin{aligned}\frac{d(Pe^{\lambda t})}{dt} &= \lambda Pe^{\lambda t} = a\left(Pe^{\lambda t}\right) + b\left(Qe^{\lambda t}\right) \\ \frac{d(Qe^{\lambda t})}{dt} &= \lambda Qe^{\lambda t} = c\left(Pe^{\lambda t}\right) + d\left(Qe^{\lambda t}\right).\end{aligned} \tag{9.24}$$

Next, we solve for λ. We do this in two ways: (1) using algebraic substitutions (for simplicity and clarity) and (2) using matrices (for generality). For the first method, we simplify and rearrange Eqs. 9.24 to form

$$\begin{aligned}\lambda P &= aP + bQ \\ \lambda Q &= cP + dQ.\end{aligned}$$

This leaves us with two equations and three unknowns, a condition that usually strikes fear in the hearts of practical biologists. Obviously, we cannot solve for P and Q without a defining relation on at least one of the unknowns. We can do this for λ by solving both of the above two equations for P/Q:

$$\frac{P}{Q} = -\frac{b}{a-\lambda} \tag{9.25}$$

$$\frac{P}{Q} = -\frac{d-\lambda}{c}. \tag{9.26}$$

Equating the right-hand sides of Eqs. 9.25 and 9.26 and simplifying, we obtain the second-degree polynomial

$$\lambda^2 - (a+d)\lambda + (ad - bc) = 0,$$

which determines λ.

This is the *characteristic* or *eigenvalue* equation associated with Eq. 9.22. It defines a relation between one of our unknowns (λ) and the two parameters of the model. The characteristic equation has two roots, λ_1 and λ_2, called the *eigenvalues*. A result from linear algebra allows us to assert that the general solution to $x(t)$ and $y(t)$ is the sum of expressions involving both eigenvalues:

$$\begin{aligned} x(t) &= P_1 e^{\lambda_1 t} + P_2 e^{\lambda_2 t} \\ y(t) &= Q_1 e^{\lambda_1 t} + Q_2 e^{\lambda_2 t}, \end{aligned} \qquad (9.27)$$

where P_i and Q_i (for each $i = 1, 2$) are products of a constant determined from initial conditions and a vector associated with each eigenvalue called the *eigenvector*. The eigenvector is a value for variable x and variable y that satisfies Eq. 9.23. Note that the overall dynamics will be dependent on both λ_i. This result indicates that the solution is a sum of exponential growth terms. Note the relationship between Eq. 9.27 and Eq. 9.21 and between λ_i and r.

Before discussing the properties of different values of the eigenvalues, we will rederive the characteristic equation using matrix algebra. We do this because this method generalizes more readily to systems of more than two equations, and it is a common formulation. We begin by noting that Eq. 9.22 can be written as a product of a matrix and a column vector

$$\begin{array}{ccccc} \begin{pmatrix} \dfrac{dx}{dt} \\ \dfrac{dy}{dt} \end{pmatrix} & = & \begin{pmatrix} a & b \\ c & d \end{pmatrix} & \begin{pmatrix} x \\ y \end{pmatrix} \\ \dot{\mathbf{x}} & = & \mathbf{A} & \mathbf{x}, \end{array}$$

where $\dot{\mathbf{x}}$ denotes the vector of derivatives with respect to time. Using the solution form assumed in Eq. 9.23 and noting that

$$\lambda P e^{\lambda t} = \lambda x \qquad \lambda Q e^{\lambda t} = \lambda y,$$

Eq. 9.24 can be written in matrix form:

$$\lambda \mathbf{x} = \mathbf{A}\mathbf{x},$$

or

$$\begin{aligned}\mathbf{Ax} - \lambda\mathbf{x} &= (\mathbf{A} - \lambda\mathbf{I})\mathbf{x} = \mathbf{0} \\ &= \left[\begin{pmatrix} a & b \\ c & d \end{pmatrix} - \begin{pmatrix} \lambda & 0 \\ 0 & \lambda \end{pmatrix}\right]\mathbf{x} = 0 \\ &= \begin{pmatrix} (a-\lambda) & b \\ c & (d-\lambda) \end{pmatrix}\mathbf{x} = 0,\end{aligned}$$

where $\mathbf{I}$ is the identity matrix and is defined as

$$\mathbf{I} = \begin{pmatrix} 1 & 0 \\ 0 & 1 \end{pmatrix}.$$

Another result from linear algebra is that we can solve for λ if the *determinant* of $(\mathbf{A} - \lambda\mathbf{I}) = \mathbf{0}$, or using the rules of evaluating two-dimensional determinants:

$$\begin{vmatrix} (a-\lambda) & b \\ c & (d-\lambda) \end{vmatrix} = 0 \tag{9.28}$$

$$(a-\lambda)(d-\lambda) - bc = 0 \tag{9.29}$$

$$\lambda^2 - (d+a)\lambda + (ad - bc) = 0, \tag{9.30}$$

which is our old friend the characteristic equation and, when solved, gives us the eigenvalues. This completes the second derivation of the characteristic equation.

Now we consider the system dynamics for some possible values for λ_i (Eq. 9.27). First, note that the λ_i, being the roots of a second-degree polynomial (Eq. 9.30), can be solved using the quadratic formula; this implies that the eigenvalues can be either real or complex numbers. If they are real, then the dynamics (Eq. 9.27) will be dominated by the largest of the two eigenvalues and will be *essentially* the same as in the single-variable case. If the largest eigenvalue is negative, the system is stable because the dynamics will be the sum of two negative exponentials. If the largest eigenvalue is positive, the system is unstable. More complex behavior is possible depending on the value of the second λ. For example, if $\lambda_2 < 0 < \lambda_1$, then the system is a saddle (hyperbolic) point (Fig. 9.9f). See Percival and Richards (1982) for more examples.

Things get interesting if the eigenvalues are complex numbers and are written as $z = \alpha + \beta i$, where α and β are constants. We must interpret the meaning of Eq. 9.27 when this is the case. It turns out that if the eigenvalues are *distinct* (not numerically identical) and if one of the eigenvalues is complex, then they both are complex and are *complex conjugates* of each other. That is, $\lambda_1 = \alpha + \beta i$ and $\lambda_2 = \alpha - \beta i$. This means that Eq. 9.27 is

$$x(t) = P_1 e^{\alpha t} e^{i\beta t} + P_2 e^{\alpha t} e^{-i\beta t} \tag{9.31}$$

$$y(t) = Q_1 e^{\alpha t} e^{i\beta t} + Q_2 e^{\alpha t} e^{-i\beta t}. \tag{9.32}$$

Furthermore, another result from linear algebra is that if the eigenvalues are complex conjugates of each other, then so are the eigenvectors. In particular, P_2 is the complex conjugate of P_1, and similarly for Q_2 and Q_1.

At first glance, Eq. 9.32 appears bizarre, producing imaginary dynamics, but an amazing result (called *Euler's formula*) from the analysis of infinite series states

$$e^{i\beta t} = \cos(\beta t) + i\sin(\beta t).$$

The solution for $x(t)$ and $y(t)$ becomes, after some more algebra,

$$x(t) = P_1' e^{\alpha t}\cos(\beta t) + P_2' e^{\alpha t}\sin(\beta t) \quad (9.33)$$

$$y(t) = Q_1' e^{\alpha t}\cos(\beta t) + Q_2' e^{\alpha t}\sin(\beta t), \quad (9.34)$$

where the primes indicate that these new constants are derived from our earlier Ps and Qs, but have different values. This is a remarkable result since it tells us that the long-term dynamics of a system of linear differential equations without a forcing function will be a sum of sines and cosines. Cycles can be produced by these simple models in two or more dimensions, whereas they could not be produced in systems with a single state variable.

Several special cases of the other constants have important consequences for stability. If $\alpha = 0$ and $\beta \neq 0$, the solution is a sum of a cosine and sine function with constant amplitude. Therefore, a perturbation of the equilibrium will cause undamped oscillations (neutral stability, Fig. 9.9a). If $\alpha > 0$, the oscillations grow exponentially and the solution is unstable. If $\alpha < 0$, the oscillations are damped and the solution is stable. Thus, by calculating the eigenvalues we can decide the stability of a set of linear equations in a manner analogous to the single-variable case.

Nonlinear Equations

The above analysis is wonderful if we have linear differential equations, which we almost never do in biology. Consequently, the last problem is to convert a nonlinear equation to a linear equation so that neighborhood stability analysis can be performed. Basically, we wish to define a new function of the *deviations* of the system following perturbation from the equilibrium. Suppose we have a system of differential equations in variables y_1, y_2, y_3, and so on. We further assume the system is at equilibrium y_1^*, y_2^*, y_3^*; by definition $dy_i^*/dt = 0$. Let $X_i = y_i - y_i^*$, the deviation of the system from its equilibrium; X_i^* is the origin when $y_i = y_i^*$. To show the linearization method for X_1, we begin by perturbing the equilibrium point by an amount x_1:

$$\frac{d(X_1^* + x_1)}{dt} = f(X_1^* + x_1), \quad (9.35)$$

where $f()$ is the model differential equation for y_1. Our problem is that we already know we cannot usually solve equations such as these when $f()$ is nonlinear, so we approximate the function with a *first-order Taylor series* (Section 9.2.2). To linearize a differential equation of a single variable, the

Taylor series approximation at the equilibrium is

$$\frac{d(X^* + x)}{dt} = \underbrace{f(X^*)}_{\text{zero-order}} + \underbrace{x\frac{\partial f}{\partial x}}_{\text{first-order}} , \tag{9.36}$$

where x is the perturbation from equilibrium. It is also true that

$$\frac{d(X^* + x)}{dt} = \frac{dX^*}{dt} + \frac{dx}{dt}.$$

Moreover, at the equilibrium both the rate of change of X^* and function $f(X^*)$ are zero, so we have the first-order approximation

$$\frac{dx}{dt} = x\frac{\partial f}{\partial x}. \tag{9.37}$$

When the function has two arguments x and y, the approximation is

$$\frac{dx}{dt} = x\frac{\partial f}{\partial x} + y\frac{\partial f}{\partial y}. \tag{9.38}$$

Since the derivatives are evaluated at X^*, they have a definite, single numerical value which is constant.

Given this extremely powerful tool for approximating nonlinear functions with linear ones, we can now apply the eigenvalue method to evaluate stability characteristics *in the local neighborhood of the equilibrium.* Since we have transformed the problem from studying dynamics of the state variables to studying dynamics of *deviations* from the equilibrium, the eigenvalue will tell us only about system behavior relative to the equilibrium. Bearing in mind that our approximation is valid only for small neighborhoods, if the deviations decrease, the system is locally stable; otherwise, the system is not locally stable.

To illustrate this for a typical model, suppose we have three differential equations from which we form the linear approximations to the deviations from equilibrium

$$\begin{aligned}
\frac{dx_1}{dt} &= f_1(x_1, x_2, x_3) \approx x_1\frac{\partial f_1}{\partial x_1} + x_2\frac{\partial f_1}{\partial x_2} + x_3\frac{\partial f_1}{\partial x_3} \\
\frac{dx_2}{dt} &= f_2(x_1, x_2, x_3) \approx x_1\frac{\partial f_2}{\partial x_1} + x_2\frac{\partial f_2}{\partial x_2} + x_3\frac{\partial f_2}{\partial x_3} \\
\frac{dx_3}{dt} &= f_3(x_1, x_2, x_3) \approx x_1\frac{\partial f_3}{\partial x_1} + x_2\frac{\partial f_3}{\partial x_2} + x_3\frac{\partial f_3}{\partial x_3}.
\end{aligned}$$

We can write this set of equations as a matrix

$$\dot{\mathbf{x}} = \mathbf{J}\mathbf{x}$$

$$= \begin{pmatrix} \frac{\partial f_1}{\partial x_1} & \frac{\partial f_1}{\partial x_2} & \frac{\partial f_1}{\partial x_3} \\ \frac{\partial f_2}{\partial x_1} & \frac{\partial f_2}{\partial x_2} & \frac{\partial f_2}{\partial x_3} \\ \frac{\partial f_3}{\partial x_1} & \frac{\partial f_3}{\partial x_2} & \frac{\partial f_3}{\partial x_3} \end{pmatrix} \begin{pmatrix} x_1 \\ x_2 \\ x_3 \end{pmatrix}$$

The 3×3 matrix ($\tilde{J}$) is called the *Jacobian* and represents the linear system which we will analyze for stability using the tools described above. This is the matrix $\tilde{A}$ we used earlier. In particular, we must calculate the roots of the characteristic equation which is obtained by substituting in parameter values and equilibrium values. Evaluating

$$|\mathbf{J} - \lambda\mathbf{I}| = \mathbf{0}$$

produces, in this case, a third-degree polynomial.

An example should make this useable. Every introductory ecology text claims that one of the possible dynamics associated with the Gause competition equations is *indeterminant exclusion*, or unstable coexistence. This condition depends on a particular choice of parameters and was illustrated in Fig. 9.11d. Below, we demonstrate that a particular parameter set which graphical, qualitative theory claims will be unstable is, indeed, quantitatively unstable according to neighborhood stability analysis. The Gause Eqs. 9.13 and 9.14 have the following Jacobian for deviations from the equilibrium defined by n_1^* and n_2^*:

$$\begin{aligned} \frac{\partial f_1}{\partial n_1} &= r_1 - \frac{2r_1n_1^*}{K_1} - \frac{r_1n_2^*\alpha}{K_1} & \frac{\partial f_1}{\partial n_2} &= \frac{-r_1n_1^*\alpha}{K_1} \\ \frac{\partial f_2}{\partial n_1} &= \frac{-r_2n_2^*\beta}{K_2} & \frac{\partial f_2}{\partial n_2} &= r_2 - \frac{2r_2n_2^*}{K_2} - \frac{r_2n_1^*\beta}{K_2}. \end{aligned} \tag{9.39}$$

Notice that the above Jacobian depends not only on the parameters but also on the equilibrium values of the state variables. To complete the analysis, we must evaluate the determinant of the Jacobian assuming a specific set of parameter values, for example,

r_1	α	K_1	r_2	β	K_2
0.05	0.2	200	0.05	6.0	1100 .

Using these parameters, there are four equilibria for species 1 and 2 (respectively): (0.0, 0.0), (200.0, 0.0), (0.0, 1100.0), and (100.0, 500.0) The latter pair of points represents the interesting equilibrium where both species are present. It is this one about which we wish to evaluate stability.

Using the equations for the elements of the Jacobian matrix (Eq. 9.39), we substitute the parameters and equilibrium values

$$\tilde{J} = \begin{pmatrix} -0.025 & -0.005 \\ -0.1363636 & -0.022772727 \end{pmatrix}.$$

Next, we construct the characteristic equation by evaluating

$$\begin{vmatrix} (-0.025-\lambda) & -0.005 \\ -0.1363636 & (-0.022772727-\lambda) \end{vmatrix} = 0$$

to get

$$\lambda^2 + 0.04772727\lambda - 0.000113636 = 0.$$

The roots of this polynomial are

$$\begin{aligned} \lambda_1 &= 0.0460379, \\ \lambda_2 &= -0.093765171. \end{aligned}$$

Since the largest eigenvalue is positive, we conclude the system is not stable. This accords with the classical, graphical interpretation of the parameter values and nullclines (Fig. 9.11d). Moreover, since λ_2 is negative, we have a saddle point (Fig. 9.9f), that is, a ridge along which the system will converge to the equilibrium. This ridge is sometimes called a *separatrix*, since it separates two domains of attraction with equilibria at K_1 and K_2.

Finally, for simple systems (i.e., five or fewer state variables) there is a short cut to stability analysis. As shown above, the sign of the largest λ determines the character of stability, and the sign depends on the roots of a polynomial (Eq. 9.30). The roots are determined completely from the coefficients of the polynomial contained in the elements of the Jacobian matrix. It is possible, therefore, to ascertain the sign of the eigenvalue simply by inspecting the constants of the matrix. These relationships have been codified in several stability criteria. Two of the more important of these are the criteria of *Routh* and *Hurwitz*. A complete description of these methods with solved problems is in DiStefano et al. (1967), but the clearest, most useful summary is in May (1973, p. 196). As an example of the method, consider a general characteristic equation for a system of m state variables

$$\lambda^m + a_1\lambda^{m-1} + a_2\lambda^{m-2} + \cdots + a_m = 0, \tag{9.40}$$

where a_i are the coefficients of the polynomial and are based on model parameters and state variable equilibrium values. A system of two state variables ($m = 2$) will be stable if and only if $a_1 > 0$ and $a_2 > 0$. If $m = 3$, then the system will be stable if and only if $a_1 > 0$, $a_3 > 0$, and $a_1a_2 > a_3$. With these rules, stability can be determined without actually having to find the roots of a polynomial. May (1973) lists the rules for $m = 1, \cdots, 5$.

Précis on Stability Analysis

Stability analysis is an elegant, but limited, tool. For most equations, we must settle for a local analysis, and it is often difficult to determine the relationship between the mathematical analysis and real-world disturbances. The analysis holds only for small neighborhoods around the equilibrium, so that the linear approximation of a system may indicate instability in a nonlinear system that has a stable limit cycle (Fig. 9.9b). If the analysis indicates local stability, then the system is generally "globally" stable for large regions of state space, but a locally unstable approximation may not be globally stable. Moreover, it is not always possible to find a closed form solution for the equilibria. Nullcline analysis addresses the same questions, but graphically. It has great heuristic power, but is difficult to perform for more than three state variables. Recent study of nonlinear equations with more complex behavior (e.g., chaotic) has developed alternative methods [e.g., bifurcation analysis, graphics, and Lyapunov exponents (analogous to λ)]. The interested reader should consult Chapter 17 and more advanced texts (e.g., Baker and Gollub 1990). Overall, stability analysis is one of several tools available for understanding model behavior to be used where appropriate.

9.4 Exercises

1. If the coefficient of variation for variable x is 0.2 and that for variable y is 0.35 and x and y are uncorrelated, calculate the mean and variance of the function $z = xy$ for the following pairs of x and y: (0,0), (3,3), (3,1).
2. Which of the functions in Table 9.2 show error compensation (error amplification)? Does it depend on the values of x and y?
3. Consider the disease model:

$$\frac{dN}{dt} = a - bN,$$

   where a is the rate of disease infection and b is the rate constant of cures. Write an equation for the equilibrium. Is the equilibrium stable?
4. Solve for the equilibria and nullclines for the yeast model of Chapter 4. Are the equilibria stable? Suppose sugar is continuously dripped into the system at a rate I. Find the equilibria of this system and determine if they are stable.
5. Draw the vectors of change in the four sectors of Fig. 9.7 as we did in Fig. 9.11. Prove algebraically that the directions are as you drew them.
6. Perform a local stability analysis on the population logistic equation: $dN/dt = rN(1 - N/K)$.
7. Show that n_1 decreases above the n_1 isocline as was done for n_2 in Fig. 9.10.
8. Use local stability analysis to show that Case III (Fig. 9.11) of the Gause competition model is stable. Use the parameters given in the text, except let $K_2 = 800$ and $\beta = 3$.

9. Fill in the algebraic steps between Eqs. 9.32 and 9.34. [Hint: let $P_1 = r + is$ and $P_2 = r - is$ and remember that $cos(-x) = cos(x)$ and $sin(-x) = -sin(x)$.]
10. Calculate var(z) for: x^3, $x^2 + y$, $x^2 + y^2$.
11. Write the error propagation equation for the extinction model (Eq. 9.4) assuming the parameters are correlated.
12. Given the algorithm for the Monte Carlo error analysis of the extinction model described in the text, is it legitimate to compare the simulated confidence interval with that of the analytical approach?
13. Perform an individual parameter perturbation sensitivity analysis on the extinction model (Eq. 9.4). For each parameter, repeat the analysis using three perturbation levels: 2%, 10%, and 20% of the mean. Rank the parameters by sensitivity. Compare your conclusions to the Monte Carlo error analysis.

Chapter 10

Stochastic Models

10.1 There's Nothing Like a Random World

Random errors could be assigned ... to the basic behavior of the system, but the value of doing so is questionable. ... [they] *are likely to produce only additional confusion. It is not the purpose of* [computer simulation] *to remind us of the normal state of affairs.*

— Walters and Bunnell (1971)

One man's mean is another man's Poisson. — JWH

In biology, it is always difficult to predict, especially the future. This difficulty increases as one moves outside the realm of tightly controlled biochemical and physiological systems to the behavior of whole organisms or to the dynamics of populations, ecosystems, or the global environmental system. One reason for this difficulty is that biological systems (like many other systems) are subject to apparently random fluctuations. That is, either the state variables themselves or the parameters are perturbed at random times and by random amounts. We will not discuss the philosophical problem of whether this is an *inherent* characteristic of biological systems or whether if we had complete information, the apparent randomness would disappear. The fact remains that our degree of certainty will only in special cases be sufficient to eliminate what appears to us as random changes. For all the other systems, we must acknowledge that predictions can be wrong simply because the real system (not the modeled one) is subject to unknown perturbations.

Repeatedly simulating random models allows us to estimate characteristics of the probabilistic model response (e.g., the distribution's dispersion and central tendency). This process is called *Monte Carlo* simulation. There are three broad areas in which probabilistic models and Monte Carlo simulation are useful in biological simulation:

1. *Statistical Hypotheses:* Sometimes we wish to test a null hypothe-

sis for which there is no easy equation to compute the test statistic. One example is the use of "null" models in biogeography. This field is frequently plagued by small sample sizes (sometimes $n = 1!$). For example, many biogeographers are interested in knowing if the occurrence of species on a set of islands in an archipelago is caused by competitive interactions between the species. The data consist of a single matrix of zeros and ones in which species are rows and islands are columns. If element a_{ij} is 1, then species i was found on island j. One approach to estimate the probability of this matrix is to generate a large number of random matrices. The frequency of occurrence of the matrix in question in this sample of random matrices estimates the probability. This can then be used in statistical tests. This type of application is known as resampling, and two popular methods for performing this test are *bootstrapping* and *jackknifing.* These topics are beyond the scope of this book, but introductions can be found in Noreen (1989) and Crowley (1992), where the basic methods applied to ecology and evolution are reviewed.

2. *Differential and Difference Equations:* The dynamics of continuous and discrete time systems can be made stochastic by randomly varying the parameters or state variables of the system. Monte Carlo simulation of these equations produces statistical distributions at different times in the solution. There are many situations where random effects can be important. For example, we may be primarily interested in a biological process such as individual growth which is affected by an abiotic factor such as temperature. We could construct a detailed and elaborate meteorological model that predicts temperature fluctuations from first principles. Or, we could simply assume that these fluctuations are drawn randomly from some probability distribution the defining parameters of which can be estimated from observations. In population dynamics models, we can adopt an even more abstract approach. To account for random changes in populations, we could build a model of population growth in which the birth and death of individuals in a small interval of time is a random process from some assumed probability distribution. In such an abstract model, we would not even have to represent probabilistically an external factor such as temperature. Finally, in models of animal movement, choice of direction for the next step could be the result of complex decisions based on the internal states of the individual. These may result in apparently random movements. However, we usually do not have access to the internal states, and so have no recourse but to approximate the movement process as a series of random choices.

3. *Markov Processes:* The dynamics of systems that can occur in only

a finite number of states (e.g., the letter grades students receive at the end of the term) can be modeled by assigning probabilities to transitions between the states. In this way, the system randomly walks through the states. Monte Carlo simulation is one method of estimating the probability that a system will, at some moment in time, be in a particular state.

When faced with random variation in systems, modelers have two fundamental choices to make. They can either ignore these random changes and model mean behavior, or they can incorporate randomness by constructing stochastic models and couching predictions in terms of probable outcomes. In this chapter, we will discuss the following topics: (1) the mechanics of generating and using computer-generated random numbers in simulation, (2) simulating stochastic differential equations, and (3) Markov chains.

10.2 Random Numbers

When casino croupiers spin a roulette wheel and roll the ball, they are using a physical device to draw a random number from a probability distribution. Gamblers have an interest not only in the particular number selected, but also in the underlying probability distribution. The distribution influences the average slot in which the ball comes to rest. Factors that influence the distribution of values include the number of slots, the distribution of value indicators (e.g., red or black, even or odd) in the slots, the balance of the wheel, the qualities of the axle and stopping mechanisms (e.g., age, rigidity, smoothness, etc.), gravity, and a host of other physical phenomena. Choice of a particular final resting slot at a particular time is influenced by the underlying distribution, of course, but also by the force exerted by the croupier on the wheel and the ball, the slot at which the ball is released, and the current physical conditions (e.g., atmospheric conditions) of the room.

The particular slot at which the ball finally comes to rest is essentially impossible to predict. This does not mean that roulette wheels violate physical laws of mechanics or thermodynamics or are somehow being dominated by the influences of quantum mechanical effects. The reason for unpredictability is that a large number of unknown physical events are interacting in complex ways. The combinations of events and interactions are so large that from the human perspective of limited knowledge, the outcome is unpredictable.

When we incorporate random numbers into computer programs, we are faced with the problem of using a digital device to mimic the outcome of physical phenomena. This is a real conundrum, since one of the most cherished characteristics of computers is that they are able to repeat calculations faithfully, that is, they are deterministic. We must, therefore, design algorithms that, although based on nonrandom mathematics, give the appearance of being random.

In roulette, we want the wheel to be fair in the sense that the ball has an equal chance to land in any slot. That is, if we had numbered the slots 1 to 72 (or whatever) and spun the wheel and ball many times, the frequency of trials in which the ball stopped in any particular slot would be 1/72 for all slots. Such a frequency distribution would be *uniform*: all slots have an equal probability of being chosen. In biological models, we do not always want such a simple distribution. At times, we want to select numbers (also called *deviates*) from normal, exponential, gamma, or other distributions. Or, we may wish to choose numbers from *empirical* distributions: those that are obtained from empirical observations and that may not be possible to describe using simple mathematical equations. Thus, our algorithms must work with any distribution. It turns out that for a large class of distributions, if we can generate random numbers from a uniform distribution, then it is a rather simple matter to use these numbers to obtain a deviate from the desired distribution.

10.2.1 Generating Uniform Random Numbers

When we use random numbers, we tend to need a lot of them, and so we are interested in generating *sequences* of numbers, all of which can be said to come from the same population (i.e., the same probability distribution with identical characteristics: mean, variance, etc.). This suggests methods that use *recursive* equations, so that the last number produced is used to calculate the next. To emphasize the deterministic origins of the numbers, we call them *pseudo-random numbers.*

Because these sequences need to be long, we put a premium on speedy algorithms. This means that the methods must use operations that are easy for the computer to perform. The methods, then, must rely as much as possible on integer arithmetic, and not on floating point operations. One such operation that lies at the heart of many algorithms is the `mod` or modulus arithmetic operation. `(y mod x)` produces the integer remainder obtained by dividing y by x.

To see how this operation can produce sequences that appear random, consider the following recursive function:

$$x_{n+1} = x_n^2 \bmod 31417.$$

If we begin with $x_0 = 123$, we generate the following sequence of numbers: 123, 15129, 13796, 5430, 15754, 25633, 26968, 891, 8456, 30261, 16822. This set illustrates several attributes of pseudo-random sequences. First, there is clearly no apparent pattern to this sequence; it is not obvious what number follows 16822. So, the remainder of a division (`mod`) of one moderately large number by another moderately large number does produce a sequence with little pattern. Second, although without pattern, the sequence is deterministic. If the starting point had been not 123 but 13796, the next number generated by this new sequence would be 5430. Also, if we repeated the sequence on a different occasion, the sequence would be the same. But equally significant, if we had started the sequence at 124, the sequence would have been quite different. Third, the sequence will eventually repeat: at some point we will again produce the number 123. All subsequent numbers will then be the same as those produced when we started from 123. Fourth, were we to continue the calculations until we had many thousands of numbers, we could apply statistical tests (e.g., goodness-of-fit) to determine if the population from which this sequence was drawn was indeed a uniform distribution. Statistical verification of the adequacy of a particular generating function is a surprisingly difficult task (Kleijnen and van Groenendaal 1992).

There are several critical characteristics of good algorithms. (1) They should produce long sequences before repeating. (2) They should be fast. (3) They should reproduce the major components of the desired distribution (mean, variance, skew, distribution at the tails, etc.).

Almost all modern compilers provide a built-in function that returns a random number from a uniform distribution. In `C`, this function is called `int rand(void)`. Although it varies among compiler manufacturers, the *linear congruential* method is most commonly used. It is the recursive function:

$$U_{i+1} = (aU_i + c) \bmod m, \tag{10.1}$$

where a, c, and m are machine-dependent constants chosen to produce a good fit to a uniform distribution. For example, on an IBM mainframe computer, $a = 314{,}159{,}269$, $c = 453{,}806{,}245$, and $m = 2^{31}$. If $c = 0$, it is a *multiplicative congruential* method. Modern implementations now frequently use $m = 2^{32} - 1$. For most compilers in which the longest integer is 32 bits, the period is close to $2^{32} \approx 4 \times 10^9$. As Press et al. (1992) point out, this is not the best method, and they define some alternatives that do not depend on machine-specific parameter values. One of these is a *shuffling* approach that gives a period of about 2×10^{18}, which is a number even larger than the national debt in pennies. They also provide, in a passage (p. 276 ff) notable for its entertainment value, many cautions and much good advice on using vendor-supplied pseudo-random generators.

In C, `rand()` produces a sequence of integers that ranges between 0 and $2^{31} - 1$. If we desire random numbers to vary between 0 and 1.0, we simply divide (using floating point operations) the random integer by the largest integer represented by the compiler (e.g., $2^{31} - 1$). Since it is a recursive function, the initial random seed (U_0 in Eq. 10.1) must be supplied before a sequence can be produced. The `C` function `void srand(unsigned int)` takes a user-supplied seed as an argument and initializes the generator used by `rand()`. On some compilers, other functions [e.g., `randomize()`]

initialize the generator with a system-supplied seed, usually derived from the current system clock. It is good practice, however, to treat the seed as any other simulation parameter and read and write it along with other data needed to initialize a simulation run. This will be crucial for debugging code when it is necessary to duplicate exactly the conditions of a run.

10.2.2 Generating Normal Deviates

Once we have a method for generating random numbers from the uniform distribution, we have the basic tool for obtaining numbers from virtually any other distribution we wish. As we will see below, there are some standard methods for generating equations that sample from nonuniform distributions. One that is especially effective is the inverse function of the cumulative distribution. Unfortunately, this method does not work on one of the most important distributions: the normal. Consequently, many other algorithms have been developed for this special distribution. One of the best of these is the Box–Muller method. This approach involves combining two uniform random numbers [U_1, U_2, obtained from 2 separate calls to the uniform generator `rand()`] to produce two random numbers from a normal distribution (z_1, z_2) having a mean of 0 and a standard deviation of 1.0 [i.e., $N(0,1)$]:

$$z_1 = \sqrt{(-2\ln(U_1))}\cos(2\pi U_2)$$

$$z_2 = \sqrt{(-2\ln(U_1))}\sin(2\pi U_2).$$

Bratley et al. (1987) caution that the Box–Muller method in combination with a linear congruential uniform generator produces correlated pairs of normal deviates; so, they too, recommend using a more complicated uniform generator.

To convert a standardized random deviate (z_i, above) to a deviate from another normal distribution with standard deviation s and mean m, use $y = sz + m$.

10.2.3 Inverse Cumulative Methods

A very powerful and general procedure for generating formulae for sampling from distributions is to use the inverse of the cumulative distribution. The conceptual basis of this approach can be illustrated by applying it to the problem of sampling from an empirical distribution.

Suppose we wish to use as a random driving variable the set of temperature values measured at some place. Our data might appear as in Fig. 10.1a: number of days at which a given category of temperatures occurred. From this we can create the relative frequency distribution or the *probability density* (Fig. 10.1b, the fraction of all days in the sample at the given temperature). This is the distribution of temperatures from which

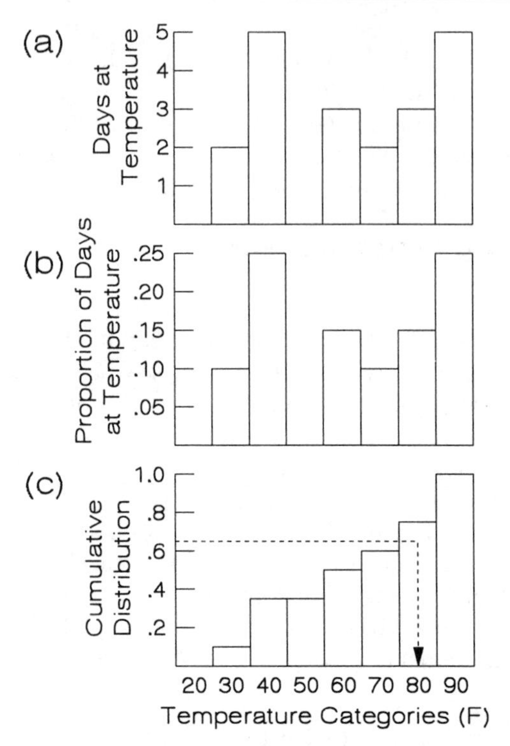

Figure 10.1: Frequency distributions of observed temperatures: (a) raw frequencies, (b) relative frequencies, and (c) cumulative distribution. The arrow indicates the random temperature generated after selecting a random uniform number 0.65.

we wish to sample. To put it another way: if we used some particular algorithm to generate many different temperatures, we would like the final distribution of those simulated temperatures to resemble this distribution. To accomplish this, we form the cumulative distribution (Fig. 10.1c), which is the sum of the relative frequencies from the lowest temperature category to the highest temperature category.

The y-axis of the cumulative distribution ranges from 0.0 to 1.0, and the difference in height between two adjacent bars is equal to the height of the relative frequency distribution at that category (Fig. 10.1b). Missing categories (0 relative frequency) become a bar of the same height as the category to the left (e.g., 50° F in Fig. 10.1c).

To sample from the distribution, first obtain a random number between 0.0 and 1.0 from a uniform random number generator. Interpret this as a point on the y-axis, and follow a line to the right until a histogram bin is encountered. The random deviate is the mid-point of the bin. The arrow in Fig. 10.1c shows the sequence. Categories that were very frequent in the original data have tall histogram bars in the cumulative distribution, and relatively many uniform random points will intersect this bar. Categories for which there were no observations in the original data are never sampled by this method. The only random number generator we need is the uniform generator supplied with the compiler. Notice that the initial width of the categories determines the resolution of the deviate generated. In the temperature example shown, we can generate temperatures only at 10° intervals.

The same method can be applied to standard probability functions if their cumulative distributions can be algebraically inverted. We illustrate the method by deriving an equation to sample random turning angles for moving insects using the wrapped Cauchy distribution. This distribution is roughly shaped like the normal distribution, but has a thicker distribution at the tails and is constrained to values between $\pm\pi$ (since these are the bounds on movement angles). The probability density function (pdf) of the wrapped Cauchy distribution is (Batschelet 1979)

$$f(\phi) = \frac{1-\rho^2}{2\pi[1+\rho^2-2\rho\cos(\phi-\theta)]}, \tag{10.2}$$

where ϕ is the angle in radians, ρ is the measure of concentration, and θ is the mean angle in radians. ρ and θ are the parameters of the distribution. Note that when $\rho = 0$, $f(\phi) = 1/2\pi$: a uniform distribution over $\pm\pi$.

The cumulative distribution function (cdf) is the integral of the pdf (Hodgman et al. 1955)

$$\begin{aligned} F(\omega) &= \frac{1-\rho^2}{2\pi}\int \frac{dx}{1+\rho^2-2\rho\cos(\omega)} \\ &= \frac{1}{\pi}\arctan\left(\frac{1+\rho}{1-\rho}\tan\left[\frac{\omega}{2}\right]\right) + C, \end{aligned} \tag{10.3}$$

where ω is $(\phi-\theta)$ (the deviation from the mean) and C is the constant of integration. Since $f(\phi)$ is symmetric and unimodal, $F(0) = C = 0.5$.

To obtain a formula to sample from $f(\theta)$ (Eq. 10.2), note that $F(\omega)$ varies from 0.0 to 1.0. Replace this value with a uniform deviate $[U(0,1)]$ and solve for the desired Cauchy deviate (ϕ) by inverting Eq. 10.3 and solving for ϕ:

$$F^{-1}(\omega) = \phi = \theta + 2\arctan\left(\frac{1-\rho}{1+\rho}\tan\Big(\pi(U(0,1)-0.5)\Big)\right). \tag{10.4}$$

We do not use this approach for the normal distribution because it does not have an equation for the cumulative distribution that can be inverted. However, even if the distribution does not have an inverse that we can write as a single equation (as above), we can still use the inverse method on theoretical (nonempirical) distributions. Simply create a discrete form of the pdf by discretizing the categories (as was done with the temperature categories), then form the discretized cumulative distribution (as if it were an empirical distribution) and apply the same method for determining the histogram bin that corresponds to the random point on the y-axis.

This approach works for the normal distribution, but since there are better approximate methods such as Box–Muller, it is not used.

10.2.4 Methods for Other Distributions

The inverse cumulative and rejection methods are general approaches that apply to many common distributions. However, efficient specialized algorithms for most of the standard distributions have already been designed. Some can be found in numerical software packages [e.g., the International Mathematical and Statistical Library (IMSL), Numerical Algorithm Group (NAG), Mathematica, etc.] or in more advanced texts (e.g., Bratley et al. 1987; Kleijnen and van Groenendaal 1992). These include, for example, the Cauchy, log-normal, exponential, gamma, F, and Weibull continuous distributions, and the binomial, Poisson, hypergeometric, and negative binomial discrete distributions.

10.2.5 Multivariate Distributions

Often we wish to use deviates for several random variables (e.g., population birth and death rates to calculate the probability of extinction, Chapter 9). If the variables are uncorrelated (i.e., do not *covary*), then we can simply generate them independently and use them separately as described above. If they are correlated, then the distribution is multivariate and we can not draw the deviates independently. The method to use depends on the underlying distribution. Here, we illustrate the approach for the multivariate normal distribution. Other distributions will require different methods.

When variables covary, it means that not all possible combinations of the variables are equally likely. If x and y are negatively correlated, then pairs in which x is large and y is large will be relatively uncommon. The degree of correlation between the variables is measured by the covariance of x with y. Moreover, in a sense, the degree that x is correlated with itself is measured by the variance of x. Consequently, the sampling distribution of a function is portrayed by its *variance–covariance* matrix (Chapter 8). This square matrix must be considered when drawing deviates from a multivariate distribution.

If the distribution of n variates is normal, then the following algorithm returns a deviate for each of the variables. (1) Select n deviates ($\tilde{z}$) from the standard normal distribution using the Box–Muller method (or equivalent). (2) Convert the n standard deviates into physical deviates with the relation $\mathbf{y} = \mathbf{m} + \mathbf{Sz}$, where $\mathbf{m}$ is the vector of variable means and $\mathbf{S}$ is a square matrix derived from the variance–covariance matrix ($\mathbf{V}$). $\mathbf{S}$ plays a role analogous to the standard deviation when using univariate normal distributions, but includes factors for the covariance of the variables. The following relationship holds:

$$\mathbf{V} = \mathbf{SS}'$$

When $n > 2$, we use software to generate the Cholesky decomposition to obtain $\mathbf{S}$. When $n = 2$, we can easily do it by hand as follows. From the defining relation, we have

$$\mathbf{V} = \mathbf{SS}'$$

$$\begin{pmatrix} \sigma_1^2 & \sigma_{12} \\ \sigma_{21} & \sigma_2^2 \end{pmatrix} = \begin{pmatrix} s_{11} & 0 \\ s_{21} & s_{22} \end{pmatrix} \begin{pmatrix} s_{11} & s_{21} \\ 0 & s_{22} \end{pmatrix},$$

where the $\mathbf{S}$ matrix on the right is the transpose of the $\mathbf{S}$ matrix on the left. From these we can derive, using the rules of matrix multiplication:

$$\begin{aligned} s_{11} &= \sqrt{\sigma_1^2} \\ s_{21} &= \sigma_{12}/\sqrt{\sigma_1^2} \\ s_{12} &= 0 \\ s_{22} &= \sqrt{\sigma_2^2 - (\sigma_{12}^2/\sigma_1^2)}. \end{aligned}$$

where σ_i^2 is the variance of the ith variable and $\sigma_{ij} = \sigma_{ji}$ is the covariance of i with j. With $\mathbf{S}$ defined as above, we can convert the n standard normal deviates into deviates of each of the needed variables $\mathbf{y}$.

10.3 Applications to Differential Equations

Stochastic differential equations (SDEs), like partial differential equations, use mathematics that is very difficult for most biologists. There are enough counterintuitive and just plain confusing aspects associated with modeling and simulating these equations that it is best to seek the advice and consent of a bona fide mathematician who specializes in this area. Nevertheless, having issued this caveat, we will now naively proceed to discuss how to do it!

Randomness may be implemented in differential equation models in the initial conditions, driving variables, parameters, or on the state variables directly. Making state variables random is not common, as it is always possible to achieve the same effect by randomizing the rates (e.g., through the parameters). The most common application is to randomize driving variables and parameters.

There are two different concepts of biological stochasticity: *environmental* and *demographic* stochasticity. These concepts have been discussed under these names primarily in ecology, but they apply to other areas of biology as well. Environmental stochasticity refers to random variation in systems modeled as populations or compartments. For example, we may have random variation in the per capita growth rate of a population or in the rate constants of a chemical reaction. The dynamic variables (e.g., populations, chemical concentration) are continuous quantities, and environmental stochasticity alters these continuous variables randomly.

Demographic stochasticity refers to random variation in the occurrence of events affecting the state of an individual. For example, an ecological population can be viewed as being composed of an integer number of individuals that undergo at least two important processes: birth and death. We can model random variation in an individual's state by assuming there are probabilities associated with an individual giving birth or dying within some small, finite time interval. For example, if the organisms in question give birth to only one offspring at a time, we might assume that the probability of having one offspring in Δt is r, and that the probability of no offspring is $1 - r$. We take a similar approach to mortality. These probabilities may ultimately be caused by chance encounters with a fertile mate or a predator. The important point is that the biological process affecting individuals either occurs, or not, according to random events. The concept of demographic stochasticity can be generalized to any particle-based system where the interest is in the discrete states of individual particles. For example, cancer cells are known to reverse their evolved resistance to chemotherapeutic drugs. This phenomenon was modeled by Kimmel and Stivers (1994) as a *branching random walk* in which the life span of cells and the numbers of gene copies in the progeny were demographically stochastic in our terminology.

To incorporate stochastic events in parameter values, we use a differential equation in which the parameters at time t are affected by random deviates from some distribution (e.g., normal). For example, a stochastic density-independent population model might be

$$\frac{dX}{dt} = r_t[N(\mu, \sigma^2)]X, \tag{10.5}$$

where $r_t[N(\mu, \sigma^2)]$ means that r is a random deviate from a normal distribution with mean μ and variance σ^2. Thus, r_t is no longer a constant, but changes randomly in time. Ludwig (1974) surveys other simple stochastic population models.

The first thing we notice is that Eq. 10.5 incorporates randomness *additively*. r_t can be written as the mean of r plus a random deviate with mean 0 [i.e., $r_t = \overline{r} + N(0, \sigma^2)$]. Alternatively, we can incorporate randomness *multiplicatively*: $r_t = \overline{r}[1 + N(0, \sigma^2)]$. In this model, we are adding a random fraction of $\overline{r}$ to itself. In at least one application to questions in community ecology, the results depend on which formulation is used (Turelli 1981). The two models make different assumptions about how randomness affects the system, but in the absence of discriminating experiments, it is not clear which form to favor in any given case.

Problems of analytical solutions aside, repeated simulation of these equations involves the following steps.

1. Determine the probability distribution to use for the parameter and estimate the descriptive statistics (mean and variance).

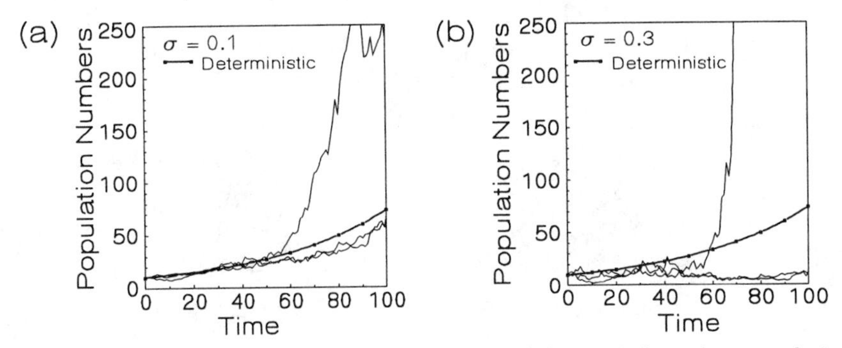

Figure 10.2: Two sets of three random sequences of density-independent population growth using additive normal variation of r. (a) Standard deviation of $r = 0.1$, (b) standard deviation of $r = 0.3$.

2. Inside the simulation loop, sample the distribution and use the resulting random deviate as the parameter value (e.g., r_t) in the differential equation.
3. Save the resulting dynamics in an array for post-simulation statistical analysis.
4. Repeat steps 2 and 3 for a large number of times to obtain a set of *Monte Carlo replicates* on which to do statistics. The size of "large" depends on the question being addressed, the underlying variability of the biological process, and the amount of time and money available to answer the question. (Monte Carlo simulations can require a great deal of computer and modeler time.)
5. Perform statistical analysis on the resulting random dynamics.

Figure 10.2 shows two sets of three random sequences of random population numbers based on the density-independent model where the standard deviation of the intrinsic rate of increase r is 0.1 (Fig. 10.2a) and 0.3 (Fig. 10.2b). To standardize the comparison, the two sets of sequences of random numbers used identical random seeds.

In analyzing random system dynamics, we can ask several different questions. First, we might ask: What is the nature of the state variable values of a single system subject to environmental stochasticity? To address this using computer simulation, we would simulate a single system and collect the variable values over a long period of time, and then statistically analyze these values for their mean, median, variance, and distribution. Second, we could ask: What is the nature of the statistical distribution of an *ensemble* of systems, where each is subject to similar environmental stochasticity? To address this, we would make multiple simulations each with random dynamics and collect the ensemble averages at a sequence of points in time. The above two questions are quite different. A third type of question is: What are the stability properties of the random dynamics? The mathematical analysis of this question is much more difficult and problematical

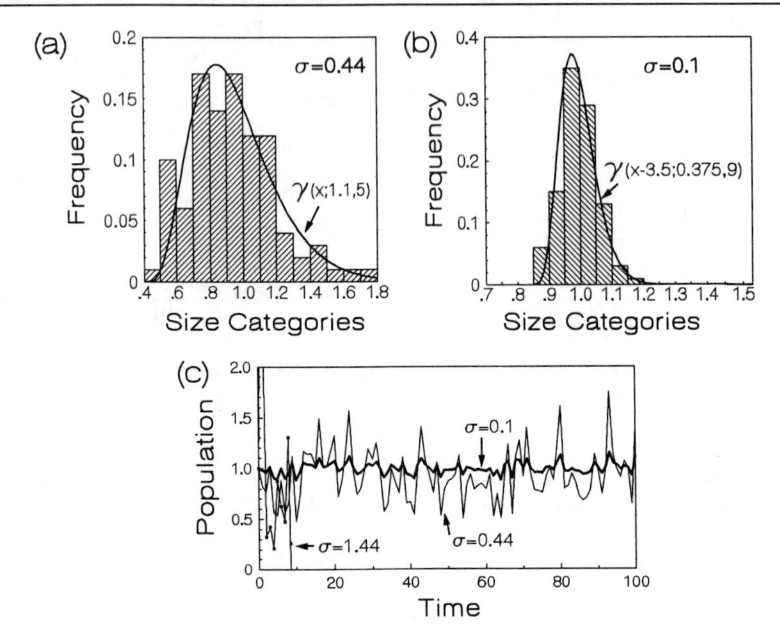

Figure 10.3: Three realizations of Eq. 10.6 using three different standard deviations of variation around K. The histograms of population size for 100 time steps are shown for (a) $\sigma = 0.44$ and (b) $\sigma = 0.1$. $K = 1.0$. Superimposed are curves for the gamma distribution with parameters fitted by eye. In (c) are shown the dynamics corresponding to the above and $\sigma = 1.44$, which causes early population extinction.

than the analogous question for deterministic systems (Chapter 9).

To illustrate these concepts, we consider an example analyzed by May (1973). We make the logistic equation stochastic as follows. The deterministic form is

$$\frac{dV}{dt} = rV\left(1 - \frac{V}{K}\right).$$

For simplicity, we eliminate r by dividing both sides by r/K:

$$\frac{dV}{dt(r/K)} = V(K - V).$$

Defining a new time variable $\tau = t(r/K)$ and an additive model of stochastic variation in K, the final equation is

$$\frac{dV}{d\tau} = V(K + N(0, \sigma^2) - V). \tag{10.6}$$

For this analysis, May (1973) was interested in the statistical properties of a single population. Three examples with different degrees of variation are shown in Fig. 10.3. When $\sigma = 1.44$, the variation is so great that the population is driven to extinction. In the cases when the populations persist (e.g., $\sigma = 0.44, 0.1$), the frequency of population sizes fits a gamma distribution (Fig. 10.3, top). This is useful information since it allowed May to describe the conditions for the population to have a non-zero equilibrium: $K > 1/2\sigma^2$. Knowing this allows us to place bounds on the probability that a population will go extinct (but see Goodman 1987).

However, as mentioned, the derivation and analysis of stochastic differential equations are problematical, and May's example is no exception. Some of the problems and assumptions of May's formulation have been discussed by Turelli (1977) and Feldman and Roughgarden (1975). A few of the problems identified were: (1) scaling time by r/K and letting K be random implies that r and K are correlated with respect to environmental stochasticity, and (2) assuming an additive stochastic model suggests a greater effect of randomness on per capita change than on K per se. Altering these assumptions gave different calculations of species extinction probabilities and the maximum amount of resource overlap between species that permits competing species to coexist (Turelli 1977). The take-home message is that equally likely assumptions will give different results, so care is required when working with these equations.

10.4 Markov Processes

A *Markov process* is a probabilistic model of system dynamics when the system variables possess only a finite number of possible states. So, assume that a system is described as being in one of a finite number of states at each time t. System dynamics are a sequence of these states (as they are in differential equation models). The rules that describe the changes can be either deterministic or probabilistic; normally, we are interested in the latter form. In addition, there are two possible ways of viewing the system. First, we may think of the system as an individual object, in which case the system visits the various states sequentially. For example, suppose a person is described as "walking on the sidewalk", "talking with a friend," or "withdrawing money from a bank." The dynamics of this individual can be generated by hypothesizing that there is a certain probability that the person will change state from "walking" to "talking" and another probability from "talking" to "banking," and so on for other combinations of transitions. Depending on random events, the person might have this sequence of states: "walking," "talking," "walking," "banking," "talking," "banking," "walking," etc. In this interpretation, we can calculate the long-term frequency distribution of an individual's states.

In a second interpretation, we may envision the system to be an ensemble of individuals that are not explicitly modeled, but each of which is viewed as changing state randomly. We interpret the system dynamics as occurring when a fraction of the individuals moves between states. In this case, our concern is with the relative proportion of individuals in all of the states, not with the sequences of the individuals. For example, we have a set of field plots characterized by their dominant plant species and a set of rules that predict which species will next dominate the plot given the current dominant species. Since we have a set of plots currently dominated by species A, a set dominated by species B, and so on, our model will predict what fraction will go from being dominated by A to being dominated by another species.

Both approaches can be described with the same mathematics. The central concept is the *transition matrix.* A transition matrix is a special case of a probability matrix which is an $n \times n$ matrix in which all elements are non-negative, and the elements in the rows sum to 1.0. For example,

$$\mathbf{P} = \begin{pmatrix} \frac{1}{4} & \frac{1}{4} & \frac{1}{2} \\ 0 & 1 & 0 \\ \frac{1}{3} & \frac{1}{3} & \frac{1}{3} \end{pmatrix}. \tag{10.7}$$

Two important facts of these matrices are: (1) If $\mathbf{P}$ and $\mathbf{Q}$ are probability matrices, then $\mathbf{PQ}$ is a probability matrix. (2) If $\mathbf{P}$ is a probability matrix, then there is a row vector bft such that

$$\mathbf{t} = \mathbf{tP}. \tag{10.8}$$

In other words, multiplying the vector and the matrix returns the vector unchanged. Obviously, if $\mathbf{P}$ is the identity matrix, the statement is true, but it is also true for other, more interesting values of $\mathbf{P}$. This is not true for all $\mathbf{t}$. If $\mathbf{t} = [0.2, 0.40.4]$, the multiplication shown in Eq. 10.8 is not the input vector, but rather $\mathbf{t} = [0.183, 0.583, 0.233]$.

10.4.1 Biological Applications of Markov Processes

We will define a transition matrix $\mathbf{P}$ to be a probability matrix such that the rows and columns are the states of the system. If the system is viewed as an individual, then the element p_{ij} is the probability that the system will change from state i to state j. If the system is viewed as an ensemble of individuals, then the elements are the fractions of individuals changing from state i to j. Below, we list, without proof, some basic facts about these matrices. Excellent elementary discussions can be found in Grossman and Turner (1974) and Hillier and Lieberman (1980).

Since $\mathbf{P}$ is the probability of moving from the current state to the next state, it is convenient to call this the *one-step transition probability matrix* (Hillier and Lieberman 1980). $\mathbf{P}$ multiplied by itself ($\mathbf{P}^{(2)}$) is the *two-step transition probability matrix* and represents the probabilities of moving from state i to state j in two steps. $\mathbf{P}^{(n)}$ is defined similarly for n steps.

Let $\mathbf{p}$ be a row vector of probabilities that an individual is in state i. In the ensemble interpretation, it is the fraction of individuals in state i. Then, we can form a recursive equation to generate the probability distribution in the next time step as

$$\mathbf{p}_{t+1} = \mathbf{p}_t\mathbf{P}. \tag{10.9}$$

If $\mathbf{P}$ is composed entirely of elements that are constant and independent

Table 10.1: The Markov transition probability matrix for a deer moving among water, grass, and sleeping areas. The powers of the matrices are indicated by the superscripts: 1, 2, 4, 16, 32, 64.

$\mathbf{P}^1$			$\mathbf{P}^2$		
0.600000	0.200000	0.200000	0.460000	0.270000	0.270000
0.250000	0.500000	0.250000	0.337500	0.362500	0.300000
0.250000	0.250000	0.500000	0.337500	0.300000	0.362500
$\mathbf{P}^4$			$\mathbf{P}^{16}$		
0.393850	0.303075	0.303075	0.384754	0.307623	0.307623
0.378844	0.312531	0.308625	0.384529	0.307743	0.307728
0.378844	0.308625	0.312531	0.384529	0.307728	0.307743
$\mathbf{P}^{32}$			$\mathbf{P}^{64}$		
0.384616	0.307692	0.307692	0.384616	0.307692	0.307692
0.384615	0.307692	0.307692	0.384616	0.307692	0.307692
0.384615	0.307692	0.307692	0.384616	0.307692	0.307692

of p_i and if $\mathbf{p}_{t+1}$ depends only on $\mathbf{p}_t$ (i.e., not on previous $\mathbf{p}_{t-m}$, where m is a positive integer), then $\mathbf{P}$ is a *Markov transition matrix.* In this case, Eq. 10.9 describes a *Markov process.* Sometimes this is referred to as a *linear, first-order Markov process* to allow for the existence of more complicated models of the same general type.

As stated above, for a given $\mathbf{P}$, there is a $\mathbf{p}$ such that in Eq. 10.9, $\mathbf{p}_{t+1} = \mathbf{p}_t$. This is, basically, an equilibrium of the probability distribution of the system, and $\mathbf{p}_t$ is called the *fixed probability distribution.* The fixed $\mathbf{p}$ can be computed from

$$\mathbf{p} = \mathbf{p}_0 \mathbf{P}^{(n)},$$

where n is large. This states that a system will eventually reach an equilibrium probability distribution if sufficient iterations are run.

As a simple example, suppose an organism such as a deer, in its daily movement, probabilistically visits three habitats: water, grass, and a sleeping area. The locations of the deer in the three habitats are the three states of the deer. Table 10.1 shows the $\mathbf{P}^{(n)}$ Markov transition matrix for $n = 1$, 2, 4, 16, 32, 64. Notice that the matrix converges. The rows of the $\mathbf{P}^{(64)}$ matrix are the fixed probability distribution. To verify that this is so, multiply the initial probability vector $[0.5, 0.5, 0.0]$ times $\mathbf{P}^{(64)}$ to determine the long-term trajectory of the probability distribution.

10.4.2 Simulating Markov and Transition Matrix Models

The assumptions of a Markov process are biologically unrealistic. In particular, the current state of the system will often influence the transition probabilities, so that $\mathbf{P}$ will not be constant. Further, many biological systems have a "history" in the sense that events in the past influence current processes, so the assumption that $\mathbf{p}_{t+1}$ depends only on $\mathbf{p}_t$ is often false. One approach to relaxing these assumptions is to use *semi-Markov processes* applied to compartments (e.g., spatial position) in which the probability of leaving increases the longer an object has been in the compartment. Matis et al. (1992) give some analytical results when the probability distribution is a gamma function.

In other cases of relaxing the original Markov assumptions, the simple analytical results discussed above may not be possible, and computer simulation will be necessary. Simulating a Markov chain is not difficult. The process can be simplified if the rows of the original transition matrix are converted to cumulative distributions. The rows denote the current state; the columns denote the new states. Given the transformed transition matrix ($\mathbf{P}'$), the algorithm is:

1. Assign an initial state to the system ($s_{i,t}$).
2. Obtain a uniform random deviate (U_t).
3. For row $s_{i,t}$, determine the column ($s_{j,t+1}$) such that $p_{ij} < U_t \leq p_{i(j+1)}$, where p_{ij} is the upper bound of the cumulative distribution for the transition from state i to state j.
4. j is the new $s_{j,t+1}$.

Once this basic structure is in place, it is possible to relax the assumptions of linear, first-order Markov chains. One relaxation is to constrain state visitation by the current state or previous transitions. For example, if the deer has visited the sleeping area (**S**) three times consecutively, then the probability of a transition from **S** to **S** can be dynamically reduced to 0. This hypothesis relaxes both the assumption of no historical effects and the independence of transition probabilities and current state.

10.5 Exercises

1. If at time t_0 the deer is equally likely to be found in all places, in which place is the deer most likely to be found after one time step?
2. Verify that the fixed probability distribution for the deer movement model is obtained regardless of the initial probability distribution.
3. Interpret the fixed probability distribution for the deer movement model if we had used the *ensemble* interpretation of the system.
4. If matrix 10.7 is a transition matrix, interpret the meaning of the second row ([0,1,0]). What is the equilibrium vector of probabilities for this matrix?
5. Simulate random temperature by drawing 100 samples from the empirical distribution of temperatures shown in Fig. 10.1.
6. Simulate the density-dependent population model with r a normal deviate. Is there a long-term average? Does the distribution of these averages fit any simple probability distribution?

7. The pdf of the exponential distribution is $f(x;\lambda) = \lambda\exp(-\lambda x)$. Derive the inverse cdf and devise an algorithm to sample from the pdf.
8. Construct and run a computer program to randomly place points (x-y pairs) uniformly in a circular region with radius a. Your algorithm should not have to throw away any tentative x–y pairs. Test the correctness of the results.
9. Below are modified tree replacement data from Horn (1975). The rows are current dominant canopy species in a stand and the columns are the percent sapling species under the canopy. Assuming that the sapling species of today become the canopy species of tomorrow, what is the equilibrium composition of the forest?

	PERCENT SAPLINGS				
CANOPY	RO	HI	TU	RM	BE
Red Oak	12	12	12	42	22
Hickory	14	5	10	53	18
Tuliptree	12	8	10	32	38
Red Maple	11	25	4	17	31
Beech	13	27	8	19	33

10. Write a program to simulate weather as random temperature values. Assume the cosine function for temperature (Chapter 4) is the mean, and add a random component from a normal distribution with constant variance. How reasonable is the assumption of constant variance? What is a simple alternative?
11. One of the four possible outcomes to the classical Gause competition equations described in Chapter 4 is an unstable equilibrium. This is sometimes referred to as "indeterminant" competition because it is difficult to predict the outcome of a laboratory system with normal stochastic fluctuations started near the separatrix. In other words, if a stochastic system is started at the unstable equilibrium, it could drop into either of the two basins of attraction. Can sufficiently large random fluctuations prevent either species from excluding the other, resulting in a system that remains near an unstable equilibrium? Test this idea by simulating the competition equations with parameters α_{ij} and K_i chosen so that an unstable equilibrium exists. In separate simulation analyses, introduce randomness in these two ways: (1) random fluctuations in population numbers (e.g., disturbances) and (2) random fluctuations in all the parameters. For each of the two analyses above, examine several levels of randomness. Use a probability distribution of your own choosing, but consider using a log-normal.

Part II

APPLICATIONS

Chapter 11

Photosynthesis and Plant Growth

11.1 Introduction

PLANTS represent some of the most difficult biological systems to model. There are major problems with choosing the appropriate spatial, temporal, and biological scales. In addition, in terrestrial vascular plants, a major component of the organism, the root system, is not easily available for study or inspection. In this chapter, we will not provide a complete overview of models of photosynthesis and plant–water relations, but rather choose a few examples from these fields to illustrate some of the problems and progress that has been made. In examining these systems, we will illustrate several important principles developed in *Part II*. These include (1) the effect of scale and biological levels of organization on model structure, (2) nullcline analysis and bifurcations, (3) the control of processes by multiple factors in biochemical networks, (4) the use of mean resistance for multiple control in hydraulic models, and (5) multiple flow variables in plant growth models.

11.2 Cellular-Level Photosynthesis

Although the biochemical pathways involved in photosynthesis are relatively well known, there is still wide variation in the set of models for this process. Some of the discrepancy is due to different objectives and scales used to describe plants. In the first model we will examine, a model of steady-state levels of carbon assimilation was desired. The central biological question addressed by this model is: What effects do light intensity and the concentrations of CO_2 and O_2 have on the net rate of plant CO_2 uptake? Another approach focuses on the dynamics of stomata, but ignores most of the biochemical details. This model addresses the question: Can the mechanisms of water flow within leaves explain cycles in transpiration?

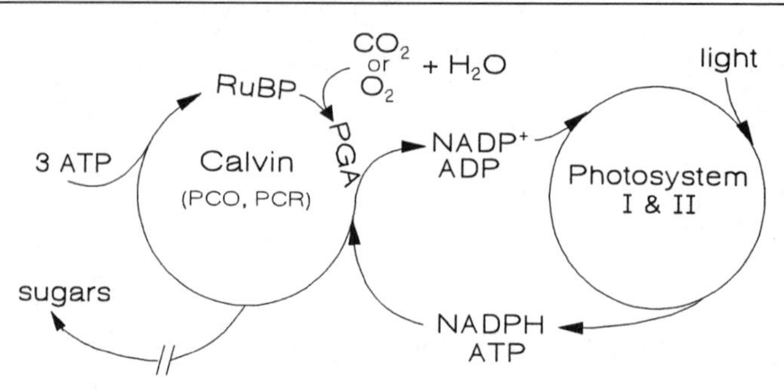

Figure 11.1: Two important biochemical cycles involved in photosynthesis are the production of ATP and NADPH using light energy and photosystems I and II and the Calvin cycle wherein RuBP is produced and the pathway to the ultimate production of sucrose is initiated. RuBP may be either carboxylated with CO_2 (PCR) or oxidized with O_2 (PCO).

A third model, describing plant growth, uses a high-level of description with few mechanistic details. The question this model addresses is: How does atmospheric CO_2 concentration affect the distribution of plant resources to shoots and roots?

11.2.1 Photosynthesis Biochemistry

An extremely simplified view of the important biochemical reactions associated with photosynthesis is shown in Fig. 11.1. In brief, light energy as photons interacts with two photosystems to produce ATP and NADPH. These compounds are required to convert (ultimately) phosphoglycerate (PGA) to sucrose and ribulose 1,5-biphosphate (RuBP) . RuBP and CO_2 are used to create more PGA, completing the cycle. Many other details are omitted, but for the purposes of the leaf-scale photosynthesis model described below, the diagram shows components relevant to two important biochemical conditions that determine the rate at which carbon is fixed. First, the production of PGA from RuBP may be limiting. Second, the regeneration of RuBP from PGA using NADPH produced in the Calvin cycle may be limiting. In the former case, the amount of extracellular O_2 and CO_2 will influence carbon fixation rates, while in the latter case, light levels will determine fixation rates.

A key step in the formation of PGA from RuBP is the carboxylation of RuBP: photosynthetic carbon reduction (PCR). Carboxylation is facilitated by the enzyme RuBP carboxylase-oxygenase (Rubisco) whose active site accepts one molecule of CO_2 and uses the carbon atom to form two molecules of PGA. The active site, however, also accepts O_2 in a reaction that *oxidizes* RuBP and starts a pathway called photorespiratory carbon oxidation (PCO) or *photorespiration* that *releases* CO_2 and thereby defeats the carbon fixation cycle. CO_2 and O_2, therefore, compete for the active site on Rubisco. When O_2 is successful, two molecules of RuBP result in the release of one carbon atom from the system. In addition (Farquhar et al. 1980), oxidation directly produces one molecule of PGA and one molecule of PGIA (phosphoglucolate). One mole of PGIA results in 0.5 mole of PGA, requiring the use of 0.5 mole of ATP. So, oxidation of RuBP results in 1.5 mole of PGA. Carboxylation produces 3 mole of PGA.

11.2.2 Carbon Assimilation

Based on differential equations of the rate kinetics of the biochemical constituents contained in a more complex version of Fig. 11.1, Farquhar and von Caemmerer (1982) derived a general, steady-state model of carbon metabolism for C_3 plants. Their derivation had two purposes: understand the mechanisms of carbon assimilation, and simplify the relationships to allow experimental test with quantitative data. Needless to say, photosynthesis is a very complex system to model, requiring details of biochemistry we do not have space to introduce. However, to give a flavor for biochemical modeling, we will give a partial derivation of some of the results more fully covered in Farquhar and von Caemmerer (1982).

In very simplified terms, the rate of CO_2 assimilation is the difference between input (carboxylation) and output (respiration):

$$A = f(V_c) - R_d, \tag{11.1}$$

where A is the assimilation rate, and R_d is *respiration*, the rate at which C is used by the plant's photosynthesizing machinery. R_d is the rate at which CO_2 is lost from cells inside the leaf due to cellular metabolism; the plant cells behave as do animal cells in this regard. The function $f(V_c)$ is the rate of CO_2 uptake and incorporation into stored products such as sugar. R_d is an important modeling problem in its own right, but we will focus on $f(V_c)$ in the following and assume that R_d is a measurable constant.

To begin, we sketch the general plan of attack by noticing that Fig. 11.1 shows two cycles affecting C assimilation and sugar production. In biochemical pathways such as this, the rate of a reaction (e.g., sugar production) is often limited by the slowest step in the pathway. This suggests that a suitable modeling approach is to write equations for the rates of all of the major biochemical steps, then invoke the Law of the Minimum to determine the overall reaction rate. This is basically the strategy that Farquhar and von Caemmerer (1982) used. In their development, they dealt with most of the known facts of all of the major steps; here, we will focus on only a subset, being guided by Farquhar and von Caemmerer's insights about those which are especially important. The two most important steps are the conversion of RuBP to PGA and the regeneration of NADPH and RuBP from the photosystems. So, our simplified problem really comes down to analyzing the case when RuBP is plentiful and the case when it is in short supply.

When RuBP is plentiful, the rate of PGA formation depends on the supply of CO_2 and competing O_2. That is, the rate of RuBP carboxylation is

$$V_c = W_c = \frac{V_{\text{cmax}} C}{C + K_C(1 + O/K_O)},$$

where C is the partial pressure of CO_2, O is the partial pressure of O_2, V_{cmax} is the maximum rate of CO_2 carboxylation, K_C is the half-saturation constant for carboxylation in the absence of O_2, and K_O is the half-saturation constant for oxygenation. We use W_c in this context to denote the rate of carboxylation when RuBP is saturating. This equation should be familiar as having the general form of the Michaelis-Menten equation. In addition, it is an example of the modeling tool wherein a primary rate equation (the Michaelis–Menten effects of CO_2 on carboxylation) is modified to incorporate a second influencing factor (O_2) by transforming a constant (half-saturation, K_C) into a function of the second factor. As a result, the equation is not exactly a Michaelis–Menten relation.

The rate at which O_2 competes with CO_2 to *oxidize* RuBP is also similar to a Michaelis–Menten relation:

$$V_o = \frac{V_{\text{omax}} O}{O + K_O(1 + C/K_C)},$$

where K_O is the half-saturation constant for oxidation in the absence of CO_2, and V_{omax} is the maximum rate of CO_2 oxidation. The ratio of these two rates will be a useful variable later:

$$\phi = \frac{V_o}{V_c} = \frac{V_{\text{omax}} O K_c}{V_{\text{cmax}} C K_o}. \tag{11.2}$$

Once we have V_c and V_o, we can complete Eq. 11.1 as

$$A = V_c - 0.5V_o - R_d, \tag{11.3}$$

where the factor 0.5 is due to the stoichiometry of the reaction: oxidation of 1 mol of RuBP releases 0.5 mol of CO_2 (Farquhar and von Caemmerer 1982).

Bearing in mind that this is a steady-state model, the next step is to relate the parameters to quantities that can be estimated in the laboratory. The first of several simplifications involves the concept of compensation points. The *compensation point* of an environmental variable that influences photosynthesis is that level at which the rate of respiration and photorespiration equals photosynthetic rate so that *net carbon fixation is zero* (the steady-state condition). Since the rate of photosynthesis depends on both CO_2 and light levels, there are compensation points for both environmental variables. While recognizing that light effects are important, the Farquhar–von Caemmerer model focused on CO_2 as the limiting variable. Consequently, the CO_2 compensation point is the one of primary interest. Since the rate of photosynthesis is determined by the competition of CO_2 and O_2 for Rubisco active sites, there is a CO_2 compensation point (Γ_*)

even when $R_d = 0$:

$$\Gamma_* = \frac{0.5V\text{omax}K_C O}{V\text{cmax}K_O}, \tag{11.4}$$

where Γ_* is the value of C at which Eq. 11.2 is 0. This follows from the fact that in Eq. 11.3, when R_d and A are 0, $\phi = 2 = 2\Gamma_*/C$. The equation for Γ_* also implies that Γ_* is a linear function of O. From Eqs. 11.2 and 11.3 it follows that

$$\Gamma_* = V_O C/2V_c. \tag{11.5}$$

When $R_d > 0$, the compensation point is derived from the same operations as above to give

$$\Gamma = \frac{\Gamma_* + (R_d/V\text{cmax})K_C(1 + O/K_O)}{1 - R_d/V\text{cmax}}.$$

In this case, as above, the compensation point is a linear function of O.

When RuBP is not saturating, carbon fixation rate depends on the rate at which RuBP is regenerated by interactions of the Calvin cycle and the photosystems (Fig. 11.1). The rate of RuBP regeneration can be limited by the rate of ATP formation in the PCR cycle. For each mole of RuBP and sugar that is formed, the PCR cycle uses 2 mol of NADPH and 3 mol of ATP. From this stoichiometry, ATP is consumed to form RuBP at the rate

$$a = 3V_c + 3.5V_o,$$

where a is the consumption rate of ATP and equals the sum of the rates of carboxylation and oxidation, respectively, weighted by their use of RuBP. Oxidation uses an additional 0.5 mole of ATP in converting PGIA to PGA. We assume that ATP is in steady state so that production in PCR equals consumption to regenerate RuBP. For each molecule of ATP produced, three protons are liberated, so

$$a' = 3a = 9V_c + 10.5V_o = (9 + 21\Gamma_*/C)V_c,$$

where a' is the proton liberation rate. Recall that when carbon fixation is in steady state and $R_d = 0$, $V_o = 2\Gamma_* V_c/C$. Finally, two protons in a water molecule cause one electron to move through photosystems I and II, so that

$$J = 0.5a' = (4.5 + 10.5\Gamma_*/C)V_c,$$

where J is the rate of electron transport.

Now, V_c is unknown and J can be estimated empirically, so rearranging

$$V_c = J' = \frac{JC}{4.5C + 10.5\Gamma_*}, \tag{11.6}$$

because we assume here that the rate is not limited by sites for ADP phosphorylation. Farquhar and von Caemmerer (1982) relax this assumption.

We can now put all of this together. The net rate of CO_2 assimilation is

$$A = V_c - 0.5V_o - R_d, \tag{11.7}$$

or

$$A = V_c(1 - \Gamma_*/C) - R_d, \tag{11.8}$$

from Eq. 11.5, where V_c is the rate of carbon fixation when either RuBP is saturated (W_c) or when irradiance and electron transport limits RuBP regeneration (J'). To determine the ultimate rate, we use Liebig's Law of the Minimum (Chapter 4):

$$V_c = \min(W_c, J').$$

So, to summarize, if RuBP is saturating

$$\begin{aligned} A &= V_c(1 - \Gamma_*/C) - R_d \\ &= V\mathrm{cmax}\frac{C - \Gamma_*}{C + K_c(1 + OK_o)} - R_d. \end{aligned} \tag{11.9}$$

If RuBP regeneration is limited by irradiance

$$A = J\frac{C - \Gamma_*}{4.5C + 10.5\Gamma_*} - R_d. \tag{11.10}$$

The only step left is to give an empirical equation for J, the potential electron transport rate for the formation of ATP. This electron production rate depends on light levels. There are several alternative formulations of this rate, but a recent one (Evans and Farquhar 1991) is

$$J = \frac{I_2 + J\mathrm{max} - \sqrt{(I_2 + J\mathrm{max})^2 - 4\Theta I_2 J\mathrm{max}}}{2\Theta},$$

where

$$I_2 = \frac{I_0}{2}(1 - f)(1 - r),$$

and where I_0 is incident radiation, J is potential electron transport rate, I_2 is irradiance absorbed by Photosystem II, f is a factor to correct for spectral imbalance of light, r is reflectance and transmittance from the leaf to photosynthetically active radiation, and the factor 2 accounts for the effect of Photosystem I on electron flow. J_{max} and Θ are empirically estimated, with the former being the maximum electron transport rate and the latter being a shape parameter.

A in Eqs. 11.7 and 11.8 represents the instantaneous rate of CO_2 assimilation and can be used in a differential equation of carbon flux in a plant. Moreover, A will vary with fluctuating light levels and internal carbon concentration. It will also be influenced by temperature, for which the reader should consult Farquhar and von Caemmerer (1982). Since A is experimentally measurable, the model can be validated directly. Figure 11.2 shows comparisons of Eqs. 11.9 and 11.10 with data from two species of wheat. See Evans and Farquhar (1991) for more details. Clearly, this

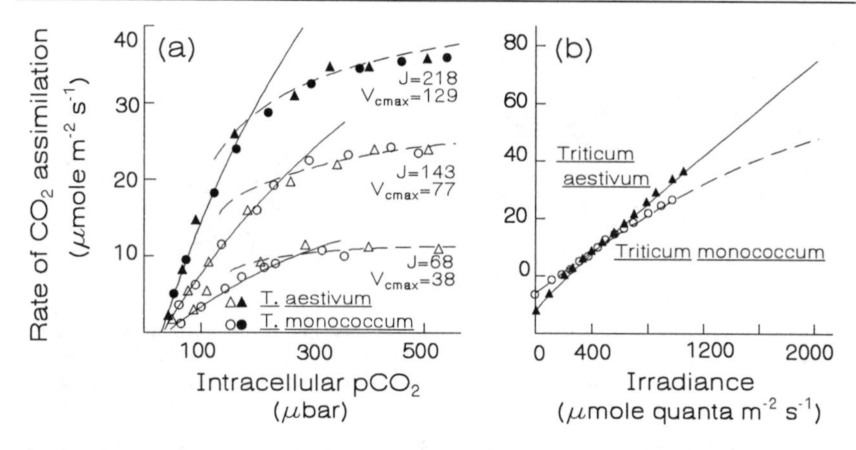

Figure 11.2: Empirical tests of the Farquhar–von Caemmerer–Evans model of photosynthesis for *Triticum aestivum* and *Triticum monococcum*. (a) Effects of CO_2 at three light intensities on CO_2 assimilation and (b) effects of light level on CO_2 assimilation. Symbols are observations; solid lines are model output for Eq. 11.10, and broken lines are Eq. 11.9. The intersection point of the model lines is the shift from CO_2 limitation to light limitation. (From Evans and Farquhar 1991, Figs. 1-2A and 1-3. © 1991 Crop Science Society of America, Inc. Reprinted with permission Crop Science Society of America, Inc., publisher.)

model gives a good fit to the data, but more importantly, it has a solid theoretical foundation in the biochemical pathways and likely limiting factors that influence electron flow and biochemical kinetics.

11.3 Leaf-Level Photosynthesis

The Farquhar–von Caemmerer model is a model of CO_2 assimilation based on intracellular CO_2 and light levels. A key process is the production of PGA from RuBP in the presence of CO_2 and water (Fig. 11.1). Therefore, understanding the processes affecting the levels of CO_2 and H_2O in a leaf is crucial for a complete mechanistic description of photosynthesis. One of the critical processes involved is the magnitude and duration that *stomata* (i.e., leaf surface pores) are open for the interchange of water and CO_2. In this section, we construct a model of the dynamics of stomata in order to better understand the hydraulic mechanisms of photosynthesis.

11.3.1 Basics of Plant–Water Relations

Before diving into the model description, we very briefly describe the central concepts needed to think about water movement in plants. The basic physical system to consider is a series of water compartments connected by semipermeable membranes. Figure 11.3 shows two such compartments under two different conditions. On the left is a case where the solutions (cross-hatching) are isotonic (all solutes in the same concentration), but at the moment in time shown a higher pressure head exists on the right

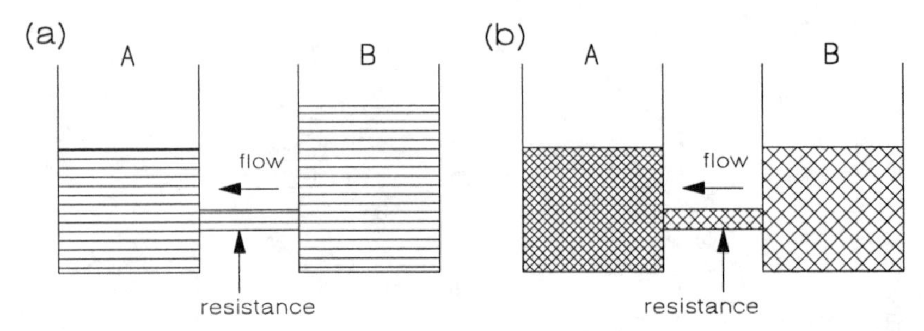

Figure 11.3: Two causes of water movement between chambers separated by a semi-permeable membrane. (a) Hydrostatic pressure differences, (b) osmotic potential differences. In both cases, A receives water from B.

compartment than on the left. In the system to the right, the pressure head between the two chambers is equal, but the solution is more concentrated on the left than on the right. The left chamber is *hypertonic* with respect to the right; it has a higher osmotic pressure than the chamber on the right, and water flows from right to left to eliminate the pressure difference.

Osmotic pressure is the amount of hydrostatic pressure that must be applied to the hypertonic chamber to offset the water flow that would occur because of differences in ionic concentration between two chambers. It is estimated using *van't Hoff's Law*

$$\pi \cong iRTc,$$

where R is the ideal gas constant (0.08314 atm·liter/g·mol· K), T is degrees Kelvin, i is the number of ions formed when a single solute molecule dissociates in the solution medium [e.g., NaCl (table salt) has $i = 2$], and c is the molar concentration of the solute (moles of solute per liter of solution). In differential equations, as a matter of convenience, π is given a positive sign.

Both hydrostatic and osmotic pressure play major roles in determining the flow of water in the living tissue of both plants and animals. When the concept of hydrostatic pressure is applied to plant cells, we use the term *turgor pressure*, denoted P . Turgor pressure is measured relative to a particular cell; it is the pressure exerted on the cell by its wall. Due to the fact that cell walls can only stretch to a limit, we will refer to cell characteristics defined when the cell is at *full turgor*. This condition occurs when the cell contains its maximum amount of water. Under constant conditions, cellular π and P combine to produce a net "proclivity" of water to move into or out of the cell. This proclivity is called *water potential* and denoted Ψ and is defined as $\Psi = P - \pi$ (π is positive). If π is large relative to P, water will tend to move in. If P is greater than π, water will leave the cell.

In addition, most living tissue is elastic, so that as water flows into a cell, the pressure increases as the cell wall or membrane expands. Thus, there can be nonlinear relationships between the flow of water due to osmotic pressure and the subsequent changes in hydrostatic pressures.

In addition to these two forms of pressure determining the rate of water flow, the membrane itself will slow down molecular movement. This can be thought of as a *resistance* (as in electrical resistance) similar to friction. The resistance of a pathway to water flow is usually an empirically determined constant that depends on the properties of the medium or membrane through which the water flows. Resistance suggests the measure of a force that prevents a flow from occurring. Consequently, it is often more convenient to use *conductance*, the mathematical inverse of resistance. This quantity is commonly used in models in a multiplicative expression to portray the quantity of fluid that flows from point A to point B.

Resistance used in this context of a physical flow means that compartments or chambers arranged in series or parallel permit simple rules for calculating overall resistance in the network. If the compartments are in series (e.g., soil, roots, stem, leaves, atmosphere), the overall network resistance (soil to atmosphere) is the sum of the resistances:

$$R_n = R_r + R_s + R_l + R_a,$$

where the subscript denotes the terminal compartment of the component flow.

Alternatively, the compartments could be in parallel, for example, water flowing along a branch to an apical cluster of leaves. In this case, the pathways (i.e., the leaves) are "competing" for the flowing material, and the overall network resistance can be computed using the fact that the inverse of the network resistance is the sum of the inverses of each component flow:

$$\frac{1}{R_n} = \frac{1}{R_1} + \frac{1}{R_2} + \frac{1}{R_3} + \ldots,$$

where the indices indicate compartments (e.g., individual leaves). This formulation should be familiar from Chapter 4, where we presented it in the context of multiple limiting factors.

With these two concepts of pressure and resistance, we can understand the basics of quite a few models in plant physiology.

11.3.2 Stomata Dynamics

As indicated, a knowledge of the rate at which water evaporates (i.e., *transpires*) from a leaf is essential to understanding photosynthesis. Transpiration (E) is defined as

$$E = g_s w,$$

where w is the difference in water vapor potentials inside and outside the leaf, and g_s is the conductance of the stomatal aperture (i.e., the ability of water to flow through the pore). Conductance is the key variable to model and one important modeling approach is to assume that the stomate is in

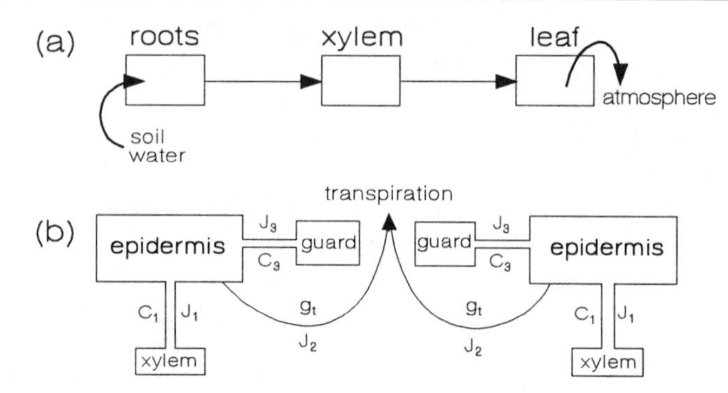

Figure 11.4: (a) Idealized view of water flow in a plant showing compartments for roots, stem, and leaf. (b) Idealized view of a stomate showing flows among xylem, epidermis, guard cells, and the atmosphere.

steady state so that simple empirical relations can be derived. Ball et al. (1987) produced one notable model:

$$g_s = kAh/C,$$

where g_s is conductance, h is relative humidity, A is net photosynthesis, C is partial pressure of CO_2 and k is an empirical constant. This simple, but useful equation, however, ignores recent developments in the short-time-scale dynamics of stomatal movement and conductance. In this section, we derive a non-steady-state model of single stomate dynamics. The model contains both hydraulic and biochemical controls of stomate opening and closing.

We use here a simplified view of a leaf shown in Fig. 11.4a. Stomate dynamics are determined by the relative pressures of the guard and epidermis cells (Fig. 11.4b). These pressures are determined by the volumes of the two cell types, and these, in turn, are determined by flows of water between the cell types and a source of xylem water (Fig. 11.4b).

Thus, three flows of water (J_1, J_2, and J_3) determine the volumes of two cell types: guard (V_g) and epidermal (V_e)

$$\frac{dV_g}{dt} = J_3 \tag{11.11}$$

$$\frac{dV_e}{dt} = J_1 - J_2 - J_3, \tag{11.12}$$

where V_g and V_e are constrained to non-negative values.

J_1 is the flow of water from xylem to epidermal cells

$$\begin{aligned} J_1 &= C_1(\Psi_x - \Psi_e) \\ &= C_1(\Psi_x - P_e + \pi_e), \end{aligned} \tag{11.13}$$

where Ψ_x is the water potential of the xylem (i.e., plant roots and stem), Ψ_e is the water potential of the epidermis tissue, P_e is the turgor pressure of the epidermal cells, and π_e is the osmotic pressure of the epidermal cells. This flow has the simple form of water flow models described above: conductance of a pathway for water multiplied by the difference in water potential between two sites. Notice that Eq. 11.13 is an example of negative feedback by an extrinsic limit (Ψ_x, see Section 4.3.3).

Cell pressure is a linear function of the ratio of current volume to the volume at full turgor, scaled by the cell wall modulus and pressure at full turgor. Cell wall modulus is the inverse of wall elasticity, and in reality varies with turgor pressure, which we ignore for simplicity. Thus, we have

$$P_e = \epsilon_e \left(\frac{V_e}{Va_{e,ft}} - 1 \right) + P_{e,ft}, \tag{11.14}$$

where ϵ_e is epidermal cell wall modulus, $Va_{e,ft}$ is the volume at full turgor of epidermal cells in a finite leaf area, and $P_{e,ft}$ is the pressure of the epidermal cells at full turgor. P_e is constrained to non-negative values. Cell wall modulus is the inverse of wall elasticity, and in reality varies with turgor pressure, but we ignore this for simplicity.

Epidermal cell osmotic pressure (π_e) is determined by the concentration of solutes from van't Hoff's Law as

$$\pi_e = \frac{N_e RT}{V_e}, \tag{11.15}$$

where N_e is moles of solutes, R is the gas constant, and T is temperature in degrees Kelvin.

J_2 represents evaporation from epidermal cells and is proportional to the difference of internal (c_s) and atmospheric (c_a) water vapor pressure:

$$J_2 = g_t(c_s - c_a), \tag{11.16}$$

where J_2 is transpiration and was denoted as E above. This equation is another example of negative feedback by an extrinsic limit (C_a, see Section 4.3.3).

Total conductance, g_t, is a combination of two conductances, one from the epidermal cell surfaces to the guard cell (g_e) and the other from the guard cell to the atmosphere (the boundary layer conductance, g_b). The latter quantity is assumed, for our purposes, to be a constant for a given leaf and environmental conditions. g_e, however, is a function of the current guard cell aperture opening:

$$a = b_0 + b_g P_g - b_e P_e \tag{11.17}$$

$$g_e = k_1 a, \tag{11.18}$$

where a is the guard cell aperture and constrained to be greater than zero; b_0, b_g, b_e are empirically determined constants, and k_1 is a proportionality constant relating aperture to conductance. The inverse of conductance is analogous to electrical resistance, and we combine stomatal conductance and boundary layer conductance by adding this series of two resistances:

$$\begin{aligned} \frac{1}{g_t} &= \frac{1}{g_e} + \frac{1}{g_b} \\ g_t &= \frac{g_e g_b}{g_e + g_b}. \end{aligned} \tag{11.19}$$

The last flow, J_3, represents the flow between guard and epidermal cells. This is determined by the differences between the water potentials of guard and epidermal cells:

$$J_3 = C_3(-P_g + \pi_g + P_e - \pi_e), \tag{11.20}$$

where C_3 is conductance between guard and epidermal cells, and the P_i and π_i are hydrostatic pressures and osmotic pressures for guard ($i = g$) and epidermal ($i = e$) cells, respectively.

Guard cell pressure is defined analogously to epidermal cell pressure:

$$P_g = \epsilon_g \left(\frac{V_g}{Va_{g,ft}} - 1 \right) + P_{g,ft}, \tag{11.21}$$

where the variables for guard cells are similar to those for the epidermal cells.

Guard cell osmotic pressure is similar to that for the epidermal cells, with the exception that solute concentration may be a direct function of time (to simulate metabolism) and is biochemically controlled by pressure of the epidermal cells and diffusion of water from guard cells:

$$\pi_g = \frac{N_g(P_e, N_g, t)RT}{V_g}. \tag{11.22}$$

To describe the effects of ion diffusion and epidermal pressures on N_g, we need another differential equation that is a function of N_g and P_e. For simplicity, we assume the following linear relationship:

$$\frac{dN_g}{dt} = sP_e - r(N_g - N_{gmin}), \tag{11.23}$$

where s and r are empirical constants, and N_{gmin} is the minimum concentration of solutes maintained by normal cell metabolism. This equation hypothesizes that guard cell ion production is stimulated by high epidermal pressure (P_e) and that ions decay from the guard cell in proportion to the excess ion concentration above normal. Biochemical control of stomate opening can be eliminated by setting s and r to zero. This permits the study of the relative importance of hydraulic compared to biochemical controls, and is an example of the use of alternative models and hypotheses.

Combining these equations for the J_i, we have

$$\begin{aligned} \frac{dV_g}{dt} &= C_3 \Bigg[\left(- \left(\epsilon_g \left(\frac{V_g}{Va_{g,ft}} - 1 \right) + P_{g,ft} \right) \right) + \frac{N_g RT}{V_g} \\ &\quad + \left(\epsilon_e \left(\frac{V_e}{Va_{e,ft}} - 1 \right) + P_{e,ft} \right) - \frac{N_e RT}{V_e} \Bigg] \end{aligned} \tag{11.24}$$

$$
\begin{aligned}
\frac{dV_e}{dt} &= C_1 \left[\Psi_x - \left(\epsilon_e \left(\frac{V_e}{Va_{e,ft}} - 1 \right) + P_{e,ft} \right) + \frac{N_e RT}{V_e} \right] - g_t(c_s - c_a) \\
&- C_3 \left[\left(- \left(\epsilon_g \left(\frac{V_g}{Va_{g,ft}} - 1 \right) + P_{g,ft} \right) \right) + \frac{N_g RT}{V_g} \right. \\
&+ \left. \left(\epsilon_e \left(\frac{V_e}{Va_{e,ft}} - 1 \right) + P_{e,ft} \right) - \frac{N_e RT}{V_e} \right] \qquad (11.25) \\
\frac{dN_g}{dt} &= sP_e - r(N_g - N_{gmin}). \qquad (11.26)
\end{aligned}
$$

As described in Chapter 9, nullcline analysis can yield insight into qualitative dynamics and stability properties of a model. We now give the nullcline equations and graphs for a particular set of parameters. Asterisks denote equilibrial values of the variable. After setting all three differential equations to 0 and simplifying, we have the nullcline equation for N_g

$$
N_g^* = \frac{s(m_{ve}V_e^* + i_e)}{r} + N_{gmin}, \qquad (11.27)
$$

where

$$
\begin{aligned}
m_{ve} &= \epsilon_e / Va_{e,ft} \\
i_e &= P_{e,ft} - \epsilon_e.
\end{aligned}
$$

The nullcline for V_g is

$$
\begin{aligned}
0 &= V_g^{*2}(-m_{vg}V_e^*) + V_g^*(m_{ve}V_e^{*2} + i_e V_e^* - i_g V_e^* - B) \\
&+ RTV_e^* \left(\frac{s}{r}(m_{ve}V_e^* + i_e) + N_{gmin} \right), \qquad (11.28)
\end{aligned}
$$

where

$$
\begin{aligned}
m_{vg} &= \epsilon_g / Va_{g,ft} \\
i_g &= P_{g,ft} - \epsilon_g.
\end{aligned}
$$

And the nullcline for V_e is

$$
V_g^* = \frac{q_3 V_e^{*3} + q_2 V_e^{*2} + q_1 V_e^* + q_0}{q_6 V_e^{*2} - q_5 V_e^* - q_4}, \qquad (11.29)
$$

where

$$
\begin{aligned}
q_0 &= B(g_b + k_1(b_g i_g - b_e i_e)) \\
q_1 &= \Psi_x(g_b + k_1(b_g i_g - b_e i_e)) \\
&- i_e(g_b - k_1(b_e i_e - b_g i_g)) - Bk_1 b_e i_e) + \frac{w}{C_1} g_b k_1 (b_e i_e - b_g i_g) \\
q_2 &= m_{ve}(-\Psi_x k_1 b_e - g_b + k_1(2b_e i_e - b_g i_g) + \frac{w}{C_1} g_b k_1 b_e) \\
q_3 &= m_{ve}^2 k_1 b_e
\end{aligned}
$$

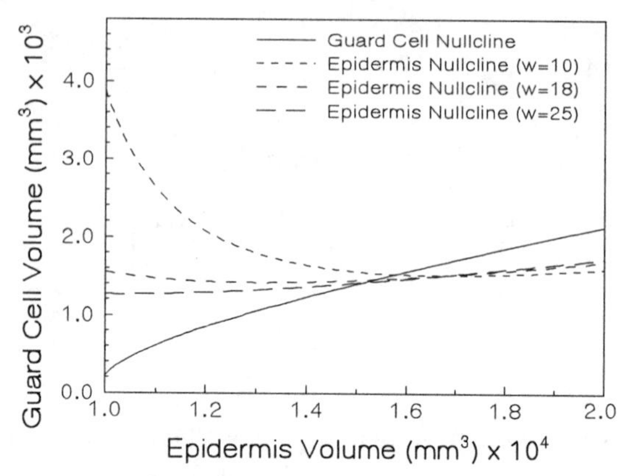

Figure 11.5: Nullclines for V_g and V_e for three water vapor deficits (w =10, 18, and 25 bars).

$$\begin{aligned} q_4 &= Bk_1b_gm_{vg} \\ q_5 &= \Psi_x k_1 b_g m_{vg} - i_e k_1 b_g m_{vg} - \frac{w}{C_1} g_b k_1 b_g m_{vg} \\ q_6 &= m_{ve} k_1 b_g m_{vg}, \end{aligned}$$

and where $w = (c_s - c_a)$ and $B = N_eRT$.

Equation 11.28 is solved for V_g as a function of V_e using the quadratic formula. The positive root produces negative V_g and is ignored. The qualitative dynamics and stability properties of the equations can be visualized by plotting the nullclines for V_g and V_e in the state space. Figure 11.5 shows the shape of the V_g and V_e nullclines for three levels of water vapor deficits. For clarity, the N_g nullcline is not shown. Equilibria exist at the intersection of the curves. Note that if only vapor pressure deficit (w) is altered (as shown here), the equilibria fall along the approximately linear V_g nullcline. Thus, steady-state epidermal volume (and pressure) is linearly related to steady-state guard cell volume. For the cases presented here, the nullclines with $w = 10$ bars indicate a stable equilibrium and those with $w = 25$ bars indicate a stable limit cycle.

Figure 11.6 illustrates this by superimposing the nonlinear dynamics for the three cases using parameters listed in Table 11.1. This shows that relatively small changes in the position of the equilibrium (Fig. 11.5) can produce dramatically different dynamics. Rand et al. (1981) proved that the similar model by Delwiche and Cooke (1977) exhibits a Hopf bifurcation from a fixed point to a limit cycle. This cycle in their model can be stable or unstable, depending on parameters. The numerical results of Fig. 11.5 are consistent with this mathematical analysis.

The existence of oscillations in stomatal conductance is well established. For example, Cardon et al. (1994) showed whole-leaf oscillations with a period of approximately 30 minutes and an amplitude of approximately 60 mmol·m^{-2}·sec^{-1}. These are approximately equal to the values produced by

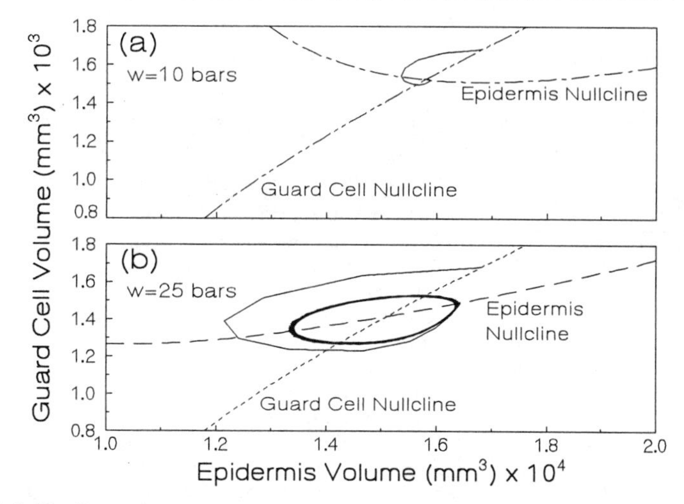

Figure 11.6: Nonlinear dynamics and nullclines for V_g and V_e for two water vapor deficits [10 (a) and 25 (b) bars].

the model using the parameters in Table 11.1 when $w = 25$ bars (Fig. 11.7).

11.4 Plant Growth

As we will discuss in more detail in Chapter 16, models and observational studies have a particular scale of space, time, and biological organization. The above models of photosynthesis apply to low levels of organization: biochemical and tissue or leaf. While it may very well be possible in principle to apply the detailed photosynthesis models over long enough time periods to model the growth of complete plants, this is not a useful endeavor. We therefore also need models at the level of the whole organism that describe how plant biomass changes with plant maturity. Such models would have important applications to agriculture as a means of predicting plant size as a function of soil moisture, fertilizer, or weather. In the following sections, we describe a few of the more widely used approaches.

11.4.1 Growth of Total Plant Biomass

One class of models treats the individual as a homogeneous black box in terms of weight of biological material. Three such empirical approaches to growth are described here.

Logistic If we think of an organism as being composed of a population of cells of fixed size, then the density-dependent model of population growth will describe an organism. This theory states that as the number of cells increases, the amount of resources for cell division decreases, reducing or-

Table 11.1: Parameter values used in the model of stomata dynamics.

		INITIAL CONDITIONS
V_e	1×10^4	Epidermis volume/m^2 leaf
V_g	1×10^3	Guard cell volume/m^2 leaf
N_g	1.017×10^3	Guard cell ion content/m^2 leaf
		PARAMETERS
b_0	0.0	Stomate aperture when cell pressure is 0.0
b_e	1.0	Effect of P_e on stomate aperture
b_g	1.0	Effect of P_g on stomate aperture
C_1	15.0	Xylem conductance
C_3	1.0	Epidermis–guard cell conductance
c_a	20.0, 12.0, 5.0	Atmospheric water vapor pressure
c_s	30.0	Leaf internal water vapor pressure
ϵ_e	50.0	Epidermis wall modulus
ϵ_g	50.0	Guard cell wall modulus
g_b	3.0	Atmospheric boundary layer conductance
k_1	0.10	Stomate aperture effect on conductance
N_e	678.4	Epidermis ion concentration
N_{gmin}	508.26	Guard cell minimum ion concentration
Ψ_x	0.0	Xylem water potential
$P_{e,ft}$	10.0	Epidermis pressure at full turgor
$P_{g,ft}$	15.0	Guard cell pressure at full turgor
R	0.08319	Gas constant
r	0.19675	Decay rate of ions in the guard cell
s	10.0	Effect of P_e on guard cell ion production
T	298	Temperature in degrees Kelvin
$V_{e,ft}$	4.2×10^{-5}	Volume of an epidermis cell at full turgor
$V_{g,ft}$	4.2×10^{-6}	Volume of a guard cell at full turgor
		NULLCLINE VARIABLES
i_e	-40.0	Epidermis full turgor pressure − ϵ_e
i_g	-35.0	Guard cell full turgor pressure − ϵ_g
m_{ve}	0.002976	$\epsilon_e/V_{\mathrm{e,ft}}$
m_{vg}	0.02976	$\epsilon_g/V_{\mathrm{g,ft}}$
w	10.0, 18.0, 25.0	Difference in external and internal humidities

ganism growth rate. The differential equation for this is the familiar logistic:

$$\frac{dW}{dt} = \mu W \left(1 - \frac{W}{W_f}\right), \tag{11.30}$$

where μ is the growth rate constant and W_f is the final weight. This simple equation has an analytic solution (France and Thornley 1984):

$$W = \frac{W_0 W_f}{W_0 - (W_f - W_0)\mathrm{e}^{-\mu t}}, \tag{11.31}$$

where W_0 is the initial weight. This is the classical sigmoid curve where the maximum rate of growth occurs when the plant is one-half W_f.

Gompertz Instead of hypothesizing that the cell division rate declines with increasing numbers of cells, we can assume that the rate simply de-

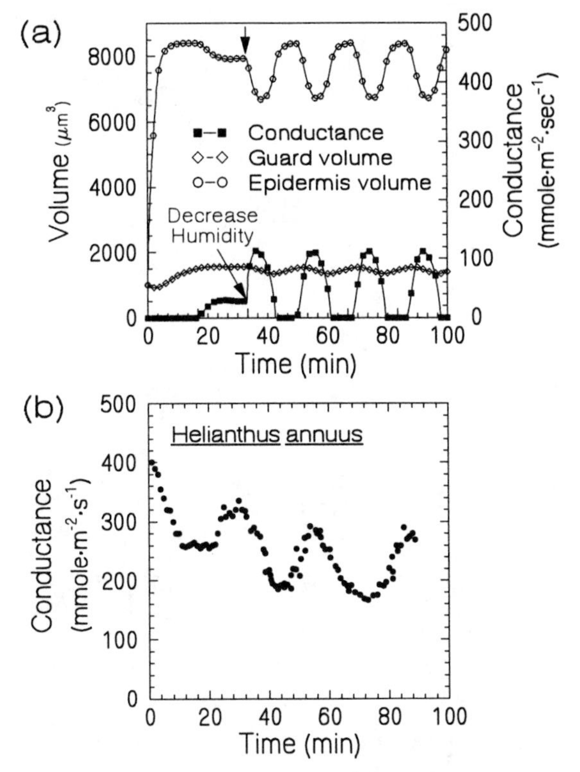

Figure 11.7: Oscillatory stomatal dynamics in experiments and models. (a) Model results using Eqs. 11.24-11.26 and parameters listed in Table 11.1. (b) Whole leaf stomatal conductance in *Helianthus annuus* measured in a gas exchange chamber. (From Cardon et al. 1994, Fig. 1a. © 1994 Blackwell Science, Ltd. Reprinted with permission Blackwell Science, Ltd, publisher.)

clines with time. This produces the Gompertz equations:

$$\frac{dW}{dt} = \mu(t)W \tag{11.32}$$

$$\frac{d\mu}{dt} = -D\mu, \tag{11.33}$$

where D is a decay constant. The latter equation does not depend on W, so it can be integrated directly: $\mu = \mu_0 \mathrm{e}^{-Dt}$. Substituting (France and Thornley 1984),

$$\frac{dW}{dt} = \mu_0 W \mathrm{e}^{-Dt} \tag{11.34}$$

$$W = W_0 \exp(\mu_0(1 - e^{-Dt})/D). \tag{11.35}$$

Chanter The Chanter model assumes both the time dependence of μ used in the Gompertz equation and the resource limitation of the logistic (France

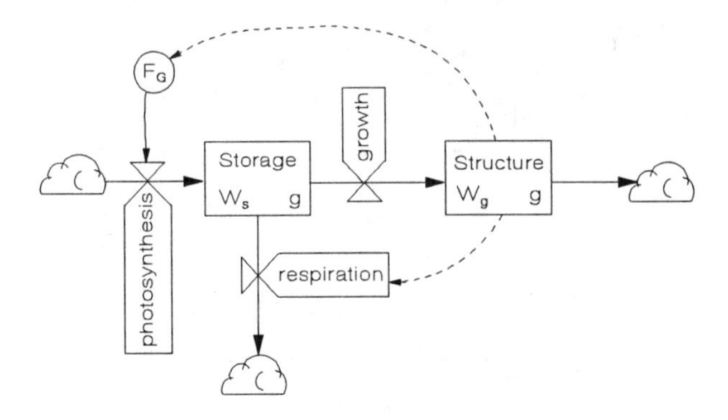

Figure 11.8: Forrester diagram for growth of a winter lettuce plant.

and Thornley 1984):

$$\frac{dW}{dt} = \mu W \left(1 - \frac{W}{B}\right) e^{-Dt} \tag{11.36}$$

$$W = \frac{W_0 B}{W_0 + (B - W_0)\exp\left(-\mu(1 - e^{-Dt})\right)/D}. \tag{11.37}$$

Several other phenomenological models are described in France and Thornley (1984). Models of the above type are often used as components in more complex models.

11.4.2 Whole–Plant Model

A simple example of the use of the Gompertz model is a model of winter lettuce growth dynamics by Sweeney et al. (1981). The Forrester diagram of this problem is shown in Fig. 11.8. In this simple model, there are two state variables for a pool of storage material (e.g., g C: W_S) and a pool representing the structure of the plant (W_G). The latter is typically interpreted as leaf area:

$$\begin{aligned} \frac{dW_S}{dt} &= \theta F_g(t) - E/Y_G \\ \frac{dW_G}{dt} &= E = W_G \mu e^{-Dt}, \end{aligned} \tag{11.38}$$

where Y_G is a conversion factor relating grams of substrate to grams of structural component, θ is another conversion factor relating grams of CO_2 fixed by photosynthesis to grams of growth substrate (W_S), and D is an empirically determined parameter that describes the effect of time on the reduction of growth rates. μ, F_g, and E are functions described below. F_g is the amount of CO_2 fixed during photosynthesis; it depends on the amount of light available, the amount of leaf area present to intercept the light, and a time decay to describe environmental and plant morphological changes over time:

$$F_g(t) = A_e P_g(t),$$

where A_e is the effective leaf area and $P_g(t)$ is the gross photosynthetic rate of a leaf.

Leaf area is a hyperbolic function of current plant size:

$$A_e = h^2 \left(1 - e^{F_G W_G / h^2}\right),$$

where h is the planting distance between plants and F_G is a proportionality constant that relates the current size of the plant to actual leaf area. A_e is asymptotic to the maximum area that the plant can expose to the sun without overlapping other plants.

Following the relation originally proposed in Monsi and Saeki (1953, and later discussed in Monsi et al. 1973), gross photosynthetic rate depends on light intensity and time:

$$P_g(t) = \frac{\alpha I(\tau C - \beta)}{\alpha I + \tau C} e^{-D_p t},$$

where α is the light utilization efficiency, I is instantaneous rate at which light strikes a unit area of the earth's surface, C is atmospheric CO_2 concentration, τ is CO_2 conductance from the atmosphere to the plant, and β is a constant loss of CO_2 to respiration during photosynthesis. Time (t) is measured in days. The maximal rate of photosynthesis decays exponentially over time by an amount D_p per day. Sweeney et al. (1981) further assumed that light levels are sufficiently low that the photosynthetic rate is restricted to the nearly linear portion of the low light portion of the curve, so that, approximately,

$$P_g = \alpha' I e^{-D_p t}.$$

This formulation is an example of the modeling principle to elaborate a process by converting a constant (maximum photosynthesis rate,: α') into a variable, which, in this case, depends on time [$\exp(-D_p t)$]. Combining these two limiting processes (light and carbon fixation), the control of photosynthetic rate by leaf area exposed to light and the biochemical rates of photosynthesis is

$$F_g(t) = h^2 \left(1 - e^{F_G W_G / h^2}\right) \frac{\alpha I(\tau C - \beta)}{\alpha I + \tau C} e^{-D_p t}.$$

Notice that this equation employs the multiplicative method of determining overall process rate as the product of two separate controlling mechanisms (leaf area and biochemical rates).

Sweeney et al. (1981) incorporated environmental effects into the growth equations by allowing temperature to influence the parameters μ, D, and D_p (hence, they are not truly constants). The authors used the Q_{10} method of incorporating temperature effects. The Q_{10} of a biochemical process is the amount that the rate of the process is increased when temperature is raised 10°C above a reference level (usually 20°C). The model of the

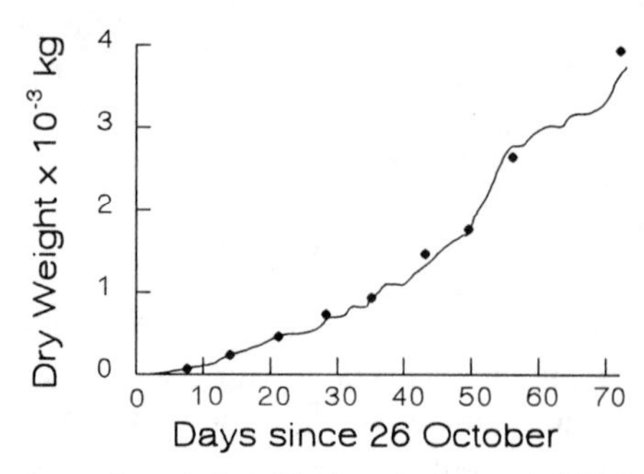

Figure 11.9: Comparison of predicted (line) and observed (filled circles) above-ground plant parts in winter lettuce. (From Sweeney et al. 1981, Fig. 2. © 1981 Academic Press, Ltd. Reprinted with permission Academic Press, publisher.)

temperature (T) effect on the rate (R) is:

$$R(T) = R(20)Q^{(T-20)/10}.$$

Q is experimentally estimated by performing the appropriate experiments at a series of temperatures and solving for Q in the above equation. Both the decay parameters (D and D_p) and the metabolic rates (μ) possess their own Q_{10} parameters.

If the basic time scale of the model is 1 d, then the complete model is

$$\begin{aligned} \frac{dW_S}{dt} &= \theta h^2 \left(1 - e^{F_G W_G/h^2}\right) \alpha' J_t e^{-D_p(T_t)t} \\ &\quad - \frac{1}{Y}\mu(T_t)\left(\frac{W_S}{W_S + W_G}\right) e^{-D(T_t)t} \\ \frac{dW_G}{dt} &= W_G \mu(T_t)\left(\frac{W_S}{W_S + W_G}\right) e^{-D(T_t)t}, \end{aligned} \tag{11.39}$$

where T_t is a time series of environmental temperatures, and J_t is the daily rate of light flux. With appropriate adjustment of some of the empirical parameters, this model fits typical field data well (Fig. 11.9).

Because of its empirical accuracy, such a model can be used to design planting regimes of lettuce. For example, the effects of planting distances (h) and planting timing on final biomass can be investigated. Both variables can be chosen to maximize lettuce size using evolutionary optimization techniques (Chapter 19).

11.4.3 Partitioning Resources to Organs

While the above model of plant growth does well for a particular crop and planting environment, it relies heavily on empirical data. It incorporates relatively few mechanistic details. In particular, it fails to distinguish growth dynamics in two major types of plant structures: roots and shoots.

The *partitioning* of nutrients and photosynthetic by-products is an important direction of plant growth modeling. This model problem also illustrates the use of submodels when a system comprises flows of several conserved quantities. Thornley (1972) described an early model that is still widely used today. Here we describe an extension of Thornley's model by Dewar (1993) that incorporates another important mechanism of nutrient translocation between roots and shoots: Münch flow.

The plant is described as having two *substrate* compartments in both roots (r) and shoots (s, aboveground plant material). The two compartments are pools of labile carbon and nitrogen, denoted as $W_{C,r}$ and $W_{N,r}$ for root compartments and $W_{C,s}$ and $W_{N,s}$ for shoot compartments. The model plant also has two *structural* compartments representing the amounts of C and N contained in anatomical structures in roots (e.g., roots and root hairs) and shoots (e.g., stems and leaves) (W_r and W_s). There are a total of six state variables and differential equations:

$$\frac{dW_s}{dt} = W_s \left[k_s C_s N_s \left(\frac{1 - \Psi_s}{\Psi_c} \right) \right] \tag{11.40}$$

$$\frac{dW_r}{dt} = W_r \left[k_r C_r N_r \left(\frac{1 - \Psi_r}{\Psi_c} \right) \right] \tag{11.41}$$

$$\frac{dW_{C,s}}{dt} = \underbrace{[\sigma_C W_s]}_{\text{photosynthesis}} - \underbrace{\left[\frac{C\text{av}(C_s - C_r)}{R_{ph}} \right]}_{\text{Münch flow}} - \left[f_C \frac{dW_s}{dt} \right] \tag{11.42}$$

$$\frac{dW_{N,s}}{dt} = [\lambda \sigma_N W_r] - \left[\frac{N\text{av}(C_s - C_r)}{R_{ph}} \right] - \left[f_N \frac{dW_s}{dt} \right] \tag{11.43}$$

$$\frac{dW_{C,r}}{dt} = \left[\frac{C\text{av}(C_s - C_r)}{R_{ph}} \right] - \left[f_C \frac{dW_r}{dt} \right] \tag{11.44}$$

$$\frac{dW_{N,r}}{dt} = [(1 - \lambda)\sigma_N W_r] + \left[\frac{N\text{av}(C_s - C_r)}{R_{ph}} \right] - \left[f_N \frac{dW_r}{dt} \right], \tag{11.45}$$

where

$$C_s = W_{C,s}/W_s \tag{11.46}$$

$$C_r = W_{C,r}/W_r \tag{11.47}$$

$$N_s = W_{N,s}/W_s \tag{11.48}$$

$$N_r = W_{N,r}/W_r \tag{11.49}$$

$$C\text{av} = f_s C_s + f_r C_r \tag{11.50}$$

$$N\text{av} = f_s N_s + f_r N_r \tag{11.51}$$

$$A = \left(\frac{1}{W_s} + \frac{1}{W_r} \right) \tag{11.52}$$

$$R_{ph} = r_{ph} A \tag{11.53}$$

$$\Psi_s = \Psi_r - E[r_{xy} A] \tag{11.54}$$

$$\Psi_r = \Psi_{\text{soil}} - E[r_{sr}/W_r]. \tag{11.55}$$

The parameters are defined in Table 11.2. C_i and N_i ($i = s, r$) are the relative concentrations of C and N in roots and shoots. C_{av} and N_{av} are average concentrations of substrate C and N in the plant. E (Eqs. 11.54 and 11.55) is transpiration. R_{ph} is the phloem resistance to sap flow between shoot and root (a sum of two resistances in series). A is the sum two resistances in series. Ψ_r is the xylem water potential between the soil and the root; Ψ_s is the water potential between the root and the shoot. The quantity in brackets in Eq. 11.54 is xylem water flow resistance between root and shoot, and the quantity within brackets in Eq. 11.55 is the resistance between soil and root. Two important auxiliary variables are the fraction of structural dry matter in the roots $[f_r = W_r/(W_s + W_r)]$ and shoots $[f_s = W_s/(W_s + W_r)]$. The shoot:root ratio (f_s/f_r) is a third auxiliary variable used to summarize the overall state of the plant.

The relative growth rates of structural C and N are contained within brackets in Eqs. 11.40 and 11.41. They are based on the mass action principle in which both C and N are required for a chemical reaction.

Equation 11.42 describes the dynamics of shoot substrate carbon using one input and two outputs. Shoot C increases by the first term on the right in brackets, which represents the amount of C derived from photosynthesis. All of this C contributes to the substrate C stored in the shoot. The second term on the right is Münch flow whereby shoot C is transported via diffusion to the roots according to the concentration gradient of C between the shoots and roots. The third term represents the amount of C substrate that is used to produce the C used in plant structure. Similar output components exist in the flow of substrate N (Eq. 11.43). Since all N is taken up in the roots, the input of substrate N to shoot storage is that fraction, λ, of the N absorbed that is subsequently transported in xylem to the shoots via transpiration.

Substrate C is added to roots by Münch flow in the phloem (first bracket pair in Eq. 11.44) and so depends on the gradient of C. A fraction (f_C) of the increase of structural C in the roots is taken from the substrate C (second bracket pair). Similar processes add and remove substrate N from the roots. In addition, root substrate N has a source directly from root N absorption (first bracket pair in Eq. 11.45).

While the model needs many more validation efforts, a change in one of the parameters indicates that the model is qualitatively accurate. The mass-specific rate of carbon fixation by the shoot component (σ_c) measures the efficiency by which atmospheric C is assimilated per unit of photosynthetic material. This efficiency is dependent upon many factors, for example, the concentration of CO_2 in the atmosphere. Whatever the mechanism, does the model respond correctly when this parameter is doubled? The answer appears to be yes. Dewar (1993) allowed the model to reach an equilibrium in its root and shoot C substrate, then doubled σ_c (Fig. 11.10 arrow). Immediately, C_s increased rapidly, and a short time later a smaller

Table 11.2: Parameters and initial conditions in the lettuce model. DM = kg of dry matter.

	INITIAL CONDITIONS	
W_s	Shoot structural DM	1.0 DM·m^{-2}
W_r	Root structural DM	1.0 DM·m^{-2}
$W_{C,s}$	Shoot C substrate DM	0.15 DM·m^{-2}
$W_{C,r}$	Root C substrate DM	0.05 DM·m^{-2}
$W_{N,s}$	Shoot N substrate DM	0.03 DM·m^{-2}
$W_{N,r}$	Root N substrate DM	0.03 DM·m^{-2}
	PARAMETERS	
σ_C	C shoot fixation	0.15 kg (DM·d^{-1})
σ_N	N shoot uptake	0.05 kg (DM·d^{-1})
σ_W	Shoot transpiration	15.0 kg (DM·d^{-1})
r_{ph}	Phloem resistance	0.5 (d^{-1})
r_{xy}	Xylem resistance	10.0 ($m^2 \cdot$ d^{-1})
r_{sr}	Soil–root resistance	1.0 (m$^2 \cdot$ d^{-1})
Ψ_{soil}	Soil water potential	-100 (J · kg^{-1})
$k_s(k_r)$	Shoot(root) growth	500 (d^{-1})
Ψ_c	Critical water potential	-1500 (J · kg^{-1})
f_C	C content	0.45 (unitless)
f_N	N content	0.03 (unitless)

increase in root substrate C (C_r) occurred due to Münch flow. As this process removed C from the shoots, C_s decreased to a new equilibrium but higher than the previous one (Fig. 11.10a). Since the new equilibrium of shoot C substrate was relatively higher than the new equilibrium for root C substrate, C_s is larger than C_r. By Eqs. 11.42 and 11.44, $W_{C,s}$ will decrease relative to $W_{C,r}$. This resulted in smaller equilibrium values of W_s relative to W_r; hence, root structure will increase relative to shoot structure (Fig. 11.10b). This is qualitatively similar to experimental manipulations of ambient C levels (Dewar 1993).

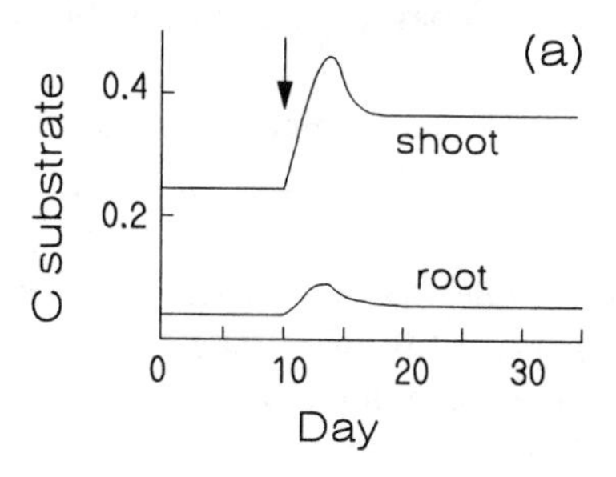

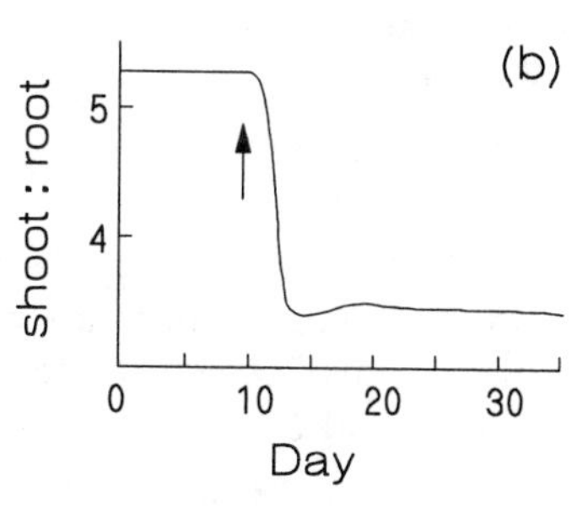

Figure 11.10: Response of a carbon–nitrogen allocation model of plant growth to increased C fixation efficiency. The model was allowed to equilibrate, then photosynthetic efficiency was doubled on day 10 (arrows). (a) Rapid increase in shoot C (C_s) is followed by an increase in root C (C_r) via Münch flow. (b) Rapid decrease in proportion of structure in shoots relative to roots (f_s/f_r). (From Dewer 1993, Fig. 4a and 4c. © 1993 Blackwell Science, Ltd. Reprinted with permission Blackwell Science, Ltd., publishers)

1.5 Summary

This chapter has illustrated a variety of modeling techniques and scales that occur in theoretical plant physiology. These examples were chosen to describe a variety of solutions and approaches. The steady-state assumptions used in the biochemical model of photosynthesis is characteristic of many models of biochemical mechanisms in plants and animals. In contrast, the dynamic model of stomata is an application of standard dynamical analysis tools (e.g., flows of conserved quantities, nullclines) that are not commonly applied to this scale in plants. As illustrated here, models at higher levels of biological organization (see Chapter 16) can achieve accurate descriptions of whole plant growth, but lose mechanistic detail. These models gain, however, potential practical applications as aids to optimizing agricultural practice.

1.6 Exercises

1. Study each of the models in this chapter and identify principles that were discussed in *Part I.* Are there other general principles contained in these models that were not mentioned earlier?
2. Write and solve a model that adds pests to the whole crop model. Assume that insects prefer young leaf material to old leaf material.
3. Write a computer program that simulates the model of stomatal dynamics. Use it address the question: How does the value of epidermis cell wall modulus affect the dynamics? Does this quantity interact with guard cell wall modulus?
4. Do the equilibrium results of the Dewar model agree with the model of Farquhar and von Caemmerer (1982) and data of Evans and Farquhar (1991)?
5. Simulate the Dewar plant model and find parameters that will cause oscillations.
6. Derive the nullcline equations for the stomate model Eqs. 11.27 – 11.28.
7. Draw a Forrester diagram for Dewer's model Eqs. 11.40 – 11.45.
8. Perform a first-order (analytical) error analysis of photosynthesis rate when RuBP is saturating based on Eq. 11.9 for all the parameters and C and O. Calculate the 95% confidence interval.

Chapter 12

Hormonal Control in Mammals

12.1 Hormonal Regulation

THE hallmark of vertebrate physiology is the fine control of physiological states by negative feedback systems. For this to be effective, there must be mechanisms to turn off operating processes and to turn on dormant processes. This requires that there be bodywide communication among system components that signals the state of operating processes. The coordinated interaction of the central nervous system and *hormones* is one of the most important mechanisms by which negative feedback is achieved. Hormones are chemicals that are transported long distances via the blood and that are capable of turning on and off processes occurring at the site of hormone action. This chapter describes a mathematical model of one of these feedback systems that causes the level of glucose in the blood to be regulated within relatively narrow bounds.

The model illustrates a number of principles developed in *Part II*. First, it demonstrates the trade-offs required in model construction to balance mechanistic realism against mathematical simplicity and the need to minimize data requirements. Because the model is relatively complex, this chapter also illustrates the utility of Forrester diagrams for model exposition. As the model structure is explicated, principles of quantitative model formulation are revisited when we introduce a new, flexible mathematical function for representing nonlinear biological processes. Finally, the use of models to address interesting, practical questions is illustrated here by investigating the effects of eating on the blood sugar levels of diabetic and obese medical patients. These simulations demonstrate the potential practical value of mathematical models for patient diagnosis and treatment.

12.2 Glucose and Insulin Regulation

It's inconvenient to have to eat continuously, Superbowl Sunday not withstanding. Eating, while generally an enjoyable experience, can interfere

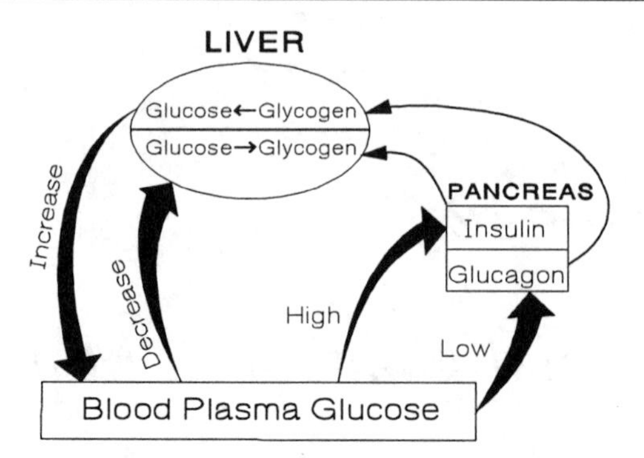

Figure 12.1: Blood plasma glucose is regulated by insulin and glucagon. When glucose concentration in the bloodstream is high, insulin production is stimulated, which results in the storage of glucose as glycogen in the liver. When blood glucose is low, glucagon converts liver glycogen to glucose, which is then added to the bloodstream.

with other worthwhile activities, such as changing channels or escaping from predators. Moreover, with the exception of a few ungulates and laboratory mice in feeding experiments, a predator's prey rarely cooperates by being continuously available for consumption. But the cells of most mammals, require a continual supply of energy, although a few organisms, such as hummingbirds, are capable of entering a physiological state called *torpor* in which their metabolic demand is reduced to extremely low levels during the night. In particular, maintaining functioning of the central nervous system and the ability to perform rapid, energy intensive muscular reactions require carbohydrates (Berne and Levy 1993). From the stomach's point of view, then, it is necessary to have a storage capacity into which glucose can be sequestered immediately following eating and that can later be resupplied to the body as needed between meals. The plasma glucose control system is an intricate and elegant mechanism for storing and releasing carbohydrates.

As Fig. 12.1 diagrams, a negative feedback system for regulating the concentration of glucose in the blood plasma has evolved. This system involves the coordinated activity of blood chemistry, the pancreas, and the liver. Although the details of glucose regulation are replete with biochemical and genetic details, the basic story is simple to tell (Guyton 1986, Raven and Johnson 1992, Berne and Levy 1993).

After a meal, sugars and other complex carbohydrates are broken down into glucose which crosses the stomach wall and enters the bloodstream. As the blood perfuses tissues and cells in its circuit through the circulatory system, plasma glucose is passed to the cells according to the cell's internal demands at the moment. Glucose cannot be stored in the blood for long periods because of the effects it has on other physiological systems. So, if the momentary supply of plasma glucose exceeds the demand, special glucose "detectors" stimulate cells in the pancreas called the *islets of Langerhans* to produce *insulin*. This is a hormone that attaches to the surface of cells and stimulates the cells to absorb glucose, that is, remove it from the bloodstream. Muscle and liver cells are especially sensitive to insulin and much of the glucose is stored there. Once inside the storage cells, glucose is transformed to *glycogen*, a relatively inert starch-like molecule similar to glucose. When the glucose detectors are switched off after the concentration of blood glucose falls, another set of cells in the islets of Langerhans secretes *glucagon* into the blood. This hormone is carried to the glycogen-storing cells of the liver or muscles and reconverts glycogen to glucose.

In a normal human, there is enough glycogen in the liver to maintain appropriate blood glucose concentrations for 10 hours without eating. After that, other noncarbohydrate molecules are converted to glucose to keep up with the demand of the nervous and muscular systems. A normal individual weighing 70 kg has about 91.5 mg of glucose per 100 ml of blood plasma, 11 μmole insulin Units per 1 ml (μU/ml) of plasma, and 75 pico-grams of glucagon per 1 ml (pg/ml) of plasma. [International Units (U) of a substance is the amount that produces a specific quantitative result in a bioassay. The physical amount depends on the substance and the nature of the bioassay.)]

The regulation of these normal levels can fail for two main reasons. If the body cannot secrete sufficient levels of insulin, then glucose is not removed from the blood and increases to dangerous levels. As a result, a cascade of chemical and physiological reactions occur that decrease blood pH to 6.8 or less. Among the many physiological reactions that are impeded by low pH is the affinity of hemoglobin for O_2. Low blood pH means that less O_2 is carried to vital organs, and death can result. This is a disease known as *Type I diabetes mellitus.* Alternatively, there may be sufficient insulin, but too few insulin receptors on the glycogen-storing cells (in the liver and muscle). Without receptors, these cells cannot detect the presence of insulin and the consequent need to stimulate the absorption of glucose for storage. This is known as *Type II diabetes mellitus* and also results in dangerously high levels of glucose in the blood. There are other clinical conditions that are correlated with abnormal insulin dynamics. For example, obese individuals have high rates of insulin production. Also, intense physical exertion reduces the rate of insulin secretion.

Mathematical models of the glucose–insulin system are valuable both for providing theoretical insight into the mechanisms of diabetes and as a diagnostic tool. In the latter case, a model is constructed that is based on easily measured patient quantities (e.g., body weight and normal plasma glucose concentration) and that can be driven by perturbations that correspond to standard medical diagnostic procedures (e.g., oral ingestion of a known amount of glucose). Depending on the subsequent dynamics of blood glucose for a patient with a given baseline concentration, irregularities in insulin secretion or cell-wall reception can be detected.

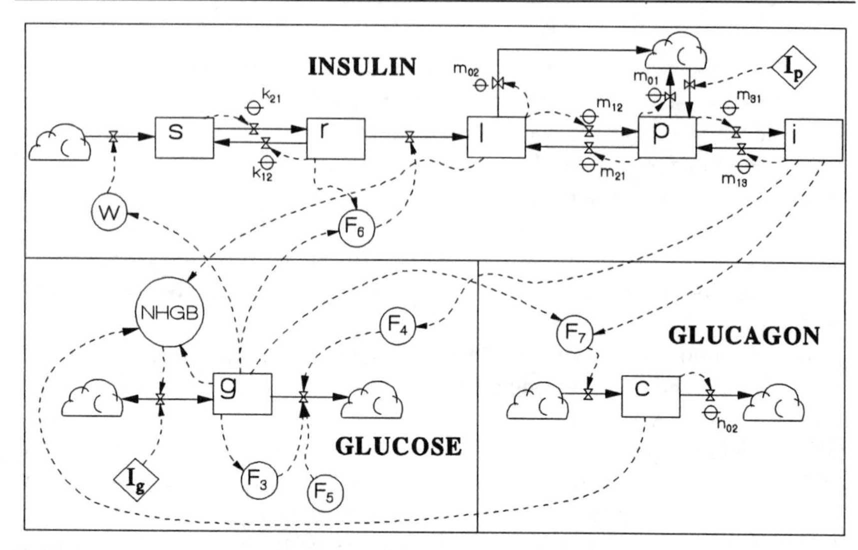

Figure 12.2: Model of the glucose–insulin regulation system based on three major subsystems: INSULIN, GLUCOSE, and GLUCAGON. For clarity, lines from parameters to rates have been omitted, and other parameters are subsumed in auxiliary functions (F_i). See Table 12.1 for variable definitions.

12.3 Glucose Model of Intermediate Complexity

Cobelli et al. (1982) developed a model of glucose regulation that has an intermediate level of complexity. As such, it incorporates many feedback loops missing from simpler models (improving its utility in diagnosis), but is simple enough to pass validation tests. A slightly simplified Forrester diagram for the model is shown in Fig. 12.2. The three submodels are shown as three parallel models in the Forrester sense (g = glucose submodel, c = glucagon sub-model, and other levels belonging to the insulin submodel: s, r, l, p, and i).

The model is semi-phenomenological since many of the important mechanisms are incorporated, but it is intended to be tailored to a particular patient. As a result, the model is parameterized so that it can be scaled around the "normal" operating conditions of the patient. This is achieved through the clever use of the hyperbolic tangent function (tanh in Fig. 12.3), which has a domain of $\pm\infty$ and range ± 1. In many of its applications in this model, the domain of the function is the *basal (baseline) plasma concentrations* of the state variables (e.g., glucose). The range of the function is the rate of production of one of the state variables (e.g., liver glucose production rate). Empirical and theoretical constants scale the maximum of tanh to appropriate biological values. In addition to being an asymptotic function, tanh is symmetric about the $x = 0$ line. In the glucose model, this fact is used so that a negative departure from normal substance levels produces a negative response. Also, by subtracting the

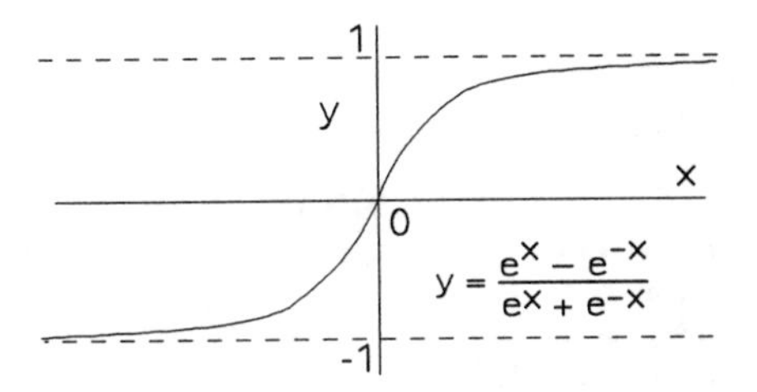

Figure 12.3: The hyperbolic tangent function (tanh) used in the glucose model.

function from 1.0 (i.e., $1 - \tanh$), we can describe a monotonically decreasing function that corresponds to a negative feedback relation between the dependent and independent variable.

12.3.1 Basic Equations

The state variables and important auxiliary variables are defined in Table 12.1. They are related by the following differential equations:

$$\frac{dg}{dt} = \text{NHGB} - F_3 - F_4 - F_5 + I_g(t) \tag{12.1}$$

$$\frac{dc}{dt} = -h_{02}c + F_7 \tag{12.2}$$

$$\frac{di}{dt} = -m_{13}i + m_{31}p \tag{12.3}$$

$$\frac{dl}{dt} = -(m_{02} + m_{12})l + m_{21}p + F_6 \tag{12.4}$$

$$\frac{dp}{dt} = -(m_{01} + m_{21} + m_{31})p + m_{12}l + m_{13}i + I_p(t) \tag{12.5}$$

$$\frac{dr}{dt} = k_{21}s - (k_{12} + k_{02})r \tag{12.6}$$

$$\frac{ds}{dt} = -k_{21}s + k_{12}r + W. \tag{12.7}$$

Many of the physiological processes depend on the concentrations of the primary variables in the model. As a consequence, the following concentrations are defined based on the absolute quantities of the state variables and the volumes of tissues in which they are confined. Associated with each state variable, we define: $\bar{g} = g/V_b$, $\bar{p} = p/V_p$, $\bar{l} = l/V_l$, $\bar{i} = i/V_i$, $\bar{c} = c/V_b$, where V_b is the volume of blood and extracellular or interstitial fluids (0.2 of body weight divided by blood density), V_p is the volume of plasma (0.045 of body weight divided by plasma density), V_l is the volume of liver (0.03 of body weight divided by liver density), and V_i is the volume of interstitial fluid (0.10 of body weight divided by interstitial fluid density). Each of these concentrations will be standardized by subtracting a patient's *basal* (or normal) concentration of the substance from the dynamic concentration. For example, the standardized glucose concentration

Table 12.1: Variables used in the glucose-insulin model. Functional relationships are diagrammed in 12.2. U = international units.

STATE VARIABLES	
c	glucagon in plasma and interstitial fluids (nU)
g	glucose in plasma and extracellular fluid (mg)
i	interstitial fluid insulin (μU)
l	liver insulin (μU)
p	plasma insulin (μU)
r	releasable pancreatic insulin (μU)
s	stored pancreatic insulin (μU)
AUXILIARY VARIABLES	
NHGB	Net Hepatic (liver) Glucose Balance ($F_1 - F_2$)
F_1	Liver glucose production rate
F_2	Liver glucose uptake rate
F_3	Renal (kidney) glucose excretion rate
F_4	Peripheral system (muscles) glucose use rate
F_5	Non-peripheral system (central nervous system and red blood cells) glucose uptake rate
F_6	Insulin secretion rate
F_7	Glucagon secretion rate
I_g, I_p	Glucose, insulin ingestion rate
W	Insulin synthesis rate

is $\Delta\bar{g} = \bar{g} - g_{\text{basal}}$; the standardized concentrations of the other substances are defined similarly. Also in the following, doubly or triply subscripted letters are constants. The notation for parameters is taken from Cobelli et al. (1982). The details of the model for each subsystem follow.

12.3.2 Glucose Subsystem

The glucose subsystem is described by Eq. 12.1. Glucose in the plasma and extracellular fluid is produced by the liver and the stomach. Net glucose production by the liver is NHGB (net hepatic glucose balance) and is the difference between liver glucose production and uptake. Glucose production rate (F_1) is limited by three factors: the standardized concentrations of glucose, liver insulin, and plasma glucagon. Glucagon stimulates the formation of glucose from glycogen (G_1); both liver insulin (H_1) and plasma glucose (M_1) reduce glucose levels. These effects are combined multiplicatively (see Section 4.3.6 on multiple limiting factors). Similarly, liver glucose uptake rate (F_2) is a multiplicative combination of the negative effects of liver insulin (H_2) and the positive effects of glucose concentrations (M_2). These hypotheses are combined as

$$\begin{aligned}
\text{NHGB} &= F_1 - F_2 \\
F_1 &= a_{11} G_1 H_1 M_1 \\
G_1 &= 0.5[1 + \tanh(b_{11}(\Delta\bar{c} + c_{11})] \\
H_1 &= 0.5[1 - \tanh(b_{12}(\Delta\bar{l} + c_{12})] \\
M_1 &= 0.5[1 - \tanh(b_{13}(\Delta\bar{g} + c_{13}))]
\end{aligned}$$

$$\begin{aligned} F_2 &= H_2 M_2 \\ H_2 &= 0.5[1 - \tanh(b_{21}(\Delta\bar{l} + d_{12})] \\ M_2 &= a_{221} + a_{222} 0.5[1 + \tanh(b_{22}(\Delta\bar{g} + c_{22}))]. \end{aligned}$$

G_1 is the positive effect of standardized glucagon concentration on glucose production, H_1 is the negative effect of liver insulin, and M_1 is the negative effect of glucose. H_2 is the negative effect of liver insulin on glucose uptake and M_2 is the positive effect of glucose on glucose uptake.

There are three other major losses of plasma glucose: kidney excretion, uptake by fatty tissue and muscles, and uptake by the blood cells and nerves. F_3 is the renal (kidney) excretion rate of glucose:

$$\begin{aligned} F_3 &= M_{31} M_{32} \\ M_{31} &= 0.5[1 + \tanh(b_{13}(\Delta\bar{g} + c_{31})] \\ M_{32} &= a_{321}(\bar{g} + c_{31}), \end{aligned}$$

where M_{31} is the negative feedback effect of deviations of glucose from the basal value, and M_{32} is the linear flow rate from the plasma glucose compartment to urine and eventual excretion.

Glucose is removed from blood plasma by being used in adipose and muscular tissue (F_4) and in the central nervous system and red blood cells (F_5):

$$\begin{aligned} F_4 &= a_{41} H_4 M_4 \\ H_4 &= 0.5[1 + \tanh(b_{41}(\Delta\bar{i} + c_{41}))] \\ M_4 &= 0.5[1 + \tanh(b_{42}(\Delta\bar{g} + c_{42}))] \\ F_5 &= M_{51} + M_{52} \\ M_{51} &= a_{51} \tanh(b_{51}(\Delta\bar{g} + c_{51})) \\ M_{52} &= a_{52}(\bar{g} + b_{52}), \end{aligned}$$

where H_4 and M_4 are positive effects of interstitial insulin and glucose, respectively, on adipose and muscle use, and M_{51} and M_{52} are the effects of positive effects of glucose on central nervous system use.

Finally, glucose and insulin are added to the plasma by means of ingestion, either intravenously or orally. $I_g(t)$ is glucose ingestion, and $I_p(t)$ is insulin ingestion. These functions of time are used for diagnostic tests. Here we will focus on IVGTT, the intravenous glucose tolerance test, which is a standard medical diagnostic test.

12.3.3 Glucagon Subsystem

In the glucagon submodel (Eq. 12.2), control of glucagon production (F_7) depends on plasma glucose and insulin concentrations. Large values of either of these two quantities result in lowered amounts of glucagon pro-

Table 12.2: Nominal parameters for the glucose-insulin model.

GLUCOSE		
$a_{11} = 1.51$	$a_{221} = 1.95 \times 10^{-3}$	$a_{321} = 1.43 \times 10^{-5}$
$b_{11} = 2.14$	$a_{222} = 5.21 \times 10^{-3}$	$a_{322} = -1.31 \times 10^{-5}$
$b_{12} = 7.84 \times 10^{-2}$	$b_{21} = 1.11 \times 10^{-2}$	$b_{31} = 20$
$b_{13} = 2.75 \times 10^{-2}$	$b_{22} = 1.45 \times 10^{-2}$	$c_{31} = -180$
$c_{11} = -0.85$	$d_{12} = 51.3$	$c_{12} = 7$
$c_{22} = -108.5$	$c_{13} = 20$	$a_{41} = 2.87 \times 10^{-2}$
$a_{51} = 1.01 \times 10^{-3}$	$a_{52} = 4.6 \times 10^{-6}$	$b_{41} = 3.1 \times 10^{-2}$
$b_{42} = 1.44 \times 10^{-2}$	$b_{51} = 2.78 \times 10^{-2}$	$c_{41} = -50.9$
$b_{52} = 4.13 \times 10^{-4}$	$c_{42} = -20.2$	
INSULIN		
$k_{12} = 0.01$	$k_{21} = 4.34 \times 10^{-3}$	$m_{01} = 0.125$
$m_{02} = 0.185$	$m_{12} = 0.209$	$m_{13} = 0.02$
$m_{21} = 0.268$	$m_{31} = 0.042$	$a_w = 0.287$
$a_6 = 1.3$	$b_w = 1.51 \times 10^{-2}$	$b_6 = 9.23 \times 10^{-2}$
$c_w = -92.3$	$c_6 = -19.68$	
GLUCAGON		
$a_{71} = 2.35$	$b_{71} = 6.86 \times 10^{-3}$	$b_{72} = 3.00 \times 10^{-2}$
$c_{71} = 99.2$	$c_{72} = 40$	$h_{02} = 0.086$

(From Cobelli et al. 1982.)

duction:

$$
\begin{aligned}
F_7 &= a_{71} H_7 M_7 \\
H_7 &= 0.5[1 - \tanh(b_{71}(\Delta\bar{i} + c_{71}))] \\
M_7 &= 0.5[1 - \tanh(b_{72}(\Delta\bar{g} + c_{72}))],
\end{aligned}
$$

where H_7 is the negative effect of interstitial insulin on glucagon production and M_7 is the negative effect of glucose.

12.3.4 Insulin Subsystem

Finally, the insulin submodel is the most complex, having five compartments described in Eqs. 12.3–12.7. Most of the rate dynamics, however, are linear, donor-controlled relationships. Parameters of these relationships (e.g., m_{ij}, k_{ij}, and a_{ij}) are not verbally defined, but have values shown in Table 12.2. The only exceptions are the rates of insulin production and secretion in the pancreas. As Fig. 12.1 indicates, insulin is formed in the pancreas and is transported to the liver, where it stimulates the conversion of glucose to glycogen. Pancreatic insulin is assumed to occur in two forms: a nonlabile, stored form produced by the pancreas at rate W:

$$W = 0.5a_w[1 + \tanh(b_w(\Delta\bar{g} + c_w))],$$

and a form that Cobelli et al. (1982) called a "promptly releasable" form which is secreted from the pancreas at rate F_6,

$$F_6 = 0.5a_6[1 + \tanh(b_6(\Delta\bar{g} + c_6))].$$

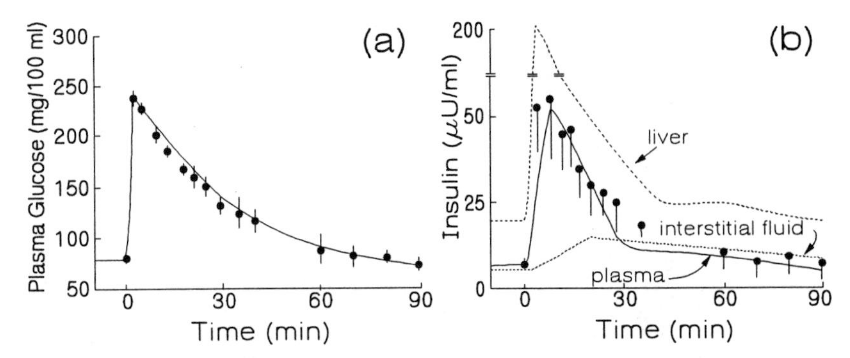

Figure 12.4: Simulated and average observed glucose (a) and insulin (b) responses of a normal individual following an intravenous pulse of glucose (IVGTT). Error bars are ± 1 standard error, $n = 5$ patients. In (b), the solid line is plasma insulin (p), the dashed line is liver insulin (l), and the dotted line is interstitial insulin (i). Points are observations. (Reprinted by permission of the publisher from Cobelli et al. 1982, Figs. 6 and 7. © 1982 by Elsevier Science, Inc.)

12.3.5 Normal Simulations

Table 12.2 lists the nominal parameters for a normal patient. The time courses for glucose and insulin, following the diagnostic glucose tolerance test (IVGTT), are shown in Fig. 12.4. Note that both plasma glucose and plasma insulin (solid lines) return to normal levels in about 90 minutes following the pulse of glucose. In Fig. 12.4b, note that liver and plasma insulin increase almost immediately with the glucose pulse.

A second kind of test applies repeated pulses of glucose at intervals less than that needed to clear the previous pulse from the system. Think of this as a way of simulating glucose injection during TV commercials. If repeated and increasing pulses are administered, glucose levels do not simply decay exponentially as they did following a single IVGTT (Fig. 12.5). A "hump" following pulse 6 develops which greatly delays the recovery of the system.

12.3.6 Diabetic Simulations

Parameters for diabetic individuals are listed in Table 12.3; functions F_2 and H_2 are replaced with constants. The response of a diabetic to the IVGTT is shown in Fig. 12.6. The basal level of glucose concentration is much higher than in a normal individual, and insulin levels are much lower. Consequently, the recovery period is much longer than in a normal individual. The system requires almost twice as long to return to basal conditions.

12.3.7 Obesity Simulations

The parameters appropriate to an obese subject are given in Table 12.3. Insulin and glucose responses to the IVGTT are shown in Fig. 12.7. Note

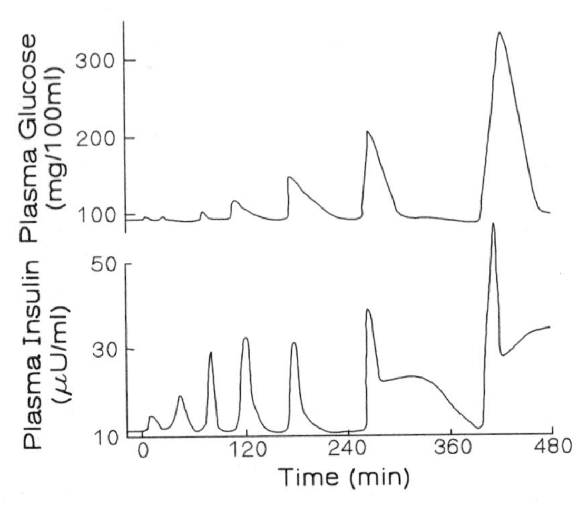

Figure 12.5: Simulated response of glucose (above) and insulin (below) from a subject given increasingly larger doses of intravenous glucose. At intervals of 0, 30, 70, 110, 180, 270, and 400 minutes, a 70-kg subject was administered 0.5, 1.0, 2.5, 5, 10, 20, and 40 grams of glucose. (Reprinted by permission of the publisher from Cobelli et al. 1982, Fig. 15. © 1982 by Elsevier Science, Inc.)

the nearly normal response of glucose but the abnormal hump in the insulin decay curve.

12.4 Summary

This model epitomizes a broad class of biomedical models that have quite good success. Part of this success is due to the fact that some (not all) model parameters are fitted to the patient being simulated. But part of the success is that we have a good understanding of these systems. This may be one class of biological models that can and have been used for diagnosis and prescription (i.e., product design). Other mammalian regulatory subsystems (e.g., cardiovascular) have been similarly studied.

Table 12.3: Parameters for diabetic and obese subjects. All other parameters as in Table 12.2.

Diabetes	Obesity
$F_2 = 0.037$	
$H_4 = 0.0012$	$H_4 = 0.0012$
$b_{42} = 7 \times 10^{-3}$	$b_{42} = 7 \times 10^{-3}$
$c_{42} = -40.47$	$c_{42} = 40.47$
$b_w = 4.5 \times 10^{-3}$	$m_{02} = 0.13$
$b_6 = 5 \times 10^{-3}$	$b_6 = 0.5$
$c_6 = -363.55$	$c_6 = -3.64$
$c_w = -306.25$	

(After Cobelli et al. 1982

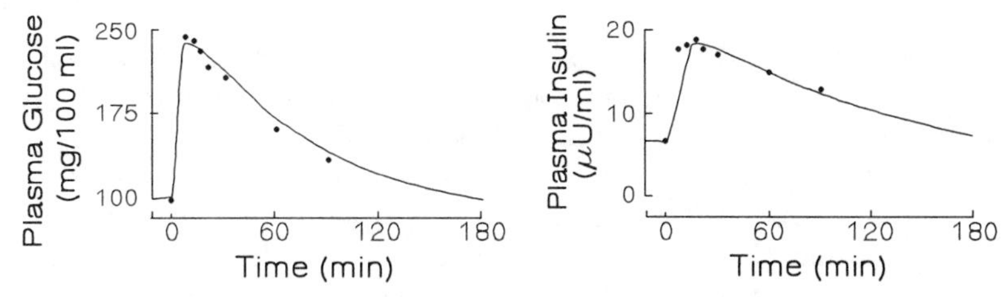

Figure 12.6: The response of a diabetic subject to the IVGTT test. Points are a single patient; the line is model predictions. Parameters as in Table 12.3. Note the slow recovery period. (Reprinted by permission of the publisher from Cobelli et al. 1982, Fig. 21. © 1982 by Elsevier Science, Inc.)

12.5 Exercises

1. Code the Cobelli model of glucose regulation and attempt to reproduce Figs. 12.4a and b. Also plot glucagon concentrations and rates of liver uptake of glucose. Discuss the results in light of the Forrester diagram. Where does the majority of glucose go?
2. Why does the "hump" in insulin concentration develop in Fig. 12.5? Why is there a similar hump for obese persons in Fig. 12.7?
3. Using the parameters for the diabetic subject, administer the sequence of glucose pulses described in Fig. 12.5. Compare to a normal subject.
4. Simulate the glucose infusion diagnostic test by adding glucose not as a pulse (as in the IVGTT), but as constant input spread out over 60 minutes, to simulate a meal. In your model, administer 25 g to a 70-kg subject over a 60-minute period. Plot plasma glucose and insulin. How do these dynamics compare to the IVGTT? Explain the dynamics in terms of mechanisms included in the model.
5. Another disease of glucose regulation is *hyperinsulinism* (Guyton 1986). This is the opposite of diabetes mellitus in that overproduction

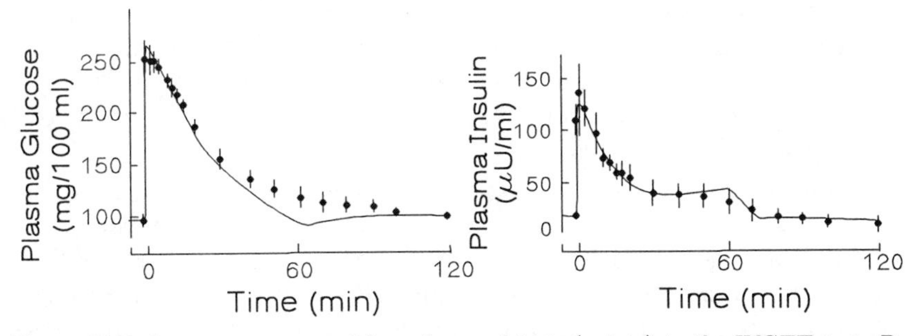

Figure 12.7: Average response of four obese subjects (points) to the IVGTT test. Parameters as in Table 12.3. Error bars are ± 1 standard error, $n = 4$. The solid line is the model prediction. (Reprinted by permission of the publisher from Cobelli et al. 1982, Fig. 23. © 1982 by Elsevier Science, Inc.)

of insulin drives down the plasma glucose concentrations (*hypoglycemia*). Since the central nervous depends almost exclusively on plasma glucose for energy, low concentrations of glucose (about 70 mg/100 ml) will begin to produce erratic behavior and loss of motor control. In severe cases of hypoglycemia, when plasma glucose falls below 20 – 50 mg/100 ml, the patient becomes convulsive and eventually falls into a coma. This suite of symptoms is called "insulin shock." A short-term treatment is to supply the patient with large concentrations of intravenous glucose.

(a) Simulate hyperinsulinism by adjusting the appropriate parameters in Table 12.2. As a first guess, try increasing a_6 in F_6, but other adjustments may be necessary. Your new model should terminate the patient when plasma glucose falls below 20 mg/100 ml.

(b) Attempt to resuscitate your dying patient by administering glucose intravenously. How much do you have to add in order to prevent death?

6. A normal patient should be able to recover from insulin shock. Simulate rapid ingestion of insulin administered as 0.10 U/kg body weight over 2 minutes. Observe the momentary hypoglycemia that resulted. Did your subject die? Repeat with an obese subject.

7. Simulate the glucose, insulin, and glucagon dynamics resulting from a normal, diabetic, and obese subject consuming an average bowl of vanilla ice cream.

8. Review Chapter 2 and write an objective statement for the glucose model.

9. Design a validation study using profile analysis of the glucose model applied to obese patients. Since we do not have the values for the individual patients, we can not use Fig. 12.7 directly. As an exercise, simulate the patient's values by estimating the variance of patient response at each time as graphically portrayed in Fig. 12.7. Knowing this, draw random values from a normal distribution for each time value with a mean as indicated in the figure. You will need to determine from the requirements of profile analysis the number of "patients" to use in your simulated study.

10. How much total insulin is produced by all of the viewers of a typical Super Bowl game? Assume there are 50 million viewers worldwide, and each viewer consumes 0.25 bags of chips ("crisps," if you're from the UK) during *each* commercial and *after each* touchdown scored by either team. Using your favorite brand of chip, calculate the glucose content, assuming that 100% of the carbohydrates are in the form of glucose. How much more insulin would there be if the viewers also ate guacamole and sour cream dip and a six-pack of beer?

Chapter 13

Populations and Individuals

13.1 Populations

A POPULATION is a set of organisms of the same species living in a particular place and time. This simple definition begs the question of how to define "species," since the traditional criterion that it be composed of "interbreeding" organisms is difficult or impossible to apply in many cases. Nevertheless, the definition works for most purposes. The key idea, to which we will return below, is that populations are composed of interacting individuals. An operational definition of the concept of *ecological community* is more elusive, however. One anonymous, but cynical, wag defined it as a set of populations about which it is interesting to speak. There is a frighteningly important element of truth in this definition. And we could accept it, provided community ecologists were never boring. This not always being the case, we content ourselves with the more typical definition: the set of co-occurring and interacting populations in a place. In practice, the set of populations and relations studied is often confined to specific taxa and ecological processes.

In this chapter, we describe some of the elementary models of populations and communities. In so doing, we will again return to the principles developed in *Part I.* In particular, we examine more complex nullcline analysis using mechanistic models of competing species. We introduce the concept of individual-based models and revisit stochastic models in the form of demographic stochasticity and time to extinction. Finally, we encounter again the problem of model validation in testing simple, alternative predator–prey models with laboratory experiments; we will also use bioenergetic models to predict and test size distributions of fish in lakes.

The central questions that these models address include: (1) Can population dynamics be predicted from the bioenergetics of individuals? (2) What is the simplest model needed to describe accurately predator–prey dynamics in simple aquatic microcosms? (3) How does predator learning affect predator–prey cycles? (4) Can pesticides effectively control insect pest outbreaks?

13.1.1 Populations Without Age Structure

We have already introduced, through examples in *Part I*, density-independent and density-dependent population growth. We will not repeat that now, but rather will give a simple, phenomenological generalization of the models. We wish to formulate an hypothesis of population growth based on the effects that the entire population has on the reproduction of an average individual. (By average, we mean average in all respects: sex, weight, age, and so on.) In density-independent models, the relation is a straight line with zero slope; in the density-dependent logistic model, it is a straight line with negative slope. To generalize the biological hypothesis that increased population size always decreases per capita birth rate, we could use a nonlinear relation such as Richard's equation as illustrated in Chapter 4.

A more dramatic departure is a phenomenon called the *Allee effect* in which two processes are operating: decreases in per capita birth rate due to competition, and increases in per capita birth rate with increases in population numbers due to increased chances of encountering mates at low population density. If our aim is simply to describe this relation, we can use any functional form that possesses a maximum and that can be scaled to biologically realistic numbers. Two candidates from Chapter 4 are the maximum function and the Blumberg function. The former, being a product of two separate subfunctions, has the advantage that each subfunction and its parameters can be associated with the two biological processes (mate location and competition).

Here is one possible phenomenological model of population growth using the Allee effect (Wilson and Bossert 1971):

$$\frac{dN}{dt} = rN\left(\frac{K-N}{K}\right)\left(\frac{N-M}{N}\right), \tag{13.1}$$

where M is a lower threshold below which the per capita rate is negative. Above the threshold the per capita rate increases to a maximum then decreases to 0 at $N = K$. The importance of the Allee effect will become apparent in Section 17.8.6, where we discuss chaos.

13.1.2 Populations with Discrete Age Structure

These models, because of their nonlinear structure, can fit many data sets (Berryman 1991), but being general, they do not satisfy our desire for more mechanistic explanations. One point in which they fail to capture basic biological mechanisms is their assumption that all individuals are equal. All individuals, of course, are not equal and everyone eventually grows old and dies. Individuals differ because of their age and other physiological and ecological variables often correlated with age (e.g., the effect of age or size on energy demands, foraging efficiency, running speed, etc.). The simplest model of an age structured population is one analogous to the density-independent finite difference model. As a simple example, assume the population has four age classes, only the oldest reproduces, and at each time step the fate of an individual is either to die or to live and become one time step older (i.e., advance to the next age class). So, the fate of age class i is

$$\begin{aligned} N_{i,t+1} &= N_{i,t} - d_i N_{i,t} - (s_i) N_{i,t} + s_{i-1} N_{i-1,t} \\ &= s_{i-1} N_{i-1,t}, \end{aligned}$$

where d is the fraction dying, and s is the fraction surviving *and* aging 1 time interval (so that $d = 1 - s$). In addition to survival, the youngest age class increases by the addition of individuals through reproduction. This is modeled as f_i, the average birth rate per female of age i. Since all individuals within an age class are considered equivalent, the net number of newborn individuals from females of age i is $f_i N_i$. The total number of newborn individuals is the sum of the contributions of all reproductive age classes. With this, the complete set of equations for all age classes is

$$N_{0,t+1} = f_3 N_{3,t} \tag{13.2}$$

$$N_{1,t+1} = s_0 N_{0,t} \tag{13.3}$$

$$N_{2,t+1} = s_1 N_{1,t} \tag{13.4}$$

$$N_{3,t+1} = s_2 N_{1,t}, \tag{13.5}$$

assuming only age class 3 reproduces.

We can extend this idea to distinguish between sexes as well. In that case, we use separate equations for males and females, but must handle male and female reproduction differently since the numbers of male and female babies are not independent of the numbers of adult females *and* males. One possibility is to assume that reproduction is limited by females (i.e., there is always an overabundance of interested males) and that the sex ratio (r) of babies is constant (e.g., 1:1 in many populations, $r = 0.5$). This leads to

$$\begin{aligned} N_{0,f,t+1} &= r f_3 N_{3,f,t} \\ N_{0,m,t+1} &= (1 - r) f_3 N_{3,f,t} \\ N_{1,f,t+1} &= s_0 N_{0,f,t} \\ N_{1,m,t+1} &= s_0 N_{0,m,t} \\ N_{2,f,t+1} &= s_1 N_{1,f,t} \\ N_{2,m,t+1} &= s_1 N_{1,m,t} \\ N_{3,f,t+1} &= s_2 N_{1,f,t} \\ N_{3,m,t+1} &= s_2 N_{1,m,t}, \end{aligned}$$

where r is the fraction of females in the population. In this simple model, we assume that the survival rates are the same for males and females.

13.1.3 Matrix Approach

Equations 13.2–13.5 are a system of linear equations. As we have seen in earlier chapters, such a system can be written in matrix notation (Leslie 1945):

$$\begin{pmatrix} N_0 \\ N_1 \\ N_2 \\ \vdots \\ N_m \end{pmatrix}_{t+1} = \begin{pmatrix} f_0 & f_1 & f_2 & \dots & f_m \\ s_0 & 0 & 0 & \dots & 0 \\ 0 & s_1 & 0 & \dots & 0 \\ \vdots & & & \dots & 0 \\ 0 & 0 & 0 & \dots & 0 \end{pmatrix} \begin{pmatrix} N_0 \\ N_1 \\ N_2 \\ \vdots \\ N_m \end{pmatrix}_t \tag{13.6}$$

where f_i are the net fecundities of older age classes and s_i are survivorships.

This is a multidimensional version of density-independent population growth, where each age class becomes an axis of the state space. It has its special form with zeros in most positions of the matrix because aging is a unidirectional process. However, a set of equations analogous to Eq. 13.2–13.5 and a matrix analogous to Eq. 13.6 can be constructed using size classes as the property defining the axes of state space. In this case, since it is possible to lose body mass or to make large weight gains, there may be nonzero values in those positions that were zero in the age class model. For example, a nonzero fraction of the individuals in each size class can lose weight and become an input to smaller size classes. (See the exercises for an example that uses real data for a stage structured plant population model.) This cannot occur in age structured models, providing that dormant ages are not modeled (see Werner and Caswell 1977).

Since the Leslie matrix describes a set of simultaneous linear equations, the mathematical properties of the Jacobian matrix used in linear stability analysis (Section 9.3) also apply here. In particular, there is an eigenvalue (λ) such that

$$\mathbf{L}\mathbf{N}_t = \lambda \mathbf{N}_t. \tag{13.7}$$

This states that the numbers of individuals in each age class at $t+1$ are a simple proportion, λ, of the numbers at t, because $\mathbf{L}\mathbf{N}_t = \mathbf{N}_{t+1}$. The proportion is the finite rate of increase of the population, and $\lambda = e^r$, where r is the instantaneous rate of increase of the population. When Eq. 13.7 is true, the proportions of individuals in each age class are constant and the vector of proportions of each age class is called the *stable age distribution.* We solve for λ using the same techniques employed when determining stability of a set of linear differential equations: solve the characteristic equation that results from evaluating the following determinant:

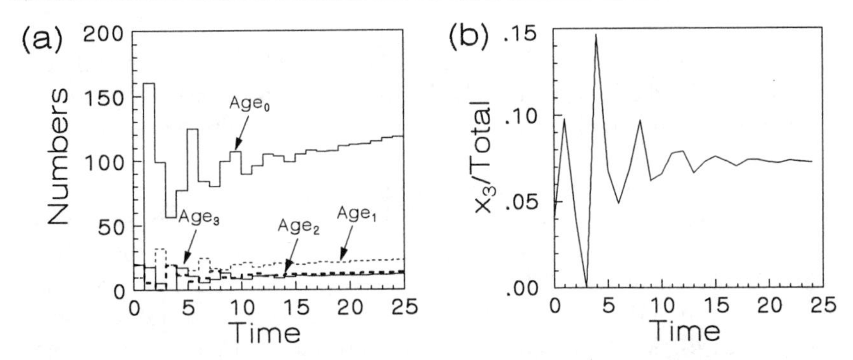

Figure 13.1: Dynamics of an age structured population model. (a) Values for age classes, (b) dynamics of the proportion of age class 3 to the total population.

$$\begin{vmatrix} (f_0 - \lambda) & f_1 & f_2 & \cdots & f_m \\ s_1 & (0-\lambda) & 0 & \cdots & 0 \\ 0 & 0 & (0-\lambda) & \cdots & 0 \\ \vdots & & & \cdots & 0 \\ 0 & 0 & 0 & \cdots & (0-\lambda) \end{vmatrix} = 0. \tag{13.8}$$

Although this is a linear model, it has more than one state variable, and therefore exhibits more complicated behavior than the age independent, density-independent model. In particular, the age dependent model can show transient or sustained oscillations. To illustrate the former behavior, we simulate a population with the following L matrix (Eq. 13.9):

$$L = \begin{pmatrix} 0 & 1 & 3 & 4.5 \\ 0.2 & 0 & 0 & 0 \\ 0 & 0.6 & 0 & 0 \\ 0 & 0 & .9 & 0 \end{pmatrix}. \tag{13.9}$$

Figure 13.1a shows the time evolution of ages 0–3. Note the distribution of individuals over the age classes: younger ages are more abundant than older ages. Figure 13.1b demonstrates the development of the stable age distribution. The ratio of numbers in age 3 relative to the total numbers in the population is initially variable, but rapidly approaches a constant. The other age classes behave similarly. Caswell (1989) gives a more complete and rigorous treatment of this class of population model.

13.1.4 Individual-Based Population Models

While the age or stage specific models illustrated above are an improvement in realism for populations, we can make the models even more realistic by modeling individual organisms explicitly. We introduced these models in Chapter 3 in the context of particle models. The terminology for this class of models is still unsettled; they are variously called *individual-based* or *individual-oriented* models (Metz and de Roos 1992). Here, we will refer to them as individual-based models (IBMs) and keep the name age specific or stage specific for models that lump individuals into discrete ages or size categories.

Individual-based population models in one form or another have existed for some time. At their core is the stochastic birth–death process which ultimately is based on random walks or Markov chains using probability theory developed in the 19th century (Ludwig 1974). Prior to computer simulation, the major mathematical results were limited to rather special biological cases. With the advent of computer simulation, however, these models produced results for more interesting biological systems. In the early computer era, Gatewood (1971) was a pioneer using individual-based models of human demography and epidemiology. The National Micropopulation Simulation Resource at the University of Minnesota continues to develop models and simulation environments for IBMs (e.g., Ackerman et al. 1993). Many theoretical ecologists have also adopted this approach based on the early applications of D. DeAngelis, H. Shugart, and M. Huston at the Oak Ridge National Laboratory (Huston et al. 1988). This is now an exciting field with applications to both theoretical and applied problems in population and community ecology of plants and animals (e.g., DeAngelis and Gross 1992; Judson 1994).

The essence of this method is to follow the fates of all the individuals in the population as they proceed through their lives, however small and insignificant they may be. A population is composed of all these little events, and if the condition and state of each individual is known, then the state of the population can be generated simply by summing the set of individuals with similar states (e.g., all those alive, or of similar size, or of the same sex, etc.) The motivation for this is the hypothesis that small variations among individuals can have dramatic effects on the ultimate state of the population. To simply illustrate the impact that individual properties can have on population structure consider the case of plants growing and occupying the soil surface. The individual property of interest is distance to the nearest neighbor (Huston et al. 1988). Assume that growth rate is inversely proportional to the proximity of neighbors and that growth stops when two individuals touch. Figure 13.2 shows two scenarios of initial plant spatial location: uniform and random. A random dispersion results in an unequal final size distribution of individuals because isolated individuals have little competition and become large. A uniform initial dispersion produces uniform final sizes.

In animal populations, the events that determine an individual's fate are essentially those also faced by humans as they daily live out their lives. As with humans, animals in individual-based models are born, seek food usually in the form of individual particles, avoid predators and other forms of death, find mates, have babies, and ultimately succumb to old age or a

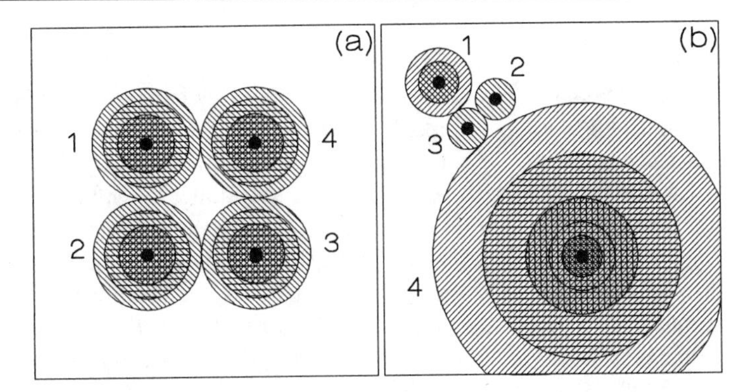

Figure 13.2: An individual's spatial location can dramatically affect the ultimate plant population size distribution. Concentric circles represent a series of growth events of four plants whose sizes are represented by the circle areas. (a) Plants are uniformly distributed, all grow at the same rate, and all achieve the same size before growth stops. (b) Plants are randomly placed; those far from others (plant 4) achieve large sizes; those with close neighbors (2 and 3) are stunted; the population size distribution is uneven.

violent end. Again like humans, what happens on a day to day basis to an individual animal is largely a stochastic process. Perhaps we failed to find a mate today, but tomorrow is always another day, and hope springs eternal in the hearts of those whose fate is in the hands of a random number generator.

Although this picture brings forth a rather grim picture of an individual's fate, its connection with our own observations of human lives is part of the appeal. Model structure and parameters are based on observations of individuals, and the natural variation among individuals and the stochastic nature of their fates can be incorporated directly and easily using this modeling approach. But there is a downside as well. In most applications, the equations of the processes are too complex for analytical solution, and computer simulation is necessary. This is nothing new for this book, but in IBMs this can mean following the fates of hundreds or thousands of individuals, each capable of being in many different states. Further, when IBMs are applied to questions in population ecology, the models describe birth and death processes. As a result, the numbers of individuals that must be simulated increase over time, possibly exponentially. This can create a huge computational burden, but Rose et al. (1993) have developed an algorithm using a fixed number of individuals that closely approximates a model that allows the numbers tracked to increase. Moreover, because IBMs are stochastic, we must simulate the system many times to determine the expected outcome. Consequently, the use of IBMs involves a trade off between analytic tractability and realism.

The main utility of IBMs is that they do not use population averages of parameters to generate population dynamics. IBMs will, therefore, be especially useful in systems in which individuals interact and behave so that a simple average does not represent the overall behavior of the population. There are three classes of circumstances in which this can occur (DeAngelis and Rose 1992). (1) When populations are small, such as founder populations on islands, there is a good chance that random sampling will select a nonrepresentative sample from the larger population. This will affect population dynamics by biasing parameters of growth, predator avoidance, and so on. It also exacerbates demographic stochasticity and increases extinction probability. (2) When populations exist in temporally variable environments, the fates of individuals will be altered by random events that may dominate the behavior of the population. (3) When individuals are not randomly mixed within the population, chance encounters (e.g., mating) can alter population dynamics. Populations will not be randomly mixed if there is spatial heterogeneity that affects movement or if social structure (e.g., social hierarchies) prevents some individuals from freely mating with others.

One system where IBMs have been especially useful is the simulation of the size distributions of fish populations. IBMs are useful here because fish consumption of prey and avoidance of predation are very sensitive to the size of the individual, and random encounters of fish with their prey dominate daily rates of food intake. The basic computational flow of one fish population IBM (Madenjian and Carpenter 1991) is shown in Fig. 13.3. Many other population IBMs are similar, but details of movement or the effects of predation on the target population will change with the system studied (e.g., Folse et al. 1989, Hyman et al. 1991). The target fish population in this example is young-of-the-year (YOY) walleye (*Stizostedion vitreum virtreum*) in Wisconsin lakes. Population characteristics are generated by the sizes of individuals (bottom of Fig. 13.3) which are determined by individual daily growth rates.

Growth rates of individual walleye are computed by the number and sizes of prey consumed each day. The time scale of this model is one season, so reproduction is ignored. Walleye death occurs when energetic intake is so low that starvation occurs. The daily number of prey (e.g., bluegill or perch) encountered by an individual fish is a random deviate from a Poisson distribution. Each prey fish encountered is given a size from a normal distribution. If the ratio of prey size to walleye size is less than a threshold, the prey is consumed. In this way, a fraction of the prey's energetic content based on walleye energetic conversion efficiency is added to the walleye biomass. New walleye size is calculated after all prey are consumed and is a power function of the walleye length. So, based on the above, the finite difference equation for the biomass of walleye individual i

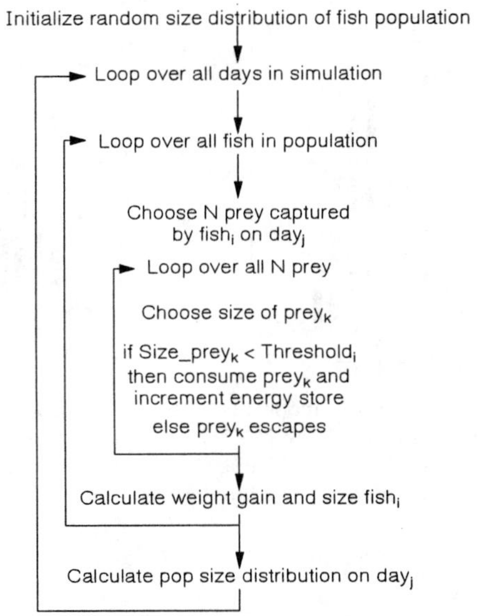

Figure 13.3: Flow chart of a fish population IBM. (From Madenjian and Carpenter 1991, Fig. 1. © 1991 Ecological Society of America. Reprinted with permission of the publisher.)

is

$$W_{i,t+1} = W_{i,t} + \begin{cases} \alpha C_{i,\max} & \text{if } M_{i,t} = 0 \\ \min \begin{cases} g \sum_{k=0}^{M_{i,t}=P(\lambda)} \beta L_k(N(\mu,\sigma))^\delta \\ C_{i,\max} \end{cases} & \text{otherwise,} \end{cases}$$

where $W_{i,t}$ is the current weight of individual i, g is the efficiency of converting prey biomass into walleye biomass, and $M_{i,t}$ is the number of fish prey consumed on day t and is drawn from a Poisson distribution (P) with mean λ. L_k is the length of the kth prey and is drawn from a normal distribution ($N(\mu, \sigma)$), and β and δ are constants that convert prey length to prey biomass. $C_{i,\max}$ is the maximum specific consumption rate for individual i and is computed as

$$C_{i,\max} = aW_{i,t}^b F(T),$$

where a and b are empirical constants and $F(T)$ is a temperature (T) response function. Walleye cannot consume more than $C_{i,\max}$, even if $M_{i,t}$ were so large as to permit greater consumption. If $M_{i,t}$ is by chance zero, then it is assumed that walleye can obtain a small fraction (α) of their maximum consumption rate from alternative prey species. Since walleye length plays an important role in determining the threshold size at which large prey escape predation, walleye biomass is converted to length by

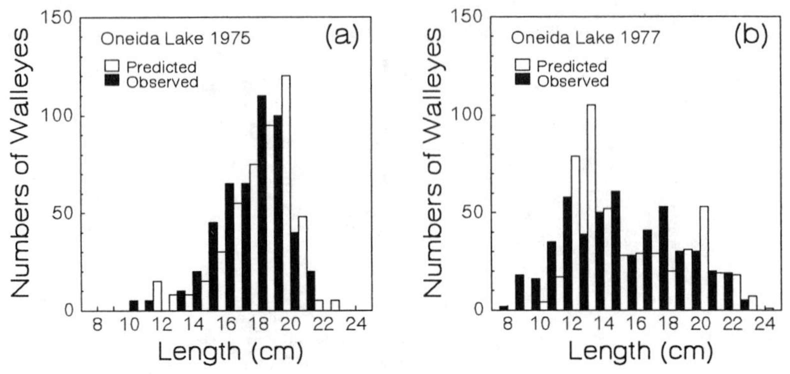

Figure 13.4: Comparison of an IBM model and observations of a YOY walleye population in Lake Oneida, Wisconsin in 1975 and 1977. Bars indicate predicted and observed fish numbers in discrete length categories. (From Madenjian and Carpenter 1991, Fig. 2. © 1991 Ecological Society of America. Reprinted with permission of the publisher.)

inverting the empirical relation

$$W_{i,t} = \gamma L_{i,t}^{\nu}.$$

It should be apparent that this model is essentially a single equation, which is based on a few simple facts of walleye behavior and energetics. The model fits empirical size distributions quite well (Fig. 13.4). The accuracy of the fit to the 1975 data (Fig. 13.4a) was obtained by fitting a parameter to these data. Encounter rate (λ, the mean of the Poisson distribution) was adjusted until the predicted walleye size distribution fit the data in 1975. The model run for 1977 was not, however, adjusted in this way, but was corrected only for differences in *mean* prey availability between the two years. This is legitimate since the prey populations were not explicitly modeled. Notice also that this validation effort did not attempt to compare two time series, as is common. Instead, validation was based on a *derived measure* (Chapter 8): the frequency distribution of the states (i.e., sizes) of individuals at a point in time.

This model is an example of a particularly simple IBM that, nevertheless, performs quite well. A slightly modified version was used for management purposes to predict body loads of PCBs (polychlorinated biphenyls) in Lake Trout (Madenjian et al. 1993). Other fish IBM models (e.g., Adams and DeAngelis 1987; DeAngelis et al. 1991) have used more complex behavioral and physiological bases to determine the sizes of prey encountered. These models are based on fish size, *reactive distances* of individual fish to prey of a given size, distance, and turbidity conditions. The theoretical implications for population dynamics of these models have also been investigated. In one example, Adams and DeAngelis (1987) found that in an IBM of bass feeding on shad, if both spawned at their normal times, over the season bass consumed 19% of the shad. If, however, bass were delayed for some reason, they consumed only 6% of the shad. The reason for this is due to the fact that the delay permitted shad to grow relative to bass and escape predation by exceeding the size threshold for successful bass attack (Fig. 13.3). We would expect, based on this significant reduction in shad mortality, strong natural selection for shad to emerge early and grow quickly. This has perhaps occurred to some extent, but because of the natural limitations to this process by the timing of winter thaws, shad are not able to push emergence very far back into early spring. Nevertheless, the IBM with its basis in individual variation gives us another approach to investigate evolutionary questions. Among others, Johnson (1994) and Toquenaga et al. (1994, see Chapter 19) have done work in this area.

In summary, IBMs, as with all the modeling approaches described in this book, are not a panacea to apply without thoughtful consideration. Their analytical intractability can sometimes prevent our seeing the broad patterns of population dynamics because of the complex details of the fluctuations of individuals. Nevertheless, if the system has small numbers, is temporally stochastic, or is nonuniformly mixed, then IBMs are another tool for our toolbox.

13.2 Interactions in Simple Communities

13.2.1 Mechanistic Models of Competition

In previous chapters, we have seen several applications of the simple two-species competition models. In particular, the simple equations based on the Lotka–Volterra models are especially amenable to analysis and study (e.g., nullclines and neighborhood stability). The major problem with these equations is the absence of a mechanistic basis. They do not distinguish *interference competition* (i.e., one organism actively inhibiting another organism from using a resource) from *exploitative competition* (i.e., no active inhibition, but consumption of a single resource by two organisms). Schoener (1976) provided a better mechanistic basis for these two basic ecological interactions. Population growth of a species in the presence of a competitor is determined by two components that correspond to the two types of competition. For two competing species,

$$\frac{dN_1}{dt} = R_1N_1\left(\left[\frac{I_{01}}{N_1+\beta N_2}\right] + \left[\frac{I_{E1}}{N_1} - \gamma_{11}N_1 - \gamma_{12}N_2 - C_1\right]\right) \quad (13.10)$$

$$\frac{dN_2}{dt} = R_2N_2\left(\left[\frac{\beta I_{02}}{N_1+\beta N_2}\right] + \left[\frac{I_{E2}}{N_2} - \gamma_{22}N_2 - \gamma_{21}N_1 - C_2\right]\right) \quad (13.11)$$

where the parameters are defined in Table 13.1. The right-hand side of each equation has two terms in square brackets. The first bracketed expression represents exploitative competition; the second represents interference competition. Both terms describe the amount of energy available to an individual for reproduction. R_i converts available energy to numbers

Table 13.1: Parameters for Schoener's mechanistic competition model.

C_i	Density-independent maintenance and replacement cost of an individual of the ith species
I_{Ei}	Rate of net energy input into the ith species of resources *exclusive* to that species
I_{0i}	Rate of net energy input into the system that is useable by both species in terms of energy to ith population
N_i	Numbers of individuals of species i
R_i	Efficiency to convert 1 unit of energy consumed by species i into new individuals of species i
β	Ability of species 2 to obtain energy relative to species 1
γ_{ij}	Energetic cost to species i of interference interactions with species j, $j = i$ is intraspecific interference costs and $j \neq i$ is interspecific costs
p_{Ei}	I_{Ei}/γ_{ii}
c_i	C_i/γ_{ii}
g_{ij}	γ_{ij}/γ_{ii}

of offspring per capita. Available energy after exploitative competition occurs is described by uniformly apportioning energy among all individuals weighted by their competitive ability. Individuals of the same species (i) have equal weight; individuals of the other species are weighted by β. If $\beta < 1$, an individual of species 2 is not as efficient at harvesting resources as individuals of species 1. The effect of interference competition is to subtract the energetic costs of behavioral interaction (γ_{ij}) and maintenance costs (C_i) from the energy input (I_{Ei}) per individual (N_i).

Using the basic approach of nullcline analysis developed in Section 9.3, Schoener (1976) found the following nullclines for species 1 and 2, respectively:

$$\begin{aligned} N_2 &= \frac{p_{E1}}{2g_{12}N_1^*} - \frac{c_1}{2g_{12}} - \frac{N_1^*}{2\beta} - \frac{N_1^*}{2g_{12}} \\ &\quad + \left[- \left(\frac{p_{E1}}{2g_{12}N_1^*} + \frac{c_1}{2g_{12}} - \frac{N_1^*}{2\beta} + \frac{N_1^*}{2g_{12}} \right)^2 + \frac{p_{01}}{g_{12}} \right]^{1/2} \\ N_1 &= \frac{p_{E2}}{2g_{21}N_2^*} - \frac{c_2}{2g_{21}} - \frac{\beta N_2^*}{2} - \frac{N_2^*}{2g_{21}} \\ &\quad + \left[- \left(- \frac{p_{E2}}{2g_{21}N_2^*} + \frac{c_2}{2g_{21}} - \frac{\beta N_1^*}{2} + \frac{N_2^*}{2g_{21}} \right)^2 + \frac{\beta p_{02}}{g_{21}} \right]^{1/2}, \end{aligned}$$

where the asterisks indicate equilibrium population values and the parameters are defined in Table 13.1. These equations are clearly nonlinear and more complex than the nullclines developed for the Lotka–Volterra–Gause equations in Section 8.4.2. Nevertheless, they can be plotted in the phase space. Figure 13.5 shows the nullclines, the multiple equilibria, and their stability properties for two relations between cost of interference (g_{ij}) and

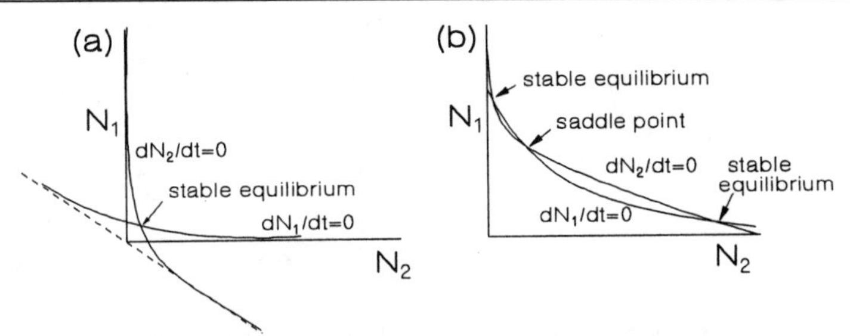

Figure 13.5: Nullclines for energy-based two-species competition model for (a) $g_{12} < \beta$; $g_{21} < 1/\beta$ and (b) $g_{12} > \beta$; $g_{21} > 1/\beta$. (From Schoener 1976, Figs. 5 and 6A. © 1976 Academic Press, Inc. Reprinted with permission of the publisher and author.)

energy gained by interference (β). Case (a) has a single stable equilibrium and corresponds with Case III of the Gause model (Section 9.3). Case (b) is analogous to Case IV, but the nonlinear nature of the nullclines permits additional stable equilibria. Appropriate choices of other parameters (e.g., K_i) permit the Schoener model to produce other nullcline relationships analogous to Gause Cases I and II. This example makes two main points: (1) the apparently simple idea to put competition on an energetic basis has resulted in nullclines that are algebraically complex; other energetic assumptions might have resulted in nullcline equations that could not be solved as these were; and (2) these more realistic mechanistic equations result in much more complex and interesting dynamics that we can now explore experimentally.

13.2.2 Predation in Simple Communities

One system where simple theory has been experimentally tested is predator–prey dynamics. Here we develop some extensions to the classical theory and examine some tests. The classical Lotka–Volterra equations and their null-clines were presented in Section 8.4.1. This model hypothesizes that predators are never satiated and that prey growth rate is density-independent. To relax these strong assumptions, we use a Type 2 functional response analogous to the Michaelis–Menten relation for biochemical reactions and a linear density-dependent function:

$$\frac{dV}{dt} = rV\left(1 - \frac{V}{K}\right) - \frac{aT_TVP}{1 + ahV} \tag{13.12}$$

$$\frac{dP}{dt} = c\frac{aT_TVP}{1 + ahV} - dP, \tag{13.13}$$

where a is the encounter rate, h is the handling time, T_T is total time available for foraging, c is a conversion factor between victims consumed and new predators created, and d is predator per capita death rate.

The nullcline equations are left as an exercise, but depending on parameters they produce curves such as those in Fig. 13.6. The prey nullcline

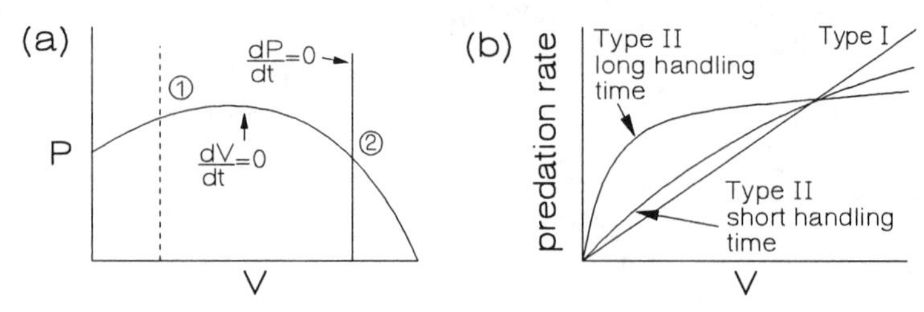

Figure 13.6: (a) Nullclines for a predator-prey model with density dependence and Type II functional response. The hump is created by intraspecific competition reducing prey growth rates at large prey numbers [point 2 in panel (a)]. At point (1), individual Type II predators with long handling times are more efficient at low prey numbers compared to predators with short handling times or Type I predators [panel (b)]. At both points 1 and 2, fewer predators are required to balance prey growth.

curve is "humped-shaped"; the predator nullcline is a vertical line. Besides the equilibria that occur when either V or P or both are zero, two interior equilibria are also shown in the figure. Equilibrium 1 is a locally unstable point, but globally stable to a limit cycle; equilibrium 2 is stable.

It is important to understand the biological reasons for the shape of the prey nullcline. By definition, the prey nullcline is the set of points (V, P) such that the prey's absolute growth rate is zero. Based on Eq. 13.12, this rate is a combination of both predation and intraspecific competition. Therefore, at a given V, the nullcline defines the number of P needed to keep V in equilibrium. If the nullcline is near the V axis, then the absolute growth rate of V is small and only a few P are needed to consume the added V. This situation occurs, for example, when V is near K, the carrying capacity.

When more complex predator behavior is incorporated, more complex system dynamical behavior arises. A good example is the analysis of the plant–herbivore interaction between spruce trees and the spruce budworm (*Choristoneura fumiferana*). The budworm is a major pest in the eastern North American forests. Since about 1750, the budworm has shown fairly regular episodic outbreaks about every 40 years. At their peak, budworm densities can be as high as 150 insects per m^2 (Royama 1984).

May (1977) provides a nice synthesis of work originally done by C.S. Holling, D.D. Jones, D. Ludwig, and others (Ludwig et al. 1978; Jones 1979). The key to the dynamics in their models is the fact that the predation rate of a single predator responds to prey density by a Type 3 functional response, which is sometimes indicative of a predator that has some form of learning (but see Taylor 1984). The shape of this relation is sigmoidal, so that at very low prey density, the predator consumes very few prey. The predator does not increase its consumption rate in proportion to increases in prey density until moderately high prey density is present. The biological mechanism might be that the predator is not efficient until it forms a *specific search image*, which does not occur until it has encountered sufficient numbers of prey. At very high prey density, the predator's predation rate is flat, and a further increase in prey density does not increase predation rate.

The equations of this model are, following May (1977),

$$\frac{dN}{dt} = rN\left(1 - \frac{N}{K(S)}\right) - \frac{\beta N^2}{N_0(S)^2 + N^2}P \qquad (13.14)$$

$$\frac{dS}{dt} = \rho S\left(1 - \frac{S}{S_{\max}}\right) - \eta N, \qquad (13.15)$$

where P is the number of predators attacking budworm larvae (assumed fixed), N is the number of spruce budworm, S is the amount of spruce leaf area available to attack. r is the maximum per capita rate of increase of budworm, and β is the rate at which budworm larvae encounter spruce leaves. $K(S)$ is the budworm carrying capacity and depends on amount of spruce biomass; it is assumed that $K = \kappa S$, where κ is the efficiency at which budworm convert spruce leaves into new budworm larvae. $N_0(S)$ is a variable that defines the shape of the predator's (P) functional predation response to budworm density and is defined as the density at which the predator saturates (May 1977). It is analogous to the half-saturation constant of the Michaelis–Menten relation. This shape variable is proportional to the amount of budworm resource available: $N_0 = \eta S$, where η is the fraction of N that consumes S. A plausible biological mechanism for this hypothesis is that spruce trees inhibit the predator's ability to find and attack budworm by allowing budworms to be more uniformly dispersed in space. Since budworms must live on trees, the budworm population will be more highly aggregated on individual trees when fewer trees are present than when trees are dense. For many predators, aggregation increases attack rates (Taylor 1984). For the spruce leaf area (Eq. 13.15), ρ and $S_{\max}$ are the intrinsic rate of increase and carrying capacity, respectively. The amount of spruce leaves being consumed is proportional to the number of budworms (η).

Solving for the nullclines explicitly is difficult in this case. Instead of doing this, we will use a graphical method. We begin with the easy case of S by referring to Fig. 13.7. Equation 13.15 has two components: a factor producing positive density-dependent growth and a factor describing population decrease (consumption by N). The net rate will be zero where these two terms are equal. To find these points, we plot the two functions together (Fig. 13.7a). Equilibria exist where population increase (parabola) equals population decrease (horizontal lines). Since consumption rate depends on the level of N, we plot several budworm population levels. It is assumed that the budworm population changes continuously, so that equilibria exist between the levels shown (e.g., between N_2 and N_3). The spruce nullcline is obtained by plotting the equilibria points in the N–S phase space (Fig. 13.7b). This is a parabola, since the consumption

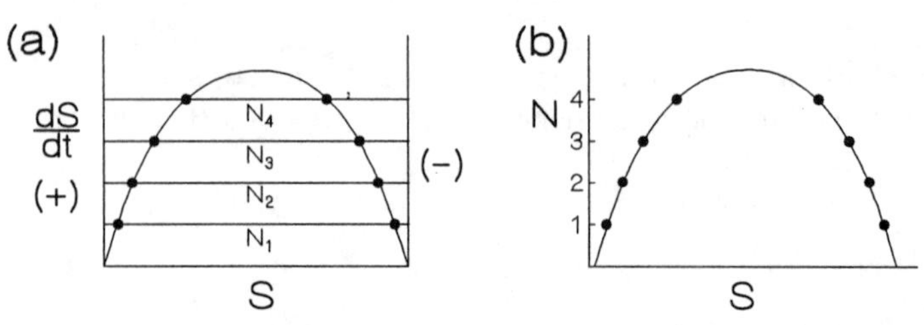

Figure 13.7: Graphical derivation of the spruce tree nullcline. (a) Density-dependent growth of spruce (parabola, positive growth on left vertical axis) plotted with budworm consumption (horizontal lines, negative growth on right vertical axis). Four levels of the budworm population are shown (arbitrary numerical scale). The intersection points are equilibria. (b) The set of equilibria, plotted at their respective values of N and S, result in a "hump-shaped" parabola in the phase space. Below the curve, S increases; above the curve, S decreases.

function intersects the growth function at two points.

The budworm nullcline is obtained using the same method, but the equations are more complex, so we first describe how the components of the dynamics change with S. Figure 13.8a shows the logistic growth rate at four levels of S that determine four different carrying capacities for N (Eq. 13.14, left component in parentheses). Figure 13.8b depicts the predation rate on the budworm population as a function of spruce numbers. Note that the two processes (growth and predation) respond in opposite directions to increasing S. When the two sets of curves are superimposed, the equlibria can be determined as a function of N and S. This is done for three levels of S in Fig. 13.9a. These curves illustrate the effects of the nonlinearities and the fact that increasing S *increases* N growth but *decreases* predation on N. These properties cause the number of equilibria

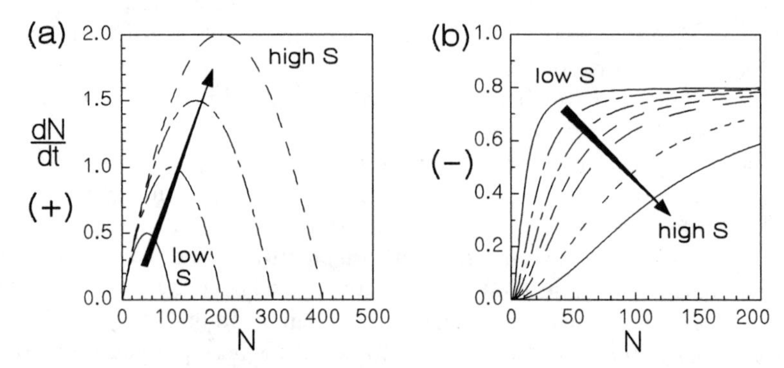

Figure 13.8: Graphical depiction of budworm growth rates as affected by spruce tree numbers. (a) Budworm density-dependent growth rate when carrying capacity is proportional to spruce tree numbers (Eq. 13.14). The arrow indicates increasing numbers of spruce trees. (b) Rate of budworm consumption by a predator population fixed at level P (Eq. 13.14), where the predator's functional response depends on spruce tree numbers (increasing in the direction of the arrow). Note the vertical scales of (a) and (b) are different.

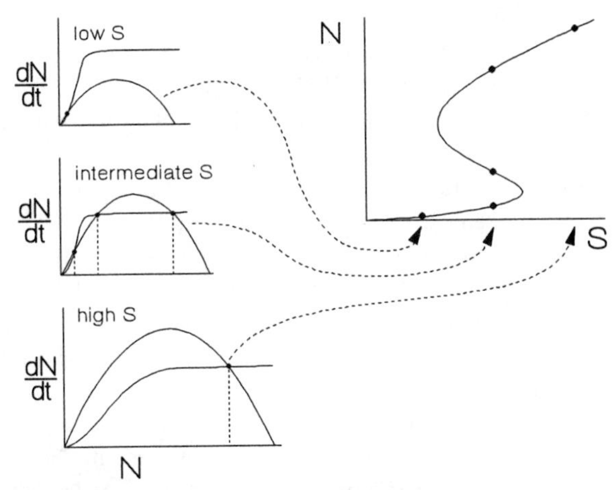

Figure 13.9: Nullcline for budworms based on three spruce tree densities. (a) Growth and predation rates superimposed for three spruce levels. At low S, there is a single intersection at low N. At intermediate S, the two curves intersect at three levels of N. At large S, there is again a single intersection point. (b) The resultant N nullcline when a continuum of S levels are considered. The intersections in (a) become the points on the nullcline curve. To the left of the nullcline, N decreases; to the right, N increases.

to change from one intersection at low S to three at intermediate levels, and back to one again at high S. At two special values of S (not shown), there are just two intersection points.

When both nullclines are superimposed (Fig. 13.10), the resulting dynamics can be a stable limit cycle, depending on parameters. The oscillations have large amplitude, and therefore the system alternates between budworm dormancy and epidemic outbreaks. Other choices of parameters can show stable equilibria at high values of N.

This graphical analysis of low dimensional mathematical models has the advantage that alternative parameters and functional forms can be applied without defining the specifics of a particular mathematical equation. Thus, we would expect the same qualitative results as long as the functional forms were roughly similar to those shown in Figs. 13.7 and 13.8. An example

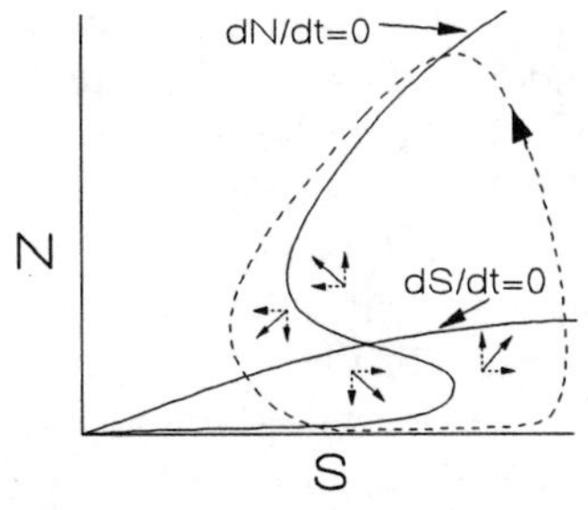

Figure 13.10: Nullclines for the spruce–budworm model. (From May 1977, Fig. 7. Reprinted with permission from *Nature*, © 1977 Macmillan Magazines Limited.)

of this is the incorporation of additional budworm mortality due to applications of pesticides. As May (1977), Ludwig et al. (1978), and Yodzis (1989) argue, if such an additional source of density-independent mortality is added to the budworm equation, its S–shaped nullcline is "straightened out." This alters the dynamics from a limit cycle to a stable equilibrium. Consequently, this analysis suggests that while pesticides cannot eliminate budworms, they can remove the outbreaks and produce a system that always has budworms at moderately high levels. It is a social decision whether permanently moderate levels are better than short periods of devastatingly high levels. Of course, this is an extremely simple model of complex biology upon which to base such a system design decision. [See Royama (1984) for a dissenting view.] Nevertheless, this elegant example of model simplification and generalization has captured, in the form of stable limit cycles, one of the main qualitative dynamical features of episodic insect outbreaks. However, as we have argued in earlier chapters, alternative models must also be evaluated.

13.2.3 Testing Predation Models

We have advocated in Chapters 4 and 8 the comparison of alternative models as an important component of the validation process. A recent rigorous example of this is by Harrison (1995), who modeled the elegant laboratory experiments of Luckinbill (1973). Laboratory experiments in small, homogeneous containers are notorious for being unstable: either the predator is too efficient and drives the prey to extinction and then goes extinct itself, or the predator is not able to find sufficient prey to survive and the prey grows to its carrying capacity in the absence of the predator. In nature, there are several mechanisms by which this instability is circumvented but that are absent in the simple containers of laboratory experiments: the prey has a refuge in which the predator cannot forage, the predator numbers are limited by other predators, or subtle prey niche requirements exist that enhance or reduce prey growth.

The conceptual framework of Luckinbill's experiments was to use the nullclines of simple predation models to predict the experimental conditions in which the prey and the predator could survive together for long periods. The nullclines are derived from Eqs. 13.12–13.13.

Figure 13.11 illustrates how changes in the parameters of the equations affect the stability of the dynamics. Figure 13.11a is meant to represent parameters and nullclines for a typically unstable laboratory experiment. In Fig. 13.11b, the predator nullcline is shifted to the right, for example, by decreasing the searching rate. In Fig. 13.11c, the carrying capacity of the system for the prey is decreased while the predator's parameters are unaffected. This stabilizes the system since it produces a reduced prey growth rate that is available to support predators. Consequently, the predators consume less, grow more slowly, and are relatively unimportant to prey

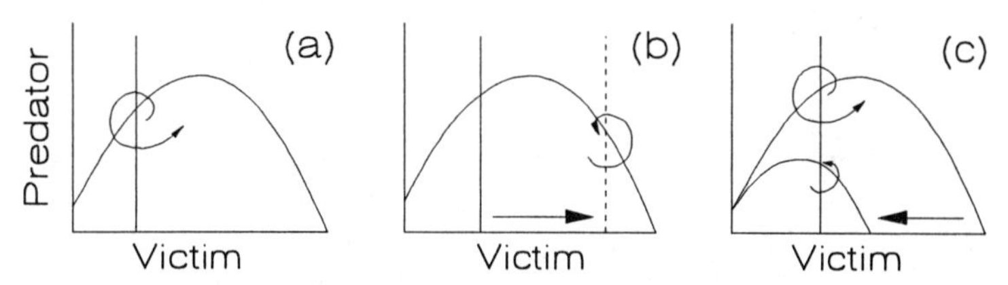

Figure 13.11: Nullclines that enhance stability in predator-prey models. (a) Simple predator-prey models predict instable dynamics if the predator nullcline intersects the prey nullcline to the left of its maximum. (b) The dynamics are stabilized if the predator is less efficient (e.g., decreased search rate) so the intersection is to the right of the peak. (c) The dynamics can also be stabilized if prey nullcline is moved relative to the predator's nullcline (e.g., decreased prey carrying capacity K).

dynamics compared to the intraspecific competition. This latter relation is also significant if looked at from the other side of the coin. If a system such as shown in Fig. 13.11c is at a stable equilibrium (small K), then increasing the carrying capacity, such as by adding nutrients, will *destabilize* the system. Since this may cause the prey to go extinct, it appears that adding nutrient, usually considered to be beneficial to the prey, will, in the long run, be bad for the prey. Rosenzweig (1971) first brought this possibility to our attention and called it the *paradox of enrichment.*

Luckinbill attempted to exploit these nullcline relationships by experimentally manipulating the foraging and growth parameters so as to shift the nullclines to the stable configuration. He used as prey the microorganism *Paramecium aurelia* and as predator the voracious ciliate *Didinium nasutum.* The two species were grown together in 6 ml of medium in which supplies of bacteria were introduced as food for *P. aurelia.* The medium was replenished approximately every 2 days so that in the absence of predators, *P. aurelia* grew as predicted by the logistic equation. In the setup just described, *D. nasutum* quickly consumed all of its prey and itself went extinct, usually within a matter of hours.

To stabilize the system, Luckinbill attempted to manipulate the searching efficiency of *D. nasutum* by forcing it to swim more slowly. He cleverly accomplished this by adding water-soluble methyl cellulose to the medium which greatly increased the viscosity of water, but did not harm the organisms. Naturally, this slowed down both the predator and the victim, but in this case, it slowed down the predator more than the prey. This manipulation did increase the time to extinction, but was not sufficient to permit long term coexistence. This was a step in the right direction, but apparently the prey growth rates needed to be manipulated (Fig. 13.11c). To do this, he reduced the amount of bacteria in the medium. By itself, this also increased persistence time, but not indefinitely. It was only when he simultaneously slowed the predator foraging rate and slowed the prey's growth rate that he was able to achieve indefinite coexistence (Fig. 13.12).

These experimental results qualitatively agree with the basic predictions of simple predator–prey theory and many would interpret the results

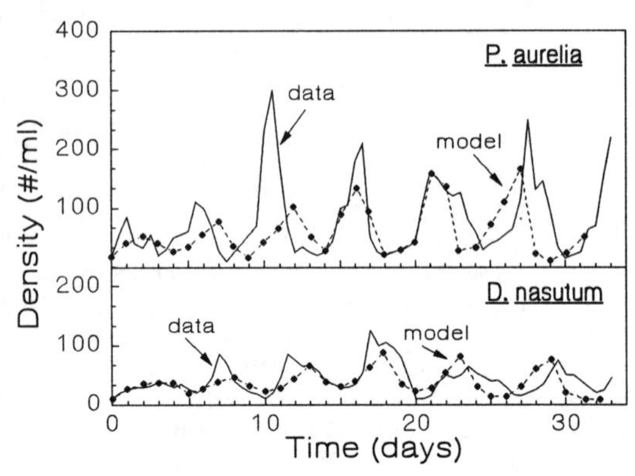

Figure 13.12: Graphical comparison of model predictions and laboratory data for prey (top: *Paramecium aurelia*) and predator (bottom: *Didinium nasutum*). The experimental conditions used one-half strength food concentration for the prey and slowed foraging rates of predators using methyl cellulose. The model predictions were based on Eqs. 13.16. (Data redrawn from Luckinbill 1973, Fig. 5; model results redrawn from Harrison 1995, Fig. 9e. © 1973 and 1995 Ecological Society of America. Reprinted with permission of the publisher.)

as validation of the model. Harrison (1995), however, was skeptical and attempted a more quantitative validation of the model. He did this by statistically comparing the data to a family of models. He examined a continuum of 11 models that ranged from the "standard" model (Eqs. 13.12–13.13) at one extreme, to models with complicated functional responses and time lags. Harrison used two of Luckinbill's data sets. Harrison used a short 18-day experiment to adjust the model parameters for minimum error. He also compared the model to the longest 33-day experiment. In this comparison, no parameters were adjusted except those associated with the controlled experimental conditions (e.g., carrying capacity controlled by food levels). Figure 13.12 graphically shows the degree of fit of one of the best models that Harrison compared to the long data set.

The model shown in Fig. 13.12 was:

$$\begin{aligned}\frac{dx}{dt} &= \rho x(1 - x/K) - \omega\left(\frac{x}{\phi + x}\left[1 - (1 + \nu x)\mathrm{e}^{-\nu x}\right]\right) y \\ \frac{dZ}{dt} &= \sigma\omega\left(\frac{x}{\phi + x}\left[1 - (1 + \nu x)\mathrm{e}^{-\nu x}\right]\right) y - \delta Z, \\ \frac{dy}{dt} &= \left(\lambda\frac{Z}{y} - \gamma\right) y,\end{aligned} \tag{13.16}$$

where the symbols are defined in Table 13.2. x is prey numbers, y is predator numbers, and Z is an energy storage compartment. Energy consumed by predators is stored in Z until it is used for reproduction. This creates a cascade of energy that introduces natural time lags in the predator population dynamics. Small δ corresponds to short lags, large δ to long lags. To simplify and eliminate the parameter λ, Harrison (1995) rescaled Z to

Table 13.2: Variables and parameters in best model to fit *Paramecium–Didinium* predator-prey dynamics.

		VARIABLES
x	15 (#/ml)	Prey density (initial conditions)
y	6 (#/ml)	Predator density (initial conditions)
Z	90.45 (kC)	Total energy in all predators
		PARAMETERS
ρ	3.02 (t^{-1})	Prey net reproduction
K	898 (numbers)	Prey carrying capacity
ω	9.74 (prey/pred·t)	Maximum predation rate
ϕ	54.3 (shape)	Type 3 shape
ν	0.0983 (shape)	Type 3 shape
σ	9.15 (unitless)	Proportion prey consumed that is stored
δ	1.78 (t^{-1})	Energy expenditure rate
λ	—	Reproduction rate relative to energy
γ	1.78 (t^{-1})	Predator death rate

$z = \lambda Z$, so λ was not estimated.

The overall index of model fit was the sum of squared differences between the 18-day data and model at each datum sampling point. For Eqs. 13.16, the sum of squares was 29,231. The sum of squared deviations for the standard model (Eqs. 13.12–13.13) was 236,137; for the best model (not shown), it was 25,439. The best model was similar to Eqs. 13.16, but added a time lag in prey growth. The two improved models show nearly an order of magnitude improvement in accuracy over the standard model, but the price we pay for this is more parameters to estimate: the standard model has five, the model of Eqs. 13.16 has nine, and the best model has ten. (While λ did not need estimating, the initial condition for z was required.) Assessing the trade off of accuracy against model complexity is usually subjective. Harrison (1995) clearly felt that the cost of four parameters needed to gain an order of magnitude improvement over the standard model was worthwhile. However, he concluded that adding one more parameter to reduce the sum of squares by only an additional 4000 was a high price to pay. This is a situation where we could usefully incorporate measures of model complexity in our evaluations (Section 8.2.5, Spriet and Vansteenkiste 1982).

Harrison (1995) did not really follow our protocol for validation of multiple models outlined in Chapter 8 to the letter. As he emphasized, his was an exercise in curve fitting using a family of models. Nor did he attempt to test formally the hypothesis that one model (e.g., Eq. 13.16) was statistically better than the simpler ones. Problems of repeated measures and other statistical assumptions probably would have made this effort problematical. Nevertheless, this is an excellent illustration of the power of the approach of multiple working hypotheses that can lead to new insights into the role of different biological processes (e.g., time delays). It is also an example of reasonably accurate predictions of simple laboratory predator–prey experiments by relatively simple equations. We will see another example of an important test of laboratory population dynamics in Chapter 17 when we examine chaos and nonlinear dynamics.

13.3 Exercises

1. Werner and Caswell (1977) developed a stage-structured model for teasel (*Dipsacus sylvestris*) with the following stage definitions

$$\begin{array}{ll} x(1) & \text{seeds} \\ x(2) & \text{dormant seeds (yr 1)} \\ x(3) & \text{dormant seeds (yr 2)} \\ x(4) & \text{rosettes } (< 2.5 \text{ cm}) \\ x(5) & \text{rosettes } (2.5 - 18.9 \text{ cm}) \\ x(6) & \text{rosettes } (> 19.0 \text{ cm}) \\ x(7) & \text{flowering plants} \end{array}$$

   The matrix was

$$\begin{pmatrix} 0 & 0 & 0 & 0 & 0 & 0 & 503 \\ .430 & 0 & 0 & 0 & 0 & 0 & 0 \\ 0 & .970 & 0 & 0 & 0 & 0 & 0 \\ .010 & .021 & .005 & 0 & 0 & 0 & 0 \\ .036 & .003 & 0 & .190 & .253 & 0 & 0 \\ 0 & 0 & 0 & .070 & .105 & .150 & 0 \\ 0 & 0 & 0 & 0 & 0 & .002 & .517 \end{pmatrix},$$

   where the upper row corresponds to seed production by flowering plants (503 seeds·plant^{-1}·yr^{-1}), and the remaining elements (L_{ij}) are the fractions of stage j that become stage i in the next year.

   (a) Starting with an initial distribution of 100 seeds (only), simulate this population for 40 years. Plot the numbers of seeds, flowering plants, and all rosettes over time. Plot the proportion of flowering plants to all stages over time. Does a stable distribution for this stage develop? Explain what you observe.

   (b) Use matrix manipulation software to estimate λ and r from Eq. 13.8. Does this agree with your simulations? Compare your calculations with the values reported in Werner and Caswell (1977) for population "L."

2. Write a computer program to simulate density-dependent population dynamics with and without the Allee effect. Summarize the differences.

3. Write finite difference equations and the matrix form for the following situation. A plant population has three size classes (0, 1, 2). Sizes 1 and 2 can reproduce: each individual of 1 produces three offspring

and each of 2 produces 4 offspring. Each individual of size 0 can either grow to size 1 or 2, or stay the same size. The average fraction doing each of these is 0.8, 0.1, and 0.1, respectively. Fractions of size 1 can shrink, grow, or stay the same size, i.e., 0.1, 0.7, and 0.2, respectively. Size 2 can shrink to size 1 or stay the same size: 0.05, 0.95.

4. For the following age-based projection matrix, compute the eigenvalues and r using Eq. 13.8:

$$\begin{pmatrix} 0 & 2 & 3 \\ .5 & 0 & 0 \\ 0 & .2 & 0 \end{pmatrix}$$

5. Using Eq. 13.7, derive a simple equation that computes λ, *assuming* the population has achieved a stable age distribution. Calculate λ for the matrix in exercise 4 using as the stable age distribution: $N_0 = 10$, $N_1 = 20$, $N_2 = 40$.
6. Convert the equations for Z and y in Eq. 13.16 to unitless form to eliminate two parameters.
7. Write a program to simulate the IBM of Madenjian and Carpenter (1991). Use parameters that they provide. Investigate the effects of stochastic variation on population dynamics by plotting the population size over time for multiple runs with different starting random number generator seeds.
8. Write differential equations and derive the nullclines for the following scenario. In the absence of any predators, a prey population grows in a logistic manner. When present, the predator consumes prey according to a Type 1 functional response, converts prey to new predators at a rate c, and a constant fraction of predators die at each moment of time. How many equilibria are there, which are stable, and which are unstable? Perform a local stability analysis according to the methods described in Chapter 8. Simulate the equations using a wide variety of parameters and starting conditions. Do the simulations agree with the stability analysis?
9. DeAngelis (1992) hypothesized a simple predator–prey–nutrient recycling model in which detritus was assumed to decompose instantaneously. In the equations below, N is the prey and X is a consumer

$$\begin{aligned} dN/dt &= I_n - r_n N - r_1 NX/(k_1 + N) + d_1 X \\ dX/dt &= r_1 NX/(k_1 + N) - (d_1 + e_1)X. \end{aligned}$$

   (a) Give a verbal description of each of the components and parameters in the above equations.
   (b) Derive and plot the nullcline equations.
   (c) Qualitatively evaluate the stability properties of the possible

equilibria.

10. Using the parameters below, simulate and compare the standard Luckinbill model with Harrison's model Eq. 13.16 against Luckinbill's 32-day data set. In Harrison's parameterization, the standard model is:

$$\begin{aligned} dV/dt &= \rho V(1 - V/K) - \omega V/(\phi + V) \\ dP/dt &= \sigma V/(\phi + V) - \gamma P. \end{aligned}$$

The parameter values used were:

ρ	K	ω	ϕ	σ	γ
1.85	898	25.5	284.1	12.40	2.07

11. Based on the curves in Fig. 13.9, draw the two sets of curves that produce exactly two equilibria at two special values of S.
12. Incorporate pesticide applications into the spruce–budworm model (Eqs. 13.14–13.15) by adding another mortality term to the budworm: $-pN$. Based on the graphical argument shown in Fig. 13.9, show why pesticides applied to budworms are likely to straighten out the budworm nullcline. What effect will this have on system stability?

Chapter 14

Chemostats

14.1 Chemostats and Simple Population Dynamics

A CHEMOSTAT is an experimental chamber (Fig. 14.1) in which the dynamics of small, usually asexually reproducing organisms are studied under controlled laboratory conditions. While it is not a requirement, chemostats are typically maintained in a *steady-state* condition. A steady-state chemostat consists of a growth chamber into which a constant concentration of nutrients are pumped at a constant rate. Organisms are introduced into the chamber and allowed to take up nutrients and grow. Both the growth medium and the microorganisms are removed from the chamber at a constant rate in order to maintain a constant volume. The purpose of this arrangement is to permit the microorganisms to grow in constant abiotic (nutrient) conditions. These systems have applications in research laboratories for physiological studies, in industry as a method to produce large quantities of chemical by-products useful in research and medicine (e.g., enzymes), and in sewage treatment plants. Chemostats are not common in nature, but they are sometimes closely approximated in aquatic upwelling systems such as those located off the western coast of South America. The biological questions that models of chemostats can address include: (1) What is the effect of temporal variability on the outcome of competition? Is it likely that high species diversity in ecological communities is maintained by temporal variability? (2) Can chaos arise in simple predator-prey models?

Because of this constancy in the physical conditions and their practical importance, chemostats have been extensively and successfully modeled. Recently, these models have been reviewed (Smith and Waltman 1995). Here, we use the models as good examples of several principles developed in *Part I.* (1) We will apply the basic techniques of quantitative model formulation to compartment models with time-varying parameters. This model will be used to examine the effects of temporal variability on com-

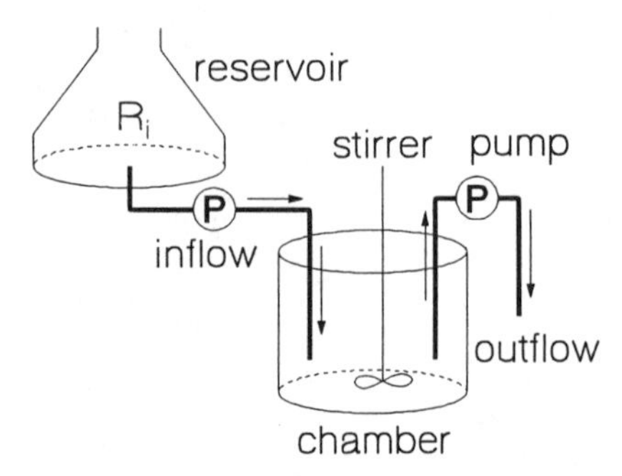

Figure 14.1: Diagram of a typical chemostat. Nutrients are pumped (P) at a rate D from a reservoir with concentration of R_i into the growth chamber containing organisms. The contents of the chamber are pumped out at the same rate.

petitive interactions. (2) Model simplification (Section 4.4) is illustrated by converting the model to a dimensionless form and by using a conservation equation. (3) In Chapter 8, we advocated the importance of investigating model reliability as well as model adequacy. In this chapter, we describe a chemostat model that is tested in an experimental setting (i.e., pulsed nutrients) for which it was not designed. (4) Finally, we also explore more advanced forms of model analysis using techniques developed for general nonlinear dynamical systems. We will see that simple models of predator–prey interactions in chemostat can produce very complicated dynamics.

14.1.1 Monod Model

A model of the chemostat is a system with two components: a nutrient (resource or substrate) measured in grams or moles and a population whose growth is limited by the substrate measured in numbers.

The classical model for the population in this system is the *Monod* equation,

$$\frac{dN}{dt} = N(\mu - D), \tag{14.1}$$

where N is the number of cells in the chamber, D is the death or dilution rate. μ is the steady-state growth rate assuming that growth follows the Michaelis–Menten saturation curve,

$$\mu = \mu_{\max}\left(\frac{R}{K_R + R}\right), \tag{14.2}$$

where R is the nutrient concentration, $\mu_{\max}$ is the maximum growth rate, and K_R is the half-saturation constant. It is evident from Eq. 14.1 that the population size will be in steady state only if $\mu = D$, which depends on the nutrient concentration.

The substrate flows into the chemostat at rate D with concentration R_i and is removed at rate D with a concentration equal to the current concentration R in the vessel. The substrate is taken up at a rate proportional to the growth rate of the population. The complete system of two coupled differential equations is

$$\begin{aligned}\frac{dR}{dt} &= D(R_i - R) - \frac{\mu_{\max}}{Y}\left(\frac{R}{K_R + R}\right)N \\ \frac{dN}{dt} &= -DN + \mu_{\max}\left(\frac{R}{K_R + R}\right)N,\end{aligned} \tag{14.3}$$

where D is dilution rate of the chemostat, R_i is the concentration of substrate in the input reservoir, $\mu_{\max}$ is the maximum per capita growth rate of the consumer (cells·cells^{-1}·time^{-1}), and K_R is the Michaelis–Menten half-saturation constant. Y (*yield*) is the amount of substrate required to produce one consumer individual; it converts growth of consumer to amount of substrate removed. This basic model has simple expressions for the equilibria and nullclines for N and R, and their determination is left as an exercise.

14.1.2 Droop Equation

The greatest limitation of the Monod model is that the amount of nutrient supplied in the growth medium does not accurately reflect the amount of nutrients available for growth. The latter is better viewed as being dependent on an internal storage pool of the limiting nutrient (recall Harrison 1995 in Section 13.2.3). The Droop equation describes population dynamics when such a mechanism is incorporated (Rhee 1980). The population model is as in Eq. 14.1, but μ is a function of the internal pool:

$$\mu = \mu'_{\max}\left(1 - \frac{k_q}{Q}\right),$$

where Q is the internal concentration of the resource also known as the cell quota. k_q is the subsistence (or minimal) cell quota and $\mu'_{\max}$ is the maximum growth rate at infinite cell quota. The cell quota is the external nutrient uptake rate (v) divided by the growth rate (μ):

$$Q = \frac{v}{\mu},$$

where

$$v = v_{\max}\left(\frac{R}{K_v + R}\right).$$

R is the external nutrient pool concentration (in the chemostat), K_v is the half-saturation constant for nutrient uptake, and $v_{\max}$ is the maximum rate of nutrient uptake. By combining the above equations and assuming the system is in equilibrium, it is possible to derive a simple equation for the half-saturation constant for cell growth (Rhee 1980):

$$K_R = \mu'_{\max} k_q K_v / v_{\max}. \tag{14.4}$$

Table 14.1: Values used in Burmaster's model. Nominal values and ranges are shown in brackets.

INITIAL CONDITIONS		
N	$0.8[0.5 \rightarrow 1.5]\text{cell} \cdot \text{L}^{-1}$	Number of cells
R	$1.0[0 \rightarrow 2.0] \times 10^{-6}$ M	Inflow nutrient concentration
PARAMETERS		
$\mu'_{\max}$	$1.03 \cdot \text{day}^{-1}$	Maximum growth rate
k_q	$7.02 \times 10^{-16}\text{M} \cdot \text{cell}^{-1}$	Minimal cell quota
$v_{\max}$	$4.68 \times 10^{-16}\text{M} \cdot \text{cell}^{-1} \cdot \text{min}^{-1}$	Maximum rate of nutrient uptake
K_v	0.51×10^{-6} M	Nutrient uptake half-saturation
D	$0.5[0 \rightarrow 1.0]\text{day}^{-1}$	Dilution rate
R_i	$1.0[0 \rightarrow 2.0] \times 10^{-6}$ M	Inflow nutrient concentration

Thus, the empirically measured half-saturation parameter of the Monod growth model can be derived from mechanisms of nutrient uptake.

14.1.3 Success of the Models

Both the Monod and Droop models use a Michaelis–Menten relationship which is based on a quasi-steady-state assumption (Chapter 4). This means that the models were not designed to accurately portray short-time-scale, transient dynamics. The models have been shown in many experiments to successfully predict the equilibrium conditions, and Burmaster (1979a) has shown that the Monod and Droop models are equivalent at steady state.

A natural question to ask is: How good are the models in predicting variable conditions? Burmaster (1979b) constructed a Droop model for the growth of a single-cell algal (*Monochrysis lutheri*) in a chemostat into which he could experimentally inject nutrients to produce rapid changes in the operating conditions of the chemostat. From independent experiments, he estimated the model parameters (Table 14.1).

Burmaster performed three different kinds of perturbations: stepped changes and pulses in the influent nutrient concentration, and stepped changes in the dilution rate. A step perturbation is one in which a variable is jumped to a new value and held there; a pulse is an instantaneous, one-time addition of nutrients. His main interest was to predict cell numbers in the growth chamber. He found good agreement when the system was subjected to a step up or down in R_i (Eq. 14.3, Fig. 14.2). The model predicted a step-down in dilution rate, but not a step-up (Fig. 14.3). The response to a pulse in R_i was not predicted by the model (Fig. 14.4). Since a step function produces alterations that persist, the cells have an extended period to adapt to the new conditions, and the steady-state model does reasonably well. In a pulse, the cells experience the new conditions only briefly, but the model does not contain detailed biochemical mechanisms to mimic the transient dynamics of the real cells. Burmaster (1979b) proposed that a desirable modification to the basic equations to better describe the tran-

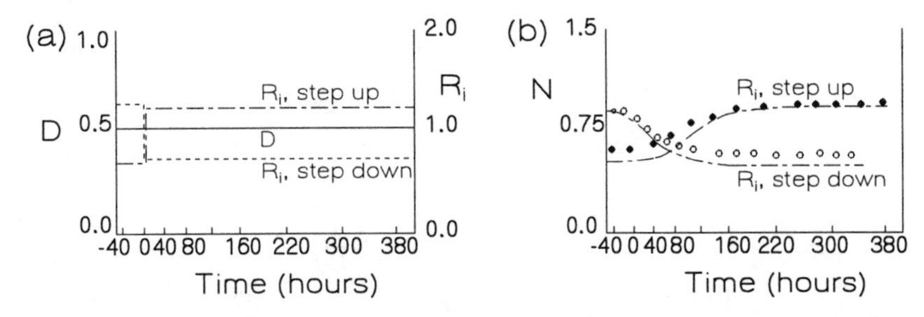

Figure 14.2: Chemostat model response to stepped changes in nutrient concentrations in the inflow (R_i). (a) Constant dilution rate (D) and experimental step-up and step-down of nutrient concentration in the influent of a chemostat. (b) Number of cells predicted (line) and observed (points) following the step perturbations. (From Burmaster 1979b, Figs. 5, 6. © 1979 Elsevier Science, B.V. Reprinted with permission of the publisher.)

sient conditions was to add a time delay in the cell division equation so that cell dynamics are described as

$$\frac{dN}{dt} = \mu'_{\max}\left(1 - \frac{k_Q}{Q(t-\tau)}\right) - DN. \tag{14.5}$$

We leave it as an exercise for the student to examine the consequences of this proposal.

14.2 Competitors in Chemostats

Classical chemostat theory makes a very elegant and clear prediction of the outcome of competition between two consumers of a nutrient in a chemostat. The Monod equations for the two competitors are

$$\begin{aligned} \frac{dR}{dt} &= D(R_0 - R) - \sum_{i=1}^{2} \frac{N_i \mu_i R}{Y_i(K_i + R)} \\ \frac{dN_1}{dt} &= \frac{\mu_1 R}{K_1 + R} N_1 - DN_1 \end{aligned} \tag{14.6}$$

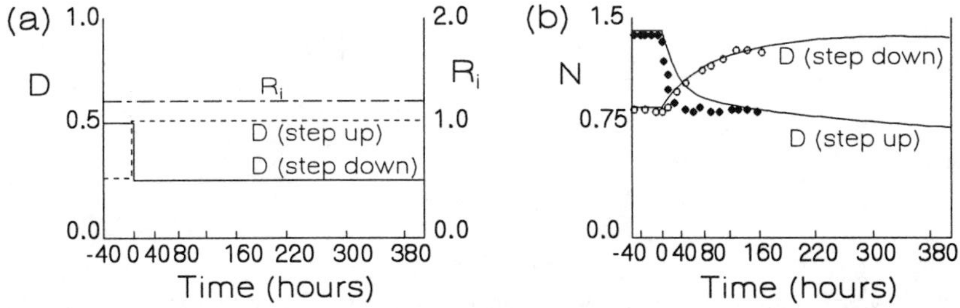

Figure 14.3: Chemostat model response to stepped changes in dilution rates. (a) Constant influent concentration (R_i) and experimental step-up and step-down of dilution rate (D). (b) Number of cells predicted (line) and observed (points) following the step perturbations. (From Burmaster 1979b, Figs. 7, 8.© 1979 Elsevier Science, B.V. Reprinted with permission of the publisher.)

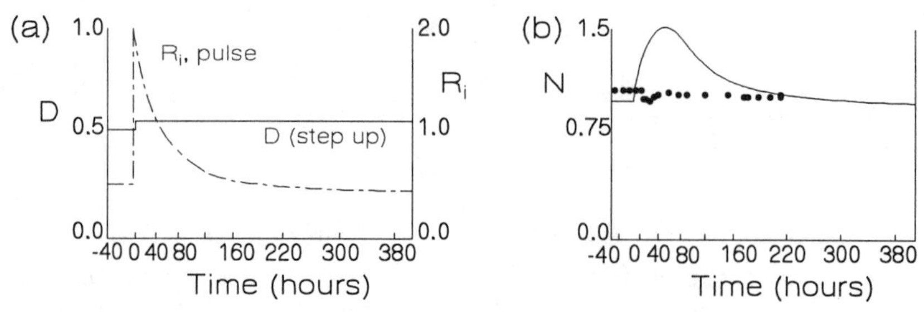

Figure 14.4: Chemostat model response to pulsed nutrient concentration in the inflow.(a) Dilution rate (D) and pulsed influent concentration (R_i). (b) Number of cells predicted (line) and observed (points) following the pulse. (From Burmaster 1979b, Fig. 9.© 1979 Elsevier Science, B.V. Reprinted with permission of the publisher.)

$$\frac{dN_2}{dt} = \frac{\mu_2 R}{K_2 + R} N_2 - DN_2,$$

where D is dilution rate, R_0 is inflow nutrient concentration, and Y_i is the yield in units of number of cells of species N_i produced for each grams of R consumed. μ_i is the maximum growth rate of species i, and K_i is the half-saturation constant for growth. From the perspective of organisms growing in the vessel, the dilution rate is equivalent to a form of mortality, so that D can also be considered the per capita death rate. The net per capita growth rate of both species, in terms of the Michaelis–Menten growth relationship and death rate (D), is plotted in Fig. 14.5. R_i^* is the value of the resource at which species i is at equilibrium and has the value

$$R_i^* = \frac{DK_i}{\mu_i - D}. \tag{14.7}$$

If the resource falls below R^*, the population numbers will decline. R^* depends on species-specific parameters as well as the dilution rate of the chemostat (D). If two species compete in a chemostat, that species which has the lowest R^* will win. It is possible for two species to have different

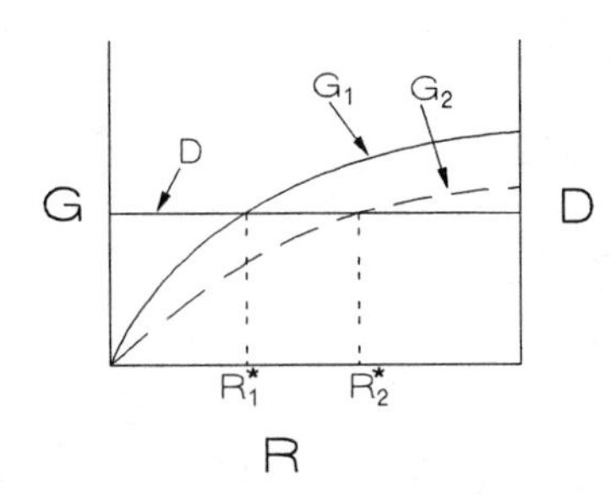

Figure 14.5: Equilibrium resource levels for one dilution rate and two growth functions. Left axis (G) is the positive per capita growth rates, and the right axis (D) is the dilution or death rate. Each growth curve represents a competing species. Equilibrium population numbers for species i occur at the resource level at which the growth curves intersect the dilution curve at R_i^*. The superior competitor is that which possesses the lowest R_i^*.

Table 14.2: Measured parameters corresponding to the basic equations for two competing species of bacteria in a chemostat.

Exp	Strain	Y_i	K_i	r_i	R_i^*	Winner
1	A	2.5×10^{10}	3.0×10^{-6}	0.81	2.40×10^{-7}	
	B	3.8×10^{10}	3.1×10^{-4}	0.91	2.19×10^{-5}	A
2	C	6.3×10^{10}	1.6×10^{-6}	0.68	1.98×10^{-7}	
	D	6.2×10^{10}	1.6×10^{-6}	0.96	1.35×10^{-7}	D
3	C	6.3×10^{10}	1.6×10^{-6}	0.68	1.98×10^{-7}	
	E	6.2×10^{10}	0.9×10^{-6}	0.41	1.99×10^{-7}	—

K_i and μ_i, but have identical R_i^* and therefore be able to coexist. The crucial attribute of this prediction is that it is based on mechanisms operating on the individual population level: the prediction uses data obtained from individual species growing in isolation. In the Lotka–Volterra–Gause competition model, to predict the outcome one must estimate the interaction parameters (α and β) by observing the two species together. This is also true of "energy-based" mechanistic competition models that do not model resources explicitly (see Section 13.2).

The smallest R^* rule is well established in steady-state chemostats (e.g., Tilman 1977). A complete test, however, must also show that if two species have different Michaelis–Menten parameters that produce identical values of R_i^*, the two species will coexist. Hansen and Hubbell (1980) demonstrated this in a system using bacterial strains competing for the amino acid tryptophan. Table 14.2 lists the parameters of Eq. 14.6 measured by Hansen and Hubbell (1980). Experiment 1 used $R_0 = 1.0 \times 10^{-4} \text{g} \cdot \text{liter}^{-1}$, and $D = 6.0 \times 10^{-2} \text{h}^{-1}$. Experiments 2 and 3 used $R_0 = 5.0 \times 10^{-4} \text{g} \cdot \text{liter}^{-1}$, and $D = 7.5 \times 10^{-2} \text{h}^{-1}$. Their results are consistent with predictions: Strain A won in Experiment 1, strain D won in Experiment 2, and neither species dominated the system in Experiment 3. These results are not surprising given that the Michaelis–Menten relationship is based on a quasi-steady-state assumption. Nevertheless, this is a good example of a validated model in population ecology.

But in light of Burmaster's (1979b) results and the rarity of constant conditions in nature, it is important to know if similar simple rules will predict competitive outcomes in non-steady-state chemostats. Grover (1990) performed simulations of periodically perturbed chemostats with two competing algae species. For comparison, he defined an *opportunist* species as one with *relatively* large maximum growth rate (high Michaelis–Menten asymptote) and large half-saturation constant and a *gleaner* as a species with a small maximum growth rate and a small half-saturation constant. In steady-state systems, gleaners have a lower R^* and therefore should outcompete opportunists. Using Eqs. 14.6, Grover (1990) modeled the input concentration as a sine function: $R_0 = \overline{R} + b \sin(\omega t)$. Since total algal growth will depend on the total amount of resource supplied to the system, it is important to choose parameters so that a constant amount of resource

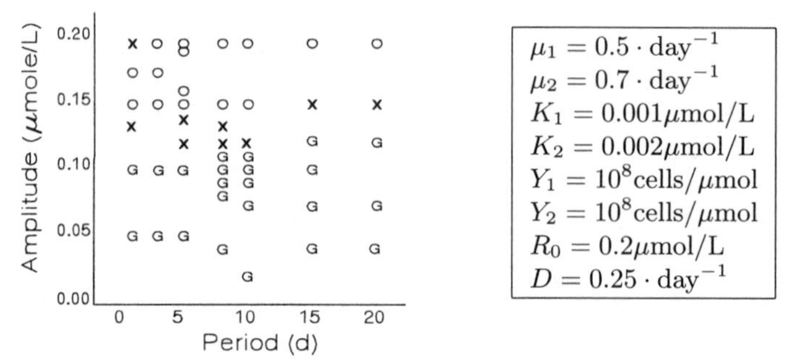

Figure 14.6: Predicted outcome of competition between a gleaner and opportunist in a chemostat under periodically pulsed resource inputs. G = the gleaner dominates the system, X = coexistence, and O = the opportunist dominates the system. (From Grover 1990, Fig. 5. Reprinted with permission of the University of Chicago, publisher. © 1990 by the University of Chicago.) The table on the right shows the parameters used in Eqs. 14.6.

is used for all simulations. To control this, Grover chose the amplitude of the sine function according to the formula

$$b = \frac{\overline{R}DT}{1 - \exp(-DT)} - \overline{R},$$

where T is the period between pulses.

His models predicted (Fig. 14.6) that gleaners would dominate the chemostat for all periods if the amplitude is less than 0.10 μmol/liter. This is consistent with the definition of a gleaner. When the amplitude is relatively small, the system acts like a steady-state chemostat and the lowest R^* (a gleaner) will dominate. Above an amplitude of 0.1, a narrow region of coexistence is predicted. Further increases in amplitude produce conditions that favor the opportunist. Interestingly, coexistence also appears at very high amplitudes, if the period is short. Grover reviewed the few empirical studies that relate to this theory, but could not find conclusive support or falsification. The relatively narrow band of conditions that permit coexistence casts doubt on the hypothesis that high species diversity in natural phytoplankton communities is maintained by temporal variability. It may be that other candidates such as spatial heterogeneity or niche partitioning are more important. The results shown here do not give a definitive answer but are an interesting step in the right direction.

14.3 Predators in Chemostats

The above models show that extrinsic temporal variability can complicate the predictions of simple models. Next we ask the question, Can extrinsic variability, when coupled with intrinsic cycles, produce even more complicated dynamics? A three-species model shows that it does.

The model for a three-trophic level system in a chemostat follows the basic form of Michaelis–Menten and predator–prey models developed earlier. These equations have been studied extensively, but a new analysis by Kot et al. (1992) and Pavlov and Kevrekidis (1992) from the perspective of nonlinear dynamics is especially instructive. Kot et al. (1992) begin with the following equations

$$\begin{aligned}\frac{dS}{dt} &= D(S_i - S) - \frac{\mu_H}{Y_H}\frac{SH}{K_H + S} \\ \frac{dH}{dt} &= \mu_H \frac{SH}{K_H S} - DH - \frac{\mu_P}{Y_P}\frac{HP}{K_P + H} \\ \frac{dP}{dt} &= \mu_P \frac{HP}{K_P + H} - DP,\end{aligned} \tag{14.8}$$

where S is the concentration of nutrient (substrate), S_i is the inflow concentration of nutrient, D is the dilution rate, H is the concentration of herbivore, μ_H is the maximum growth rate of H, K_H is the half-saturation constant of H feeding on S, Y_H is the yield of H per unit of S consumed, P is the predator concentration, μ_P is the maximum growth rate of P, K_P is the half-saturation constant of P feeding on H, and Y_P is the yield of P per unit of H consumed.

The analysis of this system is simplified by first converting Eqs. 14.8 to a dimensionless form using the following substitutions (see Section 4.4.2):

$$\begin{aligned}x &\equiv S/S_i \\ y &\equiv H/(Y_H S_i) \\ z &\equiv P/(Y_H Y_P) \\ \tau &\equiv Dt.\end{aligned} \tag{14.9}$$

When substituted into Eqs. 14.8, these produce

$$\begin{aligned}\frac{dx}{d\tau} &= 1 - x - \frac{Axy}{a + x} \\ \frac{dy}{d\tau} &= \frac{Axy}{a + x} - y - \frac{Byz}{b + y} \\ \frac{dz}{d\tau} &= \frac{Byz}{b + y} - z,\end{aligned} \tag{14.10}$$

where $A = \mu_H/D$, $a = K_H/S_i$, $B = \mu_P/D$, and $b = K_P/(Y_H S_i)$. This little maneuver has decreased the number of parameters from 8 to 4.

The second simplification is to reduce the number of state variables by noting that the total quantity of material in the system is conserved (see Section 4.4.3). So,

$$x_\tau + y_\tau + z_\tau = 1 + (x_0 + y_0 + z_0)e^{-t}, \tag{14.11}$$

which simply states that, at any moment in time, all of the material is stored in the sum of the state variables, and that over time this is a simple exponential decay from the initial total quantity to 1.0. Obviously, if any

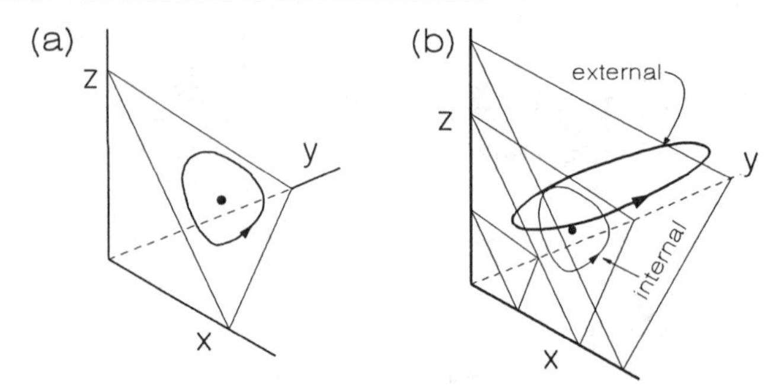

Figure 14.7: Phase space motions in the predator-prey chemostat model. (a) Internally generated limit cycle and fixed point on a plane in the phase space of the unforced system. (b) The limit cycle internal oscillations interact with an external forcing function to produce complex dynamics in phase space. (Reprinted by permission of the publisher from Kot et al. 1992, Figs. 3 and 4. © 1992 Elsevier Science, Inc.)

two of the state variables are known, the third can be computed from Eqs. 14.11.

This results in two ODEs:

$$\frac{dy}{dt} = \frac{ACy}{a+C} - y - \frac{Byz}{b+y} \quad (14.12)$$

$$\frac{dz}{dt} = \frac{Byz}{b+y} - z, \quad (14.13)$$

where $C = 1 - y - z$. These equations behave more or less like other two-dimensional predatory–prey models we have discussed earlier. They may either exhibit simple, stable equilibria or limit cycles. There are three equilibria: (1) only S present, (2) S and H present, and (3) S, H, P present. Equilibrium (3) has a Hopf bifurcation (Section 17.1) and consequently may be represented either as a fixed point or a stable limit cycle. Depending on parameter values, the fixed point or limit cycle will lie on a plane in the x–y–z phase space (Fig. 14.7a).

Considerably more complex dynamics emerge if the double Monod system is given an input that varies sinusoidally. Kot et al. (1992) replaced S_i in Eqs. 14.8 with

$$S_i \left[1 + A \sin(2\pi t/T)\right],$$

where A is the amplitude of the sine function and T is the period in the same units as t. Now there are two mechanisms of oscillation (Fig. 14.7b): the internally generated oscillations of the limit cycle and the externally generated forcing function applied to the substrate's input concentration. The former mechanism permits the dynamics to stay on a plane in phase space (Fig. 14.7a), while the latter mechanism forces the dynamics away from this plane, as the ellipse illustrates (Fig. 14.7b.

Kot et al. (1992) found that sine functions with a high frequency ($T = 5$ hr) or an intermediate amplitude ($A = 0.6$) produced only small changes in the unforced dynamics. The effects were most pronounced in the lower

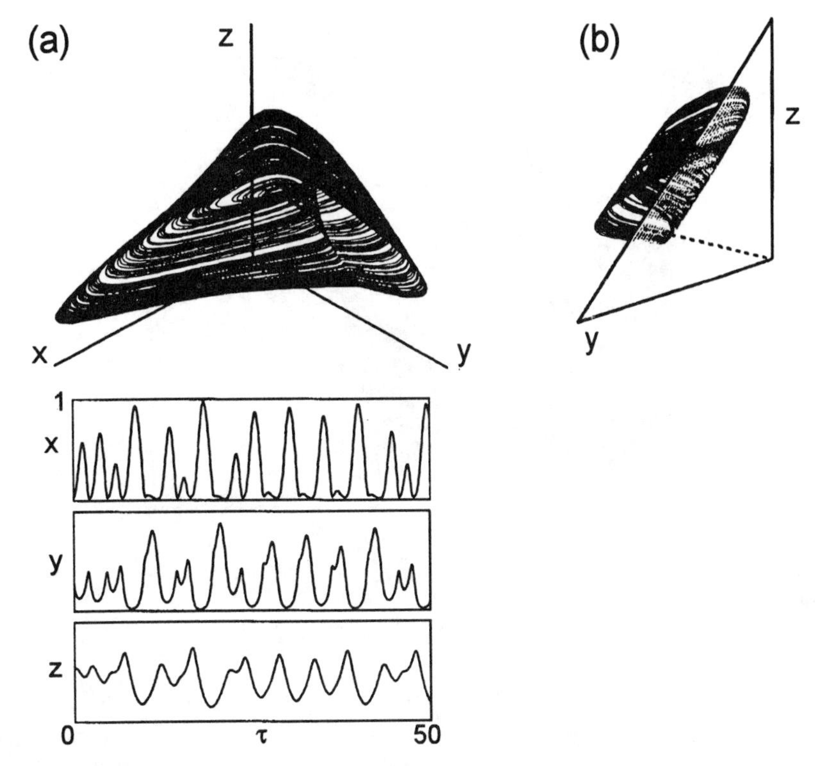

Figure 14.8: (a) The phase space portrait and time domain dynamics of the forced double Monod chemostat with large period. Note the complex, non-periodic curves in both representations. There are greater fluctuations in lower trophic levels. (b) The same dynamics as in (a)another perspective. Note that the phase space motion does not lie on a plane. (Reprinted by permission of the publisher from Kot et al. 1992, Figs. 6 and 8. © 1992 Elsevier Science, Inc.)

trophic levels, being practically undetectable in the dynamics of P. As T was increased, purely periodic solutions reappeared due to *frequency locking* where the forcing frequency matches and amplifies the intrinsic periodicity. Sufficiently large periods ($A = 0.6$ and $T = 24$ h) produced chaotic dynamics. Two pictures of these dynamics are shown. The top panel of Fig. 14.8a shows one view of the phase space motion, and the bottom panel shows the time domain dynamics for the three state variables. Figure 14.8b shows the same dynamics from a different perspective which illustrates one of the signatures of aperiodic dynamics in which it is shown that the dynamics do not lie on a plane in phase space.

A clearer picture of the complex dynamics and frequency locking properties of this forced system is found in the bifurcation diagram. Figure 14.9 shows the regions of chaotic dynamics (filled black areas) and periodic solutions (lines). This plot is the result of taking snapshots of the dynamics for many iterations by sampling at a frequency that is the inverse of the

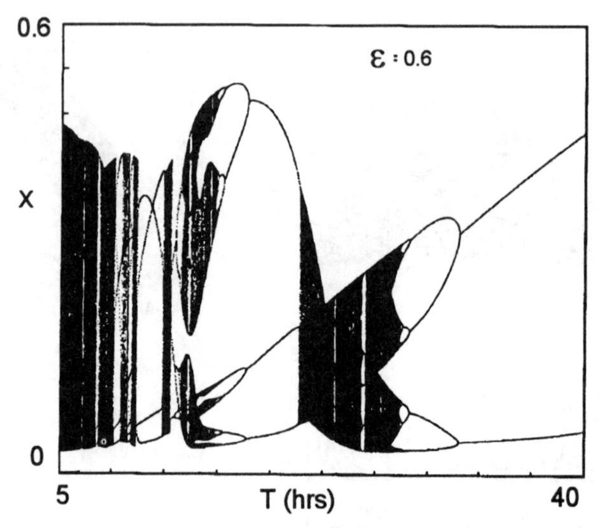

Figure 14.9: Bifurcation diagram of the force double Monod model for $A = 0.6$ and T variable. Note, as T increases, the alternation between chaotic and periodic dynamics due to resonance of the internal and external frequencies. (Reprinted by permission of the publisher from Kot et al. 1992, Fig. 17. © 1992 Elsevier Science, Inc.)

period shown on the x-axis. If the dynamics were a perfect sine wave and sampled at the frequency of the sine wave, the snapshots would always produce the same value. A curve produced by two superimposed sine waves of different frequencies will, by this sampling method, always produce the same two values. This elaboration of points will continue as the number of superimposed sine waves increase. Chaos, in this context, can be thought of as infinitely many sine waves superimposed, thus producing the black band of Fig. 14.9. We discuss these matters in more detail in Chapter 17.

14.4 Exercises

1. Discuss the strengths and weaknesses of Burmaster's evaluation of the Monod model.
2. For the model Eq. 14.3, write equations for the nullclines for both the substrate and the consumer. Plot the nullclines in the N *vs* S plane. Are the equations stable for all parameter values?
3. Perform a formal, local stability analysis for the model Eq. 14.3. Does it agree with the qualitative assessment using nullclines?
4. Do the following using the parameters for Burmaster's model in Table 14.1.
   (a) Simulate the model so that Burmaster's experiments can be reproduced.
   (b) Explore more drastic changes in stepped changes S_0, and pulses.

   (c) Explore Burmaster's suggestion that a time lag would improve the fit with pulses.

5. Derive Eq. 14.7.
6. Derive Eqs. 14.10 from Eqs. 14.8 and Eqs. 14.9.
7. Write a simulation model of single population growth in a chemostat in which the input nutrient concentration is pulsed using Burmaster's proposed model of Eq. 14.5. How well does it match the data presented in Fig. 14.4?
8. Write a simulation model of Grover's oscillating chemostat with two competitors. Investigate in greater detail his result that at low period and high amplitude the two species coexist.
9. Grenny et al. (1973) studied a chemostat model for a microbe whose growth is based on the amount of intracellular protein. They allowed chemostat flow rate to be a periodic pulse function, as did Grover, but they found a broader set of conditions over which coexistence occurred. Read both the Grover (1990) and Grenny et al. (1973) articles and discuss possible reasons for this discrepancy.

Chapter 15

Spatial Patterns and Processes

15.1 Dynamics in Space: New Complications

DYNAMICS in nonlinear systems can be immensely complex, as we discuss in Chapter 17. Unfortunately for simple theories, time is only one dimension relevant to physical and biological systems. At every level of biological organization, from biochemistry to ecosystems, dynamics are embedded in the three physical dimensions of space. In principle, our models should account for this fact by explicitly incorporating spatial effects in the mathematics. Naturally, we can sometimes avoid these problems by appropriately defining our objectives, but for many biological phenomena this is not an option. Examples abound: flows of chemicals (toxic, nutrients, or signals) in fluids (air or water); movements of organisms (in continuous space or among discrete patches); population growth of sessile individuals; and development of morphological structure (coat color patterns in animals, microtubules within cells).

Worse yet, it may be that physical space is not enough. In some systems, just as it is necessary to know how a state variable is distributed over space to make predictions, it may be necessary to know how a state variable is distributed over a physiological condition. If so, the physiological condition forms the basis of a "spatial" dimension that significantly affects biological interactions. For example, the age of an individual affects its death and birth probabilities. To be maximally accurate, then, age-specific population models must sometimes describe the "flow" of individuals from age to age, just as individuals in a river would flow from point to point. In these models, it is necessary to understand the distribution of individuals over the age dimension, just as in spatial models we must understand the distribution over the spatial dimension. Age becomes a variable analogous to physical space.

These observations lead one to conclude that the spatial distribution (physical or physiological) of variables is of fundamental importance. Further, as we saw in Chapter 13, mechanistic models of populations produce dynamics and insights not present in simple phenomenological models. In this chapter, we connect pattern with process.

Accepted usage takes "pattern" to be a quantity distributed nonrandomly and (usually) nonuniformly in space. An example is the patchy distribution of color in an animal's coat. In general, a pattern is simply the spatial dispersion of the observed quantity. By "process" we mean a mechanistic explanation. Models of pattern involve static descriptions of the distribution or mechanistic dynamic models that explain the pattern's existence. In addition, biologists, especially ecologists, have recently become aware of the fact that a quantity's dispersion depends on the *scale* with which it is observed. This affects both the time and space dimensions, and we will discuss this problem in the next chapter.

Among the principles developed in *Part I* that we illustrate in this chapter are (1) the use of partial differential equations, (2) reaction-diffusion equations, (3) the effects of parameters on stability properties, and (4) an individual-based, spatially explicit population model. The examples to follow use these principles to address several biological questions: (1) Is the aggregation of microorganisms into organized spatial structures a random process caused by simple diffusion in which individuals react only to their local environment? (2) Are spatial patches of insect pests and their predators caused by random movement or by the inclination of individual predators to rationally hunt in areas where they have had previous success? (3) To save the Spotted Owl from extinction, is it better to provide many small habitat reserves or a few large tracts of habitat?

15.2 Pattern and Process

Spatial pattern is the distribution of the quantity of a variable in the three dimensions of physical space. To model these patterns, we must mathematically describe how the quantity flows from point to point in space. We have already introduced the concepts and basic mathematics for flows in continuous space in Section 4.3.5. Here we apply this formulation to movements of animals. In the first example, the organism is relatively simple, and the model serves to introduce the basic equations. The second example shows how more complex organisms and behavior can be embedded in the same formalism.

15.2.1 Slime Mold Aggregation

Dictyostelium discoideum, a slime mold, has achieved fame because it is an extremely useful biological system for the experimental study of intercellular chemical signalling. *D. discoideum* is valuable because individual cells of this species have the ability to live much of their life in isolation, but when food resources become scarce, the cells move and aggregate to form a multicellular organism that produces a fruiting body that emits propagules. The cells accomplish this remarkable feat by moving toward high concentrations of the chemical signal $3',5'$-cyclic AMP. Keller and Segel (1970) wrote a classic paper that describes a partial differential equation (PDE) model of aggregation. We describe a simplified version here (Lin and Segel 1988).

We assume that we can arrange a laboratory experiment so that slime mold cells are constrained to move in one dimension only. In Section 4.3.5, we developed the basic reaction-diffusion PDE in one dimension. Amoeba population dynamics at a spatial point x are a function of diffusion, population growth, and aggregation. We assume straight off that the population is not reproducing because food has been exhausted. We will also assume linear diffusion rates as a function of amoeba concentration $[a(x,t)]$. The flux rate due to aggregation is a linear function of the concentration gradient of a chemical signal $[\rho(x,t)]$ and the current amoeba concentration at the point. Thus, amoeba dynamics are:

$$\frac{\partial a}{\partial t} = \underbrace{\frac{\partial}{\partial x} D_2 \frac{\partial a}{\partial x}}_{\text{diffusion}} - \underbrace{\frac{\partial}{\partial x} D_1 \frac{\partial \rho}{\partial x}}_{\text{aggregation}}, \tag{15.1}$$

where D_2 is a constant describing the "diffusivity" or random motion of individual amoeba cells. D_1 measures the strength of the chemical signal on cellular aggregation. The signs are specified as they are because (1) we use the convention of calculating gradients as $x_0 - x_{0+\Delta x}$ and (2) diffusion causes repulsion from "positive" amoeba gradients $[a(x_0) < a(x_{0+\Delta x})]$, while aggregation causes movement toward "positive" signal gradients $[\rho(x_0) < \rho(x_{0+\Delta x})]$. The negative sign for aggregation in Eq. 15.1 forces its value to be positive. Empirical data suggest that

$$D_1 = \delta a/\rho, \tag{15.2}$$

where δ is a scaling constant.

The chemical signal dynamics are ultimately based on biochemical reaction kinetics, but we assume quasi-steady-state conditions to simplify:

$$\frac{\partial \rho}{\partial t} = \underbrace{-\frac{b\rho}{1+K\rho}}_{\text{decay}} + \underbrace{af(\rho)}_{\text{secretion}} + \underbrace{D_\rho \frac{\partial^2 \rho}{\partial x^2}}_{\text{diffusion}}, \tag{15.3}$$

where a, b and K are constants, D_ρ is the diffusivity of the signal, and $f(\rho)$ is the per capita rate at which the signal is produced by amoebae. The expression $b\rho/(1+K\rho)$ should be recognized as a form of the Michaelis–Menten-type saturation relation. As a result, b is the maximum rate of ρ decay and K is a shape parameter for the saturation curve.

With the exception of the reaction terms (signal production and degradation), these are linear equations. Keller and Segel (1970) performed a stability analysis in which they derived the characteristic equation (Section 9.3.2) to be

$$\lambda^2 - \lambda(F - q^2 D_2) - (q^2 f(\rho_0) D_1 + q^2 D_2 F) = 0, \tag{15.4}$$

where λ are the eigenvalues, q is a constant, ρ_0 is the signal concentration at equilibrium, and $F = f'(\rho_0)a_0 - \bar{k} - q^2 D_\rho$ (where a_0 is the amoeba density at equilibrium, and $\bar{k}$ is a function of the signal degradation rate evaluated at the signal equilibrium). As shown earlier, if the equilibrium is to be stable, then the largest λ must be less than 0. This occurs, after expanding F, if

$$D_1 f(\rho_0) + D_2 f'(\rho_0) a_0 < D_2(\bar{k} + D_\rho q^2). \tag{15.5}$$

After more algebra and using Eq. 15.2, the system will be unstable if

$$\frac{\delta}{D_2} + \frac{a_0 f'(\rho_0)}{\bar{k}} > 1. \tag{15.6}$$

The two terms of the sum on the left represent two different sets of processes. The term to the left is, basically, the ratio of the ability of the amoebae to detect and respond to a gradient of ρ (as measured by δ) to the rate of random amoebae movement (D_2). The term on the right is, *roughly*, the ratio of the rate of signal production to the rate of signal degradation. If the term on the right is much less than 1 (low signal production), then the parameters of diffusion processes determine stability. Instability will occur if δ is large or if D_2 is small. That is, instability will occur when elevated signal and amoeba concentrations in space cannot rapidly become smoothed out by random motion of the signal or amoeba.

This smoothing process will be slowed if amoebae can react strongly to the presence of a signal gradient (large δ). This condition creates positive feedback as illustrated in Fig. 15.1. A patch of high amoeba density (Fig. 15.1b) will cause increased production of the signal, which will create a spatial signal gradient. This will attract more amoebae (due to large δ) and create even more signal in that region, which will attract even more amoebae to the area, and so on. The size of the peak of elevated amoebae and signal concentration in space will depend on the balance of the two forces represented by D_2 and δ. If δ is relatively large, the peak will grow. If D_2 is relatively large, the equilibrium will be stable (Eq. 15.6 not satisfied), and the peak will dissipate due to random motion of the amoebae. In conclusion, diffusion forces can cause spatial inhomogeneities to grow and become more pronounced over time.

15.2.2 Ladybugs and Aphids

The above model of organism movement is a good first approximation, but in essence it is a phenomenological model because it relies on empirical

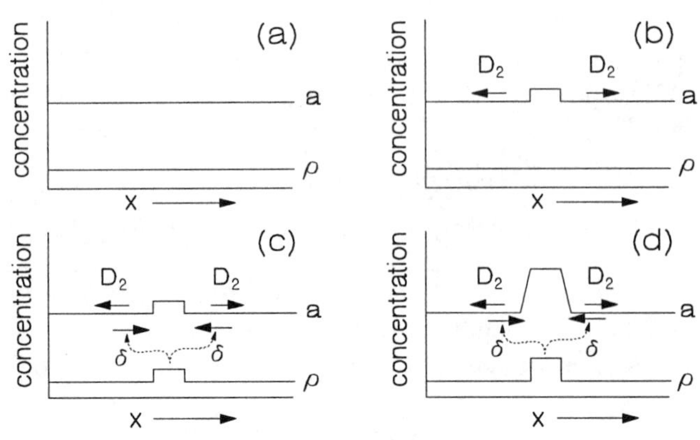

Figure 15.1: Spatial dynamics of amoebae and the aggregation signal. Four hypothetical snapshots in time of the concentration over space (x) of the amoebae population (a) and the aggregation signal (ρ). Panel (a) is the equilibrium condition. In (b), a small amount of amoeba is added in a small area. The amoeba diffusion parameter (D_2) acts to spread out the amoeba and return the system to the equilibrium. In panel (c), some short time later, the local patch of amoeba have excreted some aggregation signal which causes more amoeba to flow towards the patch according to the stimulus strength parameter δ. This produces a positive feedback which is amplified in panel (d).

estimation of the diffusivity. At extremely microscopic levels, treating organisms like physical particles is an adequate approximation. In such cases, treating organism movement as a statistical process suffices. But, when we consider larger organisms with a richer behavioral repertoire, it becomes important to relate the mechanisms of individual behavior to global movement parameters at the population level, such as diffusivity. One area where this is possible is the spatial movement of insect predators relative to their prey. Kareiva and Odell (1987) developed a mechanistic PDE model of a ladybug predator (*Coccinella septempunctata*) and its aphid prey (*Uroleucon nigrotuberculatum*). The derivation of the model is involved, so we will give a simplified version here and refer the curious reader to the original paper.

The model is composed of two coupled PDEs. The basic population equations are very similar to those used by Keller and Segel (1970), with the role of the aggregation signal played by the aphid prey and the role of amoebae played by the searching predator. The equation for the aphid (victims) is

$$\frac{\partial V}{\partial t} = \underbrace{D_V \frac{\partial^2 V}{\partial x^2}}_{\text{diffusion}} + \underbrace{\sigma_{Vn}(V, P)}_{\text{birth}} \underbrace{- \sigma_{VK}(V, P)}_{\text{death}}, \tag{15.7}$$

where V is victim numbers, and P is predator numbers. The first term on the right is passive diffusion [random movement, D_V is victim (aphid) diffusivity]. The second term is a function for net births, and the third term is a function for numbers of prey killed by predators. All of the state variables mentioned in Eq. 15.7 apply to a local point in continuous space and not to the total population sizes integrated over the entire space.

Field observations of aphid population growth dynamics in the absence of predators support the logistic, density-dependent birth model,

$$\sigma_{Vn}(V,P) = bV(V_{\max} - V),$$

where $V_{\max}$ is the maximum aphid density and was estimated from field observations; b is a growth parameter that incorporates $V_{\max}$ and was estimated from short-term field population growth experiments in which predators were excluded (Table 15.1).

$\sigma_{VK}(V,P)$ in Eq. 15.7 is the rate at which predators consume prey at a particular spatial point. Kareiva and Odell (1987) derived an expression that is related to the Holling Type 2 disc equation:

$$\sigma_{VK}(V,P) = \frac{\alpha\gamma VP}{1 + [\gamma(1-\eta)/\lambda + 1/\nu]V}, \tag{15.8}$$

where α, λ, η, ν, and γ are parameters estimated from short-term predation experiments. The parameters are defined in Table 15.1. This equation comes about because of the effect of victim numbers (V) on predator satiation (S). Satiation is a dynamic balance between the rate of increase of S (i.e., eating) and the rate at which S decreases (i.e., digestion).

In general, the number of prey consumed in unit time by a single predator is the number of prey captured and killed (predation rate) times the fraction of each individual prey consumed times the degree that a single prey increases the level of satiety (i.e., decreases hunger). As a special case in their general theory, Kareiva and Odell hypothesize that two processes affect the rate of consumption: the current level of hunger and a hyperbolic saturation of the rate of predation as prey numbers increase. The latter is basically the Holling disc equation and is represented as $V/(1 + V/\nu)$. The former, hunger, is simply $1 - S$. Both processes may limit the overall consumption rate, but how to combine them? We discussed this problem in Chapter 4, and Kareiva and Odell use the multiplicative approach so that net consumption rate is

$$C(S) = (1 - S)\frac{\gamma V}{1 + V/\nu},$$

where γ is a conversion factor.

The fraction of an individual prey that is consumed declines with satiation: $1 - \eta S$. And, lastly, each completely consumed prey increases satiation (decreases hunger) by a constant, α. If we assume that digestion decreases satiation by a constant proportion in unit time (λ), the dynamics of satiation is

$$\frac{dS}{dt} = (1 - S)\frac{\gamma V}{1 + V/\nu} - \lambda S.$$

If we assume that the acts of predation and digestion occur much more quickly than population growth and migration, then we can assume that S will achieve equilibrium rapidly for a given level of V. In other words,

$$\frac{dS}{dt} = 0 = S_0(V) = \frac{\gamma V}{\lambda + V(\gamma + \lambda/\nu)}$$

Knowing the steady state satiation level, we can solve for the predation rate per predator at constant V as the consumption rate divided by fraction of individuals consumed times the effect of consumption on satiation. This reduces to

$$K[S_0(V), V] = \frac{\alpha\gamma V}{1 + \ [\gamma(1-\eta)/\lambda + 1/\nu]V}.$$

This equation is a modified Holling disc equation in which the overall rate is the combination of two processes hypothesized to influence consumption rate. First, consumption rate will decline proportionally to the satiation level or degree to which the gut is filled $[\gamma(1-\eta)/\lambda)]$. Second, as with most predators, there are behavioral or physiological limits to consumption rates resulting in a Type 2 saturation curve. In this model, this phenomenon is parameterized by a maximum encounter frequency (ν). Since this is the rate per predator, multiplying by the number of predators gives the total death rate of victims, shown in Eq. 15.8.

Predator dynamics are more complex because both diffusion and aggregation processes are important, as are immigration and emigration. The flux equation for predators is

$$\frac{\partial P}{\partial t} = \underbrace{\frac{\partial}{\partial x} D_p(V) \frac{\partial P}{\partial x}}_{\text{diffusion}} - \underbrace{\frac{\partial}{\partial x}\chi(V)P\frac{\partial V}{\partial x}}_{\text{aggregation}} + \underbrace{\sigma_{Pa}(V,P)}_{\text{immigration}} - \underbrace{\sigma_{Pd}(V,P)}_{\text{emigration}}. \tag{15.9}$$

The loss term on the extreme right represents emigration only, but in another model could also include death processes. This function depends on both V and P in that ladybugs will stay in a patch unless aphid density is below a minimum threshold described as

$$\sigma_{Pd}(V,P) = \min\{0, A_1 P(V - A_2)\}, \tag{15.10}$$

where A_1 and A_2 are empirical constants with A_2 being the threshold. The middle term in Eq. 15.9 is immigration, but could also represent birth processes. Birth is ignored here because the time scale of the Kareiva–Odell model is short relative to the generation time of the ladybugs. Given this, field experiments demonstrated that immigration occurred at a constant rate, independent of local aphid densities.

The remaining terms in Eq. 15.9 are the now-familiar summation of diffusion and aggregation. The important feature of this model, which distinguishes it from others (e.g., Keller and Segel 1970), is that diffusivity and aggregation are mechanistically defined and estimated by individual movements. The reader should consult Kareiva and Odell (1987) for the detailed derivation, but here we repeat the intuitive description contained in their Fig. 2. Remember that the central hypothesis of the model is that area-restricted search, an individual-level phenomenon, will produce bulk population flows that concentrate predators in regions of high prey density. The critical assumption needed to achieve this is that individuals will have a greater tendency to reverse their direction of travel when they are more satiated than when they are less satiated. In other words, hungry bugs will tend to walk straight ahead; full bugs will be indecisive, moving first this way, then that way. This is a reasonable hypothesis and a plausibly adaptive strategy: if you're hungry, you're not finding food, and if you're not finding food, you should look elsewhere.

Will this reversal hypothesis cause a net flow of predators in the direction of increasing prey density? To see that this is the case, imagine a prey population that increases monotonically from left to right. Predators on the right will have relatively high reversal rates, because, according to the hypothesis, they are finding lots of prey and are relatively satiated. Predators on the left will have relatively low reversal rates because they are hungry and finding relatively few prey. Suppose that, at some given point along the prey spatial gradient, five satiated predators happen to be moving to the left and five hungry predators happen to be moving to the right. In the next time interval, most of the hungry predators will continue moving right because they have a low reversal probability. Let's say that four of the five predators continue moving right. Conversely, many of the satiated predators will reverse direction, by the reversal hypothesis. For instance, two of the five reverse and move right. As a result, in the next time interval, four predators are moving left toward lower prey density and six predators are moving right toward higher prey density. Thus, the individual mechanism produces net flow towards the higher prey density. To complete the intuitive argument and to test your understanding, suppose the prey population is low at both the right and left ends of the spatial interval. This means that the prey population has a maximum somewhere along the interval. Will the net flow be towards the right all along the interval, or will it be towards the left over some sub-interval and towards the right over a different sub-interval?

Based on algebra underlying the above intuitive argument, Kareiva and Odell derived the predator diffusivity function as

$$D_p(V) = \frac{u^2}{2R(S_0)}, \tag{15.11}$$

where u is the ladybug travel speed, and $R(S_0)$ is an empirically fit relation between equilibrium satiation (S_0) and the number of reversals per day. To avoid a complicated derivation of the psychology of satiation in ladybugs, the authors simply fit experimental observations of individual reversal rates performed at different prey densities to an empirically posited third-degree polynomial with parameters $\beta_i, i = 0 \ldots 3$. The result is the required monotonic increase in reversal probability with increasing densities of prey and, consequently, increasing levels of satiation.

The last term is the aggregation function. The critical term to define is the *preytactic* sensitivity coefficient $[\chi(V)]$ which represents the degree that the prey gradient induces area-restricted searching behavior in ladybugs. Based again on the intuitive argument, Kareiva and Odell (1987) derived

$$\chi(V) = \frac{u^2 \dfrac{dR}{dS} \dfrac{dS_0}{dV}}{R\left[2R - \dfrac{\partial S_r}{\partial S}\right]}, \tag{15.12}$$

where R is the reversal function (Eq. 15.11), S is satiation level, S_0 is equilibrium satiation level, and S_r is the rate of change of satiation. Equation 15.12 is not easy to interpret; parts of it are "...intuitively obscure, at least to us." (Kareiva and Odell 1987, p. 246). Basically, the numerator causes aggregation to increase as the speed of the predator (u) increases or as the reversal rate (R) increases with changes in victim density (acting through satiation level). The mechanism of this relation is area-restricted search: the ability of the ladybug to detect aphids and reverse direction. The denominator describes how the ladybugs will become more sensitive to aphid gradients as the local density of aphids becomes small. Overall, χ declines with increasing V.

The complete set of parameters and their values are shown in Table 15.1. Using a combination of short-term laboratory and field experiments, all of the parameters were estimated. The model was solved numerically using a standard method for PDEs. The model as a whole was tested in an independent field experiment in which 10-m strips of goldenrod were maintained in an isolated field (Fig. 15.2). At 1-m intervals, nonuniform densities of aphids were deposited. The test distribution used were two patches, at 3 m and 7 m, using two different concentrations of aphids. Uniform densities of ladybugs were deposited at each of the 1-m positions. The resulting spatial distributions of aphids and ladybugs were followed for several subsequent days. This test was repeated at several times during the season. The results of one such test replicated three times are shown in Fig. 15.2 .

Although not a quantitatively rigorous validation (Chapter 8), the model results are in good qualitative agreement with the data. In particular, two ladybug patches, centered on the aphid peaks, developed as predicted. Victim numbers on day 2 are overpredicted at 3 m, suggesting that either predation rates are higher than predicted or that aphids have movement processes (e.g., escape behavior) that were not modeled. These results demonstrate that area-restricted searching behavior at the individual level translates into patchy population distributions. This may be one mech-

Table 15.1: Parameter values and units for the Aphid–Ladybug model.

	VARIABLES	
S	Unitless ($0 \to 1$)	Satiety: fraction of gut filled
V	aphids/m	Victim (aphid) density
P	ladybug/m	Predator (ladybug) density
	PARAMETERS	
α	8.00	Aphids killed at predator consumption rate
A_1	0.0095 m/d	Ladybug emigration scaling
A_2	107.0 m/d	Ladybug emigration threshold
β_0	1.7115 reversals/d	Empirical parameter for reversals
β_1	45.3098 d^{-1}	Empirical parameter for reversals
β_2	-180.172 d^{-1}	Empirical parameter for reversals
β_3	272.991 d^{-1}	Empirical parameter for reversals
b	3.76×10^{-6} m/d	Aphid population growth rate
D_v	0.02 m^2/d	Aphid diffusivity
η	0.9866	Fraction aphid not consumed at satiety level
γ	0.018632 m/d	Maximum predator consumption rate due to empty gut
λ	2.3384 d^{-1}	Ladybug excretion rate
ν	711.2 m^{-1}	Maximum ladybug encounter frequency
σ_{Pa}	0.5 m/d	Ladybug immigration at low aphid density
u	5.87 m/d	Ladybug movement speed
$V_{\max}$	50,000 aphids/m	Maximum aphid population

anism by which spatial heterogeneity is maintained in opposition to the homogenizing effects of pure diffusion. This model has become a classical example of the ability of PDE models to represent individual behavioral processes that produce spatial pattern. Moreover, models of this kind can also be used to investigate practical questions of predator control of prey pests.

15.2.3 Other Continuous Applications

Many spatial models concern the development and persistence of patterns in space (e.g., patchy dispersion of prey and predator). Kareiva and Odell (1987) give a useful synthesis of the fundamental processes necessary for pattern to arise in space. They observe that spatial pattern requires and will almost always develop when there exists (1) the short-range, fast *activation* of a signal that can increase in the absence of other forces (e.g., aphid populations that can increase independently of predators), (2) a long-range, slow *inhibition* of the signal (e.g., ladybug predation), and (3) a *positive relation* between the strength of the signal and growth rate of the inhibition (in effect a negative feedback). Patterns arise because the relatively slow movement (aggregation) of inhibitory effects (ladybugs) and its tendency to diffuse away permits the relatively fast autocatalytical growth of the signal to reach levels that can be sustained against the inhibition.

Many biological systems satisfy these conditions, although in some cases "signals" and "inhibitors" are not physically distinct objects such as at-

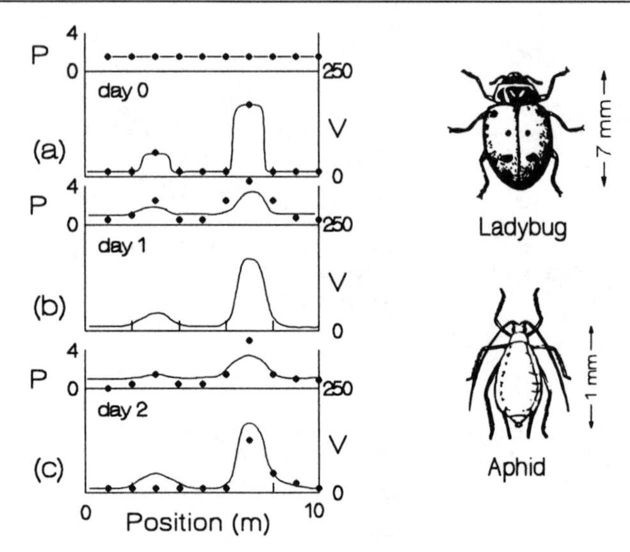

Figure 15.2: Time and space dynamics of the Kareiva–Odell ladybug–aphid movement model. Predator (P: top) and victim (V: bottom) population densities across 10 spatial positions at 1-m intervals for three time periods, averaged over three replicate experiments. (a) Initial conditions (day 0): victim patches at 3 m and 7 m and homogeneous predator population. (b) One day later; no victim data recorded. (c) Two days later. (Redrawn after Kareiva and Odell 1987, Fig. 5. © 1987 by the University of Chicago. Reprinted by permission of the University of Chicago, publisher.)

tractant chemicals and slime molds. In some cases, they are simply different rate processes acting on a single system. For example, spatial pattern in chemical toxicant flowing in fluid (e.g., a river) can arise given the proper balance between chemical production, diffusion, advection, and biotic breakdown. Other examples of systems to which continuous spatial models have been applied include water and nutrient flows in soils. Similar but less obvious examples are "flows" of pulsating blood pressure in blood vessels or of voltages in nerve cells. A great many problems in morphological development (e.g., striping or spotting patterns in animal coats) can be formulated as the production and inhibition of a chemical substance that affects coat pigment (Murray 1989).

15.3 Patches and Metapopulations

Another broad class of spatial models represents space as discrete patches. These models come in two flavors: (1) the patches are contiguous with each other, and (2) the patches may be separated by an undefined distance. In (1), the patches represent a coarse-grained discretization of continuous space, similar to the fine-grained representation used in solving PDE models of spatial flows (Kareiva and Odell 1987). This approach to spatial structure is frequently used in large-scale ecosystem models where the number of components and the complexity of their interactions require a relatively simple spatial structure in order to reduce the computational load or to match model structure with a low resolution sampling schedule. Models in class (2) are intended to describe patches of habitats or islands. Typically, the populations have two dynamical phases: within patch growth and between patch migration. When migration between patches is possible, the set of patches constitutes a *metapopulation*. Two central questions associated with metapopulations are: What factors influence the fraction of patches with nonzero population sizes?, and What factors influence the probability that the species will become globally extinct over all patches? The latter question is obviously of great concern to conservation biology and is a major factor in *population viability analysis* (Boyce 1992). We discuss models addressing these two questions below.

15.3.1 Populations of Patches

The simplest model of patch occupancy is due to Levins (1969), who viewed the occupied patches in a metapopulation as a population itself (separate from the populations that inhabited the individual patches). The variable of interest is the *fraction* of occupied patches (p) and since this variable must decline to 0 as patches become occupied, Levins chose a logistic-like relationship

$$\frac{dp}{dt} = mp(1-p) - ep, \tag{15.13}$$

where m is the migration or dispersal rate and e is the extinction rate.

This equation is formally equivalent to the density-dependent population growth model and has one stable equilibrium. If we divide the left-hand side by p and plot the extinction (e) and migration ($m(1-p)$) against p, we see how the equilibrium (p^*) is altered by the parameters (Fig. 15.3a). Setting Eq. 15.13 to 0 and solving for the equilibrium in terms of the parameters is left as an exercise.

This model basically predicts that either no patches will be occupied if $e > m$ or there will be an intermediate equilibrium that is stable. These results arise because the model assumes that extinction is independent of the fraction of occupied patches. Hanski (1991), however, reviewed a large number of studies that showed a positive correlation between the average population size (N) and the fraction occupied (p). Hanski (1991) extrapolated this fact to single-species dynamics and assumed that extinction rate declines as the fraction of occupied patches increases. The modified model is

$$\frac{dp}{dt} = mp(1-p) - e_0 pe^{-ap}, \tag{15.14}$$

where m is migration rate, e_0 is the extinction rate when no patches are occupied, and a is a shape parameter that describes the extinction rate decrease as a function of p. The exponential term is one possible im-

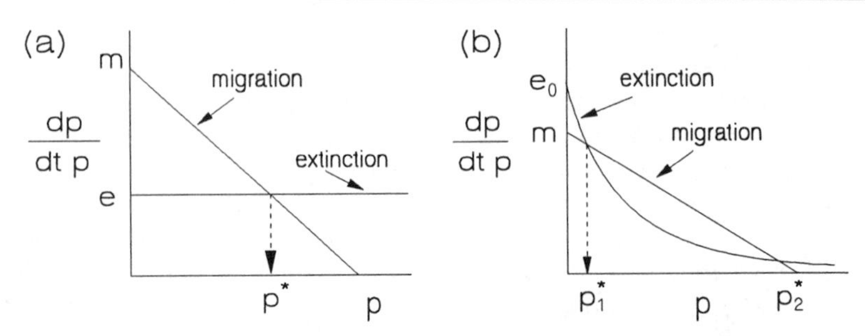

Figure 15.3: Equilibria for two patch models. (a) The original Levins model plotting the gain rate and loss rate against the fraction of occupied patches. The point of intersection is the stable equilibrium. (b) A modified model in which the rate of extinction declines with fraction of patches occupied. p_1^* is an unstable equilibrium; p_2^* is a stable equilibrium.

plementation of the Allee effect. This addition dramatically changes the nature of the model (Fig. 15.3b). If $e_0 > m$, there are now three equilibria; two are stable and the third lying between these two is unstable. As a consequence of this simple change in the assumptions, the model now predicts that there will be a threshold fraction of occupied patches, p_1^*, below which the population will go extinct globally (across the entire metapopulation). If $p > p_1^*$, the metapopulation will converge on p_2^* patches occupied.

15.3.2 Population Processes Within Patches

Levins' model of patch occupancy was phenomenological in that it did not contain any mechanisms to explain the fraction of patches occupied. Lande (1987, 1988) and Lamberson et al. (1992) have generalized Levins' model by writing explicit equations for the number of occupied sites in terms of demography and population dynamics. The results of the two sets of models are similar, but the mathematical analyses are quite different. Here, we describe the approach of Lamberson et al. (1992).

The system being modeled is the extremely controversial case of the endangered Northern Spotted Owl (*Strix occidentalis caurina*). References to the biology of the owl can be found in Dawson et al. (1987) and Lande (1988). In brief, this predator feeds high on the foodchain and is long-lived, territorial, and apparently requires large tracts of mature coniferous trees ("old-growth" forests). Single males establish territories that attract females. The males are monogamous, so that males are either single or paired with females. Juveniles must find new territories beginning in their second year. The controversy arises because the forests that the Spotted Owl inhabit are extremely valuable as lumber. Timber harvesting fragments the forest, producing small patches of habitat. Territorial birds occupying the patches create a metapopulation.

The Forrester diagram for the modeled system is shown in Fig. 15.4; there are four state variables that interact in the standard structured popu-

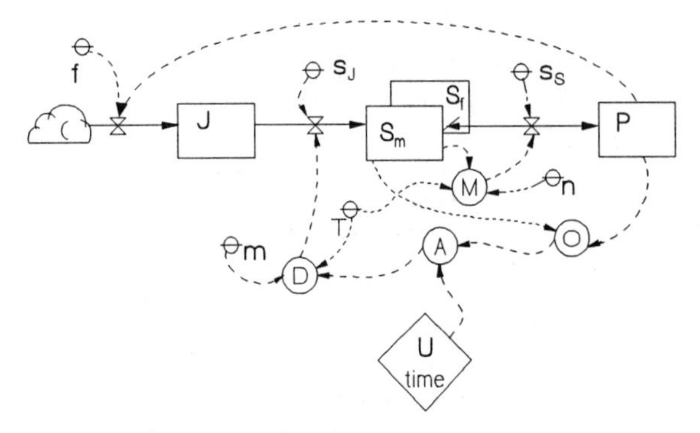

Figure 15.4: Forrester diagram of the Spotted Owl Model. See Table 15.2 for definitions.

lation age class form (see Section 13.1.2). The effects of the patch structure on global population dynamics occur through the effect of available nesting sites on dispersal mortality and mating success. This is also the mechanism by which population density affects dynamics. The driving variable (U) allows the effects of timber harvesting to be included. Notice that there is no explicit representation of the spatial positions of owls. The equations for the state variables are

$$J_t = P_t f \tag{15.15}$$

$$P_t = P_{t-1} p_s + S_{t-1} s_S M_t \tag{15.16}$$

$$S_{m,t} = 0.5 J_{t-1} s_J D_t + S_{m,t-1} s_S (1 - M_t) + p_f P_{t-1} \tag{15.17}$$

$$S_{f,t} = S_{m,t}. \tag{15.18}$$

The auxiliary equations are

$$\begin{aligned} S_t &= S_{m,t} + S_{f,t} = 2S_{m,t} \\ O_t &= P_t + S_{m,t} \\ A_t &= U(t) - O_t \\ D_t &= 1 - (1 - A_t/T)^m \\ M_t &= 1 - (1 - S_t/T)^n. \end{aligned}$$

The definitions and parameter values pertinent to the model are shown in Table 15.2. The nominal time step is 1 year. The model assumes a 1:1 sex ratio, hence the factor 0.5 in Eq. 15.17. Two important assumptions of the model should be noted. First, D_t is the probability that a juvenile will find an unoccupied patch before dying. $(1 - A_t/T)$ is the probability of not finding a patch in one "search attempt." This value raised to the number of attempts made before dying (i.e., the search efficiency, m) is the probability of dying before finding a patch. One minus the probability of dying is the probability of surviving and finding an unoccupied territory in m attempts. Second, M_t is the probability of an unmated female finding a male occupying a territory on a patch. The probability calculation uses the same logic as juvenile dispersal: $(1-S_t/T)$ is the probability of not finding

Table 15.2: Parameters and definitions for the Spotted Owl model.

Symbol	Definition	Value
STATE VARIABLES		
J	Juvenile numbers	
S	Total single (unpaired) adult numbers	
P	Paired adult numbers	
S_m	Numbers of single males	
S_f	Numbers of single females	
AUXILIARY VARIABLES		
O	Number of occupied sites	
A	Number available sites	
$U(t)$	Time varying number of suitable sites	
D	Probability of juveniles surviving dispersal	
M	Probability of female finding male	
PARAMETERS		
s_S	Fraction of single owls surviving	0.71
s_J	Fraction of juveniles surviving to single adults	0.60
p_s	Probability both individuals of a pair survive	0.88
p_f	Probability only female of a pair dies	0.056
f	Number of offspring per breeding pair	0.66
m	Unoccupied site search efficiency	*var.*
n	Unmated male search efficiency	*var.*
T	Total sites in system	1000

a mate in a single search attempt. n attempts are made before the female dies or leaves the area. The probability of finding a mate is nonlinear in the sense that it has a maximum at intermediate numbers of unmated males on territories (S_t/T). To see this, consider that as S_t/T goes to 0, M_t goes to 0; as S_t/T goes to 1, M_t again goes to 0. At intermediate values of S_t/T (e.g., 0.5), the probability is moderately large, depending on n. This produces a "hump-shaped" curve: the Allee effect. To test the significance of this assumption, Lamberson et al. (1992) also analyzed an alternative, simpler model that modeled only females and used a fixed probability of mating success.

These equations were simulated and analyzed analytically for equilibrium conditions (Lamberson et al. 1992). Simulations (Fig. 15.5) revealed three equilibria: one at zero pairs, a stable 150 pairs, and an unstable 25 pairs. An unstable equilibrium was also present in Hanski's modification of Levins' model. The same phenomenon is operating here through the effects of density on available territorial sites and mating success. This is shown in Fig. 15.6, where the solid lines indicate the single equilibrium present when density does not affect mating success, and the broken lines indicate stable and unstable equilibria resulting from the mating effects. The unstable equilibrium is pernicious in this case because it constitutes a threshold below which the population will inevitably go extinct (Fig. 15.5). Notice (Fig. 15.6) that for fixed searching efficiency the line of unstable equilibria is essentially flat for much of the abscissa. This implies that increasing the proportion of suitable habitat will not significantly reduce the extinction

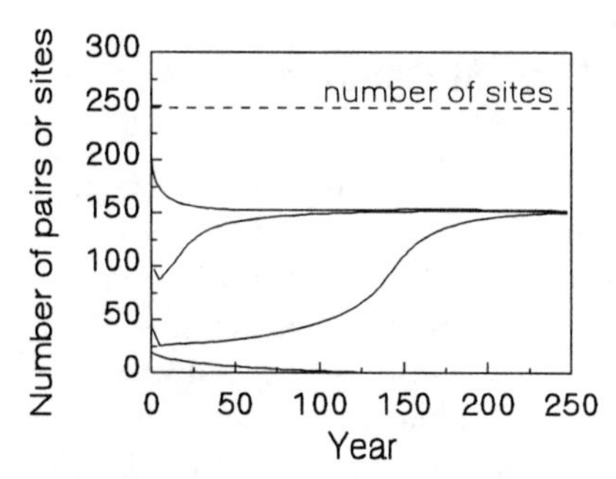

Figure 15.5: Simulation results for Spotted Owl breeding pairs. Top line is the number of suitable sites (U) in the system (25% of total). Trajectories below are scenarios with different initial conditions. Populations started with fewer than 25 pairs go extinct; those above 25 reach a stable equilibrium of 150 pairs. (From Lamberson et al. 1992, Fig. 3. © 1992 Blackwell Science, Inc. Reprinted by permission of Blackwell Science, Inc.)

threshold. Other management strategies will need to be explored.

Lamberson et al. (1992) also investigated the effects of timber harvesting on owl population dynamics through the effects it has on suitable breeding sites (driving variable U in Fig. 15.4). They assumed suitable breeding sites were reduced from 40% of the landscape to 20% at a rate of 4% per year. Numbers of pairs declined, as expected, but did not equilibrate at about 100 pairs until 15 years following harvesting cessation. They also found that if harvesting was continued until only 13% of the landscape was suitable habitat, the owl population went extinct. This is a slightly lower threshold than determined by Lande (1988).

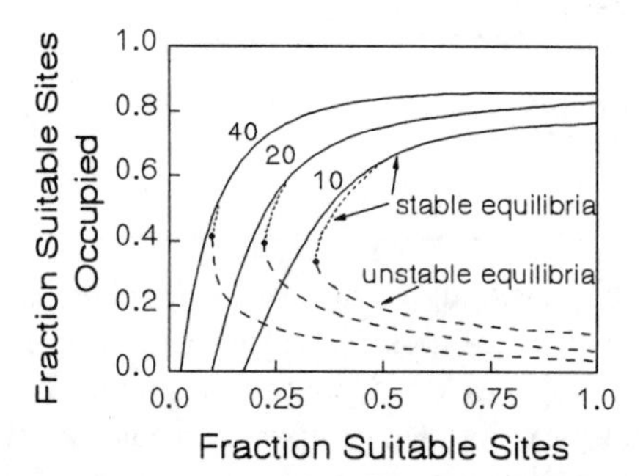

Figure 15.6: Equilibria of Spotted Owl patch occupancy as a function of percent suitable habitat available using three levels of juvenile searching efficiency for unoccupied patches (m = 40, 20, 10). The solid line is the single equilibrium resulting when mating success is independent of population levels (females do not search for mates). The broken lines show the two equilibria that result when mating success declines at low male population density (Allee effect). At each value of m, the dotted line (top) is the locus of stable equilibria; the dashed line (below) is the locus of unstable equilibria. (From Lamberson et al. 1992, Fig. 4. © 1992 Blackwell Science, Inc. Reprinted by permission of Blackwell Science, Inc.)

15.3.3 Spatially Explicit Patches

While the above model of Spotted Owls allows us to manipulate space-related parameters such as the ability of females to find distributed mates, we cannot investigate the effects of the dispersion of suitable sites. For this, we need to know the spatial location of each patch; that is, we need a spatially explicit representation of the suitable sites and a model of the explicit movement of individual owls among the sites. Such a model is an example of an *individual-based* model (IBM) discussed in Section 13.1.4. McKelvey et al. (1993), building on their aggregated patch model, constructed a spatially explicit landscape model of the Spotted Owl.

An individual-based model, when applied to population phenomena, follows the fate of a number of individuals from their birth to death. In the process, this IBM follows the individuals' movement across patches searching for suitable territories or mates. The determination of whether a particular individual will find a mate, or if it will die is the result of probabilistic rules. In this case, a set of rules for males and females operates at each time step. As an illustration, the flow of computation for females is similar to an IBM for fish population dynamics (Section 13.1.4).

```
1. Assign random spatial locations to all owls
2. Loop over time steps
3.   Loop over all females
4.     Calculate probability of predation or starvation (Pd)
5.     Choose Uniform random deviate (x)  Is x < Pd?
         Yes: KILL this female. GOTO 8
          No: Continue
6.     If female not mated, SEARCH for male in surrounding
         patches.  Male found?
         Yes: Assign to pair. GOTO 8
          No: Continue
7.     FIND new suitable patch and MOVE.
         If Out_of_Region, KILL female.
8.     GOTO 3
9.   GOTO 2
```

By repeating this basic algorithm for all females, and a similar one for males and mated pairs, the position of each individual is known as well as its current state: alive or dead, mated or single. The rules used in this model are more complex than the simple equations of the population-based model above; they are designed to incorporate more of the known behavior of the owls. For example, owls can move toward good habitat and away from poor sites; females will avoid crossing territories with mated pairs, and so on. More importantly, since space is explicitly represented, the model can predict the population effects of different spatial arrangements of good and bad sites. This permits an investigation of whether forests should be harvested to preserve one large tract of nesting habitat or to preserve many small "islands" of suitable habitat widely dispersed throughout the forest.

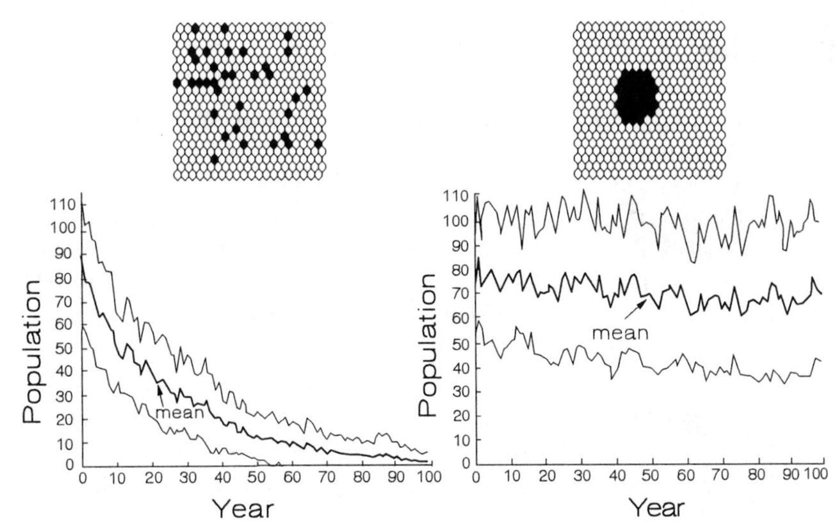

Figure 15.7: Results of 30 replicate simulations of a spatially explicit model of Spotted Owl population dynamics. The scenario on the left shows population mean and ± 1 standard deviation for 100 years when the habitat is distributed as many small patches (see insert). On the right are the results when a single large patch is present. (From McKelvey et al. 1993, Figs. 8 and 10. © 1993 Sinauer Associates, Inc. Reprinted with permission of the publisher.)

This is the basic question of biotic preserve design known as SLOSS: Is it best to create a **S**ingle **L**arge **O**r **S**everal **S**mall preserves? McKelvey et al. (1993) simulated 30 replicates of several scenarios to investigate this question. The results (Fig. 15.7) show that many small patches of suitable habitat produced smaller populations than a single large patch.

In both the earlier spatially aggregated model as well as this more complex spatially explicit individual-based model, we have a model of a system that is neither a white box nor a black box (*sensu* Karplus, 1983). The model results, nevertheless, are tantalizing pictures of possible outcomes of management strategies. Because the issue of whether to cut mature forests is so contentious with such a great deal at stake on both sides of the issue, it behooves us to evaluate carefully if this model in particular, and population ecology, in general, is mature enough to support the model's use as a prescriptive tool (Fig. 1.2). On the one hand, the model captures the basic demography and behavior of the birds using the best data available for parameter estimates. On the other hand, this simple model is a shallow caricature of a mature forest ecosystem of which the owl is a single component. Should we base decisions that will affect the profitability of a major industry, thousands of jobs, and the fate of a cherished species on such a model? Perhaps no other issue or simple model more starkly confronts us with the potential and limitations of computer simulation models to address societal conflicts.

15.4 Exercises

1. In Chapter 4 we discussed several methods to combine multiple rate-limiting processes. Which method was used in the aphid-ladybug model to describe the rate at which aphids are killed?
2. In Chapter 6, we described the method of lines (or coupled ODEs) to solve PDE models. Write the system of coupled ODEs that are appropriate for the aphid–ladybug model. Numerically duplicate Kareiva and Odell's results using this method. Compare to the original approach used.
3. Using the above solution method, or the original method, investigate the importance of ladybug movement speed (u) to the aggregation process. How important is it to measure this parameter precisely?
4. Implement and simulate the Keller–Segel model of slime mold aggregation.
5. Create a conceptual individual-based model (Section 13.1.4) of ladybug movement that incorporates the same behavioral and ecological processes as those used in the Kareiva–Odell PDE model. Compare and contrast the two approaches.
6. Using Eq. 15.13, solve for the equilibrium in terms of p, e, m.
7. Show why the equilibria in Fig. 15.3 are classified as stable or unstable.
8. Simulate the Spotted Owl model and attempt to reproduce the results of Lamberson et al. (1992). Add stochastic effects in the environment; do sufficient Monte Carlo runs to produce 95% confidence intervals at several points in time for breeding pairs and site occupancy. As one possibility, try making fecundity a normal random deviate, then repeat using survival fractions. What effect does stochasticity have on the probability that the population will survive 250 years?
9. Where in Fig. 1.3 would you position the model of Lamberson et al. (1992)? Where would you position the model of McKelvey et al. (1993)? Why?

Chapter 16

Scaling Models

16.1 Pattern and Scale

THE Kareiva–Odell model of spatial predation (Chapter 15) did well in describing the dispersion of aphids and ladybugs over a scale of 10 m^2. But, what we often want to know is: How many ladybugs will it take to eliminate aphid damage in my yard? This is a question of population densities over hundreds of square meters. In Chapter 6, we discussed how the computational difficulty of a problem increases as we increase the spatial extent and dimension of a system. We can reduce this computational explosion if we decrease the spatial resolution we use to solve the equations as we increase the extent. Unfortunately, when we reduce spatial resolution, we often lose the mechanistic basis of the fine-scale model, since the low-resolution model will not be able to represent processes that can be described only at small spatial scales (e.g., biochemical reactions, or individual animal movement). Thus, we have a conundrum: How can we incorporate mechanistic processes into models that must predict over long time and large distances? In short, how can we scale models from the small to the large? One solution is to simply use larger and faster computers, which may include massively parallel computers (Haefner 1992). But another approach is to build large-scale models that preserve the behavior of the mechanistic, small-scale models. In this context, Levin (1992) has noted "...the problem of pattern and scale is the central problem in ecology ...." Moreover, as Levin (1992) emphasized, the scale at which a pattern is observed is often much larger than the scale at which the process is studied. Because of its importance to spatial models, we will discuss some basic issues underlying the concept of scale and its implications. In particular, we will discuss the role of models in bridging the gap between process and pattern, but first a few fundamentals.

16.1.1 Scaling as Extrapolation

The essence of scaling is extrapolation. Given a measurement that depends on an independent variable (x), we want a rule or law that permits us to predict the variable for values of the x beyond those used in defining the original relationship. This is the *scaling problem*, and it has been around for a long time. Most biologists encounter it in the form of allometric relations that state that one morphological variable is a power function of another morphological variable: $M_2 = \alpha M_1^{\beta}$, where the parameters vary depending on the system to which the function is applied. For example, the weight of a mammalian heart is proportional to the total body mass: $H = 0.006 M_b^{1.0}$ (Schmidt-Nielsen 1979). This has been extended to relationships between physiological processes (e.g., oxygen consumption as a function of running speed). In physiological or organismal systems, body size is an important independent variable, as many physiological processes are simple power functions of body size (Schmidt-Nielsen 1984). This application of power functions as scaling laws has been generalized and extended to ecological relations (Peters 1983, Brown 1995).

More recently, the scaling problem has taken on new meaning with the realization that not only can properties of a system (e.g., body size) form the basis of a scaling law, but that the measurement device *itself* can determine the magnitude of the dependent variable. Mandelbrot (1977) graphically brought to our attention the fact that simple measurements such as the length of a natural object (e.g., the shoreline of an island) will depend on the basic unit of measurement used. For example, in measuring the length of the coast of England, if the length unit is 100 kilometers our estimate will be far shorter than if the length unit is 1 cm, because in the former units we skip many little twists and turns that the shorter ruler picks up. In this case, the quantity measured (Q) is related to the measurement scale (L) by a simple power law: $Q = aL^D$, where a and D are constants.

The point here is that the characteristics of the measurement device (or, more generally, the sampling regimen) will determine the result. Consequently, for many natural problems there is no one unique, special answer to questions of measurement. To distinguish this aspect of the scaling problem from physiological scaling, I refer to it as *measurement scaling*. This aspect of the scaling problem has attracted much attention in the analyses of distributed systems and models.

16.1.2 Measurement Scale

Within the context of scaling problems created by finite measurement units, two aspects of scale must be distinguished: *properties* and *dimensions*. There are three dimensions along which scale can be defined: space, time, and biological organization (Frost et al. 1988). The dimensions of space and time have the usual physical definition. So, we can speak of the time and space scale at which observations are made or to which models pertain. The dimension of biological organization relates to the biological object studied: biochemical, cellular, organismal, population, etc. We can define, for each of these dimensions, two properties: *extent* and *resolution*. Extent refers to the length or duration of the observations (e.g., number of months or years). Resolution refers to the frequency of observation or the period between observations. Applied to biological organization, extent refers to the number of levels of organization incorporated in the study; resolution pertains to the position of the system studied in the hierarchy of biological systems (e.g., cell, organism, population, ecosystem). The lower the position, the higher the resolution (e.g., molecules *vs* ecosystems). For many processes, there is a positive correlation among the dimensions and the properties. As one increases the spatial extent of analysis, the time-scale extent also increases: events that occur uniquely at a small spatial scale occur more quickly than those that exist at larger scales. For example, water percolates quickly among soil particles, but requires more time to flow from one end of a watershed to the other. Also, as one studies biological systems higher in the hierarchy of organization (e.g., whole vertebrate organisms *vs* their individual cells) processes proceed more slowly and over larger spatial distances.

Every model or observational study must be performed at a particular resolution and extent. Improper choice of scale properties can provide misleading data. To illustrate this, suppose Martians landed in a wheat field in central England and their spacecraft left a large imprint identical to the Roman letter "A" (Fig. 16.1). Suppose further that it was our job to determine the shape of the imprint, and, like the elephant and the blind men, we were restricted to ground sampling at discrete points. Without knowing the size or shape of the spaceship, we would have to choose an area from which to sample and a distance between sampling stations. If the distance between stations is too large, we might conclude the ship was shaped like the letter "Y" (Fig. 16.1a). If the size of the sampling plot was too small, we might conclude the ship was shaped like the letter "V" (Fig. 16.1b). A blind man who touched the elephant with his hand at four points that just exactly coincided with the legs of the creature would conclude he was in a forest! The same problem occurs if we sample at a single spatial point over time: we can sample too infrequently or for too short a duration to accurately describe the true dynamics (Fig. 16.1c,d).

The relative intractability of the scaling problem depends on the quantity being measured. If the quantity is a length-related measurement such as length, area, and so on, then Mandelbrot (1977) has shown how to use the fractal dimension of the process to define a scaling rule. More often, however, the quantity to scale is a biological property that covaries with spatial or temporal scale but is the outcome of complex biological subpro-

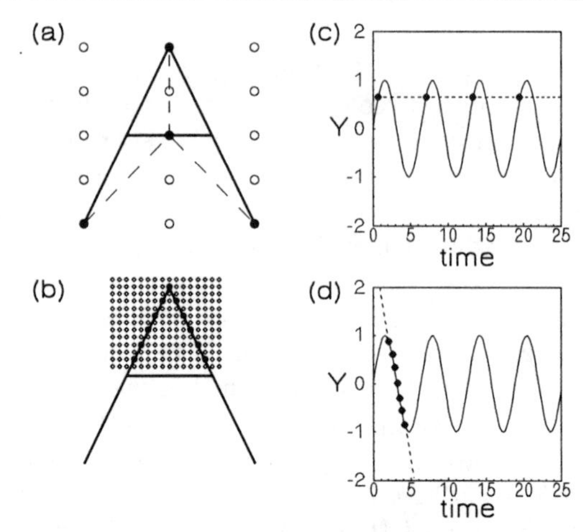

Figure 16.1: Errors encountered when improper choice of extent and resolution are used in space [(a) and (b)] and time [(c) and (d)]. The correct pattern in space is the letter "A" and in time a sine wave. The sampled points are represented by circles. Inadequate resolution of spatial sampling (a) suggests the pattern is the inverted letter "Y" (dashed line). Inadequate resolution in the time dimension means the peaks and troughs will be missed (c). Too short an extent of observations in space (b) suggests the inverted letter "V"; temporally, (d) the data would suggest a constant decrease over time.

cesses. It is not obvious how to relate scale and quantity in these cases. Almost certainly, it will not be a simple power function which is the basis of fractal dimensions. The magnitude of the problem can be glimpsed by considering the problems of describing and modeling plant photosynthesis at scales ranging from biochemistry to global primary production (Fig. 16.2). It is clear that a simple equation will not successfully predict the global consequences of humidity changes over a small area of a leaf (Section 11.3). Other approaches will be required and below we survey some of the models and scale-related issues.

16.1.3 Approaches to Scaling

There are two components to the problem of scale: identifying the important scales and producing an algorithm for relating processes across scales. The first problem has two solutions: (1) decide *a priori* what the biological levels and scales are and (2) use the patterns emerging from statistical analyses. The problem with the first solution is that the biases of the observer may influence the choice of scales. When the decision is made by an experienced observer, well informed on the spatial and temporal dynamics of the biological components of the system, often good, even optimal, results can be obtained. If we are not well informed, then we can be badly misled by a poor choice of the appropriate scales (Fig. 16.1). Alternatively, we can apply statistical techniques that have been developed over the years. We do not have space to cover any of these in great detail, but we can mention some and give a few examples.

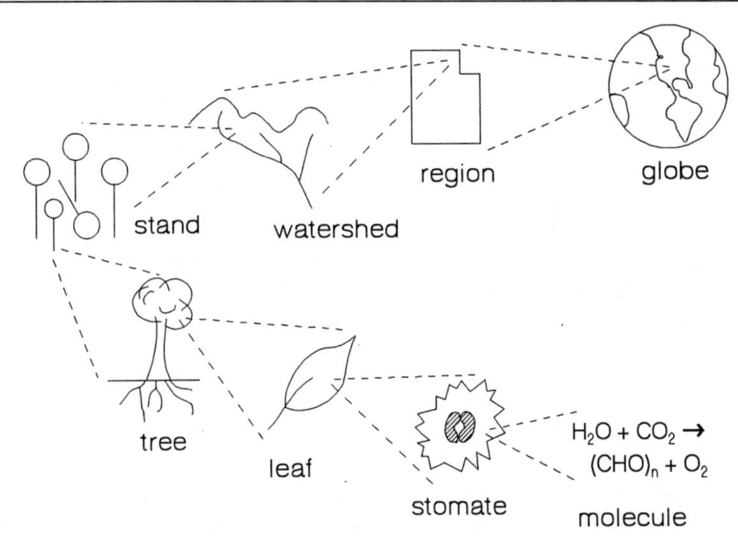

Figure 16.2: Schematic of the hierarchy of scales at which photosynthesis occurs. At the lowest (reasonable) level are molecular and biochemical processes. At the highest (reasonable) level are global processes. Each level has a characteristic space, time, and biological scale. The challenge is to find algorithms for using mechanistic knowledge at small scales to predict patterns at large scales.

16.1.4 Statistical Techniques for Scale Identification

In both space and time, the primary technique, in one form or another, is to search for correlations among sampling points. In spatial sampling, the *semivariogram* is a powerful tool (Davis 1986). Basically, this method examines the variances associated with a set of points separated by a given distance. The method calculates the variances for all distances $n\Delta h$ where Δh is a step size between samples, and n is the number of pairs of distances examined. If the quantity measured is similar for a particular $n\Delta h$, then the variance will be low. If the variance is high, there is no correspondence between the two spatial points: they are independent. Figure 16.3 shows hypothetical data along a transect (Fig. 16.3a) and the associated semivariogram (Fig. 16.3b). The latter shows that nearby points are similar (low variance) and distant points are uncorrelated. The distance at which the semivariogram reaches its maximum (3.5 m) indicates a natural spatial scale of patchiness.

A related technique is *spectral analysis* (Platt and Denman 1975, Levin 1992). In this method, the time or space series is assumed to be a summation of sine waves that combine to produce a complicated signal (see Section 17.2). For each component frequency in the signal, there will be an associated variance that is proportional to the amplitude of that frequency. A time series that is dominated by frequency x will have a high amplitude associated with x. The variance is also called *power*, so the power spectrum of a signal is a plot of the variance against frequency or, equivalently,

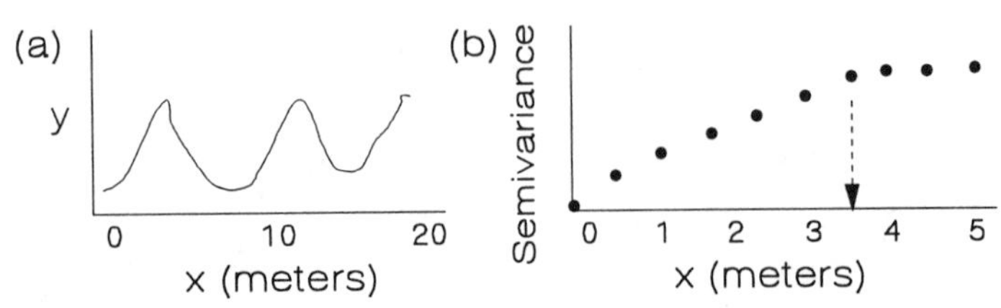

Figure 16.3: The semivariogram of a spatial transect can indicate space scales. (a) A quantity y varying across a transect of 20 m. (b) The semivariogram for the transect where the breakpoint (arrow) indicates a natural spatial scale of 3.5 m.

period.

These analyses can be used to identify characteristic length scales that explain relatively large amounts of variability in data sets. For example, from stationary meters and transects in Lake Tahoe, Powell (reported in Platt and Denman 1975) plotted chlorophyll variance and current speed variance against the inverse of distance. The power spectra of both these variables increase from short to long length scales. The two spectra have similar slopes from 100 m to 10 m. At 100 m, the chlorophyll power spectrum shows a sharp break and discontinuity where the current speed does not. This is interpreted as indicating that physical processes determine biological patches of length scale less than 100 m, but that biological processes dominate above this scale. Levin (1992) tells a similar story for phytoplankton and krill in the Southern Ocean.

O'Neill et al. (1991) have gone a step further with this type of analysis to draw broad generalizations about ecosystem structure. Using remotely sensed digital photographs, they placed 32 transects radiating from a common central point. The fraction of the landscape occurring as grassland was determined every 200 m along each of the transects and analyzed to simulate transects of different distances. For example, a short transect included only the first 1000 m from the central point. This was a small spatial scale analysis. The longest transect (large spatial scale) used the complete transect and covered 30,000 m. The variance was computed from these 32 observations and repeated for 30 total transect lengths. While there was a noisy relationship between scale and variance at extremely short scales, a striking pattern emerged for the middle to large spatial scales (Fig. 16.4). The slopes relating scale to variance changed in a stair-step fashion. O'Neill et al. (1991) interpreted this as indicating a hierarchy of processes in the system. Scale intervals labeled B, D, and F are believed to be structured by different processes with F representing events that occur on the largest spatial scale and subsuming the progressively smaller spatial scales of D and B. O'Neill et al. (1991) conclude that the pattern in Fig. 16.4 reveals *hierarchical structure* in the ecosystem, where the hierarchical levels are those operating at the spatial scales labeled B, D, and F.

This method does not tell us what the processes are that cause these

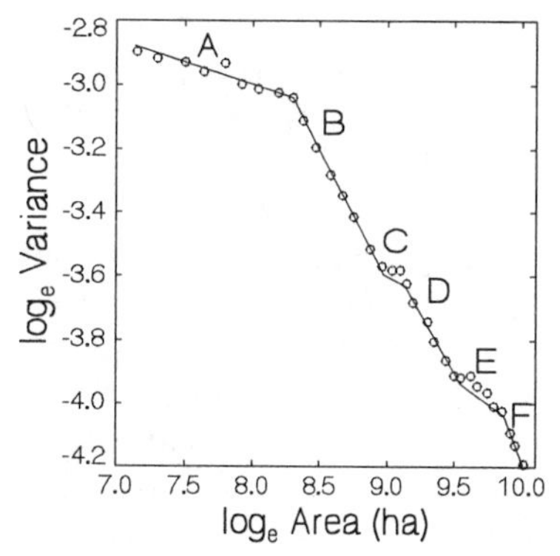

Figure 16.4: Variance in percent of space as grassland in remotely sensed images as a function of the size of the integrated area. The stair-step pattern in which steep slopes alternate with shallow slopes indicates the presence of distinct hierarchical levels. (From O'Neill et al. 1991, Fig. 5.2. © 1991 by Springer–Verlag New York, Inc. Reprinted by permission of the publisher and author.)

patterns, but it helps identify the *scale-dependent* scaling laws which Levin (1992) believes must be found. Levin and his colleagues (Levin and Buttel 1987; Levin 1992; and Moloney et al. 1992) analyzed spatially explicit simulation models of disturbance and plant dispersal using similar techniques and have found strikingly similar patterns. In this work, the discontinuities arose from spatial correlation lengths that were determined by the interaction of dispersal ability and disturbance frequency. Since the structure of the model is known, spatial correlation techniques in combination with mechanistic models provide a methodology for teasing apart the relative contributions of processes producing spatial pattern.

While there have been successes and insights from these methods, many others have been proposed and used. These include tests for randomness, fractal dimension, average patch size, autocorrelation, and others. Cullinan and Thomas (1992) compared the ability of many of these approaches to detect structure in hypothetical and real data. Not too surprisingly, they found that no one method was clearly superior and recommended that several methods be used in detecting scaling laws. Other methodological analyses can be found in Turner et al. (1989) and Milne (1991).

16.1.5 Scaling Models

A popular belief is that the world is hierarchically organized (Pattee 1973, Allen and Starr 1982, O'Neill et al. 1986), perhaps along the lines of Fig. 16.2, perhaps in some other way (Fig. 16.4). On the one hand, most modern-day biologists are sufficiently well-indoctrinated with mechanistic and reductionistic explanations to accept the idea that lower level processes (e.g., biomolecular) are the *cause* of higher level patterns. On the other hand, researchers of large-scale, global-level phenomena, such as climate change, believe that the states of higher levels of organization *constrain* low level processes. One example of this might be the belief that the global level of CO_2 influences average surface temperatures, which in turn affects enzyme reaction kinetics. There is the possibility that the myriad of fine-scale systems interact among themselves and with physical transport systems to alter physical regions beyond the small spatial scale of the lower level systems. So, one reasonable world view envisions an hierarchical system with material causation working upwards and set-point constraints operating downwards. Holding this view does not imply that phenomena exist that cannot be given mechanistic or causal explanations.

This basic philosophy has led to two different approaches toward model construction. *Bottom-up* models attempt to predict higher level phenomena using low-level processes. These models are deterministic, mechanistic, and process-based to explain high-level system performance as the outcome of systems at smaller spatial and shorter time scales (Jarvis 1993). Such explanations may be error-prone due to error propagation of knowledge beyond the scales at which it was acquired. *Top-down* models attempt to describe system behavior as the result of a phenomenological relation between system variables and an external driving variable (Jarvis 1993).

Scaling can be done for each of the three scaling dimensions. When the dimension is biological resolution, we are dealing with the errors of model aggregation. This was discussed in Section 9.2.3. Scaling across space and time are similar problems since, as a general rule, the scales are correlated: long times are associated with large spatial extent. One obvious approach is to derive a new model appropriate to the larger scale from a combination of first principles and empirical data for the new scale. The new and original models can be compared because the output of the new model is also produced by the original model, when it is iterated over as many spatial or temporal units as necessary to achieve the larger extent. We have already noted that using the original model in this way is computationally burdensome and cannot be done for routine analysis of the larger scale. It is possible, however, on reduced problems to quantify the discrepancies among models designed for different scales. New methods are being developed to address specifically the problem of scaling across space from an existing, low-level model. King (1991) identified four such approaches: (1) *lumping*, (2) *direct extrapolation*, (3) *extrapolating by expected value*, and (4) *explicit integration*. We briefly review these below.

(1) Lumping (also *calibration* in Rastetter et al. 1992) is probably the simplest and most common approach to scale changes. It involves retaining the original mathematical model, but selecting new parameter values applicable to the larger scale. An example is the "big leaf" approach to scaling from leaf-based physiological models of photosynthesis to the total photosynthesis associated with the canopy of a population of plants. (2) In direct extrapolation, the model's inherent spatial unit is replicated a sufficient number of times to encompass the larger spatial scale with appropriate

information and material flow between the units. Although this approach may be mechanistically realistic, it can be computationally impractical. (3) Extrapolating by expected value is an approach that scales local output to a wider region by multiplying the area of the large region by the expected local output. One problem of this approach is defining which of the local outputs to use or how to combine them into an aggregated variable. As we saw in the section on error propagation, a nonlinear function evaluated at the mean of its arguments is not equal to the mean of the function evaluated at a range of values of the arguments. The expected value approach assumes that local output is a random variable distributed across the landscape according to some assumed probability distribution, which is used to estimate the expected value of a local function (Rastetter et al. 1992). This has the problem that we must estimate the probability distribution given incomplete and uncertain knowledge. Rastetter et al. (1992) provide some methods to approximate the distribution. (4) Finally, explicit integration is an analytical solution that requires mathematical integration of the local function over two- or three-dimensional space. This is usually impractical because complex, nonlinear models cannot be analytically integrated (King 1991).

Of these methods identified to date, lumping and direct extrapolation are the most common. When some information on the probability distribution of model components is known, Rastetter et al. (1992) provide some tools for correcting the response to the variable input values. Their recommendation is to lump if sufficient data exist at the larger scale. Otherwise, some kind of expected value approach is needed, but this too requires data at the fine scale for as many of the contributing functional components as possible.

Scaling up, however, is only half of the problem, although it is disproportionately important due to our current uncertainty and inexperience with the concepts. We must also understand how large-scale events (e.g., global) will affect the scales at which the mechanisms operate. For example, if average global atmospheric CO_2 increases, we must be able to relate that event with changes in photosynthetic capacity at the leaf level and below. To address these issues, Jarvis (1993) and Reynolds et al. (1993) have called for a combined approach that uses both top-down and bottom-up strategies. This is not a new idea. A quarter of a century ago, it was extensively investigated and implemented in a specialized computer simulation language called FLEX (Overton 1972, White and Overton 1974). The FLEX modeling language forced the modeler to define the hierarchical structure of the system and to specify explicitly the constraints from higher levels as well as the causal mechanisms of the low levels. A fundamental concept was that every model has some *target system* that can be influenced by at least one organizational level below and one constraining or forcing level above. Models constructed in this paradigm simultaneously integrated bottom-up and top-down forces, just as Jarvis has recently suggested.

These ideas and coding efforts were the result of the U.S. International Biological Programme of the 1970s. Although theoretical ecology has moved in different directions since then, we now have a renewed need to address questions of global change and landscape ecology combining new tools with recent advances in the technology of parallel computers, individual-based models, and object-oriented simulation. Perhaps we should now reexamine some of the innovative approaches to ecosystem modeling that emerged from those early years. As we will note below, some modelers are undertaking this challenge in order to comprehend the complexity of large-scale models.

16.2 Scaling Plant Processes: Stomate to Globe

No other single system characterizes the problems of scale better than that of photosynthesis and primary production. Figure 16.2 illustrates the levels that interact to produce global pattern. Models have been constructed at each level and several have been extended across scales. We have space only to mention a few of these and direct the readers to the literature.

Stomate to Leaf One of the central modeling problems at the level of a single stomate is transpiration: rate of water loss through the guard cells. An important mechanistic hypothesis is that water flows between the guard cells and the surrounding epidermis tissue, causing the former to open and close. We have presented details of one such model in Section 11.3. The scaling problem here is to extrapolate from the single stomate to the leaf. There are several possibilities.

(1) Replicate the system of ODEs for single stomata across the entire leaf for each of several million stomata. This would require massive computer resources. For a leaf of 10 cm^2 with 100 stomata/mm^2, this requires that we solve a system of 3 million ODEs. Converting the framework to continuous space using PDEs is possible (Rand and Ellenson 1986), but loses the natural discrete form of the plant anatomy. Since most of the current technology applied to water-relations physiology cannot make measurements at the individual stomate level, the PDE approach may lose little in the way of spatial resolution.

(2) Lump groups of contiguous stomata. Nearby stomata are likely to behave similarly, so pooling them together is not likely to affect overall outcomes. One obvious choice is to pool stomata sharing a given areole (i.e., the leaf area lying within the smallest veins). Effectively, this approach scales by lumping. Other spatial averaging methods are possible, including pooling all stomata on the leaf. This could be called the Big Stomate model of the leaf and has the disadvantage that model parameters do not correspond to the physical setting of the individual stomate (Rassetter et al. 1992). Moreover, in the extreme, spatial averaging removes the ability

to incorporate recent discoveries of the spatially patchy nature of stomatal responses that may materially affect transpiration rates (Mott et al. 1993).

(3) An intermediate approach is to simplify the equations so that cellular automata (Chapter 18) can be used. This simplifies the solution but retains the spatially explicit nature of the phenomena. One possible implementation of this approach might be to create a large lattice in which each cell represents a stomate. Each lattice cell would be composed of a guard cell and epidermal tissue, both of which would switch from one of a finite number of states to another state by very simple state transition rules. For this approach to be successful, it would be necessary to demonstrate that the reduced equations are dynamically faithful, in some sense, to the original, mechanistic and physically correct equations.

Leaf to Canopy Jarvis and McNaughton (1986) and Boote and Loomis (1991) reviewed attempts to scale from the leaf to the canopy and region level. A classical approach is the level-specific models of canopy transpiration developed separately by Penman and Monteith. This equation is based on a careful analysis of the energy balance for vegetation at this scale. We will not derive the equation here, but cogent presentations can be found in Jarvis and McNaughton (1986) and Thornley and Johnson (1990). Using primarily the notation of France and Thornley (1984), the Penman–Monteith equation is

$$E = \frac{sA + c_p \rho \Delta_p g_a}{\lambda[s + \gamma(1 + g_a/g_c)]}, \tag{16.1}$$

where E is transpiration, s is the slope of the effect of air temperature on saturated vapor pressure, A is net radiation or available energy, c_p is the specific heat capacity of air, ρ is the density of air, Δ_p is the vapor pressure deficit or the difference between saturated vapor pressure of air at ambient temperature and actual vapor pressure, g_a is the conductance of water from the leaf surface through the boundary layer, g_c is the water conductance through the canopy, λ is the latent heat of evaporation of water, and γ is the psychrometric constant or $c_p P/\lambda\epsilon$, where P is atmospheric pressure and ϵ is the ratio of the molecular weights of water and air.

All of the parameters apply to the canopy level; there is no attempt to scale in the sense of King (1991) from a lower level (e.g., leaf). So while it does not address the scaling problem from lower levels *per se*, this equation (and others like it) is tremendously important for scaling from the canopy to higher levels.

Norman (1993) has made extensive attempts to scale leaf processes to canopy pattern. He compares his scaling equations with the output of complex mechanistic plant–environment (PE) models. The canopy-level models use synthetic variables such as leaf area index (LAI) and photosynthetically active radiation (PAR) to statically predict assimilation and conductance. Norman (1993) made two sets of comparisons: when light was the only control on assimilation; and when CO_2, wind speed, and other factors influenced the energy balance of leaves. Although he did not quantitatively compare the relative precision of the five scaling equations to the PE model, it appears that when light alone was limiting, stratification of the canopy into sunlit and shaded leaves produced the closest fit to the more detailed PE model. Adding multiple canopy layers did not increase the precision significantly. However, using more complex controls on photosynthesis such as temperature and canopy air vapor pressure allowed the scaled-leaf model to be a closer fit to the complex PE model (Norman 1993).

Others have also built complex energy balance models to scale up from the leaf. Most of these require an iterative procedure. The class of possible strategies was reviewed by Boote and Loomis (1991) as a continuum from simple to complex. At one extreme is the "big leaf" approach which assumes homogeneously distributed photosynthetic material over a "leaf" that has the area of the canopy. Complex models, in contrast, are those that divide the canopy into layers and model layer-specific leaf conditions for light, temperature, boundary layer thickness, and so on.

An example of this latter approach extrapolated leaf-level transpiration rate to canopy-level photosynthesis and water loss rates (Caldwell et al. 1986, see Fig. 16.5). This model assumes that (1) a leaf is internally homogeneous so that leaf assimilation rate is proportional to the product of leaf volume and assimilation rate per chloroplast, (2) a canopy possesses a fixed density of leaves measured as leaf area index (LAI) for each of several canopy layers, (3) leaves are oriented using a fixed frequency distribution of azimuth angles (usually randomly determined), (4) photosynthesis is based on steady-state chloroplast biochemistry similar to that described in Chapter 11, (5) transpiration is a simple empirical function of temperature and the water vapor difference between the leaf and atmosphere (unlike Chapter 11), and (6) the energy budget of incident radiation is balanced with regard to heat and its spectral components.

The model is driven by the global rate of shortwave radiation that impinges on the highest canopy layer. The amount incident depends on the position of the sun in the sky. Shortwave radiation strikes the canopy in two forms: direct beam and diffuse. Not all of the direct beam component is absorbed by the leaves in the top layer. Due to random placement of leaves, some radiation passes through to lower levels. Diffuse radiation is treated approximately the same. Longwave radiation is produced by scattering from surfaces and the sky itself and is dependent on temperature and the average emissivity of the surfaces. Shortwave radiation inside the canopy is either absorbed or scattered as longwave radiation.

When light enters the canopy and strikes leaves, a number of events are initiated. First, the leaf heats up; second, photosynthesis occurs so that stomata open and transpiration occurs. Transpiration, however, alters

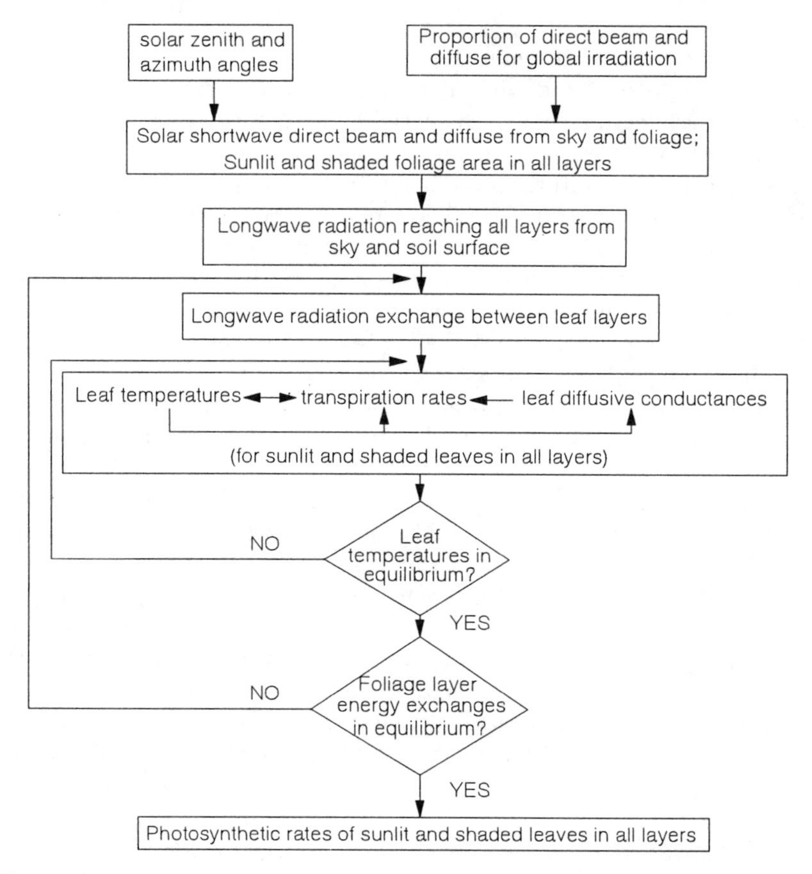

Figure 16.5: Flow of computation for a layer-specific canopy photosynthesis model. Light from all important sources strikes layers and alters leaf temperature and transpiration. Layers affect each other and two loops ensure that layer temperatures and energy exchanges are in equilibrium. (From Caldwell et al. 1986, Fig. 1. © 1986 by Springer-Verlag GmbH and Co. Reprinted by permission of the publisher and author.)

the microclimate of the leaf and thereby influences the leaf's temperature, which feeds back on transpiration rates (Section 11.3.2). Caldwell et al. (1986) therefore used an iterative procedure to alter transpiration rates until the leaf temperature within a canopy layer equilibrates (Fig. 16.5, inner loop).

In addition to these interactions within a canopy layer, there are radiation interactions between layers. Because light passes through layers and can be scattered upwards, the total radiation input to a layer can not be determined from the top to bottom layer. Moreover, longwave scattering within the canopy depends on the temperature of leaves, thereby creating interactions involving scattering between layers, leaf temperature, and leaf transpiration rates. Consequently, Caldwell et al. (1986) employed another iterative scheme to bring the dynamics of energy exchange between adjacent layers into equilibrium (Fig. 16.5).

The iterative nature of this algorithm illustrates extremely well the problems of scaling models. Both spatial and temporal scales become important when we attempt to maintain a mechanistic basis for large-scale processes. Leaf temperature and transpiration rates operate on faster scales than inter-layer energy transfer. Therefore, the model uses an inner loop to achieve equilibrium at the leaf-scale. The outer loop addresses the spatial scale component in the form of energy exchange between layers. When all iterations are complete, the canopy as a whole is in equilibrium until global environmental conditions change (e.g., daily solar azimuth or season). Once in equilibrium, the state of the canopy can be used as input to other larger-scale models or to other iterative loops to model seasonal and stand-level scales. This conceptual framework can be further extended to watersheds, regions, and beyond (Fig. 16.2).

These iterative canopy models do remarkably well in predicting physiological processes of canopies (Caldwell et al. 1986; Boote and Loomis 1991). For example, in one comparison conducted in Portugal on *Quercus coccifera* (Caldwell et al. 1986), the model fit observed net photosynthesis with $r^2 = 0.91$; validation with independent data sets had similarly high accuracy. Part of the reason for this accuracy is that the models rely heavily on empirical measurements to parameterize empirical functions and driving variables. But empirical accuracy is exactly what is required for confident extrapolation of knowledge at the small-scale of a whole leaf to the scale of the canopy.

Stand to Watershed While the above model works well for relatively closed canopies, the patchy distribution of trees at the level of a forest stand involves demographic processes that shape the structure of the canopy itself. The gap models epitomized by JABOWA, FORET, and the many variants (Botkin et al. 1977; Shugart and West 1977; Shugart 1984; Shugart et al. 1992) are stochastic, small-scale models of a plot that contains a medium number of trees. Random events determine the number of trees of the various species present on the plot that die, grow, and reproduce. These models have been used for scaling by the direct extrapolation method of replicating the plot over larger areas (Shugart et al. 1992). Used in this mode, they can address questions of community succession and the range expansion of species under different abiotic conditions (e.g., global change, Davis and Botkin 1985; Davis 1986; Shugart et al. 1992, and references therein). These models can reasonably be applied to large regions by direct extrapolation because the computations on each plot are relatively simple. In part, this is achieved by employing very simple empirical relationships for photosynthesis and transpiration. This approach, while tractable, can nevertheless require substantial computer resources.

Watershed to Region Great strides have been made in recent years modeling spatially explicit landscapes with variable soil, hydrology, plant communities, and human impacts. Several groups are working on major models that integrate ecosystem processes and individual-based FORET-like models of plant responses: Costanza et al. 1990; Band et al. 1991; Sklar and Costanza 1991, and Lauenroth et al. 1993. Most of these modeling projects use sophisticated technology ranging from remote sensing, geographic information systems (GIS), automated data collection, supercomputers, and high-end visualization hardware. The models are complex and technically difficult since they integrate the physics of water flow in saturated and unsaturated soils, surface runoff, complex hydrological sequences, climate and weather modeling on a regional scale, plant growth patterns, and animal movement. They are collectively known as process-based, as opposed to individual-based, since most of the models use a complex, system-specific lattice of discrete-space cells through which materials and biological populations flow. They therefore attempt to scale from the small grid cell to larger regions by direct extrapolation. The lowest level of biological organization used is dependent on the size of the larger region model and the computational requirements for each lattice cells.

16.3 Summary

Scaling knowledge from small to large scales remains one of the challenges of ecological modeling (Levin 1992). A variety of statistical techniques must be used within each study in order to avoid a biases view (Cullinan and Thomas 1992). Models play a crucial role in extrapolating to the level of regions from the individual or organ (leaf) level, but this requires computational power beyond the limits of current technology. Thus far, no universally applicable scaling methods or laws have been discovered. Rather, we now se the use of system-specific models (e.g., Caldwell et al. 1986) that permit extrapolation among two or three scales. This approach is likely to dominate for the near future.

Chapter 17

Chaos in Biology

17.1 Nonlinear Can Be Weird

CHAOS, the mathematical concept, was rediscovered, explicated, and applied in the mid-1960s and 70s (Lorenz 1963; May 1974) and has since then been broadly assimilated into contemporary Western culture. (Of course, the concept of social and political chaos has been well-known to even casual observers of contemporary events for a long time.) An informative, brief history of mathematical chaos was given by Holton and May (1990). Although the word is encountered frequently, as with similar overarching and broadly applicable concepts such as relativity, Darwinism, or connectionism, the concept of chaos is sometimes only vaguely understood. In this chapter, we have the very modest goal of giving a qualitative, informal exposition of some of the underlying concepts plus a few examples. Many fine books on the subject exist ranging from the popular (Gleick 1987) to the mathematical (e.g., Guckenheimer and Holmes 1990, Hilborn 1994).

To begin, we must recognize that chaos is a mathematical property of the time domain solutions of a set of equations and the parameters. Only nonlinear equations possess this property, so the study of chaos is a subset of *nonlinear dynamics.* Every well-educated student of nonlinear dynamics should have at least passing familiarity with the following core concepts:

- Bifurcations
- Attractors: fixed, cyclic, toroidal, and strange
- Lyapunov exponents and sensitivity to initial conditions
- Fractal dimensions of dynamics
- Types of models that produce chaos
- Identifying chaos in empirical data.

Below, we will address each of these in different degrees of detail. As before, we will encounter principles used in *Part I.* These include age-structured population models, stability and eigenvalues, stochasticity, limit cycles, and one-dimensional maps of finite difference equations. The biological questions we will examine include: Are biological systems chaotic? What is the best test for chaos in biological systems? What biological processes cause chaos?; Is chaos an adaptive trait? Do chaotic populations have a lower probability of extinction than nonchaotic populations?

17.1.1 Attractors

An *attractor* is a mathematical object to which a system's dynamics are eventually confined. Qualitatively, the object is the set of solutions to the dynamic equations when the system is allowed to run for a long time. There are four main types of attractors: fixed point, limit cycle, toroidal or quasiperiodical, and strange. A fixed point attractor is just a fancy name for an equilibrium point. A limit cycle, discussed in Chapter 9, is a closed curve that represents the repetitive solutions. A toroidal attractor is a surface in phase space shaped like a torus or doughnut, which may be stretched and twisted. The system solutions are confined to this doughnut-like surface, and we discuss an example of this in the model of the forced Monod chemostat system (Chapter 14). A strange attractor is a similar but more complicated surface to which the solutions are confined. The important point of the attractor concept is that the dynamics are *bounded* by being confined to the attractor structure in the long run. The choice of the word "attractor" is appropriate here, because, when a strange attractor exits, almost all trajectories will approach it over time.

17.1.2 Bifurcations

A structure *bifurcates* when it splits into two branches, as in footpaths or tree branches. The word applies to equations because we can qualitatively represent the possible solutions to an equation as a path along which we traverse, not through time or physical space, but through parameter space. The qualitative solutions of an equation bifurcate in parameter space when the number of solutions changes as a parameter is changed. The parameter being altered is called the *control parameter*. The effects of the control parameter on the qualitative dynamics are represented in the *bifurcation diagram*, which is an x–y plot with the control parameter on the abscissa and the long-term values of the state variable on the ordinate.

By qualitative dynamics, we mean, among other things, the number of different values for a state variable that the long-time dynamics produce. For example, the continuous form of the density-dependent population growth model (the logistic) has a long-term equilibrium of K, the carrying capacity. Over a long time, this equation converges on one solution value. The continuous Lotka–Volterra predator–prey model has for the prey (or predator) either one solution at the equilibrium point or two extrema when the prey (or predator) oscillates.

Now, when the prey oscillates, it literally has an infinite number of states as it moves from its maximum to its minimum. But when speaking of qualitative solutions, we ignore all of these except a finite number of points. Researchers do not completely agree on which points to plot, but the usual practice in continuous systems is to plot the local maxima (the peaks). Focusing on the peaks of a state variable's dynamics is useful because, as we will see, there are some equations that oscillate between several maxima. In the theoretical (mathematical) analysis of these equations, we are not concerned with the exact values of states that are produced (even though they are definite quantities), but only the essential feature that separates one class of equations or conditions (e.g., parameter values) from another. When the system is forced by an oscillating driving function, a slightly different analysis is used. The period of the driving function is the control parameter, and we analyze the dynamics by systematically varying the period of oscillation. In this situation, the points in the bifurcation plot are generated by taking a snapshot of the system at the frequency of the forcing function. This is done irrespective of whether or not the system happens to be at an extrema. This is the technique that Kot et al. (1992) used in Chapter 14. Producing a bifurcation diagram when finite difference equations are used is much simpler, for in that case only a finite number of points are generated. The bifurcation diagram consists of all the solution points plotted.

The algorithm for generating a bifurcation diagram is straightforward:

1. Set the initial and maximum parameter values and the number of parameter values to sample.
2. While the current parameter value is less than the maximum, do:
   (a) Run the simulation for approximately 200 time steps to allow the system to settle down to long-term behavior.
   (b) To sample, run the simulation for approximately 200 time steps.
   (c) Store the sample from the dynamics [e.g., find the peaks (continuous case), or save the current value (discrete case)].
   (d) Increment the parameter and go to Step 2.

While this is correct in principle, two caveats must be mentioned. First, the algorithm will find only one attractor. Different starting values might converge on a different attractor. Second, the number 200 is a vague rule-of-thumb, at best. Larger values may be needed; some experimentation is usually required. Obviously, when the equations are solved on digital computers (as described above), large numbers of iterations can require long computing times. Electronic analog computers (see Section 5.3.1) are much faster for this type of application, and there has been a resurgence in their use because of the interest in these dynamics. In any case, bifurcation diagrams give great insight into the dynamical structure of the equations. An important lesson from work in nonlinear dynamics is that this structure

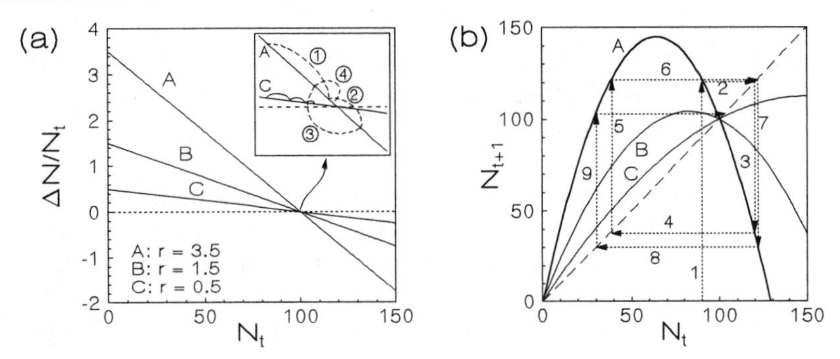

Figure 17.1: Nonlinear dynamics of the logistic map. (a) Density effects on per capita growth rate for three values of r (A, B, C). Insert: changes in the per capita rate for several iterations when $r = 0.5$ and $r = 3.5$. Note that when r is large (A), the finite time step forces $N_t > K$: the sequence of jumps (1, 2, 3, 4). When r is small (C), N_t does not over-shoot K. (b) The logistic map representation for three values of r. The dashed line is the 1:1 line. Solid curves are the density-dependent function at three values of r. Dotted lines show the history of population sizes visited when $r = 3.5$ and the initial population is 90. When r is large (A) the dynamics (points $1, 2, 3, \ldots 9$) are complicated.

can be incredibly complex. We next illustrate this complexity with a few biological examples.

17.1.3 Chaos in Finite Difference Equations

Here, we illustrate bifurcation and chaos using the standard logistic map for finite difference equations (May 1974, 1976). However, we give a slightly different slant to the equations to make a point about numerical stability in ODE solvers. The Euler approximation of the continuous density-dependent population growth equation is

$$N_{t+\Delta t} = N_t + rN_t[1 - N_t/K]\Delta t \qquad (17.1)$$
$$= N_t[1 + (r\Delta t) - ((r\Delta t)/K)N_t]. \qquad (17.2)$$

We assume for the moment that $\Delta t = 1$. The expression in square brackets in Eq. 17.1 represents the effects of population density (N_t) on the per capita growth rate of population. This relation is plotted in Fig. 17.1a for three values of r, the maximum per capita rate of increase. The insert shows how the per capita rate changes over several iterations for $r = 0.5$ (line C) and $r = 3.5$ (line A). This figure shows that when the density effect is steep (large r) for a finite time step ($\Delta t = 1$), the dynamics will overshoot K, producing negative growth rates and a population decline. The dynamics resulting from small r converge on K without oscillations. It is the steep slope coupled with a finite time step that produces the weird dynamics. As Eq. 17.2 indicates, the critical quantity is not r, but $r\Delta t$. Even if r is small, by choosing Δt too large, oscillations can be produced. This is the source of *numerical instability* in the Euler solution method. When the interest in a modeling study is not bifurcation, but simply accurate solutions of the equations, the numerical instability of the Euler method should be avoided either by choosing Δt small or by using a more robust numerical method (Chapter 6).

There is a simple graphical method to follow the dynamics of this nonlinear equation. Equation 17.1 expresses N_{t+1} as a quadratic function of N_t. This relationship is plotted as three humped-shaped curves (for three different values of r) in Fig. 17.1b along with the 1:1 line (dashed). Points on the 1:1 line signify a population at equilibrium: $N_t = N_{t+1}$. To follow the dynamics, begin at an initial point (e.g., 90) on the x-axis, then move up to the curve that defines N_{t+1}. This value becomes the new N_t; but, rather than laboriously returning to the x-axis, we move from the curve to the 1:1 line, and then up (or down) to the curve again for the second iteration of the equation. Continue this for as many iterations as desired; nine iterations are shown for $r = 3.5$ in Fig. 17.1b. The population values produced are the points of intersection of the dotted lines and the function projected onto the y-axis. Since this model contains a single state variable (N), it is called a *one-dimensional map*, or the *logistic map* for this particular model.

Consistent with Fig. 17.1a, the one-dimensional map shows that at low r (curve C in Fig. 17.1b) the system monotonically converges on the equilibrium. At slightly larger r (curve B), the population oscillates as it converges on the equilibrium. Increasing r increases the negativity of the slope of the curve at the point it intersects the 1:1 line. As r continues to increase, the slope becomes more negative until a slope is achieved for which there exists a pair of N_t and N_{t+1} such that N_t maps onto N_{t+1} and N_{t+1} maps onto N_t. This is a two-point cycle that has emerged from a bifurcation from a single solution.

This bifurcation phenomenon continues as the slope is made progressively steeper. The second bifurcation produces a four-point cycle, followed by eight-point cycles , sixteen-point cycles, and so on until the condition of *chaos* is achieved. Since this process and its ultimate endpoint is so important, we develop further the dynamical properties of the logistic map. This has been published in many other places, but it is still the best illustration. Figure 17.1 demonstrates that the dynamics resulting from one form of the logistic map (Eq. 17.1) depend on the parameters. Curve C corresponds to small r, and the dynamics are a smooth convergence on the equilibrium. Curve A corresponds to large r, and we obtain very erratic dynamics.

This transition from smooth, asymptotic dynamics to dynamics that do not appear to settle down to any simple behavior is a subject worth studying. We will do it here for a simpler version of the logistic map (May 1976):

$$y_{t+1} = ay_t(1 - y_t), \tag{17.3}$$

where a is the growth rate of the system scaled by the carrying capacity.

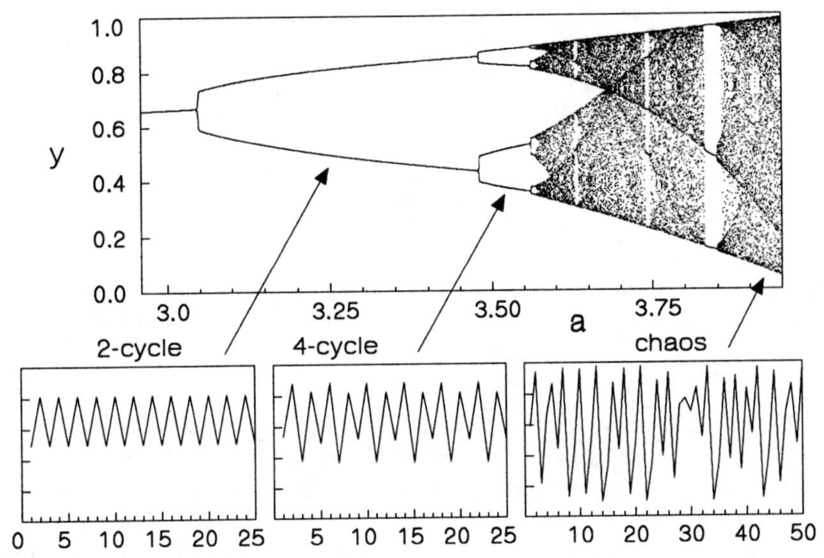

Figure 17.2: The bifurcation diagram for the simple one-parameter logistic finite difference equation. In the top panel, the parameter a is varied, and a number of solutions are plotted for each a. The dynamics corresponding to three values of a are plotted in the bottom panels.

In effect, the dynamics represent the population size as a fraction of the carrying capacity.

Clearly, the size of the parameter will determine the number of distinct solutions that the equation produces. We can study the behavior of this equation by plotting the system dynamics for many time steps at a series of parameter values. Such a plot is called a *bifurcation diagram* (Fig. 17.2).

When we construct the bifurcation diagram over a range of a values (Fig. 17.2), we see sharp jumps from one type of dynamics to another. At small $a < 3.04$, the asymptotic values are the equilibrium. The equilibrium (y^*) in this model depends on the parameter a; it is easy to show that $y^* = 1 - 1/a$. The set of stable fixed-point equilibria are represented by the slowly increasing line for $2.95 < a < 3.04$. As a increases beyond 3.04, however, the dynamics converge not on a single equilibrium value, but on a two-point cycle. The qualitative solutions have bifurcated. These dynamics are illustrated in the bottom panel of Fig. 17.2. Further increases in a cause bifurcations to a four-point cycle, an eight-point cycle, and continued proliferation of cycle periods until, at $a \approx 3.57$, the system enters a chaotic regime (May 1976). This region basically is one in which the dynamics are characterized by cycles having an infinite number of points before repeating. The complicated time courses associated with these regions (Fig. 17.2, bottom-right panel) can superficially appear to be random, but it is crucial to remember that the model (Eq. 17.3) is completely deterministic. Below, we will complicate things further by relaxing this requirement.

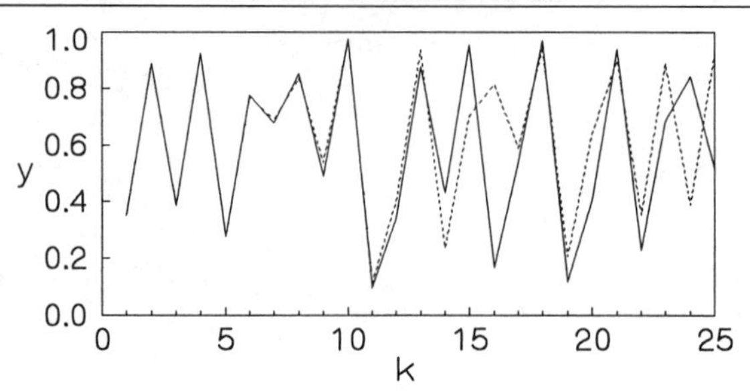

Figure 17.3: Sensitivity to initial conditions in chaotic systems based on the logistic map. Using Eq. 17.3 with $a = 3.9$, two different trajectories were started with values that differed by 0.0001. The cumulative deviation between the two sequences continually grows with time.

One of the fundamental implications of the complicated dynamics occurring in the chaotic region of parameter space is that slight differences in the starting point will produce drastically different sequences of values. This phenomenon is called *sensitivity to initial conditions.* This sensitivity is illustrated for $a = 3.9$ of Eq. 17.3 in Fig. 17.3, where two sequences are plotted that differ in their initial conditions by 0.0001. Initially, the dynamics are the same, but they diverge and never coincide, except at isolated times. Divergence is exactly the opposite of the dynamics in the region of a single stable equilibrium (e.g., $a < 3.04$). No matter where one starts, the dynamics always converge on the same equilibrium value.

The above bifurcations occurred with one state variable (e.g., population size) and were therefore one-dimensional maps. There is a similar concept in two or more dimensions when the system is continuous. A *Hopf* bifurcation is a bifurcation from a stable fixed point (equilibrium) to a limit cycle in multiple dimensions. It is more difficult to picture, but can be done in low dimensional systems. Figure 17.4 shows how the dynamics can change from a stable fixed equilibrium when the control parameter is below a threshold (p_c) to stable limit cycles confined to the surface of a cone-like structure. This phenomenon was observed in the stomate model (Section 11.2.3, Fig. 11.6, and Rand et al. 1981) when an equilibrium bifurcated into a stable limit cycle when Δw exceeded about 20 bars.

The Hopf bifurcation theorem states conditions under which this bifurcation will occur for small parameter perturbations in state space. Mullin (1993b) gives an understandable introduction with meaningful graphics. Caswell (1989) gives a slightly more technical discussion of the nature and limitations of the theorem when applied in ecology.

17.1.4 Chaos in Continuous Models

The above discussion was based on finite difference equations (except for the mention of Hopf bifurcation). It is also possible for chaos to arise in continuous systems, but the system must have at least three state variables. In Chapter 14, we saw an example in the form of the forced chemostat

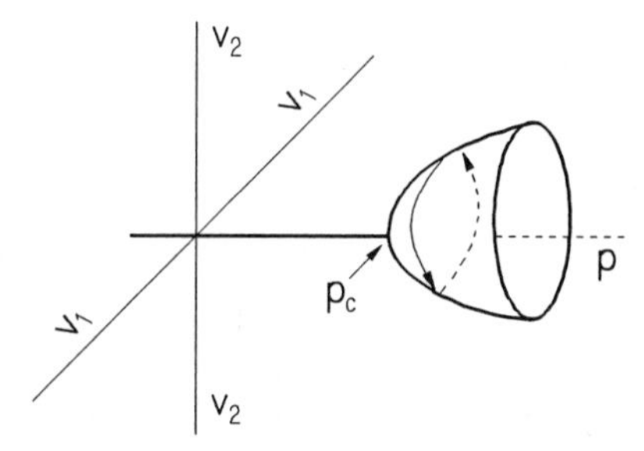

Figure 17.4: Diagrammatic view of a Hopf bifurcation. The v_1 and v_2 axes are the state variables of the system (v_1 and v_2). Below the critical value of the control parameter (p_c), the system is in a stable equilibrium, as indicated by the heavy, straight line. As the parameter p is increased, the dynamics change from an equilibrium to oscillations that are confined to the cone.

model of Kot et al. (1992). Complex, chaotic dynamics can also arise from purely endogenous interactions without external perturbations. One of the earliest examples from ecological systems was a model with two prey and one predator (Vance 1978, Gilpin 1979).

Vance's model used Lotka–Volterra relationships among two competing prey (N_1, N_2) and a predator (P):

$$\begin{aligned} \frac{dN_1}{dt} &= N_1\left[r - \frac{r}{K}N_1 - \frac{r}{K}N_2 - bP\right] \\ \frac{dN_2}{dt} &= N_2\left[r - \frac{r}{K}\alpha N_1 - \frac{r}{K}N_2 - (b-\epsilon)P\right] \\ \frac{dP}{dt} &= P\left[cbN_1 + c(b-\epsilon)N_2 - d\right]. \end{aligned} \tag{17.4}$$

Most of the parameters should be familiar by now (Chapter 9). The new parameters in this model are α (effect of an individual of N_1 on per capita growth of N_2) and ϵ (the predator avoidance advantage of N_2 relative to N_1). N_1 is a superior competitor to N_2.

Using both linearized neighborhood stability analysis and computer simulations, Vance (1978) showed that this system produces a wide range of qualitatively different dynamics depending on parameter values. Vance showed that the dynamics were especially sensitive to K (carrying capacity) and ϵ. He summarized this sensitivity in a *stability diagram* (Fig. 17.5) that graphs the boundaries of the qualitatively different dynamics in parameter space. As can be seen in the figure, different parameter combinations produce qualitatively different dynamics. The dynamics include all possible outcomes: competitive exclusion, coexistence of all populations, extinction of the predator, and aperiodicity or chaos. These results are interesting because they indicate great complexity in outcomes from simple, continuous models. Also, the stability diagram is an important descriptive tool, one that we will use again below.

Gilpin (1979) followed Vance's study with a brief note that plotted the

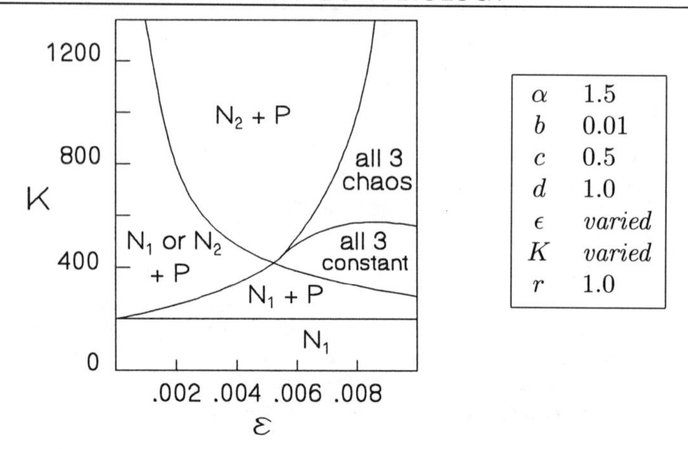

Figure 17.5: Stability diagram for the two-prey one-predator model. Parameter values are listed to the right. (From Vance 1978, Fig. 1. © 1978 by the University of Chicago. Reprinted by permission of the University of Chicago, publisher.)

dynamics in 3D phase space to reveal a strange attractor. This structure had been previously classified as *spiral chaos*, since the attractor is twisted in such a way as to resemble a spiral. This is one of the first formal analyses of a continuous ecological model that produced chaos. Gilpin's paper is worth reading both because of its historical importance for theoretical ecology, and because it contains one of the few published stereoscopic views of a strange attractor in an ecological journal. If strange attractors plotted in phase space look bizarre under normal circumstances (see Chapter 14), imagine how they look with your eyes crossed. In any case, the result demonstrates endogenous chaos in simple ecological models and the difficulty of visualizing and understanding the complex dynamics that nonlinear equations produce. It is to this second concern that we now turn our attention.

17.1.5 Signatures of Chaos

For all the youthful enthusiasm associated with the recent interest in nonlinear dynamics, it is surprisingly difficult to unambiguously determine the existence of chaos in either empirical or theoretical time series. As we will see below, the problem is even more severe in stochastic, empirical data. Nevertheless, there are philosophical reasons deeply embedded in the human psyche for wanting to make this determination. Chaotic dynamics are based on deterministic laws, but appear to be random. The underlying laws are a unifying principle that ties the incoherence of immediate sensation to regularity, constancy, and, in some sense, predictability at a deep level. We can replace the jumble of observations with a single line of mathematics (e.g., Eq. 17.3).

Dynamics produced by purely random processes, on the other hand, are nothing more than one particular sequence out of an infinity of others. Although we may be able to discover the underlying probability distribution from which the sequence of events we experience is drawn, this knowledge does not provide even the crude mechanistic explanation given by the logistic equation. An empirical, probabilistic explanation does not seem to carry the same philosophical weight as a small number of deterministic differential equations. Why? Possibly, the desire for determinism is an evolved trait, but it is hard to attach individual adaptive value to a need for an ordered universe. Predicting the future, however, is another matter. Predicting individual events (as opposed to probabilistic likelihood statements) has obvious survival value. Knowing that there's an 80% chance of rain is fine as far as it goes, but what we really want to know is whether or not to carry an umbrella tomorrow. Unfortunately, as we have just seen (Figs. 17.2 and 17.3), finding the underlying nonlinear equation of the universe will not necessarily improve our predictions, if we are operating in a chaotic region of parameter space.

So, it is not really clear philosophically why so much effort is being expended on tests to distinguish random from complex, but deterministic, dynamics. One thing we can all agree on, however, is that it is a hard problem. To illustrate this, consider the two sequences in Fig. 17.6. These sequences were generated from

$$y_{t+1} = (m + a\sin(pt/2\pi))\, r_t \tag{17.5}$$

$$y_{t+1} = (cy_t(1 - y_t))\, r_t, \tag{17.6}$$

where t is time, r_t is a sequence of uniform random deviates from the interval 0 to 1.0, and m, a, and p are the mean, amplitude, and phase of the sine function, respectively. Equation 17.5 is a random sine curve and Eq. 17.6 is a random version of Eq. 17.3. The models represented by Eqs. 17.5 and 17.6 are fundamentally different. Equation 17.5 is an empirical description based on time alone; there are no feedbacks or relationships between the system variable y. Equation 17.6 hypothesizes that a negative feedback drives the dynamics.

Which of these sequences was generated by which equation? Perhaps, for these particular models, which have very simple implementations of stochasticity, it is not so hard to tell by inspection. But more complicated models can be much more difficult, and we need alternative ways of looking at time series in our quest for underlying pattern. To briefly illustrate the possibilities and to motivate the discussion that follows, instead of using the time domain, we represent the time series by plotting y_t *vs* y_{t-1}. Figure 17.7a is this set of pairs of points for the random sine function (Eq. 17.5). Figure 17.7b is the plot for the random logistic model (Eq. 17.6). This figure makes three points. First, this type of plot reveals patterns that are not apparent from the time domain (Fig. 17.6). The random logistic model does appear to be qualitatively different from the sine function. Second, while the simple method of introducing randomness into the logistic model

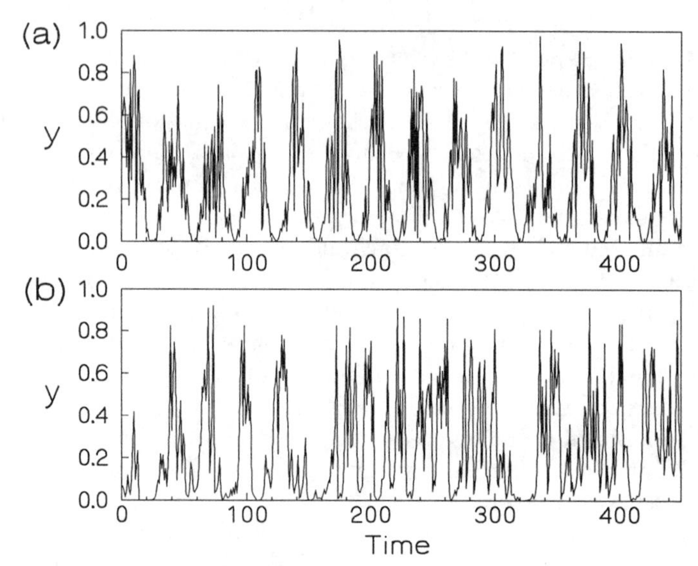

Figure 17.6: Two sequences of values. One is generated from a noisy sine function; the other comes from a noisy version of Eq. 17.3.

shows differences in Fig. 17.7, possibly some other method would destroy the pattern seen there. Third, even with this method, if the data set were restricted to a small region around $y_{t-1} = y_t = 0.5$ in Fig. 17.7b, it would be difficult to distinguish the two phase plots. In short, stochastic nonlinear difference equation models can be difficult to distinguish from random models, but a deeper analysis of the time series, for example, phase plane plots, may help.

The problem of distinguishing the dynamics of stochastic empirical descriptions from theoretical nonlinear difference equations also applies to observed time series. Figure 17.8 shows the time series from a linear random model and the time series of heart rate (beats/min) from a sleeping human. The observed data were taken from a long time series described in Rigney et al. (1994). Heart rate is calculated as the inverse of the inter-

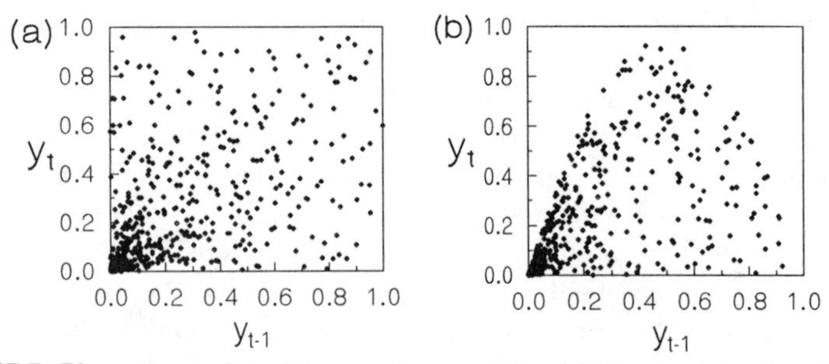

Figure 17.7: Phase space plot of two random models. (a) Random sine function. (b) Random logistic model.

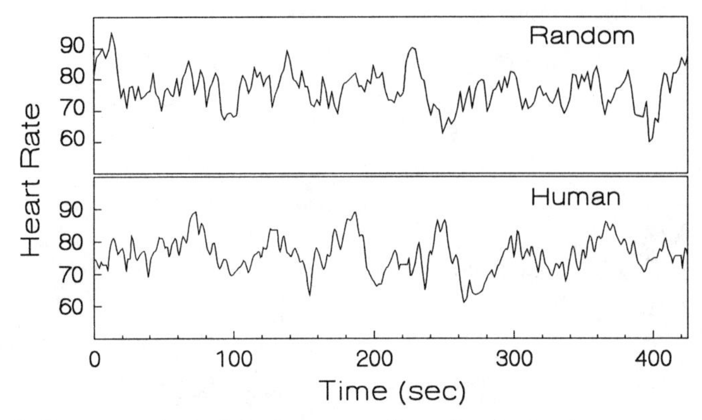

Figure 17.8: Two time series of the heart rate (beats/min) of a sleeping patient (lower panel) and a linear random model (upper panel).

val between sequential R events in the ECG record of the patient. (Read Section 18.4.2 for a description of ECGs and QRST events.) The linear model is one from a family of possible models called *autoregressive* models [$AR(M)$], where M represents the order of the model (Chatfield 1975, Gershenfeld and Weigend 1994). The general form of the family is

$$y_t = \sum_{m=1}^{M} a_m y_{t-m} + Z_t,$$

where y_{t-m} is the mth previous value of the series, and Z_t is the tth deviate from a normal distribution with a mean (μ) and variance (σ^2) estimated from the time series. The simplest model of this family is $AR(1)$, the first-order process also known as a Markov process

$$y_t = ay_{t-1} + Z_t. \tag{17.7}$$

$AR(M)$ models can, with proper choice of the parameters, provide very accurate fits to empirical data. Figure 17.8 is an example of this using $a = 0.7$, μ estimated as 22, and σ^2 estimated as 16. Even though these parameters are probably not statistically optimal, there is a remarkable, albeit superficial, similarity between the data and the model. This is the central problem for understanding empirical time series (Gershenfeld and Weigend 1994): Are the dynamics a simple, linear autoregressive process, or is there a deterministic nonlinear model that underlies the data? Is there a method for deciding?

These last two questions have not yet been answered, except that, so far, there is no "silver bullet" algorithm, index, or visualization scheme that will definitively identify the nature of the nonlinearity or if chaos is present. As a result, a number of characteristic patterns or *signatures* have been developed that suggest the existence of chaos or other patterns in models or empirical data. These fall into five general categories: (1) patterns in the time series, (2) structure in phase space (i.e., *attractors*),

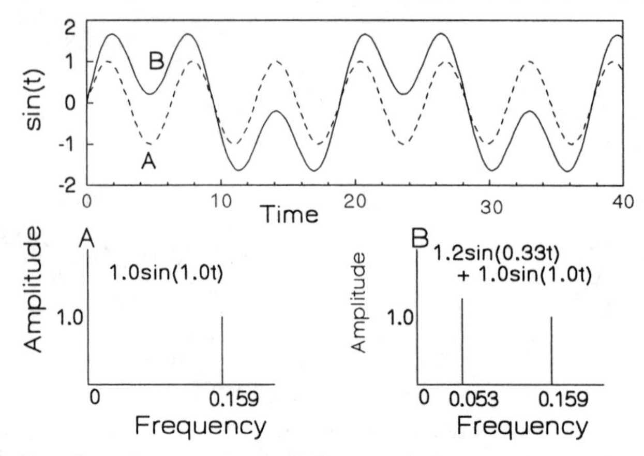

Figure 17.9: Complicated time series can be represented by translating plots in the time domain to plots in the frequency domain. Curve A: a simple sine wave with amplitude 1.0 and a single frequency of 0.159 (period 2π) is represented in the frequency domain as a single point in the amplitude-frequency space. Curve B: a more complex time series formed from the summation of two simple sine waves is represented as two points in the amplitude-frequency space, one for each component sine wave.

(3) dimensionality of the phase space structure, (4) sensitivity to initial conditions, and (5) controllability of time series.

In discussing these signatures, we will proceed in two steps. First, we will introduce the main concepts underlying these signatures; then we will discuss a few examples from biology. The reader should be aware that all of the measures and characteristics discussed below are problematical and can yield ambiguous conclusions when applied to short, noisy time series. There are many more recent and ingenious techniques being invented daily, but all have flaws (see Weigend and Gershenfeld 1994 for a summary using real world data). Consequently, this is a good point at which to remind the reader: *caveat emptor.*

17.2 Patterns in Time Series

Power spectral analysis is one approach for distinguishing chaotic dynamics from random fluctuations (see also Section 16.1.4). The idea is based on the fact that all time series can be approximated by a summation of sine waves with different amplitudes, frequencies, and phases (see Section 4.3 for a brief review). Figure 17.9a shows two time series: Curve A is a single, simple sine wave with an amplitude of 1.0 and a period of 2π (frequency of 0.159). Curve B is the result of summing two sine waves, one identical to A, the other with amplitude 1.2 and frequency 0.053. Since they are periodic, the essential features of the curves are just two numbers: amplitude and frequency. From these, we can reconstruct the dynamics. These features can be graphically presented by plotting the two values in the frequency domain (Figs. 17.9b and c).

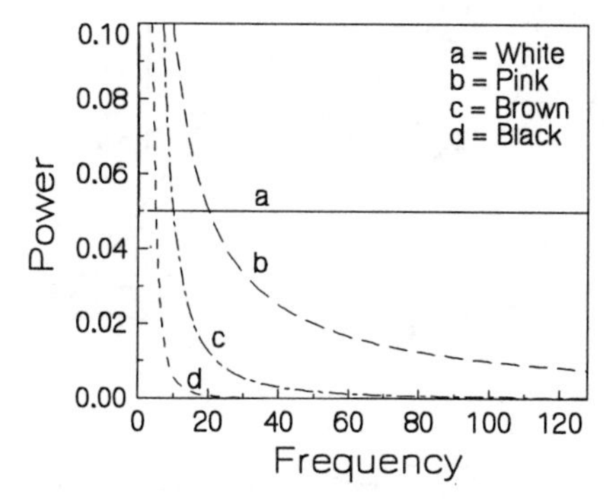

Figure 17.10: Power spectra for four classes of noise generated by the function $p = \alpha f^{\beta}$, where $\beta = 0, -1, -2 - 3$ for white, pink, brown, and black noise. Curves are scaled by α to fit on the graph.

Roughly speaking, the *power* of a particular frequency is proportional to the square of the amplitude divided by the square of the frequency. Power represents the importance of the frequency in determining the nature of the time series: frequencies with zero power (i.e., zero amplitude) make no contribution to the dynamics. Likewise, waves with large power in extremely low frequencies are important because they set the long-term trends in the series. The set of power values at all frequencies needed to approximate the dynamics to some level of precision is the *power spectrum* of the dynamics. In the simple example with only two sine waves (Fig. 17.9), the spectrum is just two points, but the concept can be extended to more complicated dynamics that require many frequencies for adequate approximation. It is also possible to represent continuous time series as possessing an infinite number of frequencies that produce a continuous function (curve) in the frequency domain.

Different physical phenomena can be characterized by a family of power spectra (Schroeder 1991). Figure 17.10 illustrates four of these. White noise is that in which all frequencies are equally likely; in color, it is an equal admixture of all wavelengths. Red light has relatively many low frequencies compared to high frequencies. Pink noise is less than red and has substantial numbers of the shorter wavelengths (i.e., broad band), but decreases with frequency relative to white noise. Brown and black noise decrease even more rapidly with frequency than pink.

All of these functions follow a power law: $p = f^{\beta}$, where p is power, f is frequency, and $\beta = 0, -1, -2 - 3$ for white, pink, brown, and black noise. On a log–log plot, the spectra appear as straight lines with slope equal to β. Brown noise ($\beta = -2$), as the name suggests, characterizes aspects of Brownian motion. Black noise ($\beta = -3$) describes natural catastrophes such as the occurrences of droughts, or floods (Schroeder 1991). Pink (or $f^{-1} = 1/f$) noise is important here because it is characteristic of the power spectra of the complex dynamics associated with chaos and strange attractors. Power spectra are not very powerful tests for chaos, but they do

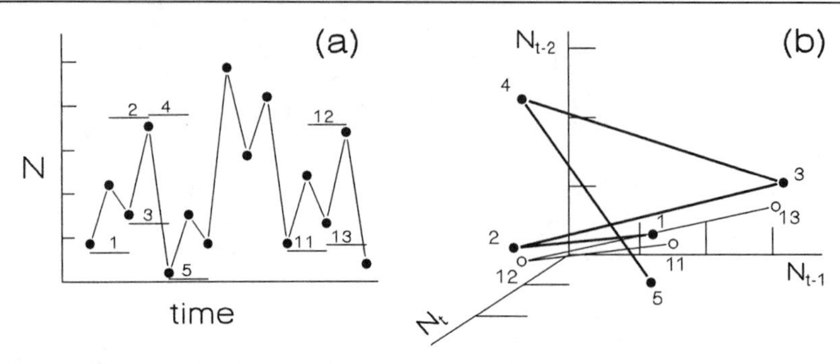

Figure 17.11: A multi-dimensional attractor can be reconstructed from a time series using sets of time lagged values. Groups of values in the time series (a) (e.g., triplets: N_t, N_{t-1}, N_{t-2}) are grouped together sequentially (numbers on horizontal lines). These are plotted as points in a three-dimensional space (b) so that repetitive sequences [triplets (1,2,3) and (11,12,13)] will appear as nearby points.

provide evidence for its existence. West and Shlesinger (1990) also review the above topics and provide additional examples from physics, psychology, and sociology.

To give a single theoretical example here, and some empirical cases later, Schaffer and Kot (1986) calculated the power spectrum of Vance's predator–prey model (Eqs. 17.4), and in a log–log plot found a strong linear relationship indicating f^{β} colored noise, which they interpreted as partial evidence for chaos. Unfortunately, the AR(1) model (Eq. 17.7) also produces colored noise (Chatfield 1975) and contains no nonlinearities.

17.3 Structure in Phase Space

As we saw in Chapter 14, phase space plots of chaotic systems are not similar to those of limit cycles. Nonchaotic systems with three state variables have orbits that lie on a plane. When the dynamics become chaotic, the attractor has greater dimensionality and the trajectories do not remain on a plane. Chaotic systems have the "signature" of complicated phase plane plots such as the spiral attractor studied by Vance (1978) and Gilpin (1979). A simple approach to detecting chaos is to plot the time series in a phase space and visually (i.e., qualitatively) determine that it looks complicated and chaos-like.

An immediate problem is that a single time series such as population numbers or heart rate does not have additional variables that can form the other axes in a phase space. This is solved by using values from previous times (Fig. 17.11). So, we plot the trajectory of the time series in a space having axes N_t, N_{t-1}, N_{t-2}. This sounds like an attempt to get something (multidimensional objects) from nothing (one-dimensional time series). Does this really help, or is it nothing more than the mathematical analog of rearranging the deck chairs on the Titanic? As it turns out, this is a very clever and useful idea. It graphically reveals subtle, repetitive

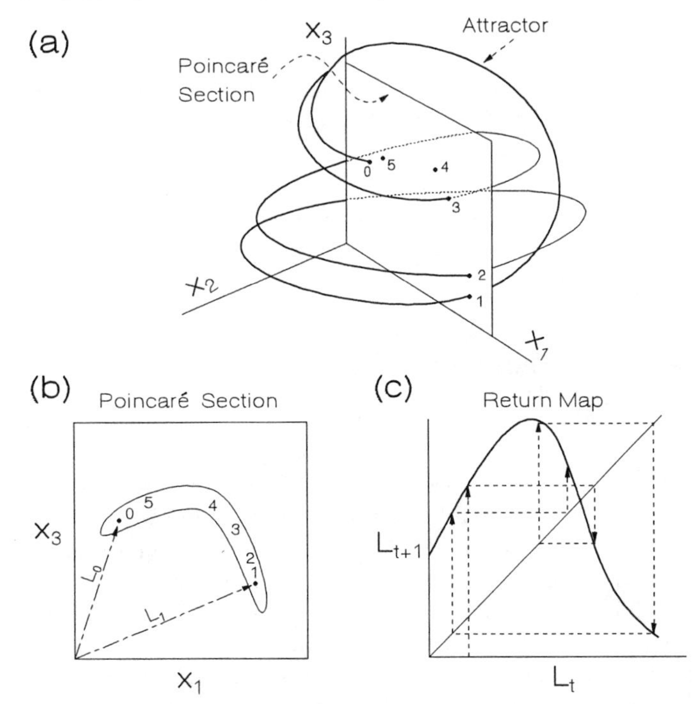

Figure 17.12: Graphical depiction of (a) an attractor for a hypothetical model showing the location of the Poincaré section and a few points (in time order) of intersection of the attractor trajectory and the section; (b) the plane of the Poincaré section, the envelope of the set of points that pass through the section, and a measure (L_i) of the relative distance of the point of intersection; and (c) the return map summarizing the iterative function that relates the distance at time t to that at time $t + 1$.

structure in the time series by causing similar *sequences* of time points to be plotted near each other (Fig. 17.11b). The method is not limited to three dimensions; when longer time lags are used we lose the graphical visualization, but other analytical tools relevant to chaos still apply.

Common practice is to use a lag of one or two time steps. A lag of two means these plots are three-dimensional and, therefore, difficult to visualize. Consequently, a standard technique is to dissect the attractor by taking a *Poincaré section* and constructing the associated *Poincaré (return) map.* A Poincaré section is a planar slice through the attractor as shown in Fig. 17.12. The lines in Fig. 17.12a are fragments of the trajectory that constitute the attractor (assuming one exists and that transient dynamics have settled down). The vertical plane shown is the Poincaré section, and the points on its surface constitute the intersection of the attractor with the section. The numbers represent the time order of the points. Figure 17.12b is a clearer portrait of the slice through the attractor, where the closed curve represents the set of points on the slice. As shown in the hypothetical example, this curve represents organized structure in the attractor; random dynamics would produce a cloud of disorganized points.

Additional attractor structure is visualized by constructing the return map (Fig. 17.12c). This is created by assigning a measure to the position of each ordered intersection point (L_i in Fig. 17.12b). A recursive map describing the relation of sequential values of this measure (Fig. 17.12c) illustrates the deterministic relations underlying the time series. In the example shown, the map represents chaotic dynamics because of the large negative slope of the map at the point of intersection with the 1:1 line (see Fig. 17.1).

17.4 Dimensions of Dynamics

As seen above, chaotic dynamics produce complex attractor structures that are bounded. Since they are bounded, they do not completely fill up the space in which they are embedded; nor, however, are they simple planar objects (see the forced double Monod in Section 14.3). Strange attractors are somewhere in between, so that one of the signatures of chaotic dynamics is the existence of objects having intermediate dimensionality. Several methods of measuring the dimensions associated with chaos and weird dynamics have been defined: correlation dimension, fractal dimension, information dimension, and the list goes on (see Farmer et al. 1983 for a review). Here, we give only a qualitative description of correlation dimension originally introduced by Grassberger and Procaccia (1983) to give the reader the flavor of the approach.

Suppose we have solutions on a strange attractor that is sampled at discrete points in time. First, choose a focal point on the attractor, and assume, for the moment, that the attractor is flat around the chosen point. Draw a circle of radius r around the focal point and count the number of solution points $[N(r)]$ also in the circle (i.e., $|x_i - x_c| < r$, where x_c is the circle center and x_i is a randomly chosen point on the attractor). Decrease the radius, say to $r/2$, and again count the number of points inside the circle. This number will be smaller than the previous number because the circle is smaller. Continue making the radius smaller and counting the interior solution points. Roughly speaking, the number of points (or "mass") inside a circle of radius r relative to the total number points in the attractor is proportional to r^D, where D is the correlation dimension (Mullin 1993a). Thus, the correlation dimension can be approximated by counting points inside circles of different radii as described above. Using linear regression on the log transform of $N(r) = cr^D$ gives an empirical estimate of the D.

This procedure does not require that the attractor be flat; that was assumed only for explanatory purposes. The attractor can be arbitrarily convoluted (i.e., multidimensional), but we do not know how convoluted it really is. For more complex attractors, we must use N-dimensional spheres in place of circles. So, a complication of the above procedure is that the answer we get will depend on the embedding dimensionality of the hyper-sphere that we use. The procedure, then, is to compute D for a series of embedding dimensions and hope that after some number of dimensions, our answers begin to converge. For deterministic chaotic time series this indeed happens (Mullin 1993a). As the embedding dimension increases, the family of curves plotted in $\log N$ *vs* $\log r$ space converges on a single linear relationship with positive slope. This can be visualized by plotting D against the embedding dimension. In chaotic systems, D quickly rises and levels off to a constant. In random time series, however, the correlation dimension usually continues to increase with embedding dimension. This is another signature of chaotic *vs* random time series. As always, though, there are types of random sequences whose correlation dimension will behave like that of a chaotic sequence.

17.5 Sensitivity to Initial Conditions

There is almost universal agreement that chaotic series have one property that distinguishes them from other dynamics. Chaotic dynamics are *bounded fluctuations that are sensitive to initial conditions* (Ellner and Turchin 1995). By this we mean that if we compare two solutions produced by a chaotic, deterministic model that differ only in slight differences in the starting values, the resulting two sets of dynamics will diverge over time (Fig. 17.3), but both will stay within a finitely bounded region of state space. This should sound paradoxical to you: how can two points move away from each other yet remain within a small region? Here we discuss two methods for quantifying the concept: Lyapunov exponents and predictive ability.

17.5.1 Lyapunov Exponents

The standard method for ascertaining that dynamics are sensitive to initial conditions is to measure the rate at which two points in phase space diverge from one another. For example, the two points labeled 1 and 2 in Fig. 17.12a diverge: 1 continues on to cut the section at 2, and 2 eventually cuts the section at 3. This divergence of nearby points on an attractor is quantified by the *Lyapunov exponent.* This quantity gets its name from its use in the one-dimensional divergence equation (Mullin 1993a):

$$d(t) = d_0 e^{\lambda t}, \tag{17.8}$$

where λ is the Lyapunov exponent and d_0 is the initial difference in initial conditions.

This quantity plays a role analogous to the eigenvalue of the characteristic equation in local stability analysis (Section 9.3.2). If $\lambda > 0$, the solutions diverge and the system is sensitive to initial conditions. An algorithm for calculating the exponent is based on the following facts. The

Lyapunov exponent for a pair of solutions of the one-dimensional map is the average of the natural logarithm of the absolute value of the derivatives of the map function at each of n solution points (Hilborn 1994):

$$\lambda = \frac{1}{n} \sum_{i=0}^{n-1} \ln |df/dx_i| , \tag{17.9}$$

where n is the iteration number and represents discrete time (t) in Eq. 17.8.

Also,

$$|df/dx_i| \approx \frac{|f(n, x+\epsilon) - f(n, x)|}{\epsilon}.$$

Equation 17.9 simply states that λ is the geometric mean ($1/n$ applied to a sum of logarithms) of the deviations (df/dx) that are calculated at progressively greater time intervals ($i = 1$ to $i = n - 1$). There are many assumptions and computational considerations involved in implementing this definition. Generally, several starting points are sampled and the average of Eq. 17.9 is the estimate of λ. The interested reader should consult Earnshaw and Haughey (1993) or the texts by Mullin (1993c) and Hilborn (1994) for details.

17.5.2 Predictive Ability

A related idea is that if the dynamics are deterministic, but sensitive to initial conditions, then we might expect that our ability to predict from past trajectories might be high for short time scales, but will become poor as we attempt to predict further into the future. Farmer and Sidorowich (1989) developed nonlinear forecasting techniques appropriate to this problem, and Sugihara and May (1990) developed a simpler version that they applied to ecological data. In the latter method, a trial time series is used to create projection rules for predicting the future some number of time steps into the future. Applying these rules to new data permits predictions τ time steps into the future that are compared to observed values. Prediction accuracy is measured by the correlation coefficient between predicted and observed datum points, where 1.0 implies perfect correlation or predictability. This method can be applied to both chaotic and white noise sequences. The effect of the size of τ, the number of future time steps for prediction, on prediction accuracy is shown in Fig. 17.13 for a chaotic series and a sine wave to which were added random values from a normal distribution. After examination of several models and data sets, Sugihara and May (1990) found that chaotic systems, unlike random sequences, showed a decrease in the correlation coefficient as prediction time increased.

17.6 Controllability of Chaos

A recent development in nonlinear dynamics is the study of the control of chaotic systems (Ott et al. 1990, Peak and Frame 1994). This is important

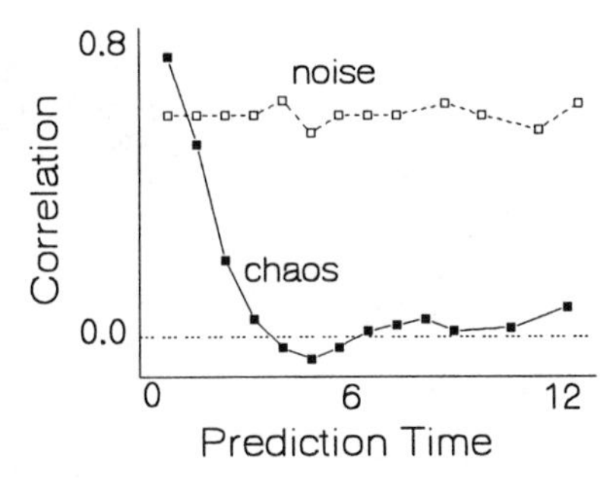

Figure 17.13: The prediction accuracy of chaotic systems (solid curve) decreases with prediction time, but not for a random sine wave (dashed line). (From Sugihara and May 1990, Fig. 2a. © 1990 Macmillan Magazines Limited. Reprinted from *Nature* with permission of the publisher and author.)

for the practical problem of the management of nonlinear systems and as another tool to identify a signature that suggests the existence of chaos. This method is based on the fact that chaotic systems can be controlled because of their underlying nonlinear deterministic structure, whereas random sequences cannot be controlled because there is no underlying structure.

We have discussed how time series produced by stochastic nonlinear recursive equations are difficult to distinguish from those produced by noisy sine functions. Although both can produce a cloud of points in the y_t–y_{t-1} plane (Fig. 17.7), there will be in the chaotic trajectory a number of subsets of points that form an alternating pattern about the 1:1 line. This pattern will be missing in the uncorrelated random sequences. Figure 17.14a illustrates the alternating pattern in question for a deterministic model. The points produced may be hidden in the cloud of points (Fig. 17.14b), but these can be discovered. If the subsets of alternating points exist, then there may be underlying determinism that we can exploit for control.

Attempts to control chaos exploit the situation in Fig. 17.14a in the following way. We will interpret "control" to mean manipulating the system so as to keep the dynamics within some finite region around the fixed

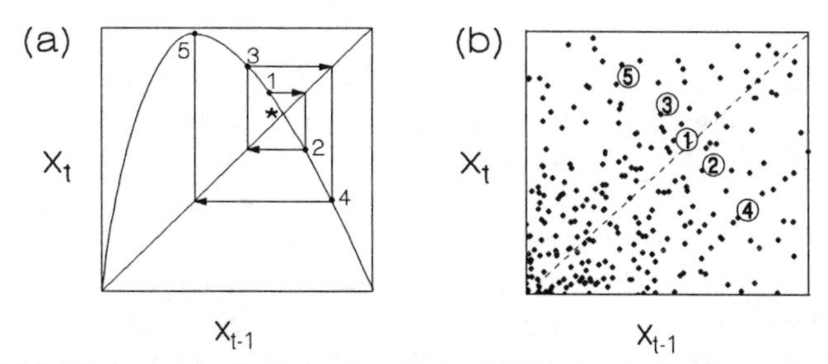

Figure 17.14: The pattern of alternating states hidden in stochastic nonlinear models. (a) Points that alternate around the 1:1 line for short segments of the trajectory in a deterministic model. (b) The same points that may be hidden in a noisy model.

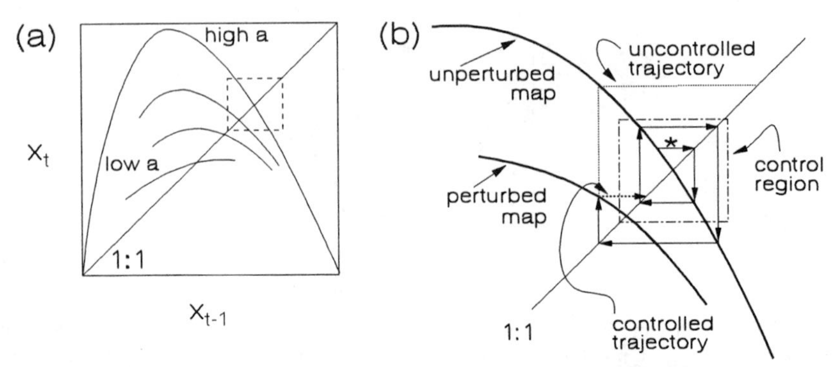

Figure 17.15: How to control chaos. (a) One complete and three fragments of maps from Eq. 17.3 with different values of the parameter a. Shallow slopes at the intersection of the map with the 1:1 line imply smaller equilibrium values and more constant dynamics. The complete map has a slope producing chaos. The dotted box is the region in which we wish to confine the dynamics. (b) An enlarged view of (a) near the fixed point. The original, unperturbed map is shown as a heavy line; the dot-dash box is the control region. The trajectory starts at * and, if uncontrolled, would quickly wander outside of the control region. By perturbing the map using Eq. 17.10, the trajectory encounters a new map with a slightly lower slope that projects the trajectory back into the control region.

point (* in Fig. 17.14a) as determined by the parameters of the equation. We further assume we can manipulate the system in time by tweaking a parameter p that interacts with the original nonlinear system so that

$$y_{t+1} = p_t a y_t (1 - y_t), \tag{17.10}$$

where p_t is our time-dependent manipulations of the control parameter. If p_t is 1.0, we have, of course, the familiar logistic equation. Now, a was the original parameter that determined the qualitative dynamics of the system as shown in Fig. 17.2. By inventing pa, we have simply defined a new time-dependent parameter for the system. If we restrict $p < 1.0$, we are effectively reducing the system parameter a. Smaller values of a were *usually* indicative of less complex dynamics (Fig. 17.2, when $a < 3.57$). This is diagrammed in Fig. 17.15a, which shows portions of the logistic map with four parameter values. Notice that the fixed point is reduced as a is reduced.

To control such a system when a of Eq. 17.10 is in the chaotic region, we set p_t to values slightly less than 1.0 whenever y_t moves outside the desired region. Figure 17.15b shows that doing this will force the system back into the box. It might occur to you that if we simply want to contain the fluctuations of y, we should just set p_t to some relatively small fixed value that will produce a permanent, stable fixed point. This will not work because our constraint on control was to keep the dynamics near the *original* fixed point, not the new one that would be produced if p_t were kept small. Figure 17.15a illustrates that a small value of the parameter of Eq. 17.3 will shift the fixed point.

17.7 Biological Models Producing Chaos

Not all models produce bifurcations and chaos. We can make a few brief generalizations concerning the biological conditions under which we would expect chaos and other weird dynamics to emerge.

Nonlinearity It should be obvious by now that while systems of linear differential equations can produce complicated and oscillatory dynamics, they do not exhibit limit cycles, strange attractors, and chaos. Nonlinear relationships, such as exemplified in the logistic map, are required. Moreover, in chaos producing models, these nonlinear relations are of the type that show strong positive feedback in one interval of the domain of the function and strong negative feedback in another interval (Berryman and Millstein 1989). By strong positive or negative feedback, we mean that the function increases or decreases rapidly for a unit increase in the state variable (the domain of the function). For complex dynamics to result, the positive feedback interval must occur at domain values that are smaller than the negative feedback interval. Again, the logistic map illustrates the relation. When population size is smaller than the peak of the hump in Fig. 17.1, positive feedback (amplification) is strong and drives the population up rapidly. This causes a sudden transition to the negative feedback interval at large population levels (to the right of the hump).

Time Delays Time lags in continuous systems can also produce complex dynamics and instabilities. A general form of these equations is:

$$\frac{dx}{dt} = f(x, \tau),$$

where τ is the time delay so that x responds not to the current value of x, but to the value τ time units in the past. Broadly speaking, time delays arise because information has a finite transmission rate through biological systems. In physiological systems, information transmission has a literal interpretation: propagation speeds of nerve impulses or diffusion rates of chemical signals (e.g., hormones). In ecological systems, time delays are often associated with age classes: events influencing a young age class (e.g., a bad winter in a habitat frequented by young individuals) will not be manifested in population growth rates until those individuals experiencing the catastrophe reach reproductive age. Usually, these systems could be modeled with explicit representations of the information transmission process. But time delay formulations can be a useful and simpler approximation to this more complex system description.

May (1973) analyzed the logistic model of population growth when time delays were present in the density dependence term, so that

$$\frac{dN}{dt} = rN_t(1 - N_{t-\tau}/K),$$

where N is the population of interest (e.g., herbivores) and $t-\tau$ is the past time influencing current growth rates. The ecological rationale for this is that it captures in one equation the dynamics of a theoretical system with two variables, one of which, for example, is vegetation and the other (N) is a herbivore. The current population growth rate depends on the number of reproductives that were produced at $t-\tau$, when the vegetation was different. May (1973) derived some simple relations that indicate how large τ must be before instability arises. He found that the system will become unstable if $r\tau > \pi/2$. He also provides some numerical simulations that indicate that the instability that arises produces a stable limit cycle.

As another example, Mackey and Glass (1977) studied the effects of time delays on a model of white blood cells:

$$\frac{dx}{dt} = -\gamma x + \beta x_{t-\tau} \frac{\theta^n}{\theta^n + x_{t-\tau}^n},$$

where x is the number of white blood cells (WBC); γ is the rate of WBC destruction; β, θ, and n are WBC proliferation parameters; and $t-\tau$ is the past time influencing the dynamics. As with May's population analysis, a Hopf bifurcation occurs at a critical time delay. See Glass and Mackey (1988) for more examples.

Compartment Cascades Certain models have an effective time delay induced by a cascading flow of material (individuals) through a series of compartments that have a single input and output. An example is age-structured population models in which each compartment represents the numbers of individuals in an age class.

Caswell (1989) studied a variant of a simple system originally examined by Guckenheimer et al. (1977). The two-age class system is

$$n_{1,t+1} = \mu n_{1,t} e^{-0.1N_t} + 2\mu n_{2,t} e^{-0.1N_t} \tag{17.11}$$

$$n_{2,t+1} = 0.9 n_{1,t}, \tag{17.12}$$

where $N = n_1 + n_2$. Even though the equation for $n_{2,t+1}$ is linear in $n_{1,t}$, it has an indirect nonlinearity on $n_{i,t-1}$:

$$\begin{aligned} n_{2,t+1} &= 0.9 n_{1,t} \\ &= 0.9 \mu n_{1,t-1} e^{-0.1N_{t-1}} + 2\mu n_{2,t-1} e^{-0.1N_{t-1}}. \end{aligned}$$

The combination of the nonlinearity and time lag produces bifurcations and chaos.

Forcing Functions External periodic perturbations of a system that also oscillates from its own internal forces have long been known to produce complex dynamics. This was the basis for the complexity of the forced chemostat system studied by Kot et al. (1992) in Section 14.3. In that study, the predator induced internal limit cycles. Perturbing these cycles by a signal whose frequency did not match that of the internal dynamics produced dynamics that were sensitive to initial conditions. When the perturbing frequency did match the internal frequencies, phase locking occurred and more regular oscillations were observed.

There can be no doubt that reasonable models of biological systems can be chaotic. It is another problem, however, to demonstrate that a real biological system is or can operate in a chaotic parameter region. This is difficult because there are superficial similarities between a chaotic time series and a random time series. Neither can be predicted for long times into the future. Both can fluctuate widely with no apparent simple period. For theoretical and practical reasons, it is valuable to determine if a given time series is random or deterministically chaotic. Unfortunately, this is surprisingly difficult to do, and all of the attempts to recognize chaotic signatures, as described above, have problems. Below, we briefly summarize some of the recent applications and results. The status of efforts to detect chaos in ecological systems was reviewed by Logan and Allen (1992) and Hastings et al. (1993). Below, we review some of the attempts to identify the signatures of chaos in empirical time series. Lastly, we describe more recent experimental manipulations of insect populations that provide evidence for chaotic behavior.

17.7.1 Power Spectra

The usual view of physiological systems, especially in the "higher" organisms such as mammals, is that they are a finely tuned, well-articulated collection of mechanisms all of which act in concordance and cooperation with one another. Proper physiological functioning is typically associated with regular dynamics. Similarly, the opposite relation is also commonly accepted: physiologically corrupt systems will result in irregular and unpredictable dynamics. This view is held by some under the name of "dynamical disease": the conditions of disease cause a breakdown of regular, coordinated dynamics to produce chaos (e.g., Glass and Mackey 1988). The opposite extreme is "chaos is healthy" (e.g., Golberger 1992), in which normal operation is thought to be irregular within bounds, but becomes more regular and cyclical during the onset of disease (e.g., heart failure). While the final assessment is far from settled, there is some evidence that chaos may be beneficial.

Power spectra, despite their shortcomings in detecting chaos, have been used extensively in the study of nonlinear heart dynamics. Goldberger and his colleagues (e.g., Goldberger and West 1987 and Goldberger and Rigney 1991) calculated heart rates (beats per minute) from interpulse intervals (Fig. 17.16). A healthy patient (Fig. 17.16a) showed irregular dynamics that could be described with a classical $1/f$ noise power spectrum. Contrary to this, a patient undergoing heart failure (Fig. 17.16b) shows oscillatory heart rate dynamics in which the power spectrum indicates the strong peak at about 0.02 cycles/second (period = 50 seconds). Other studies have found similar loss of "broad-band complexity" in the heart rates of

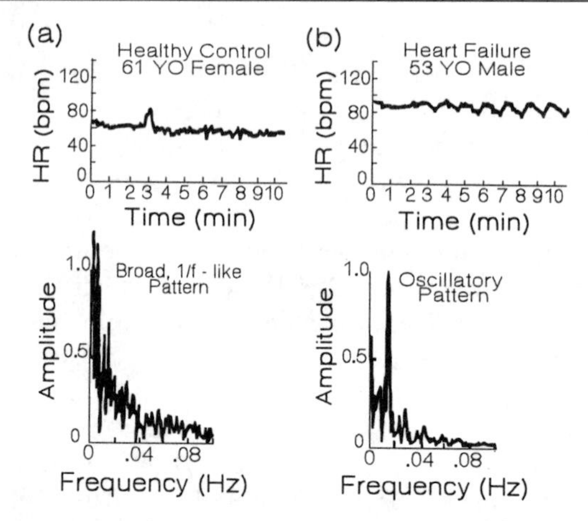

Figure 17.16: Power spectra of normal and abnormal hearts. (a) A healthy patient with irregular heart rates (above) and a power spectrum showing $1/f$ noise. (b) An unhealthy patient with cyclic dynamics and strong power peak corresponding to a period of 50 seconds. (From Goldberger and Rigney 1991, Fig. 22.10. © 1991 Springer–Verlag New York, Inc. Reprinted by permission of the publisher and author.)

older patients, the effects of toxicological stresses, and so on.

Schaffer et al. (1990) have also used this method to show that the chickenpox (apparently not chaotic) power spectra was dominated by a single frequency of 1 year, while that of the measles time series (which possessed other chaotic signatures) had several important frequencies. Schaffer (1987) cautions that when applied to short time series this method of detecting chaos can mislead.

17.7.2 Attractor Structure

Another common signature examined in the search for chaos in physiological and ecological systems is the structure of the attractor. This includes not only a graphical picture of the flows in phase space, but also the Poincaré section and return maps. For example, Hayashi and Ishizuka (1987) inserted a microelectrode into the esophageal ganglion of a marine snail so that they could both record electrical signals as well as stimulate the nervous tissue. They stimulated the nerve by passing an oscillating current across the ganglion membrane.

When a membrane experiences a periodic current whose amplitude is below a threshold specific to a particular experimental preparation, the nerve cell responds with a depolarization manifested by an increase in electrical potential across the membrane (see Section 18.4.2 for more details). This is followed by a repolarization and voltage drop back to the resting potential; a complete action potential does not occur. This response and the voltage achieved is called the "subthreshold response" (SR). When the current threshold is exceeded, an action potential occurs with a preparation specific increase in potential (AP) across the membrane. Because mem-

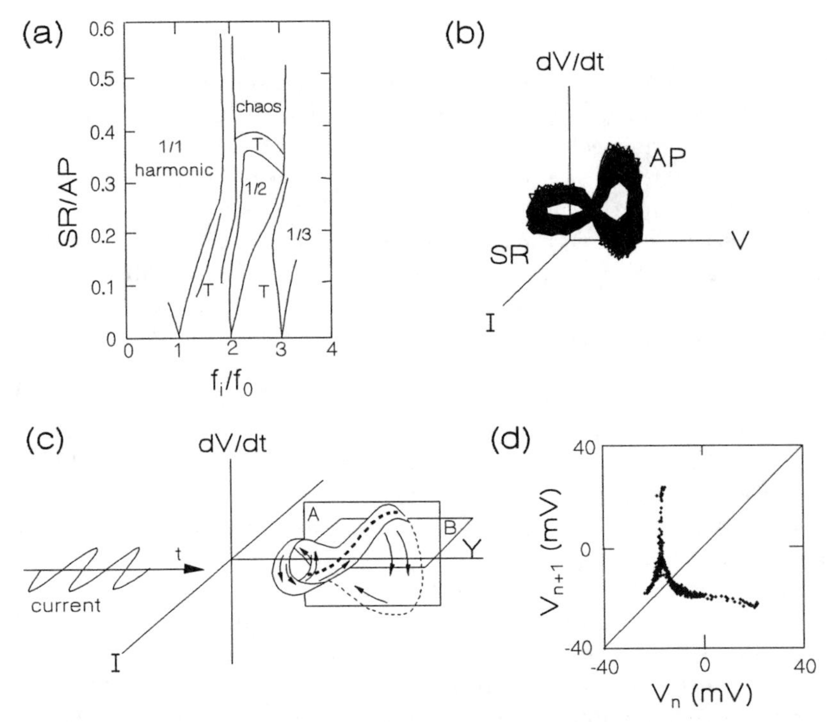

Figure 17.17: Complex dynamics resulting from externally forcing snail ganglia. (a) Stability diagram for two control parameters: f_i/f_0 (where f_i = frequency of forcing stimulus, f_0 = frequency of spontaneous firing) and SR/AP = voltage ratio of the preparation (where SR = subthreshold response voltage, AP = action potential voltage); labels indicate qualitative dynamics: $1/n$ harmonic = nerve fires once every n stimulation periods, T = transitional dynamics. (b) The attractor based on observed dynamics of the three state variables. (c) Graphical depiction of the reconstructed attractor with stimulus pattern shown to the left. (d) The return map for chaos conditions. (From Hayashi and Ishizuka 1987, Figs. 2, 4, 6. © 1987 Plenum Publishing Corporation. Reprinted with permission of the author and publisher.)

brane resistance varies among preparations, the amplitude of the applied current does not measure the intensity of the stimulus. Therefore, Hayashi and Ishizuka (1987) used the ratio of the subthreshold voltage (SR) to the action potential voltage (i.e., SR/AP) as a measure of the resistance of the membrane to the applied current fluctuations. In addition, a special feature of the snail esophageal ganglion is that, in resting conditions, this tissue spontaneously emits electrical pulses, so that an external current is a forcing function applied to an endogenous limit cycle. As discussed in Section 14.3, this can, in general, induce complex dynamics, and Hayashi and Ishizuka (1987) did indeed observe this (Fig. 17.17).

This experimental system produces large amounts of relatively noise-free data, unlike ecological systems. As a consequence, the data can be used directly to reconstruct the attractor (Fig. 17.17b). Figure 17.17c shows a diagram of the attractor in the space of membrane voltage (V), stimulus current (I), and voltage rate of change (dV/dt). Depending on the parameter values of the forcing function, this system can be driven into regular oscillations, chaos, intermittency, and random alternations (Fig. 17.17a). Again, thanks to the large data set, Hayashi and Ishizuka (1987) were able to construct an accurate return map from the Poincaré section (Fig. 17.17d). The map clearly supports the chaos hypothesis. It should be remembered, however, that these data arise from periodically forcing a stable system. The data do not represent the naturally occurring dynamics.

Ecological systems are much shorter than physiological datasets and, therefore, more difficult to analyze using these techniques. Nevertheless, Schaffer and Kot (1986) have done this for several time series of natural populations and human disease epidemics. Figure 17.18a is Ellner's (1991) representation of the attractor originally reconstructed by Schaffer and Kot. The attractor appears to be organized as expected by deterministic chaos. The Poincaré section (Fig. 17.18b) and return map (Fig. 17.18c) also show simple, nonrandom structures.

There are two major problems with this approach. First, long time and noise-free time series are required to see pattern in the section and maps. These are difficult to obtain in ecological systems. Second, Ellner (1991) obtained essentially identical qualitative graphical results when he simulated a periodic equation with noise using parameter estimates derived from the thrip time series (Fig. 17.18d–f). Thus, simple structure in time-lagged phase space is not compelling evidence for deterministic chaos.

17.7.3 Prediction of Time Series

One of the first features of complex nonlinear dynamics to be discovered in recent times was the fact that particular trajectories are sensitive to the starting point. A consequence of this is that chaotic systems, being deterministic, can be accurately predicted for a short period, but the predictions get worse over long times. In contrast, the accuracy of predictions of true, stationary, stochastic processes is independent of the time scale. As a result, we expect high correlation between observed and predicted variable values when we predict over two or three time steps; we expect the correlation coefficient to decrease if we predict over five or six time steps Fig. 17.13).

Sugihara and May (1990) applied this technique to empirical epidemic and natural population time series. They found that the decline over time of prediction accuracy varied depending on the system. Prediction of the monthly incidence of measles in New York state from 1923 to 1963 was consistent with deterministic chaos, as found by Schaffer and his colleagues. The correlation of predicted and observed measles cases declined from about 0.85 over one time step to 0.4 over seven time steps. Sugihara and May obtained a similar result for the analysis of inshore marine

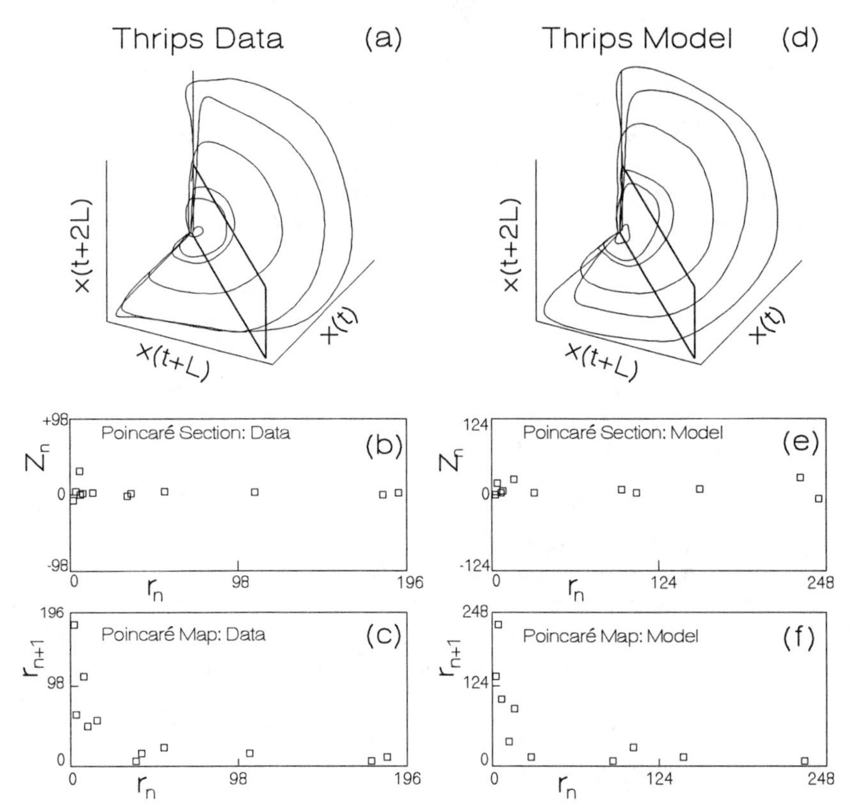

Figure 17.18: Graphical analysis of a time series of thrip population numbers and a theoretical stochastic model. (a) Reconstruction of the thrip time-lagged attractor. (b) and (c) the Poincaré section and map, respectively. (d) The "attractor" for a noisy, non-chaotic periodic function; (e) and (f) the Poincaré section and map associated with (d). (From Ellner 1991, Figs. 2 and 3. © 1991 Virginia Agricultural Experiment Station. Reprinted courtesy of Virginia Agricultural Experiment Station.)

plankton population dynamics. Their analysis for chickenpox epidemics, however, showed little evidence for chaos as the correlation coefficients varied between 0.7 and 0.8 for prediction times 1–12. This is evidence for noisy seasonal cycles, again consistent with Schaffer et al. (1990).

Like power spectra and attractor structure, this method, however, may also have little power to distinguish random series from deterministic chaos. Ellner's (1991) forecasting analysis of periodic cycles with noise showed the same decline of prediction correlation with time scale as did chaotic series.

17.7.4 Lyapunov Exponents

Much current interest in statistically identifying chaos in time series focuses on estimating the Lyapunov exponent. The original algorithms worked best when there was no noise in the signal and there were several thousand data points in the series (Ellner 1991). Because these conditions are rarely, if ever, met in many time series, an alternative method has been developed. The empirical time series is used to estimate the parameters of a time-lagged finite difference equation that relates the next point in the series to a function of previous values in the series. If a sufficiently good fit is obtained, this function is used to generate surrogate data from which an estimate of the Lyapunov exponent is obtained. Recent work has extended this to include stochastic variation. In ecology, this approach has been widely applied by William Schaffer, Peter Turchin, Steven Ellner, and others.

Ellner and Turchin (1995) calculated Lyapunov exponents for nearly 50 population time series to determine the distribution of exponents in nature. Ellner and Turchin used the time series data to estimate parameters of a convenient but biologically meaningless equation that could reproduce the original time series with statistical accuracy. They then used the model to generate "datum" points to be used in reconstructing the attractor. Their results depended on the function they used for fitting, but, in general, they found that Lyapunov exponents were less than zero in all but a few data sets. This is evidence for the absence of sensitivity to initial conditions. Moreover, most of the exponents were small negative numbers near zero, meaning that the populations were "at the edge of chaos." This is an intriguing situation that other theoretical models have also predicted. Although Ellner and Turchin found little evidence for chaos, they were quick to point out that their methods were conservative and that chaos may exist in more of the data than their methods revealed.

17.7.5 Controlling Chaos

The controllability test for chaos has recently been successfully performed in several biological systems. In a lovely set of experiments, Schiff et al. (1994) demonstrated that populations of neurons in extracted tissue of the rat brain fire in bursts separated by a chaotic sequence of intervals. They further demonstrated that this chaos could be controlled by direct stimulation of surrounding neurons. They removed and sectioned the hippocampus of rats (an area where sensory inputs are distributed to the forebrain) and perfused it with artificial cerebrospinal fluid. Under these *in vitro* conditions, the CA3 neurons continue to fire spontaneously. A recording electrode was inserted in the CA3 region and a stimulating (controlling) electrode was inserted about 1 mm away in the Schaffer collateral fibers. Input from the recording electrode in the form of the interburst firing intervals was sent to a computer system that determined when the system was diverging from an unstable fixed point. This determination was made based on sequences of states in the y_t-y_{t-1} phase space (Fig. 17.14). When this intermittent (non-periodic) condition was detected, the control electrodes delivered a single, short burst of current directly to the Schaffer fibres. The control burst instigates an action potential (see Section 18.4.2)

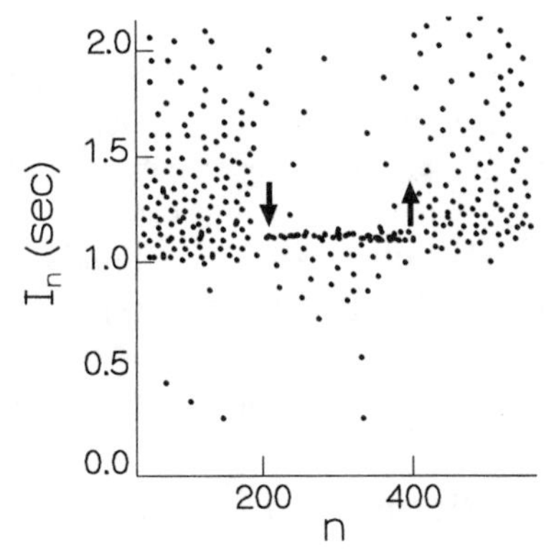

Figure 17.19: Control of chaotic sequences of intervals between bursts of neuronal activity. The x-axis is the burst number (analogous to time); the y-axis is the time interval between successive bursts. The down-arrow indicates the onset of control; points to the left of that arrow are normal intervals; to the right are the controlled intervals. The up-arrow indicates the removal of control. Control was effected by single pulses of electrical current in cells adjacent to those shown. (Loosely after Schiff et al. 1994, Fig. 3. © 1994 Macmillan Magazines Limited. Reprinted from *Nature* with permission of the author and publisher.)

that propagates into the CA3 pyramidal cells, thereby depolarizing and synchronizing a large population of these cells.

Figure 17.19 shows a rough caricature of some of their results. As shown, control using the above technique was effective in maintaining a constant interburst interval. The effect of the control was instantaneous as shown by the sharp change in dynamics. Schiff et al. (1994) also investigated several other control firing schedules. Simply firing the control electrode with a fixed period (ignoring the chaotic phase space dynamics) also reduced the scatter around the interburst interval, but not so effectively as the single-pulse, chaos-based method.

It is tempting to conclude from these positive results that interburst intervals in this *in vitro* preparation are chaotic. As evidence, the authors found subsets in the time series consistent with unstable fixed points, and they were able to control the variability of interburst interval using knowledge of the phase space. Unfortunately, as with many such previously positive tests for chaos, Christini and Collins (1995) were able to duplicate the chaos-based control result on a stochastic, nonchaotic model of neuron firing. They used the FitzHugh–Nagumo model of nerve voltage with Gaussian white noise added. Like Schiff et al. (1994), they were able to find sequences of solutions in the y_t–y_{t-1} phase space that mimicked the exponential divergence from unstable fixed points characteristic of chaos (Fig. 17.14). Christini and Collins (1995) were also able to control the dynamics using the same single-pulse firing scheduling method as Schiff et al. (1994).

These results may be disappointing to those searching for the Holy Grail of a definitive test for chaos in empirical systems. But the positive side is that we now have another tool for controlling dynamic variability, whether it is generated from nonlinear deterministic systems or stochastic systems. Moreover, there is some evidence that interburst intervals during epileptic seizures behave like those observed in the hippocampus. It is possible that these preliminary results will develop into practical medical treatment techniques. Chaos-based control techniques certainly need to be attempted on a wide variety of systems, for example, population fluctuations.

17.7.6 Experimental Population Studies

While chaos control has yet to be tried in ecology, an experimental test of nonlinear dynamical theory in population dynamics was recently performed. An elegant study of flour beetle (*Tribolium castaneum*) dynamics by Costantino et al. (1995) provides strong empirical evidence for deterministic mechanisms of complex, aperiodic population dynamics. Flour beetles inhabit large bins of ground grain such as flour (hence the name) and can be a major pest. Like many other insects, they have discrete age classes (larvae, pupae, and adults, Fig. 17.20). Adults, of course, contribute individuals to the larvae, but unlike many insects, the adults and pupae can also cannibalize smaller age classes. These two phenomena result in both positive and negative feedback mechanisms and have the potential to produce complex dynamics.

The model is a wonderfully simple system of three finite difference equations that describe an age-structured population:

$$L_{t+1} = bA_t \exp(-c_{ea}A_t - c_{el}L_t) \tag{17.13}$$

$$P_{t+1} = L_t(1 - \mu_l) \tag{17.14}$$

$$A_{t+1} = P_t \exp(-c_{pa}A_t) + A_t(1 - \mu_a), \tag{17.15}$$

where the variables and parameters are defined in Table 17.1.

Cannibalism is important in this species and is represented in the model as a decrease in the survival rates of the consumed ages in the presence of the consuming age. Adults and larvae both consume larvae; pupae do not eat larvae. Adults eat pupae, but not other adults. So, the negative exponentials in Eq. 17.13 represent the reduction of larval survival rate as adult or larvae numbers increase. The number of larval recruits increases linearly with adult numbers by the rate b. Thus, the rate equation for larvae is composed of two process rates multiplied together: reproduction and mortality. This equation should be familiar as the *maximum* function presented in Chapter 4. The equation for pupae is simple: every larva that survives predation pupates, but pupae die from causes of mortality other than cannibalism at the rate μ_l. Adults have a similar death term. In addition, adults consume pupae, so the number of pupae emerging as

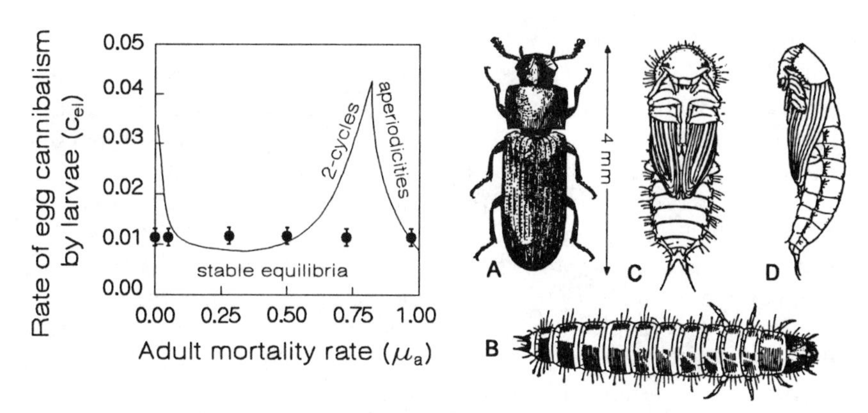

Figure 17.20: On the left is the stability diagram of a model of flour beetle population dynamics (genetic strain SS) as it is affected by rate of egg cannibalism (y-axis) and adult mortality rate (x-axis). Filled circles are experimental manipulations of adult mortality rates. Labeled regions are the qualitative dynamics as predicted by the model. (From Costantino et al. 1995, Fig. 1c. © 1995 by Macmillan Magazines Limited. Reprinted from *Nature* with permission of the author and publisher.) On the right are the three life stages of *Tribolium confusum*, a species similar to that studied by Costantino et al. (1995). A=adult, B=larva, C=pupa (ventral view), and D=pupa lateral view. (From California Agricultural Experiment Station Report 696. © 1956 California Agricultural Experiment Station. Reprinted with permission of the publisher.)

adults is the fraction not eaten by the current adult cohort. This survival rate declines as a negative exponential term with increasing numbers of adults.

The model has all the ingredients for interestingly complex dynamics; it contains (1) positive feedback in the form of larvae production by adults, (2) negative feedback at high adult densities by cannibalism, and (3) an implicit time lag in the form of the compartment cascade through the age classes. This promise of complex dynamics is well fulfilled, as Costantino et al. (1995) and Cushing et al. (1996) demonstrated. With proper choice of parameters, the model exhibits stable equilibria, two-point cycles, and

Table 17.1: Values and definitions for variables and parameters in the *Tribolium* model.

VARIABLES		
L	250 numbers	Larvae
P	5 numbers	Pupae
A	100 numbers	Adults
PARAMETERS		
b	11.68 number$\cdot t^{-1}$	Larvae recruits per adult
c_{ea}	0.011 unitless	Susceptibility of eggs to cannibalism by adults
c_{el}	$\approx$ 0.013 unitless	Susceptibility of eggs to cannibalism by larvae
c_{pa}	0.017 unitless	Susceptibility of pupae to cannibalism by adults
μ_l	0.513 unitless	Fraction of larvae dying (not cannibalism)
μ_a	varied unitless	Fraction of adults dying

aperiodicity (Fig. 17.20a).

But Costantino and colleagues did more than a simple numerical analysis of the model to produce yet another set of bifurcation diagrams. In a set of papers summarized in Costantino et al. (1995), they described parameter estimation, model validation using independent data, and experimental tests of the major predictions. To do this, they identified two parameters that controlled the dynamics: larvae cannibalism rate (c_{el}) and adult mortality rate (μ_a). In computer experiments, they numerically manipulated these parameters in order to identify the qualitative dynamics (i.e., stable equilibrium, two point cycles, or aperiodic fluctuations) that resulted from different combinations of these two parameters (Fig. 17.20). They then attempted to determine if, indeed, the real insects behaved as predicted by the model. Many other modelers have previously attempted and succeeded in experimentally validating their models, but what was new here was the attempt to push the experimental conditions to the point that qualitative dynamics changed dramatically from an equilibrium to chaos or aperiodic behavior. (Recall the discussion of model reliability in Section 8.2.)

The experimental system consisted of small laboratory containers of flour and *T. castaneum.* The environments of the containers were held constant to minimize environmental stochasticity. The populations were censused every 2 weeks by removing, aging, and counting all individuals, which were then returned to the container. Adult mortality was manipulated to coincide with different stability regions (Fig. 17.20). Adult mortality was manipulated by adding or removing adults at the time of the census.

The observed larval dynamics were gratifyingly close to the predictions (Fig. 17.21). By and large, the three theoretical kinds of dynamics were observed when the adult mortality was set to values predicted by the model. Adult numbers were slightly less consistent with expectations than larvae. This is the first experimental validation of chaotic behavior in a real population that was predicted *a priori* by a simple model. It is further significant that the complex dynamics seen were generated from endogenous interactions without external forcing (e.g., as in some neurophysiological systems: Hayashi and Ishizuka, 1987).

Now, experimentally applied levels of adult mortality are not necessarily those of natural populations, and an analysis of an unrelated *Tribolium* experiment (Dennis et al. 1995) estimated adult mortality to be in the range 0.096 – 0.148, much smaller than that needed to produce aperiodic dynamics (depending on larvae susceptibility to cannibalism: circles in Fig. 17.20). So, the generality of these results to other populations and experimental situations will need to be explicitly tested. But, the methodological framework has been developed in these studies, and we can expect to see future attempts to demonstrate similar effects in other experimental populations.

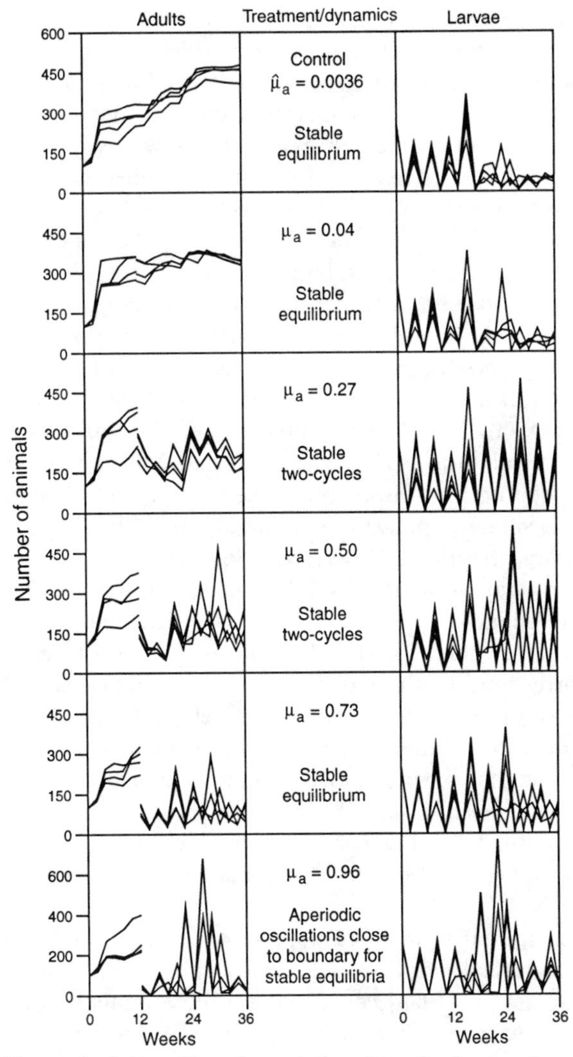

Figure 17.21: Observed adult and larval population dynamics in experimental conditions corresponding to the six filled circles in Fig. 17.20. The manipulation of adult mortality rates (μ_a) commenced on week 12. The predicted qualitative dynamics are listed in the center panel. (From Costantino et al. 1995, Fig. 3. © 1995 by Macmillan Magazines Limited. Reprinted from *Nature* with permission of the author and publisher.)

17.8 Why Is There Chaos in Biology?

At the beginning of the previous section, we mentioned that one view of physiological systems held that chaos was a healthy state. Another view holds that those systems are basically well-regulated around a stable equilibrium or limit cycle and that chaos indicates disease. We have seen some evidence that biological systems can be chaotic and somewhat less that natural unforced systems are chaotic or have complex dynamics. Why should this be so? Why is chaos rare in some systems and common in others? Are complex dynamics adaptive or merely a consequence of the multitude of constraints placed on evolving systems? Here, we explore these questions a bit further, particularly as they apply to ecological systems.

There are two primary reasons why chaos and weird dynamics might not be common in ecological systems. First, the parameter values needed to generate these dynamics are large compared to typical values. For example, in the logistic map, the onset of chaos occurs when the finite rate of increase is approximately 3.5. Most populations appear to have much smaller values. Second, in the logistic map, chaos occurs as population numbers range from very large to nearly zero (Fig. 17.2). If a natural population were to experience similar ranges, it would surely go extinct at such low population levels (Berryman and Millstein 1989). There are a number of counters to these arguments. First, even though the analysis occurs with a single population it is understood that in fact many other populations are present in the system. So, the parameter values obtained from reconstruction or fitting of single populations dynamics are those from a projection of high-dimensional dynamics (with many populations) to a single dimension. There is, therefore, no real reason to expect the fitted values to correspond to those obtained from populations truly growing in isolation. Second, as we have seen previously, the logistic map is not the only mechanism to produce chaos. It can occur in multidimensional continuous systems that have bounded attractors that do not approach one or more of the axes (zero values). So, complex dynamics can occur without a great risk of extinction. Further, chaos in these multidimensional systems may arise at biologically reasonable parameter values (McCann and Yodzis 1994).

Moreover, spatial versions of population models similar to the logistic map give theoretical evidence that chaos may reduce the chance of extinction. Allen et al. (1993) constructed a metapopulation model (Section 15.3) in which each population in a patch was subject to global and local random perturbations . Global perturbations simulated such events as region-wide weather events ("bad" years). Local perturbations were small-scale events that affected patches independently of each other. Migration occurred among patches. Allen et al. (1993) reasoned that in the absence of spatial structure (i.e., a single population in a patch), global perturbations would, of course, act on all individuals simultaneously. If a bad year occurred at the same time at which internal chaotic dynamics had driven the population to low numbers, then extinction would, indeed, be very likely. However, if the populations were in isolated patches, their complex dynamics would not be synchronized and a bad weather year would adversely affect only that subset of populations that happened to have low numbers

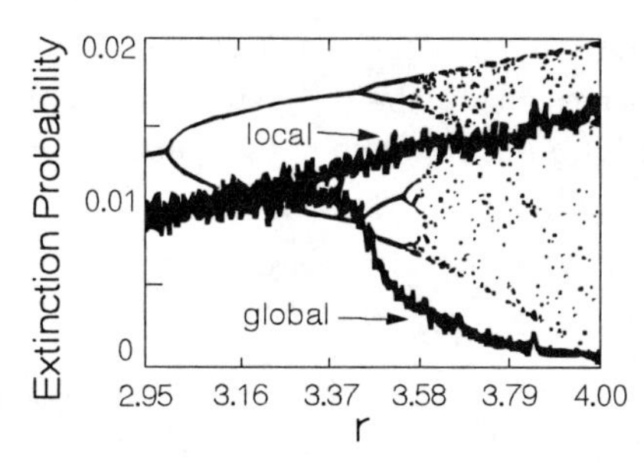

Figure 17.22: The proportion of subpopulations in a metapopulation going extinct (heavy lines) as a function of reproductive rate (r). The upper branch corresponds to the average extinction probability of local populations. The bottom branch shows the extinction probability of the species (metapopulation) as a whole. The background contains the bifurcation diagram for the non-spatial logistic map showing the increasing oscillations produced by large reproductive rate. (From Allen et al. 1993, Fig. 2e. © 1993 by Macmillan Magazines Limited. Reprinted from *Nature* with permission of the author and publisher.)

due to their chaotic dynamics. Other populations would have large numbers, would not be driven extinct, and could then colonize the patches at which extinctions had occurred. In short, chaos would desynchronize the local populations so that the metapopulation is in some sense more *adaptable* to environmental stochasticity (Conrad 1986). This idea has also been suggested to explain chaos in individual cardio- and neurophysiological systems: changing external and internal environments requires rapid change and adaptation that is more easily effected if variability exists.

To test the adaptability concept in ecological systems, Allen et al. (1993) constructed a simulation model with a variety of conditions for global and local perturbations. A representative result illustrates the potential advantages of chaos (Fig. 17.22). When the probability of a global perturbation is 0.05, then as the probability of local perturbations increases from 0.0 to 0.01, the species extinction probability gets smaller as the dynamics become more complex.

None of the arguments and studies discussed permit a clear and convincing answer to the question: Why is there chaos in biology? As noted, there is still disagreement as to the frequency of its occurrence. We will need more careful studies under natural conditions. While still important at this point, it is less convincing to show that applying an arbitrary external driving force to biological material will induce complex dynamics. It is important, however, to demonstrate that complex dynamics are produced by naturally occurring exogenous forces (e.g., seasonal weather) and extreme parameter values in the absence of external forcing functions. In addition, the adaptation of interacting biological subsystems (be they chemical transformation pathways, organ systems, or populations) involves responses to complex connections within the entire system. In ecology, the study of the evolution of chaos must include not only spatial heterogeneity, but complex evolution within the ecosystem and its complicated articulation of interacting components (Ellner and Turchin 1995). This is a fruitful area for future research.

17.9 Exercises

1. The Allee effect was discussed in Section 4.3. Its effect can be modeled as a negative per capita growth rate when the population level falls below a threshold. Draw the one-dimensional map (Fig. 17.1b) for this situation and follow the dynamics for a wide range of starting values.
2. Derive Eq. 17.9 beginning with the definition:
$$d_n \equiv |f(n, x_0 + \epsilon) - f(n, x_o)|$$
and Eq. 17.8.
3. Write a program or use simulation software to study the Vance model (Eqs. 17.4). Verify that the stability diagram is accurate. What happens if ϵ and b are pushed beyond 0.01? How does the diagram change if other parameters are varied?
4. The heart rate data of Fig. 17.8 were used in a friendly contest to test and compare different methods to identify pattern in complex time series (Weigend and Gershenfeld 1994). These data can be obtained from the following anonymous ftp Internet site:
`ftp.santafe.edu` in the directory
`pub\Time-Series\competition`.
Download the two files of human physiological data (`B1.dat` and `B2.dat`). Try a few of the techniques described above. Examine the other two variables present in the data: respiration rates and blood oxygen saturation levels. Try plotting the data in a three-dimensional phase space consisting of y_t, y_{t-1}, and y_{t-2}. Use plotting software to fit qualitatively the time series to AR(1) and AR(2) processes.
5. Draw a Forrester diagram of the *Tribolium* model (Eqs. 17.13 – 17.15). Use an auxiliary variable to represent the probability of surviving cannibalism.
6. Generate a bifurcation diagram for Eqs. 17.11–17.12.
7. As Allen et al. (1994) did for ecological systems, construct a model to test the hypothesis that aperiodic dynamics in physiological systems are beneficial. The model should show, for example, that periodic heart rates have a lower ability to respond to random environmental demands for blood flow than chaotic heart rates.

Chapter 18

Cellular Automata and Recursive Growth

18.1 An Analog and Digital World

WHEN we stroll across the quad, we feel we move through unbroken space that smoothly connects our beginning and ending points and that time flows continuously without interruption during our walk. When we pour water from one container to another, it is a continuous stream of water that flows. Yet, we know that water, at one level, is composed of discrete molecules. And we know that organisms reproduce discretely; each female produces an integer number of offspring or each asexually dividing cell results in exactly two cells. In the space and time scales of the quad, we are a distinct entity that moves, not an amorphous, diffuse electromagnetic field. Moreover, our neurons fire at discrete intervals, with finite recovery periods, and, more or less, in an on–off manner that prevents our observing the world at arbitrarily small time intervals. In this way, our senses and perceptions are digital; it is something else that makes us think reality is continuous. It could be reality itself that gives us this idea, even if we base our beliefs on incomplete knowledge. Indeed, many of our earlier models and techniques used a discrete representation that was justified as an abstraction to simplify our computations or analysis of what we believed to be the true, underlying continuous physical reality. But given the particulate nature of our perceptions of nature and the underlying discreteness of many biological processes, we have to ask: Which is the reality and which the abstraction: continuous or discrete – analog or digital? Given that the world includes the observers and that, to a certain extent, the world is as we observe it, the answer is probably "both."

Other chapters (e.g., Chapters 4 and 14) describe biological processes with spatial extent. Systems that are viewed as occupying space invite a discrete representation. Even when we use continuous mathematics such as PDEs to describe movements from place to place, to solve the equations we discretize space and time. Space becomes a grid of discrete points at which events occur. Moreover, not every model concerns a dynamical system; we also wish to model things such as biological shapes or organism morphology and anatomy. These too can be usefully represented as discrete: a plant is composed of discrete modules such as branches, nodes, leaves, etc. In this approach, we model an individual plant as a repetitive collection of these basic discrete building blocks. In this chapter, we will describe several disparate approaches and tools for the discrete perspective of biological modeling. One of the central concepts for this perspective is finite state automata, a mathematical object used extensively in theoretical computer science.

The systems and models we present in this chapter address some fundamental biological questions. (1) Does the spatial position of individual plants affect population-level phenomena such as species coexistence? (2) What are the causes of heart failure? (3) What biological forces are necessary to explain the broad patterns of plant evolution? All of these questions share the characteristic that discrete, recursive structures can be used to answer them.

18.2 Finite State Automata

Finite state automata (FSAs) are a family of mathematical constructs that, informally speaking, are defined by a finite set of states, an output alphabet, and rules that take the automaton from its current state to the next state. [This definition is a simplification, and the reader should consult Arbib (1965) or Hopcroft and Ullman (1969) for embellishment.] One of the states is designated the initial state from which the execution of the FSA begins.

When the machine changes state it produces a symbol; the dynamics of the machine are reflected in the sequence of symbols that it produces. The symbol might simply represent the last state of the machine, but there is no necessary connection between the value of the state and the symbol produced. Figure 18.1a shows a FSA that has three states and that produces either a, b, or c at the indicated state transition. The set of sequences of symbols can be considered to be the *language* that the FSA produces, and the state transition rules constitute the *grammar* that underlies the language.

A FSA can be either *determinant* (Fig. 18.1a) or *indeterminant* (Fig. 18.1b). A determinant FSA is one in which the state transition rules are such that each state goes from one state to only one other state. An indeterminant FSA has at least one state that can become one of several possible states. In this case, the transition rules must have a mechanism

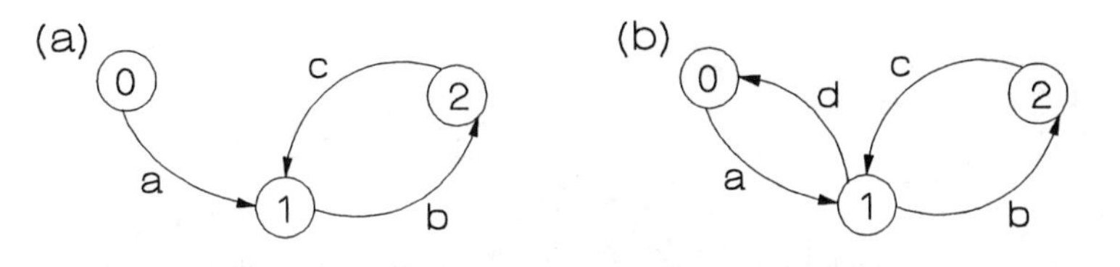

Figure 18.1: Two state transition diagrams. (a) A three-state determinant FSA and a sample of the output language. (b) A three-state indeterminant FSA and a sample of its output.

for choosing one of the alternatives. This is usually done randomly. Rules can be stated as a look-up table or as an equation in which the next state is computed from the current one (Fig. 18.2).

18.3 Cellular Automata

A cellular automaton (CA) is a spatially explicit form of a FSA. A set of cells are defined in a space; each cell is a FSA whose transition function depends on the cell's own state and those of neighboring cells. Typically, the symbolic output of CAs is the state of each cell. Later, we will discuss *L-systems*, which are another special case of a CA-like construct that has symbolic output used to describe the growth of biological structures (e.g., plants).

Consider the following simple example. We define the space to be a one-dimensional sequence of squares. Each square represents a FSA that has two states (0, 1), and the transition rules depend only on the immediate left and right neighbors of cell. The transition rule is very easy to state verbally: if the middle cell is in state 0 and has exactly one neighbor in state 1, the middle cell state changes to 1. Otherwise, the state becomes (or remains) 0. Figure 18.3 shows the spatial pattern (horizontal) that develops over time (vertical) when the rules are applied to each cell in the space. Recall that all changes in state are done "in parallel," so that the previous state of neighbors is used, not that resulting from the current changes.

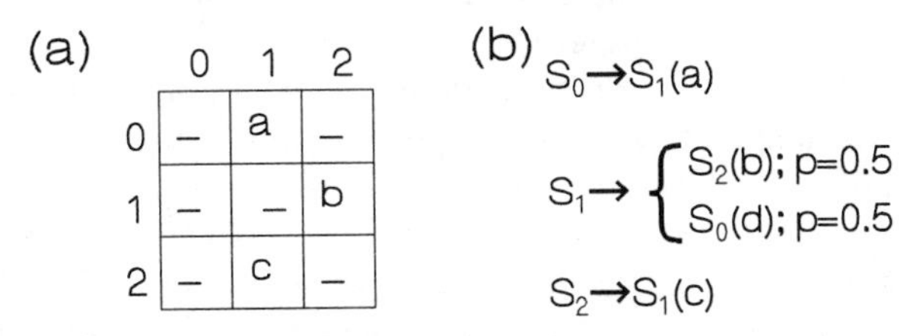

Figure 18.2: FSA transition rules stated as a table (a) for a determinant FSA, where rows are the current states; columns are the subsequent state, and table elements are the output symbols, and as probabilistic rules (b), where $S_1 \longrightarrow S_2(b); p = 0.5$ means "change state 1 to state 2 and output b with probability 0.5."

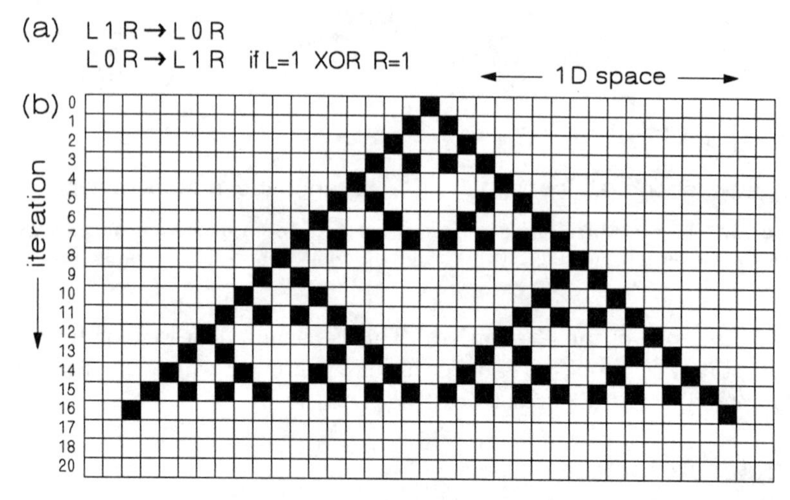

Figure 18.3: A simple example of a one-dimensional cellular automaton. (a) Transition rules for each cell (middle symbol on left-hand side) based on its state $(0, 1)$ and the states of its left (L) and right (R) neighbors. 'XOR' is logical exclusive OR; it is false if L=1 AND R=1. The first rule says that a cell in state 1 will become 0, regardless of its neighbors. The second rule states that a cell will change from 0 to 1 only if it has exactly one neighbor. If neither rule applies, the cell does not change state. (b) The spatial pattern (a row) produced over time (rows moving downward). A black cell is in state 1, and an open cell is 0.

CAs can be defined over a space with any number of dimensions and with any geometrical relationships between neighbors. Typical applications use two dimensions and define the cells on a rectangular lattice. Other lattice arrangements are possible, for example, equilateral triangles and hexagons. The number of neighbors to use can be made a property of the model in two senses. First, the model must specify if diagonal neighbors are to be included. Thus, in a rectangular grid there may be four or eight contiguous neighbors, depending on the definition. Hexagonal grids do not have this problem, but triangular grids do. Second, the model must specify if non-contiguous "neighbors" are to influence the transition functions.

In real CAs coded in computers, the size of the lattice is finite, and this creates the problem of dealing with the ends of the lattice. The last cell at each end is missing one of its neighbors (Fig. 18.3b). This is the same problem as boundary conditions in PDEs. The possible solutions include (1) make a buffer (i.e., an edge of one cell around the edge of the lattice) that has a fixed state, (2) create a special rule for the edge cells that uses a single neighbor, or (3) connect the edges together to form a surface without boundaries. In the latter solution, this converts a 1D lattice into a circle. A 2D lattice becomes a torus (i.e., a doughnut-shaped surface). See the Exercises for further details.

Another modeling consideration is the order of updating the cells. Two choices are possible: (1) change each cell immediately after its new state is computed (*asynchronous* updating), or (2) save the new state of each cell until all cells have been computed, then change all old states to new states in one operation (*synchronous* updating). The latter approach is analogous to using the entire system condition at the current time step to evaluate the state at the next time step. The advantage is that the latter method eliminates the effects of the order in which updating is done on the lattice. It does not matter where in the lattice the transition algorithm begins. The disadvantage is that one loses the "parallel" nature of state transitions in spatially distributed systems. The latter method has an implicit assumption that the time step is short compared to the time scale of the processes being simulated.

No discussion of CAs in biology would be complete without mention of John Conway's remarkable game "Life" (Gardner 1970, 1971). This is a two-dimensional CA with deceptively simple rules that produces complex and interesting behavior. A cell is "alive" if its state is 1; a dead cell has a state of 0. The transition rules are inspired by simple notions of competition and mating in real organisms as these are influenced by the number of neighbors surrounding a target cell. The density of neighbors is the number of the eight contiguous cells that are "alive." If a cell is "isolated" with one or zero living neighbors, then it dies. If a cell has exactly two living neighbors, it stays in its current state, either dead or alive. If a cell has exactly three living neighbors, then it becomes alive (either by birth or by not dying). If a cell is crowded with four or more living neighbors, it dies or stays dead.

These simple rules permit a great variety of patterns. For example, if the spatial pattern is three contiguous horizontal living cells at time t, then it changes to three vertical living cells. The vertical pattern flips back to a horizontal row of three cells, and this continues indefinitely as a limit cycle. If the three cells form three-fourths of a square, then the fourth cell becomes alive to form a square of four living cells, which do not change in further iterations: an equilibrium. All other different arrangements of three cells go extinct. Other dynamics are possible, for example, "gliders" can be created that simply translate across the lattice. Interactions between patterns can evolve. One famous example is the "eaters" and the "glider gun." The gun shoots gliders at the eater, which devours them and returns eventually to its former configuration in time to catch another glider shot from the gun. This complex "predator–prey" interaction continues indefinitely.

Finally, like all spatially explicit models, there are serious problems of visualizing and summarizing model results. The usual approach is to show the reader many interesting snapshots of the system over time. More advanced treatments attempt to characterize the statistical properties of the system by calculating a measure of entropy or spatial power spectra. For a more advanced treatment, consult Toffoli and Margolus (1987), Casti

(1992), or Langton (1992).

18.4 Applications in Biology

While the game "Life" is simple and fun, it is only a ghost of real systems. But the reader should not conclude that CAs are only video games of blinking monitor pixels. CAs are serious tools for spatial processes. Below, we examine two examples, one dealing with the ecological interactions of plants and the other modeling the spread of electrical voltage across the surface of the vertebrate heart.

18.4.1 Plant Competition

Silvertown et al. (1992) constructed a CA model of spatial competition among five species of grass in the United Kingdom. This model elegantly demonstrated that spatial configuration significantly affects competitive outcomes. The grasses involved were *Agrostis stolonifera* (A), *Holcus lanatus* (H), *Cynosurus cristatus* (C), *Poa trivialis* (P), and *Lolium perenne* (L). A 40×40 lattice was used in which each grid point (cell) could contain one of the five species. Thus, the grid spacing was approximately the size of one plant.

Each cell was updated based on the number and species of its four immediate neighbors (N, S, E, W) and random chance. Using data from a field study, Silvertown et al. (1992) determined the probabilities of replacement of a species in a cell by a neighbor crossing one of the four neighboring grid faces. Table 18.1 lists the probabilities, assuming all four neighbors belong to the species listed in the rows. To determine the new state of the cell, the number of neighboring cells occupied by each species was counted, and the replacement probabilities (Table 18.1) were weighted accordingly. Thus, if *Agrostis* was the resident species and had three neighbors of *Holcus* and one of *Poa*, the probability that *Agrostis* was replaced by *Holcus* would be $0.08 \times (3/4)$, and by *Poa* would be $0.06 \times (1/4)$. Otherwise, the cell remained as *Agrostis*. A call to a random number generator determined which of these three outcomes occurred (see Section 9.4.2). Based on the probabilities, *Agrostis* is both an aggressive invader and resistant to invasion. So, we would expect that equilibrium plant communities would be dominated by this species.

Silvertown et al. (1992) examined the effects of spatial configuration by simulating the dynamics of the cell states beginning with three different initial spatial relationships among the five species. The initial pattern at time $t = 0$, and those at two later times ($t = 100, t = 300$) are shown in Fig. 18.4. Also shown are the numbers of individuals in each species summed over all spatial cells. The results clearly show that the spatial configuration matters. Table 18.1 indicates that *Agrostis* is only a slightly better competitor than *Holcus* (0.09 *vs* 0.08). But if *Holcus* has another

Table 18.1: Probabilities that a species in a grid cell (columns) will be replaced by a species in neighboring grid cell (rows), if all neighboring grid cells are occupied by the neighbor species.

	Resident Species in Cell				
Neighbor	L	A	H	P	C
L	1.0	0.02	0.06	0.05	0.03
A	0.23	1.0	0.09	0.32	0.37
H	0.06	0.08	1.0	0.16	0.09
P	0.44	0.06	0.06	1.0	0.11
C	0.03	0.02	0.03	0.05	1.0

good competitor (e.g., *Lolium*) on the other side, the effect of these two neighbors is to cause the early demise of *Holcus.* The top row (Fig. 18.4) shows *Holcus* being eliminated by the combined action of its neighbors, followed by *Agrostis* outcompeting *Lolium* to become the dominant species. In the middle row, however, *Holcus* lies between two weaker species and is able to remain a co-dominant in the community for a longer time. Finally, the bottom row shows that *Holcus* can, through random chance, dominate *Agrostis* for a time, even if it is adjacent to the latter, if its other neighbor is a weak competitor (*Poa*).

The lesson from these simulations is that spatially explicit models can show dramatically different transient dynamics depending on the initial configuration. This phenomenon was also illustrated by the game "Life." A related lesson in the context of plant competition studies is that *diffuse competition* (competition from many species in the same habitat) can delay or alter the outcome of competition. In those situations, the spatial configuration can be as important as the quantitative effect of one species on another.

18.4.2 Excitable Tissue

Heart Basics

Everyone knows that the heart beats and blood flows. But the specific mechanisms by which this marvelously adapted structure accomplishes blood circulation are truly remarkable. Spatially explicit models help understand not only how the actions of individual muscle packets are coordinated to produce normal beating hearts, but also how disease can interfere with these mechanisms to cause heart failure. As in population and community ecology, both continuous models (e.g., Keener 1991) and discrete models (e.g., Saxberg and Cohen 1991) are applicable. Here, as illustration, we will describe a CA model of heart beating. But first, we briefly review some of these basic mechanisms involved in heart function. More detail can be found in the standard physiology texts (e.g., Guyton 1986; Berne and Levy 1993).

In mammals, the heart is a four-chambered structure composed of two

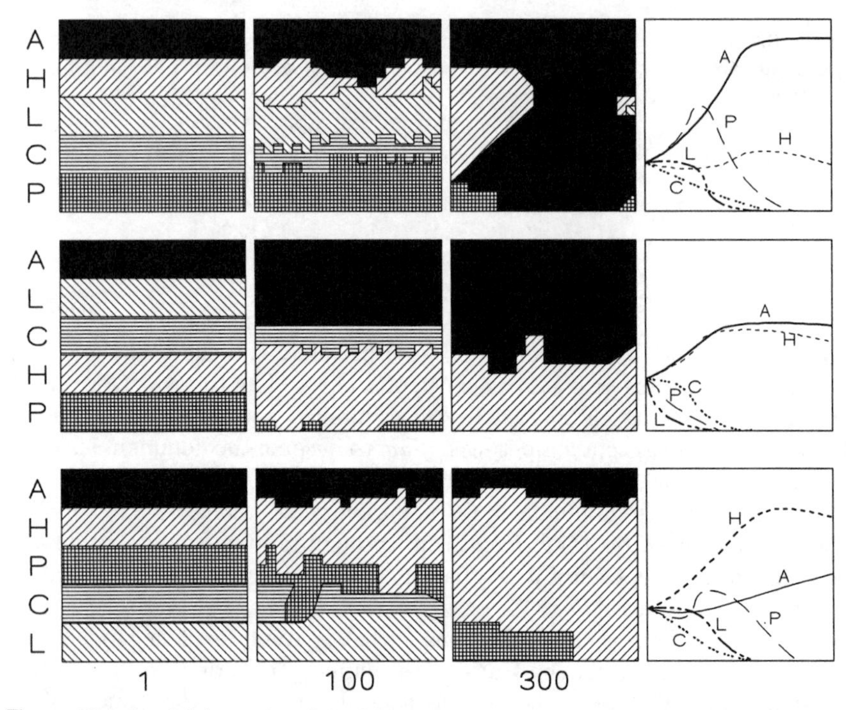

Figure 18.4: Spatial dynamics of the Silvertown cellular automaton model of plant competition. A is *Agrostis*, H is *Holcus*, P is *Poa*, C is *Cynosurus*, and L is *Lolium*. Three different initial positions of the three species are depicted in the vertical sets of panels. In each row, the spatial positions of plants are shown for $t = 1, 100, 300$. The right panel shows the numerical abundances of the species. (Redrawn from Silvertown et al. 1992, Figs. 2 and 4. © 1992 British Ecological Society. The spatial panels shown are greatly smoothed and simplified; the original color figures show much more fragmentation and intrusions at the borders.)

pairs of chambers that pump more or less in unison (Fig. 18.5). Deoxygenated blood enters the right atrium from veins, and oxygenated blood enters the left atrium from the lungs. These two chambers pump their contents to the left and right ventricles (respectively). After the ventricles have filled, the right ventricle contracts and pumps the deoxygenated blood to the lungs, and the left ventricle pumps its oxygenated blood to the rest of the body. These events are timed so that the contraction of the ventricles occurs after the atria are emptied. The contractions are produced by the rhythmic excitation of the heart's conductive system. Before discussing the specific details, we review a bit of the physiology of excitable media.

An electrical voltage is a potential for electrical charges to flow from one point to another. This is analogous to the concept of water potential that we introduced in Chapter 11. Spatial heterogeneity is implied in electrical potentials, as the concept applies only between two points that are, for some reason, electrically isolated from each other, but that have different

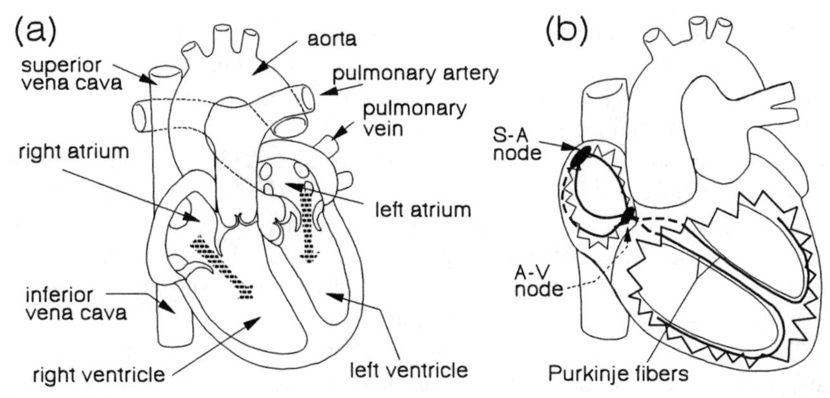

Figure 18.5: Basic anatomy of the human heart. (a) Interior view with major chambers and vessels indicated. Atrial contraction is diagrammed with blood flow indicated by shaded arrows. (b) Cut-away of an exterior view showing the location of the SA (sinoartrial) node (pacemaker) on the inside wall of the right atrium and the AV (atrioventricular) node on the interatrial septum with the Bundle of His and connecting Purkinje fibers that innervate the ventricles. Stimulus from the SA node causes atrial contraction and propagates via internodal tracts (heavy lines) to the AV node which, after a delay, causes ventricular contraction via the Purkinje fibers.

amounts of electrical charge. Think of applying a volt meter to a battery. We do not measure volts by placing both meter probes on the positive terminal or on the metallic battery case. We must put one probe on the positive terminal and the other on the negative terminal. Inside the battery, the terminals are electrically isolated from each other; outside the battery the terminals are connected by air, which, of course, does not conduct electricity. Conduction occurs only when we connect the terminals with a conductor such as the meter probes.

In a charged battery, the negative terminal has more negative charges than the positive terminal. In chemical systems, such as batteries and nerves, negative charges are electrons. Cell membranes are the barriers that separate points at different electrical potentials. As with water potential, two physical processes contribute to the *electrochemical potential* across a cell membrane. With water potential across a membrane, the processes involved are hydrostatic pressure and the relative ionic concentrations across the membrane. In electrochemical potential, the processes are ionic concentration and the electrical potentials at the two points. Electrical potential is measured relative to a fixed reference point ("ground"), just as hydrostatic pressure is measured as the "pressure head." The total electrochemical potential at a point is the sum of these two forces composed of ionic concentrations and electrical potential, just as water potential is the sum of its forces. The net flow of an ion (e.g., Na^+) between two points (e.g., across a membrane) is the difference between the two electrochemical potentials.

In excitable tissue, we are concerned with electrochemical potential across membranes. Cell membranes in animals are complex structures composed of lipids (fatty acids) and proteins. Membranes are filled with holes called *channels*. Some channels (also called "pumps") use *active transport* to move compounds through them, which requires ATP and special carrier substances embedded in the membrane. Others, which are important in excitable media, are open passages "lined" with special compounds that can close the passages at either the exterior or interior side of the membrane. One form of the latter are called "leak" channels, because they continually allow ions to leak across the membrane.

In excitable cells, a pump actively transports Na^+ to the outside of the cell and K^+ to the inside. This produces ionic concentration differences across the membrane which diffusion counteracts by moving the ions through the leak channels. The combined action of these processes results in a nonzero *resting potential*. Such membranes are then in a *polarized* state. The amount of the potential varies according to cell type, but in typical motoneurons it is −70 mV (millivolts). In typical smooth muscle such as that composing the ventricular wall of the heart, the resting potential is about −90 mV. (Just as with measuring the voltage of batteries, the sign of voltage depends on which probe is applied to which terminal; the sign of a membrane's electrical potential is, by convention, determined using the *inside* of the cell as the reference.)

When a polarized membrane is *depolarized*, ions are able to move across the membrane so as to reduce the electrochemical potential. If this process continues gradually to the point at which a tissue-specific minimum threshold potential is obtained, conformational changes in the structure of special channels occur. These changes cause an *action potential*, which is a characteristic time course of membrane potential. Figure 18.6a shows a typical action potential for the ventricle.

Hearts are special excitatory material in that they contract at regular intervals under the control of the autonomic nervous system. This regularity is generated by a system of interacting neurons and electrically conducting fibers that connect different areas of the heart. Figure 18.5b illustrates the major components. The SA (sinoatrial) node is the primary "pacemaker" that initiates the contraction sequence at the beginning of each heart beat. This collection of excitable cells depolarizes the rest of the atrium nearly instantaneously because of fast transmission of the potential wave along low-resistance pathways. The wave eventually reaches the AV (artioventricular) node that comprises excitable tissue which actually delays the wave's progress to permit the atrium to contract and fill the ventricle with blood. Once the ventricle is filled, the electrical potential at the AV node is regenerated and rapidly transmitted to the excitable tissue in the ventricles via a collection of low-resistance pathways called the *Purkinje fibers*. When this system is working properly, a coordinated set of contractions is initiated at precisely the moment the chambers of the heart are filled with blood. To function optimally two conditions must be met: (1) the major depolarizations at the SA and AV nodes must be timed to occur when the chambers are full of blood, and (2) each element of the excitable tissue in the atrial wall and the ventricular wall must be, more or less, synchronized so that coordinated contraction results in the expulsion of the blood from the chamber. If either of these two conditions is prevented from occurring, a heart beat will not occur or the contraction will not pump blood.

CA Model of Ventricular Excitation

Predicting the flow of electrical potential over the surface of the heart is an important problem because any errors in these dynamics affect normal cardiac function. This problem requires a spatially explicit approach, and standard numerical procedures to solve the appropriate PDEs have been applied (Smith and Cohen 1984). These models, while incorporating the details of heart muscular and neurological structure, are computationally intensive and, for some model objectives, may contain more detail than necessary. Being based on PDEs, they assume that conduction is spatially continuous across the heart, but evidence indicates it may actually be a discrete, discontinuous process. Modifications of PDE models can account for this phenomenon, however (Keener 1991).

To incorporate spatial discontinuity in conduction and to minimize computational requirements, Mitchell et al. (1992) created a cellular automaton model in which each grid cell of the automaton represents a homogeneous group of excitable units. Each grid cell is in one of four possible states that define the cardiac action potential: *quiescent*, *excited*, *absolute refractory*, and *relative refractory*. Figure 18.6a shows the relation of the four states to the ventricular action potential. Each grid cell is connected to its eight neighbors with which it interacts either by exciting its neighbors or being excited by its neighbors.

The rules of state change are as follows. (1) If a grid cell is Q, then it becomes E if any neighbor is E. (2) An excited cell remains in state E for EP (excited period) msec, at which point it becomes A. (3) A cell remains in A for $AP_{ij} = RP_{ij} - RRP$ seconds, where RP_{ij} = refractory period (msec) of cell (i, j), and RRP = relative refractory period (msec). At the end of AP_{ij} msec, the cell becomes R. (4) A cell remains R for RRP msec when it becomes Q, unless a sufficient number of excited E neighbors transform it to E. The number of neighbors required for this change of state depends on the amount of time the cell has been in the R state (Fig. 18.6b).

Only the ventricle is modeled. Therefore, the behavior of the SA node is ignored, and the AV node is represented abstractly as a periodic stimulation from an external driving function. The surface of the ventricle is assumed to be a cylinder created from a 50×50 matrix. The time-dependent pulses of stimulation from the atrium appear at a single point on the upper rim

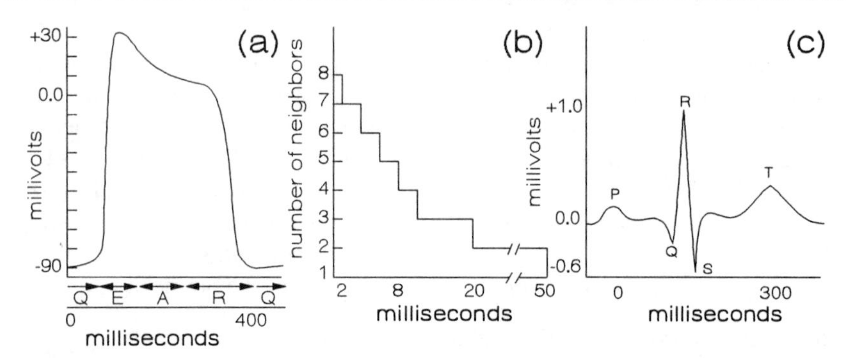

Figure 18.6: CA model of neuron states during heart excitation. (a) The four states of a CA grid cell in relation to a typical cardiac action potential. Q=quiescent, E=excited, A=absolute refractory, R=relative refractory. (b) The number of excited neighbors required to excite a grid cell in the relative refractory state. Time axis indicates elapsed time since the cell was excited. (c) Typical electrocardiogram (ECG) for a normal human.

of the cylinder (roughly equivalent to the AV node). Recent research has shown that the total refractory period (RP_{ij}+ RRP) is not identical over the surface of the ventricle. This is modeled by assigning random RP_{ij} values to grid cells at the beginning of each simulation run. The values are drawn from a normal distribution with mean MP and standard deviation SD. The parameter definitions and values are shown in Table 18.2.

Model output is presented as sequences of states of the grid cells in the matrix. This is not only cumbersome, but the states of heart grid cells are not available for most human patients. Heart dynamics in living subjects are obtained as electrocardiographs (ECG). A typical ECG is shown in Fig. 18.6c with the three components (P, QRS, and T) labeled. The model simulates the ECG based on the current electrical states of the CA (see Mitchell et al. 1992 for details). This derived output is used for validation.

The dynamics evolve on the surface of the ventricle modeled as a cylinder as shown in Fig. 18.7. Initially, all elements are in state Q. With period

Table 18.2: Definitions and values of CA variables and parameters.

VARIABLES

Variable	Definition
Q	Quiescent state
E	Excited state
A	Absolute refractory state
R	Relative refractory state

PARAMETERS

EP	Excited period (10 msec)
AP	Absolute refractory period (msec)
RRP	Relative refractory period (50 msec)
RP	Refractory period, msec deviate of N(MR,SD)
MR	Mean refractory period (250 msec)
SD	Standard deviation of refractory period (70 msec)

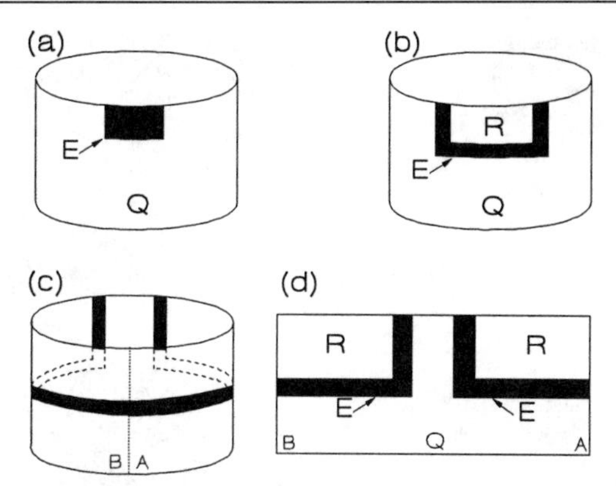

Figure 18.7: Spatial dynamics of the CA model of ventricular conduction when the ventricle is modeled as a cylinder. Three time periods are depicted in (a) – (c) showing the spread of the ventricle elements in the excited state (E) through the quiescent state (Q). Behind the E front are refractory elements (R). For ease of presentation, in (d) the cylinder is bisected along the vertical dotted line and shown as a plane.

SP a stimulation signal is simulated as arriving from the atria as a group of excited elements on the top border of the cylinder (Fig. 18.7a). The E elements excite their neighboring Q elements and a front of E elements spreads down and outward (Fig. 18.7b,c). Since an element remains as E for only *EP* msec, E elements behind the front quickly change to the absolute refractory state A. These elements then become relative refractory, but for visualization we lump the two refractory states together (Fig. 18.7b,c). Also for presentation ease, we unfold the cylinder and present the spatial dispersion of states on a plane (Fig. 18.7d).

The left set of panels in Figure 18.8 shows normal ventricle conduction during three simulated times. Shortly following the top panel, the wave fronts of E states meet in the center of the square and fuse, producing a horizontal wave front that moves downward. After it reaches the bottom of the cylinder, all of the ventricular elements are in the refractory state. They go into the quiescent state asynchronously in the middle panel because the duration of the refractory period is randomly assigned to each element. In these simulations, all of the elements go completely to the quiescent state (Q) before another stimulation arrives at the top. The bottom panel shows the E wave front several tenths of a second after the second stimulation. Since all elements are Q before the arrival of another stimulus, this pattern is repeated consistently and regularly. The ECG trace below the panels also shows reasonably normal heart behavior. The T wave is inverted from typical ECGs (Fig. 18.6c) because the model incorporates only heart surface phenomena (Mitchell et al. 1992). The P wave is missing because the atrium is not modeled. For comparisons, actual ECG traces from normal and abnormal subjects are shown in Fig. 18.9.

In addition to these simulations of normal behavior, the model is able to simulate various pathological conditions. Mitchell et al. (1992) described

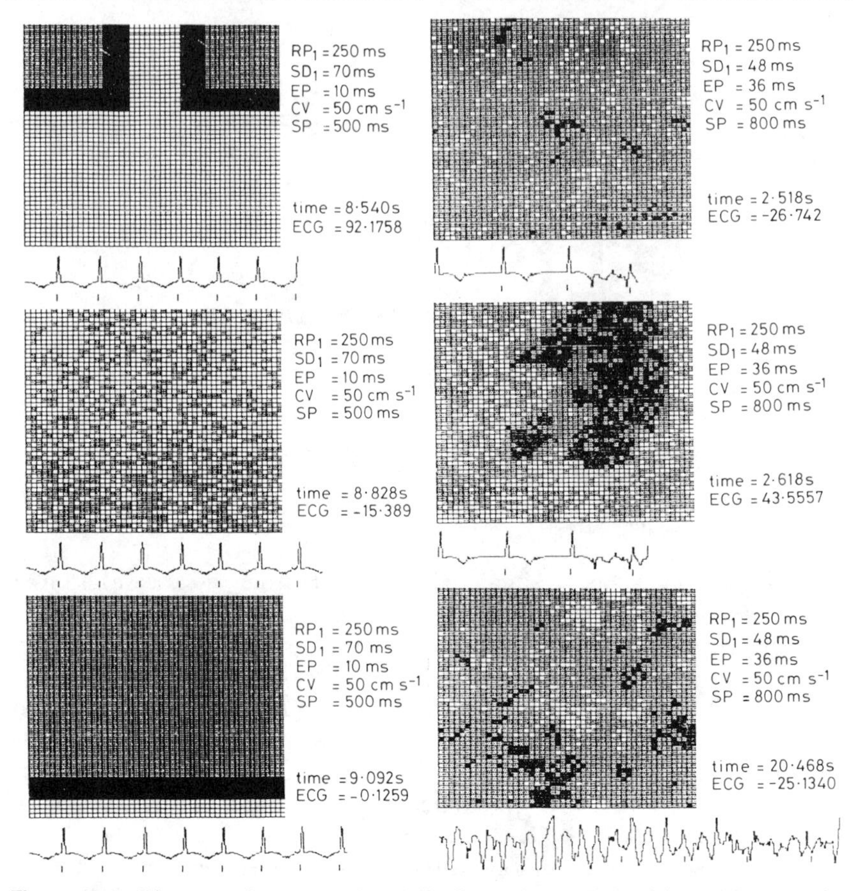

Figure 18.8: CA output for a normal and fibrillating heart. Left: CA model output for a normal heart. From top-to-bottom, $t = 0.0$, $t = 0.29$, and $t = 0.55$ sec since atrial stimulation. Below each panel is the ECG over several stimulation episodes based on the spatial patterns shown. The picture is the CA state corresponding to the ECG at the extreme right of its time trace. The bottom panel shows the state following an intervening stimulus event after the middle panel. The parameter values are shown to the right of the figure. Right: CA model output for a fibrillating heart. From top to bottom, $t = 0$, $t = 0.10$, and $t = 17.95$ sec since simulation initiation. (From Mitchell et al. 1992, Figs. 3 and 8. © 1992 Peter Peregrinus, Ltd.)

the results of pushing the system to unstable behavior by decreasing the period of stimulation events. This simulates abnormal electrical behavior in the atria and results in a variety of conduction blocks in the ventricle. The blocks are produced as the stimulation period is decreased because the heart elements do not have adequate time before the next stimulation to recover to a quiescent state. As the new wave passes over the heart, some of the elements encountered by the wave are in a refractory state and not excitable. This creates islands of unexcitable material that disrupt the synchrony of the heart elements and would prevent heart contraction. At

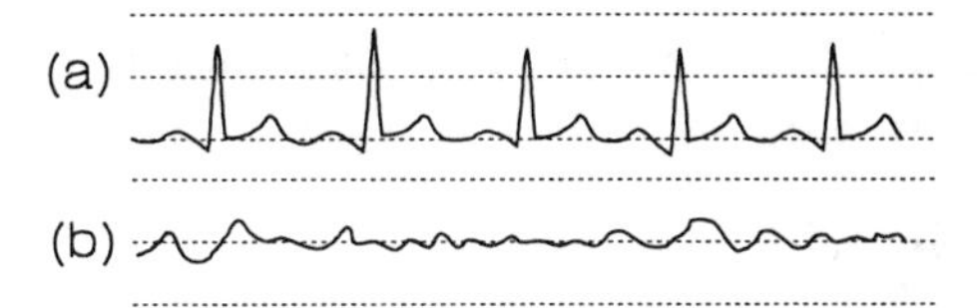

Figure 18.9: Typical human ECGs for (a) a normal heart and (b) one in ventricular fibrillation.

sufficiently short stimulation periods (e.g., $SP = 170$ msec), this imbalance of stimulation and recovery results in a 2:1 conduction block and *electrical alternans* in which every other contraction is skipped. With different stimulation periods, the model can produce ECGs other rhythm abnormalities (e.g., a 4:4 alternans which is a set of four repeating beats each with a different QRS signature in the ECG).

A more interesting case is ventricular fibrillation, a pathological condition described above. Previous models have shown that fibrillation can occur if conduction times are long, stimulation periods short, or refractory periods short. A better test of the model is to determine if fibrillation will arise and be maintained with reasonable action potential parameters and periodic stimulation rates. This model shows that fibrillation can be induced simply by the discontinuous nature of the conducting elements and the spatially inhomogeneous dispersion of refractory periods. A fibrillation episode is triggered when a small number of central heart elements (not those associated with the AV node) are externally stimulated. In the CA model, fibrillation is induced by stimulating a few of the central elements during the relative refractory state (R) of a normal heart beat.

When normal parameters are used and the heart model is stimulated in this way a persistent fibrillation episode is induced. The resultant spatial pattern of element states is shown in the panels on the right of Fig. 18.8. Note the islands of E states that propagate over the ventricle surface according to the pattern of Q and late refractory stages (R states). Note also how the normal rhythm of the ECG is disrupted early in the sequence (compare with the panels on the left of Fig. 18.9). After about 15 sec into the fibrillation episode (bottom panel), the ECG is completely irregular and shows potential fluctuations much faster than atrial stimulation. There is no coordinated wave of E states that produces a coherent contraction. This heart would be pumping essentially no blood to the brain, and death would quickly result.

In conclusion, this model is an interesting example of CA models that combines the finite state approach with the ability to study time lag effects by permitting CA elements to remain in a state for fixed time periods. Although it does not represent the complex physics of an accurate three-dimensional model, it preserves the essence of the electrical phenomena. As a result, it captures much of the qualitative behavior. Although one would not want to use such a model for the design of artificial hearts used in human patients, it is nevertheless a valuable tool for understanding a complex, spatially extended system.

18.5 Recursive Growth

A final biological example comes from a hybrid of FSAs and CAs. Like the latter, the problem is spatial: the shape of growing plants. Like FSAs, it is based on a finite state machine with symbolic output. Aristid Lindemayer (1968a, b) invented *L-systems* (hence the name) to describe the morphological development of simple organisms in space. L-systems are not finite state machines as defined above, but are represented as a *rewriting system* or *grammar*. L-systems are related to FSAs because FSAs can be shown to be equivalent to a form of grammars.

18.5.1 Definition of L-Systems

A *rewriting system* is a formal construct (or algorithm) the output of which is a string in a formal language (Chomsky 1957). There is a deep, mathematical connection between rewriting systems and finite state automata. Rewrite rules and output strings can take many forms, such as rules of grammar that produce sentences composed of English words. Lindenmayer had the great insight to represent biological structure and morphology as symbols distributed in space. Unlike other biological grammars (e.g., Haefner 1975; Dale 1980), L-systems have no nonterminal alphabet, and thus bear a strong resemblance to cellular automata. The grammar is considered to be "parallel" because each symbol is rewritten in one pass through the current structure, as is the case with asynchronous CAs. The system outputs a linear string, which by the definitions of its symbols, defines complex biological morphology.

Lindenmayer defined a hierarchy of grammars based on the complexity of the rules and on the complexity of strings that could be produced. The simplest grammar is a *D0L* grammar signifying a deterministic rewrite system in which the symbol produced for a given spatial cell depends only on the current state of the cell and not on the states of any neighbors. In other words, this grammar assumes no interactions between cells in a developing structure. Biologically, this is an overly simplified assumption, but it serves as a baseline.

18.5.2 Plant Shape as an L-System

Figure 18.10 is a simple example of a grammar that describes the margins of leaves. Since leaves are typically bilaterally symmetrical, the grammar produces strings that are symmetric around the uppermost apex. The rewriting rules capture this feature by expanding around the symbol at the midpoint of the string which represents the uppermost apex of the leaf.

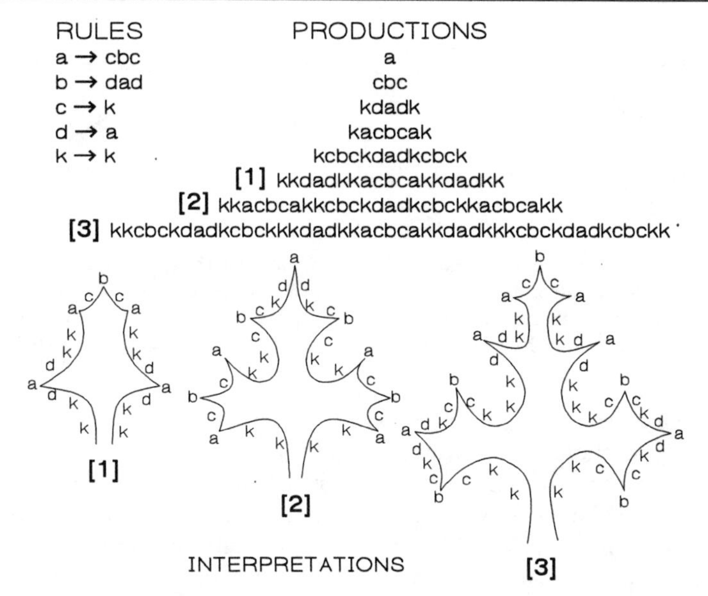

Figure 18.10: A simple D0L grammar for leaf margins. The recursive rewrite rules are listed in the upper left; the initial symbol is the letter "a." A sequence of productions is shown in the upper right. The last three productions listed are interpreted as specific leaf shapes. The symbols have the following interpretations: a and b are sharp tips (apex); c and d are lateral margins of lobes, and k is a notch between lobes. (Redrawn from Lindenmayer 1975, Fig. 1. © 1975 Academic Press, Ltd.)

By its recursive nature, the grammar produces leaves that are lobes within lobes within lobes. The grammar is nonterminating in the sense that no production ever gets to the state of all "k," so there are always symbols that can be rewritten, and the leaf grows indefinitely. The size of leaves can be incorporated by assigning to each symbol a distance along the leaf margin.

The concept of a parallel rewriting system such as that used for leaf margins can be generalized to any spatially distributed structure that is recursively generated in time. Another important example is the growth of branches, limbs, and twigs in the development of vascular plants. Early models of this problem were developed by Lindenmayer (1968b) and Hogeweg and Hesper (1974). Figure 18.11 illustrates the basic concept. Symbols represent cytological states related to the timing of cell divisions. Branching is modeled as a recursive process in which branches are hierarchically composed of nested branches. In the grammar, the nested nature of the branches is denoted by nested braces (e.g., "[⋯ [⋯] ⋯]"). Square brackets (i.e., [ ]) indicate a branch to the right of the stem, and parentheses [i.e., "( )"] denote a left branch.

This simple grammar produces only two-dimensional structures and, as with the leaf margin model, is not able to describe metric properties of growth forms. For example, the modeling approach cannot vary the angle

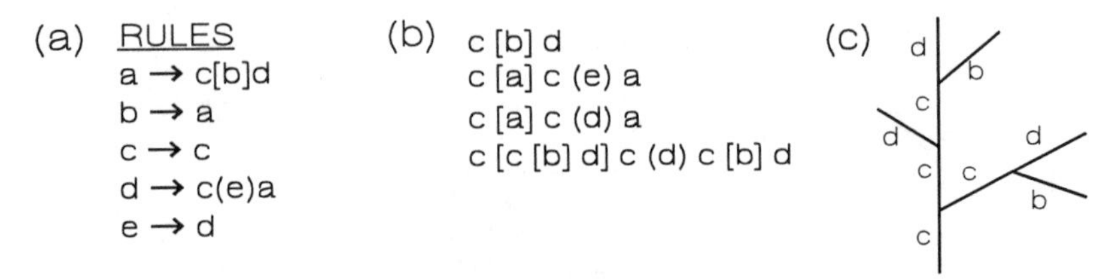

Figure 18.11: A Lindenmayer grammar of branching growth. (a) Rewrite rules, (b) several strings that result from use of the rules, (c) the interpretation of the last production in (b). Symbols represent cytological states. Square brackets indicate a branch forming to the right; parentheses indicate left branches. Brackets and parentheses may be nested.

of branching or the length of branches. Extensions to the basic formal language approach, however, provide strikingly accurate graphical simulations of a wide variety of plant forms (Prusinkiewicz and Lindenmayer 1990). This accuracy has, however, been achieved at the expense of the simplicity of the formalism. While still using L-systems, the new approach uses continuous parameters, and, consequently, follows more closely other recursive plant models not tied to a grammar perspective (Honda 1971, Fisher 1992). The best formalism to use apparently depends on the requirements for high-resolution graphical display, the complexity of the morphology simulated, and the aesthetic beliefs of the modeler (Aono and Kunii 1984). Jaeger and de Reffye (1992) give an overview of computational methods appropriate to the nongrammatical approach.

18.5.3 Plant Evolution

While the ability to capture broad, qualitative features of leaf gross morphology is interesting, the rules are, nonetheless, nothing more than formal descriptions. We have to ask, therefore, if any of these modeling approaches can provide deep insight into questions of plant evolution and design or into the ecology of plant communities.

Happily, we can indicate several applications that give us new understanding of the evolution of optimal plant design. For example, Honda and Fisher (1978), using a recursive method, simulated individual tropical trees having the basic morphology of *Terminalia catappa.* This is a tall tree having a canopy composed of horizontal tiers of three to five lateral branches. Its morphology is typical of upper canopy trees in the tropics where competition for light is intense. A reasonable prediction, then, is that the morphological parameters (e.g., number of branches per tier and branching angles) will have evolved to maximize the effective light gathering (leaf) area of the tree. The model should show maximum leaf area using parameters from real trees. Using deterministic simulations over a range of plausible parameters, Honda and Fisher (1978) and Fisher and Honda (1979) found statistically significant agreement between the model parameters that produce maximum leaf area and the parameters of many

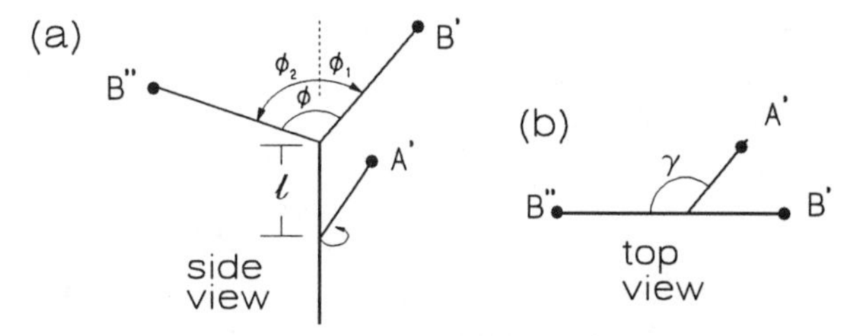

Figure 18.12: Definitions of parameters controlling plant growth and branching. (Adapted from Niklas and Kerchner 1984, Fig. 2. © 1984 Paleontological Society.)

species of tropical trees. This provides some support to the idea that trees of this form have evolved an optimal structure. The model was crucial in the argument as a tool to generate alternative trees based on suboptimal parameters.

Karl Niklas and colleagues (Niklas 1986a, 1986b, 1992, Niklas and Kerchner 1984) took a more elaborate approach to early vascular plant evolution which included multiple evolutionary constraints and interspecific competition. They used a stochastic, recursive growth function to grow leafless trees in which photosynthesis occurs in the axes (i.e., leafless stems). Four parameters determine the shape of the plant (see Fig. 18.12). (1) The bifurcation angle (ϕ) is the angle between branches that arise from a "mother axis." ϕ is composed of two subangles (ϕ_1, ϕ_2), one for each branch as measured from the angle of the mother branch. (2) The rotation angle (γ) of a bifurcation is the angle between two planes, one formed by the current bifurcation and the second formed by the previous bifurcation (Fig. 18.12). (3) The length of branch growth elements (l). (4) The probability of a bifurcation following a unit branch growth. In the simulations reported in Niklas and Kerchner (1984) and Niklas (1986b), l is fixed at 1.0 in arbitrary units, and γ and ϕ are varied systematically. The sizes of plants were limited by the maximum number of bifurcations ($N = 10$) that were allowed. By simple observation of existing plants, it is apparent that branchiness increases vertically from the ground to the terminus. Niklas incorporated this fact by forcing the probability of bifurcation to increase linearly as the plant grows. Plants are grown from an initial segment of stem. Either growth or a bifurcation to a branch occurs according to the parameters. Each of the two branches thus produced can either grow or bifurcate. The process continues in this recursive manner until the maximum levels of bifurcation (N) is achieved.

The three manipulated parameters (p, ϕ, γ) define a three-dimensional space that characterizes the shape of all possible plants. The computer algorithm generates random forms using specified parameter values. At each combination of parameters, the plant's photosynthetic efficiency and its resistance to mechanical damage are calculated from the randomly generated forms. Photosynthetic efficiency increases with plant photosynthetic surface area, but decreases due to self-shading. If two plant forms have identical total surface areas (A), self-shading will be greatest in that form that possesses the smallest area projected from the angle of the sun (θ) on to the ground (where A_p is projected area). Furthermore, for two forms with identical projected areas, that form with the largest total surface area will be the least efficient since it produces photosynthetic material that is shaded. So, photosynthetic efficiency (I_θ) at a given solar angle θ is large if projected area is large and decreases as total area decreases. Thus, $A_p(\theta)/A$, the ratio of projected to total area, is an index of photosynthetic efficiency at solar angle θ. Total photosynthetic efficiency (I) is I_θ times solar irradiance (S_p) integrated over all solar angles during a day. Or, when angles are measured in degrees,

$$I = \int_0^{90} \frac{A_p(\theta)}{A} S_p(\theta) d\theta. \tag{18.1}$$

In computer simulations, S_p is assumed to be independent of θ and set arbitrarily to 1.0. θ is varied in fixed, discrete intervals. The ratio of projected to total area at given θ is a complex function of θ, γ, and branch diameter and length. The reader is referred to Niklas and Kerchner (1984) for details of spherical trigonometry.

The second evolutionary constraint that simulated plants must satisfy is the ability to stand up under their own weight. A branch growing at some angle from vertical (e.g., in Fig. 18.12 $\phi \neq 0$) experiences compression and tensile stresses which cause the branch to bend and, ultimately, break. The *bending moment* (M) measures the tendency of a branch to bend. In early plants, M depends only on the geometric placement and size of branches so that bending moment is

$$M = \frac{\pi d^2 l^2 mg}{\gamma} sin(\phi/2), \tag{18.2}$$

where m is the mass weight of a branch, d and l are the diameter and length of the branch (respectively), g is the constant of gravitational acceleration, and ϕ is the branching angle.

The overall fitness of a generated plant is a function of photosynthetic efficiency and bending moment. Early plant species did not possess specialized tissue that reduces bending moments at horizontal angles, such as is present in modern plants. The presence of this tissue negates the importance of purely geometrical placing of branches (i.e., angles and rotations). For species with this tissue, fitness (f) is best represented by photosynthetic efficiency: $f = I$. For species lacking the reinforcing tissue, an index of fitness is the ratio of photosynthetic efficiency and bending moment: $f = I/M$. In both cases, f is a function of the three fundamental parameters: ϕ, γ, and p. Thus, the fitness of different plant shapes can be summarized by the value of f at different points in a three-dimensional space whose axes are the three parameters.

The model is stochastic because the occurrence of a bifurcation at any particular node depends on the overall branching probability (p). p was varied from 0 (highly branched plants) to 0.9 (little branching). To determine plant fitness in the parameter space Niklas and Kerchner (1984) simulated 10 plants at each of 10 choices of p, γ, and ϕ_2. (This latter parameter was used to characterize branching angle because ϕ_1 was arbitrarily held constant to reduce the parameter dimension from four to three.) Consequently, 10,000 simulations were performed for both of the fitness functions (f) tested. The average of the 10 random trials are plotted in the three dimensional parameter space.

When $f = I/M$, the most fit forms are those with large ϕ and γ (Fig. 18.13a). The probability of branching (p) has only a slight effect on the optimum. Notice that the shapes associated with the most fit forms are not those of modern plants. The oak-like and conifer-like forms have low to intermediate fitness. Indeed, some of the most fit plants seem almost to be random structures with branches going every which way. If, however, structurally reinforcing tissue is present, so that the appropriate fitness function is $f = I$, then plants with modern shapes seem to be the most fit (Fig. 18.13b). In this case, branching frequency and maximum rotation angle are important parameters (large p and low γ have low fitness), while fitness does not change much with branching angle (ϕ_2).

These results are intriguing because they make a certain sense and were generated from an obviously simplified set of assumptions. One major assumption of the model is that all positions in the parameter space are equally likely. Real evolution, however, is a stepwise, historical process. Moreover, one of the primary mechanisms by which natural selection operates is through competition between individuals of the same and different species. Niklas (1986a) addressed these omissions by starting with primitive plants (low, with little branching) and allowing them to evolve by determining the most fit neighbor in parameter space. He did this in two different ways that produced similar evolutionary trajectories. First, he ignored interspecific competition and, from the current best parameter set, computed fitness (I or I/M) for all 26 neighboring parameter locations ($3^3 - 1$). He took as the next best morphology that parameter set which had the largest fitness. By iterating this process, he traced the optimal evolutionary trajectory through the discretized parameter space such that each evolutionary step produced increased fitness. The results are shown in Fig. 18.14 (Niklas 1986a). Similar results were obtained for both fitness measures (I, and I/M).

Niklas (1986a) repeated the exercise with a more realistic model that included competition for light and space. Primitive plants were simulated to grow in a physical space in which they shaded themselves and neighboring plants. Each plant grew according to its location in parameter space, and dispersed spores into the wind. The ability of spores to disperse to uninhabited sites was a function of the height of the parent plant and the

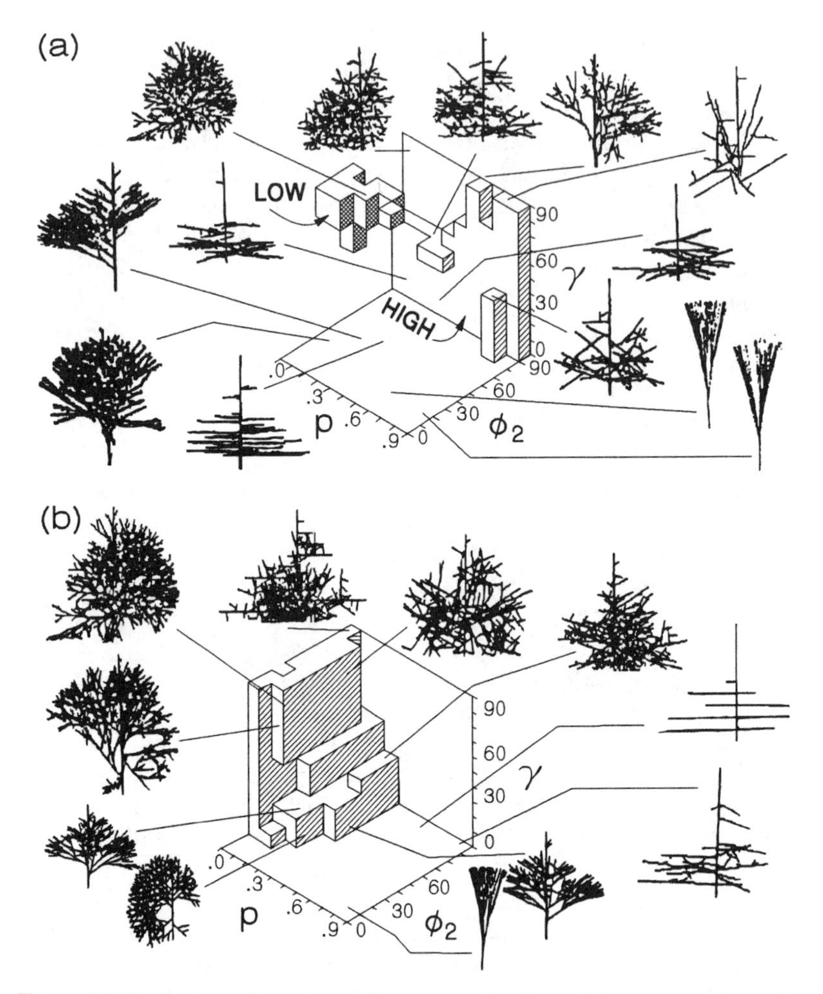

Figure 18.13: Regions of maximum fitness as a function of three parameters: branch rotation (γ, vertical direction), probability of branching (p, horizontal direction), and branch angle (ϕ_2, third direction). Representative plant shapes that are produced by various parameter values are shown. (a) The fitness function uses both photosynthetic efficiency and bending moment ($f = I/M$). The blocks of values with diagonal lines (HIGH) are parameter combinations with high fitness. The blocks of values with cross-hatching are parameters producing low fitness (LOW). (b) The fitness function uses only photosynthetic efficiency ($f = I$). The blocks of values with diagonal lines are parameter combinations with high fitness values. (From Niklas and Kerchner 1984, Fig. 13b, d. © 1984 The Paleontological Society.)

number of branch tips from which spores were emitted. Spores falling on areas shaded by any plant died. Spores that did not die, germinated and grew according to parameters slightly altered from their parents due to random mutation. This procedure was repeated for many "generations," in a fashion reminiscent of genetic algorithms (Chapter 19). Niklas (1986a)

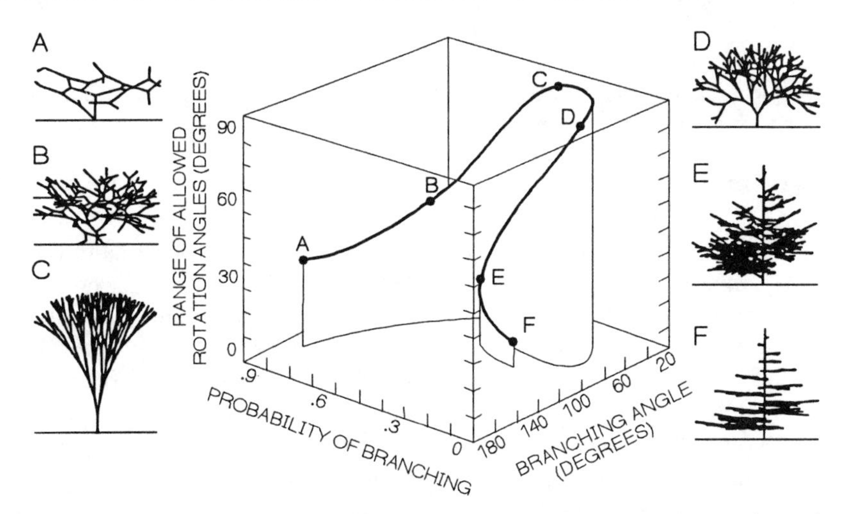

Figure 18.14: Optimal evolutionary trajectory of plant shapes. The evolution of vascular plants from a primitive form (point A) is modeled by growing plants, measuring fitness, and searching nearby points in parameter space for a more fit shape. Repeated application of this procedure produces the curve in the three-dimensional parameter space (see Fig. 18.13). Letters indicate the location of representative plant shapes. (From "Computer-simulated plant evolution" by K.J. Niklas 1986a.

found similar results in this method of simulated evolution as he did in his "direct search" of neighboring parameters. This modeling approach is interesting because it mechanistically calculates fitness from morphological parameters. The focus has shifted from system dynamics given fixed parameter values to the evolutionary dynamics of the parameters themselves. Similar models of the evolution of animal-like structures are presented in Chapter 20.

18.6 Summary

The class of models presented in this chapter differ markedly from the continuous differential equations with which we have dealt to such a large extent in this book. The CAs and recursive models described here use relatively simple and abstract rules. Consequently, statistically rigorous validation was not attempted. Moreover, the models are hard to analyze mathematically, and most of the results are obtained from computer simulation. This level of computational analysis has become possible only in the last decade because of easy access to powerful computers. As a consequence, these models require great care and attention to numerical analysis. Their strong point, however, is that most are relatively easy to formulate and, thanks to recently available computing environments and tools, easy to implement on computers. CAs, in particular, replace continuous systems (PDEs) and so it should be recognized that an explicit form of discretization is being used that may be inconsistent with physical and mathematical concepts. On the other hand, vascular plant morphology is an inherently discrete process, and discrete, recursive models are well-suited to this use.

18.7 Exercises

1. Code the Silvertown et al. (1992) cellular automaton model of plant competition, and verify their results. Examine the case where the plants are randomly distributed. Add disturbance that clears space of all species and is re-invaded probabilistically. Do different levels of disturbance support the Intermediate Disturbance Hypothesis (Connell 1978), where species richness is low at both low and high disturbance rates, but maximal at intermediate disturbance levels?
2. A torus is created from a flat surface by joining one pair of opposite edges to form a tube followed by joining the two ends to form a doughnut. Design an algorithm to do this operation on a two-dimensional rectangular lattice of CA states stored as a two-dimensional array. The algorithm should work correctly when the CA uses eight neighbor calculations. Repeat with hexagonal lattices.
3. Write the complete set of transition rules of the pattern shown in Fig. 18.3 as a look-up table. (Each row will be a triplet of 0s and 1s.)
4. For 10 iterations, simulate the CA in Fig. 18.3 when it is started with three contiguous 1s.
5. Develop the CA pattern over time for the CA shown in Fig. 18.3 with the following rule added: *If a cell in state 1 has two 0 neighbors, it remains in state 1.*
6. How do the dynamics of the game of Life change if you use a torus or absorbing barrier as the boundary condition? How to they change if you use asynchronous updating as opposed to synchronous updating?
7. Write the leaf margin production that follows [3] in Fig. 18.10.
8. Verify the nullcline equations for the stomate model. Incorporate N_g into the nullcline portrait.
9. Draw the state transition diagram for the heart CA model. Include the time lag parameters.
10. List the sequence of productions and rules used to generate the branching interpretation shown in Fig. 18.11.
11. Draw the state transition diagram for the Werner-Caswell model of teasel described in the exercises of Chapter 13.
12. Use a CA model based on Silvertown et al. (1992) to investigate the effects of arrival order of propagules on the community that develops.

Chapter 19

Evolutionary Computation

19.1 The Problem of Global Optimization

PROBLEMS to find maxima and minima (optimization) are common in biological modeling. We have already encountered them in the context of parameter estimation where we minimized the error between data and a function. Optimization also arises in models of the evolutionary process (e.g., optimal foraging) because a valuable working hypothesis when addressing questions of adaptations in organisms is that the observed traits maximize individual Darwinian fitness (i.e., reproductive contribution to future generations). A related application arises when modeling biological systems as *control systems*: systems that are able to adjust parameters in order to maintain system dynamics within some specified operating range. For example, in mammals, heart rate is increased when oxygen demand through physical exertion increases so that a constant amount of oxygen is delivered to vital organs. This can be considered to be an optimization problem since the system is "attempting" to minimize deviations of oxygen delivery rates from normal (acceptable) values. A third application arises when dynamic models must adjust flow rates of physical quantities between compartments so as to adhere to a physical law. For example, Caldwell et al. (1986, Chapter 16) modeled radiant heat absorption by leaves as part of a canopy-level photosynthesis model. Since there was no analytical solution for the heat flow into each layer of leaves in the canopy given only the input radiation, they used an iterative approximation that minimized the difference between the energy input at the top of the canopy and the total amount of energy absorbed based on a model of the effects of higher leaf levels on lower ones.

The above applications share a common feature in that they are all based on real-valued, continuous functions. That is, least-square minimization in parameter estimation, individual fitness as a function of a foraging efficiency, and balancing energy budgets all fit this pattern. When the functions to optimize are "simple," there are robust and well-studied methods (e.g., nonlinear regression). There are, however, two situations in which the methods have difficulty: *global optimization* in the presence of many local extrema and *combinatorial optimization.*

Hill-climbing methods such as Nelder–Mead simplex will usually find the closest extremum, but this may not be the global optimum. In Section 7.4, we recommended that the locality of the solution be tested by starting the simplex at several different initial parameter values. However, as we will see below, there are better approaches. Combinatorial optimization problems are those in which optima are sought that are not simple, real-valued functions. These are optimization problems, as the name suggests, whose goal is not to find parameter values, but to find the best way to combine objects together. A classical example is the traveling salesperson problem where the problem is to find the best sequence of cities to visit in order to minimize total distance traveled. Both of these aspects of optimization are hard, and new computational techniques using biological metaphors have been developed to deal with them. Foremost among these are methods based on analogies with evolution by natural selection.

New optimization techniques are useful only if they address interesting biological questions. In this chapter, the models used as examples will attempt to answer the following. (1) What are the best days to irrigate a field crop to get the highest yield and use the least water? (2) What set of parameters best predicts dry mass accretion in a cotton crop simulator? (3) To survive when competing for food, is it better for bean weevils to stay and fight with each other or to eat fast and run? (4) How can a lizard know when to chase an insect and when to pass it by? All of these questions are optimization problems that can be answered with evolutionary computation.

19.2 Optimization as Natural Selection

Based on the observation that biological evolution by means of natural selection produces organisms progressively better fit for a given environment, several computer scientists [e.g., Fogel et al. 1966 and Holland 1975 (reissued and updated as Holland 1992)] proposed an analogy between natural selection and general optimization algorithms. While there are many biological situations where we would not expect organisms to be optimally adapted to their environment, the general relationship is strong enough to encourage computer scientists. The basic analogy is that potential solutions to an optimization problem are similar to the phenotype (observable traits) of organisms, and the proximity of a potential solution to the true solution is similar to Darwinian fitness. If there are differences within a population of potential solutions, then some will be "fitter" than others and, such as biological evolution, the best potential solutions will be those that contribute the most to the next iteration of the algorithm just as more fit organisms contribute more offspring to the next generation.

A large family of algorithms uses this basic analogy and has been subsumed under the label *evolutionary computation.* The algorithms differ in their interpretations of the basic elements of the analogy and in their computer implementations. Below, we briefly survey some of the alternatives and describe in more detail one especially popular approach.

19.3 Kinds of Evolutionary Computation

Most approaches to evolutionary computation use the following general *evolutionary algorithm* (Bäck 1994). A population of N potentially optimal solutions at algorithm iteration k is denoted $\mathbf{P(k)}$.

1. **Initialize** $\mathbf{P(0)}$ with random solutions.
2. **Evaluate** the fitness of each element of the initial $\mathbf{P(0)}$.
3. **Recombine** elements of the current $\mathbf{P(k)}$ to form a new $\mathbf{P'(k)}$.
4. **Mutate** $\mathbf{P'(k)}$ to form $\mathbf{P''(k)}$.
5. **Evaluate** the fitness of $\mathbf{P''(k)}$.
6. **Select** the best of the $\mathbf{P''(k)}$ to form a new $\mathbf{P(k)}$.
7. **Repeat** step 3 with $k = k + 1$ until a stopping criterion is met.

The differences among the methods depend on the class of problems (and solutions) attacked, methods to evaluate fitness, choice of the potential solutions to retain for the next generation, and techniques to modify the current set of potential solutions to produce variation in the population. In this discussion, we include as evolutionary algorithms the following techniques: simulated annealing, evolutionary programming, evolution strategies, genetic algorithms, and genetic programming.

19.3.1 Simulated Annealing

Simulated annealing (SA) is based on an analogy with physical thermal annealing: a process used to create crystals by heating a substance to liquid and allowing it to cool. If the cooling proceeds sufficiently slowly, pure crystals will form because the individual molecules will succeed in reaching an energy minimum given the states of their neighbors. If cooling is too fast, not all molecules can orient properly before their thermal energy is removed, and imperfect crystals are the result. Imperfections are not necessarily bad; different types of metal are produced by different cooling rates.

The basic approach to SA is straightforward: (1) generate a single random solution to the problem, (2) calculate the *cost* or *quality* of the solution (i.e., "energy"), (3) if the solution is better than the previous best, accept the current solution, (4) if the solution is worse than the previous best, accept the current solution with some probability, and (5) repeat step (1) until a stopping criterion is satisfied. SA is a special case of the general evolutionary algorithm because it uses a population size of 1 and does not perform recombination among existing solutions (step 3).

The purpose of step (4) in SA is to avoid local minima by sometimes accepting poorer solutions. This allows the proposed solution to jump out of local minimum energy traps. The probability functions used vary greatly among applications. In general, the probability decreases as the control "temperature" increases (van Laarhoven and Aarts 1987):

$$Pr(k) = q_k(c) = \frac{1}{Q(c)} e^{-(\Delta C(k)/c)}, \tag{19.1}$$

where k is iteration number, $q_k(c)$ is usually called the *Boltzmann probability*, $Q(c)$ is a normalization function, c is the control constant analogous to temperature, and $\Delta C(k)$ is the difference between the costs of the current solution and the previous best. If $\Delta C(k) < 0$, the new configuration is accepted as the best. If $\Delta C(k) > 0$, the choice to retain the inferior current solution is essentially accomplished by a coin toss. If $q_k(c)$ is greater than a uniform random deviate chosen from the interval 0–1, then the current solution becomes the best solution, even though its quality is less than that of the previous best.

When c is small, $q_k(c)$ is large, causing the algorithm to accept "inferior" solutions relatively frequently. This permits the algorithm to continue searching for the global minimum when it is in the vicinity of a local minimum. To converge on a solution, however, the acceptance of inferior solutions must eventually become unlikely. This is accomplished by reducing c, analogous to cooling the medium in real annealing. In minimization problems, it is typical to choose the new control value as $c_{k+1} = f(c_k)$, where $f()$ is the cooling schedule.

The algorithm for reducing c is not specified and, generally, is chosen by a combination of intuition and iterative trials. Common approaches (van Laarhoven and Aarts 1987) include: (1) linear decrement: $f(c_k) = \alpha c_k$, where α is a number slightly less than 1.0. α can be chosen by fixing the final control value and maximum number of iterations to be performed. (2) Nonlinear decrement:

$$c_k = \left[\frac{K-k}{K}\right]^{\gamma} c_0,$$

where $k = 1, \ldots, K$. (3) Complex nonlinear decrement: van Laarhoven and Aarts (1987) discuss several approaches based on the variance of the cost at the kth iteration.

The stopping criterion is usually the "equilibrium" state. Equilibrium is achieved when the previous best solution is not replaced by the current trial solution for N iterations, where N is on the order of 20. Termination can also be specified by setting a final value for the control constant (c_k).

Applications that have used evolutionary computation include complex electronic circuit design, determination of chemical structure of molecules, and scheduling problems (factory optimization). The typical application of SA is, following the physical analogy, minimization of an error function, but it can easily be adapted to maximization problems. The convergence rate of the algorithm can be increased by choosing $Pr(k)$ (Eq. 19.1) not from the Boltzmann distribution, but from a Cauchy distribution (Ingber 1989, Ingber and Rosen 1992). These methodological details are currently the subject of intense debate and research.

A common application of SA is to optimize real-valued functions; this can be extended to complex differential equation models. An example of the latter is to optimize parameters in simulation models. Walker (1992) used SA to predict the optimal timing and volume of irrigation that must be applied to a peanut crop in order to maximize yield. He used PEANUT, a validated and well-studied simulation model of peanut crop growth that incorporates temperature, rainfall, irrigation, and soil water content to predict yields using several irrigation schedules. For each of the years 1974–1991, he used PEANUT, yearly local weather data, and SA to set the optimal irrigation schedule in the form of the volume of water applied to a standard field on 10 different days during the growing season. He used 10 irrigation days because this was the usual number used by the local growers. Since this is basically a function optimization problem applied to scheduling, Walker (1992) relied heavily on theoretical research by Bohachevsky et al. (1986). He compared the yields predicted with those obtained using a "typical" schedule employed by local growers.

An irrigation schedule was a vector of 10 days and volumes of water applied. The initial schedule was 10 equally spaced days. A new schedule was generated on the kth iteration from the previous best schedule according to the recursive function:

$$D_{k,i} = D_{0,i} + \Delta d R_i,$$

where $D_{k,i}$ is the julian date of the ith irrigation time during the kth SA iteration, $D_{0,i}$ is the same quantity in the previous best schedule, Δd is an empirically determined time step equal to 6, and R_i is a normalized uniform random deviate (Bohachevsky et al. 1986). After the 10 $D_{k,i}$ had been determined, the PEANUT model was simulated to determine predicted current yield. If the current yield was greater than the yield of the previous best schedule, $D_{k,i}$ was accepted as the new best schedule. If the current yield was less than the previous best, it was rejected if a uniform random deviate was greater than the Boltzmann probability. Otherwise, the inferior schedule was accepted. The SA algorithm terminated when the D_k schedule was rejected for 20 consecutive iterations.

Walker (1992) used as the Boltzmann probability

$$p_k = e^{-\beta |Y_m - Y_k|^g \Delta y}, \tag{19.2}$$

Table 19.1: Irrigation optimization results using SA and computer simulation. The typical irrigation schedule was determined from historical records.

Schedule	Yield (kg/ha)	Total Water (mm)
Typical	7063	154
SA	7586	138

where Y_m is the estimated maximum yield, Y_k is the yield of the current schedule, Δy is the difference between the yield of the current schedule and the previous best schedule, β is a positive scaling variable (≈ 0.85) for cooling, and g is a negative constant (≈ -1.0) that controls the shape of the cooling schedule at low temperatures.

Since g is negative, Eq. 19.2 has the standard, general form of the Boltzmann probability. In this application, the cooling schedule of the control constant is a linearly decreasing function of the current yield. The cooling schedule depends on Y_m, which is unknown but iteratively increased by small amounts as the SA proceeds to ensure $Y_m > Y_k$. As the SA approaches the global maximum, the probability of accepting an inferior schedule approaches zero.

For each year in the period 1974–1991, Walker (1992) determined the optimal schedule and compared its predictions to those produced from a typical irrigation schedule. The results, averaged over the 21 years, are shown in Table 19.1. Walker (1992) found that optimizing irrigation resulted in approximately 7% greater yields while using 10% less water compared to a typical irrigation schedule determined intuitively by local growers. The actual schedule to use varies among years and depends on the yearly rainfall, but in average rainfall years the best strategy is to begin irrigation on 13 June and repeat for nine additional irrigation episodes spaced approximately 10–13 days apart.

19.3.2 Evolutionary Programming

SA finds the optimum by randomly walking through the solution space using a single (currently best) solution. Global optimization was achieved by occasionally accepting poor solutions as the current best. The remaining evolutionary approaches we discuss differ in that they use a *population* of current solutions iterated over time and an algorithm based on the metaphor of biological reproduction, ecological relationships, and evolution. The three major approaches are *evolutionary programming*, *evolution strategies*, and *genetic algorithms*; these have been recently reviewed and compared (Bäck and Schwefel 1993; Bäck 1994; Schwefel 1995).

Evolutionary programming (EP) was invented by Fogel et al. (1966), who used the technique to estimate finite state automata transition probabilities (Section 10.4). Since many optimization problems can be cast in the framework of FSAs, this is a broadly useful technique. Moreover, the basic idea has been extended to include other problem domains that involve estimation of continuous parameters (Fogel 1994a, 1994b). The typical application is function optimization (e.g., parameter estimation, function minimization). EP follows the general evolutionary algorithm except it does not recombine the solutions (step 3).

Population variability is generated entirely by random mutations. Fogel and Stayton (1994) report accuracy and efficiency benchmarks on function minimization comparing EP with recombining methods (genetic algorithms, see below) that suggest that recombination does not improve the searching. Each element of $\mathbf{P}(\mathbf{t})$ (a "parent") generates by mutation a single offspring to form $\mathbf{P}'(\mathbf{t})$. In most applications, mutation occurs by drawing new values of solution components (e.g., continuous parameters) from an N-dimensional normal distribution. The variance from which to draw the deviate determines the amount of variability in the population of potential solutions. The control variables of the algorithm (e.g., the variances from which solution components are drawn) are variable and adaptable during runs.

Selection occurs by pooling $\mathbf{P}(\mathbf{t})$ and $\mathbf{P}'(\mathbf{t})$ and placing a subset (e.g., 10 individuals) in competition with each other. This is a stochastic form of *tournament* selection. Each solution is placed in competition with a randomly chosen subset, and the number of competing solutions that are worse than the target solution are counted. When all solutions have had an opportunity to compete, they are rank ordered by the number of wins they experienced. The new population of solutions is the best N by rank.

Fogel (1994b) demonstrated the method and compared it to genetic algorithms (GA, see below) on the problem of maximizing the total harvest of a population growing exponentially. The problem is to identify the population harvest schedule (i.e., the amounts to remove from the population at each point in time) that maximizes the total amount taken over a time interval for a population that is increasing exponentially. The schedule must be chosen so that the population is the same size at the end of the interval as at the beginning. This problem has analytical solutions of 73.23768 and 279.275275, when the duration of the population dynamics and harvesting is 20 and 45 time steps (t), respectively. Fogel found that when $t = 20$ after 1000 algorithm iterations, EP obtained the value 73.234749 and GA obtained 73.1167. When $t = 45$, EP found 214.033813 and GA found 277.3990. This shows that both of these evolutionary algorithms can get close to the optimum and that the comparative efficiency of the two methods depends on the size (duration) of the problem. Choosing the algorithm to use is not always an easy task.

19.3.3 Evolution Strategies

Evolution strategies (ES) (Bäck and Schwefel 1992, Bäck 1994, Schwefel 1995) are similar to EP except they incorporate recombination among solutions in the general algorithm (step 3). Each parent can produce more than one offspring. Most applications concern continuous function optimization. Mutation occurs by random draws from an N-dimensional normal distribution, but the methods for choosing the standard deviation differs from EP. Selection in ES also differs from EP in that tournament competition is not used. Instead, each offspring $[\mathbf{P}''(\mathbf{g})]$ is ordered by its fitness and the best N are chosen as the population in the next iteration. Solutions are described as a set of *components* (e.g., a finite number of parameter values in function minimization). Recombination occurs by swapping a subset of these components among a number of parents to produce offspring. Unlike GA, the values of the components are not affected by recombination. There are many variants on the basic ES described. Schwefel (1995) reviews many of these with function minimization benchmarks against a wide variety of functions.

19.3.4 Genetic Algorithms

Genetic algorithms (GA, Holland 1975) and their derivative, genetic programming (GP, Koza 1992b), follow the general algorithm described above, but differ from EP and ES in two ways. First, this method was designed to be applied to a broader class of problems than EP and ES. In particular, GA/GP, like SA, are useful for combinatorial optimization, although EP and ES also work on these problems. Second, the GA/GP method was derived from a close analogy with biological reproduction and evolution. Since this is one of the most important evolutionary optimization techniques (in the United States, at least), we will describe it in detail below.

19.4 Genetic Algorithms and Genetic Programming

19.4.1 The Basic Genetic Algorithm

As mentioned, the classical GA formulation follows the general evolutionary algorithm described earlier. GA is distinguished from EP and ES by its method of representing problem solutions and by its ability to address combinatorial problems. Like EP and ES, GA uses populations of potential solutions, but each individual solution is likened to a biological *chromosome* on which reside *genes*. The composite of genes on a given chromosome represents the potential solution. For example, if the problem was to estimate the parameters in a linear regression, the chromosome would be composed of two genes, one for each of the parameters that are sought. The number of genes is fixed for a particular problem. However, in principle, there can be any finite number of genes up to the storage capacity of the computer.

Genotypic and phenotypic variability in biological populations arises from many sources, but *mutation* and *recombination* among chromosomes in sexually reproducing species are two important sources. Variability is

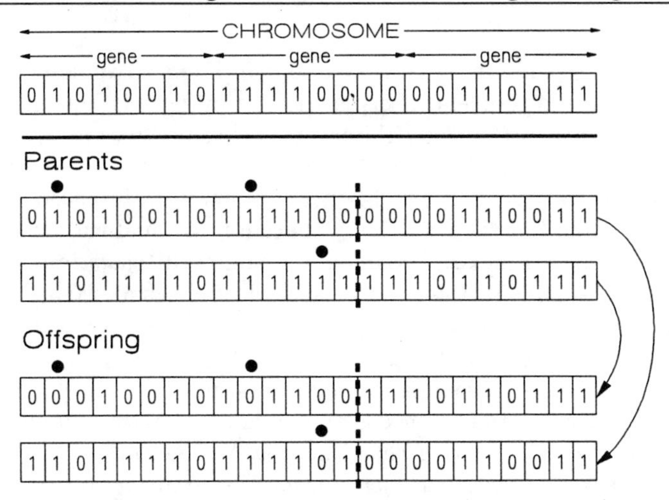

Figure 19.1: Chromosome definition and basic operations applied to bit strings used in GA optimization. Above: chromosomes as bit strings of three genes, each eight bits in length. Below: mutation is a random bit-by-bit reversal of gene values within an offspring (denoted by •); crossover exchanges portions of bit strings (to the right of the vertical dotted line) between two parents.

good in search algorithms since it is the primary way to avoid becoming trapped at local extrema. These two genetic operators manipulate the basic representation of solutions in GA: binary strings. In classical GA, every gene is represented as a string of 0s and 1s. A chromosome, being a collection of genes, is a set of binary strings. GA implements mutation as the reversal of N randomly chosen bits along the chromosome regardless of the interpretation of the gene to which the bit belongs. That is, if the bit is a 0, it is changed to a 1; if it is 1, it becomes 0 (Fig. 19.1). For a given GA run, a *mutation rate* (m) is specified as a small number (usually $\ll 0.5$). Each bit is tested for reversal by selecting a deviate from $u_i = U(0,1)$ and reversing bit i if $u_i < m$.

When mating occurs between two chromosomes, recombination occurs by exchanging portions of the binary strings between the two chromosomes (Fig. 19.1). This is called *crossover*. At a random point on the two chromosomes, portions of the binary string to the right of the crossover point are swapped between the chromosomes. As Fig. 19.1 indicates, the crossover position is the same for both chromosomes. The crossover point need not occur at a gene boundary, which is different from ES. There may be more than one crossover point, although most applications use only one. Recombination is one of the hallmarks of GA, but EP does not use crossover and some have claimed that it is not necessary (e.g. Fogel, 1994c). This is a major point of controversy in evolutionary computation; there is no easy answer, but the value of crossover appears to depend on the problem being solved.

While maintaining variability is important, it is also important that the algorithm actually progressively improve the mean value of the population to solve the problem. This is ensured by (1) biasing the choice of potential parents so that only the most fit chromosomes are parents and (2) by forcing the very best chromosome in the current generation to be present in the next generation. To effect (1), mating chromosomes are selected at random but the probability that an individual will be selected is proportional to its fitness. A common method is the *roulette wheel* (Davis 1991) that is similar to the inverse method of choosing random deviates from an empirical probability frequency distribution described in Section 10.2. When strategy (2) is implemented by forcing the best chromosome to survive into the next generation, *elitism* occurs.

Most implementations use constant numbers of chromosomes. This can be achieved by requiring that there be $N/2$ matings in a population of N chromosomes and that each mating result in two offspring. Then the next generation is formed by replacing the parents [$\mathbf{P}(\mathbf{t})$] with the children [$\mathbf{P}'(\mathbf{t})$] created through mutation and recombination. This can be generalized so that fewer than N offspring are produced and only this smaller number of parents are replaced.

The binary representation of genes permits a wide variety of problems to be attacked by this evolutionary method. This includes combinatorial problems such as the traveling salesperson zproblem, which is the problem of ordering a set of integers (cities are given arbitrary integer codes). Problems such as parameter estimation that require manipulation of noninteger (real) numbers are easy in EP and ES, but are actually somewhat cumbersome in GA. It can be done, however, by suitably coding real numbers as binary strings. A simple method is *binary coding* in which a range of real numbers is approximated by $n = 2^m$ divisions, where m is the number of bits per gene. In the binary coding method, the bit string is interpreted as binary numbers that have the usual integer interpretations: `0000` is integer 0, `0111` is integer 7, etc. For example, if a gene has eight bits, then it represents 256 different integers. A real interval from 0 to 10 (say) can be approximated by these 256 integers by assigning gene `00000000` to real 0.0, gene `00000001` to the real interval $0.0 < x \leq 0.03906$, gene `00000010` to the real interval $0.03906 < x \leq 0.07813$, and so on.

Obviously, the larger the length of genes, the better the approximation. But long genes are computationally expensive. There is another problem with the binary coding method: adjacent integers are not "adjacent" bit strings. For example, to change integer 7 to integer 8 requires that four bits be reversed. But GA mutation operates by reversing a single bit of a gene. Based on the rather loose but intuitively appealing analogy between the bit strings and biological genes, we would like a small *effect* of mutation to be caused by a small number of bit reversals. The *Gray code* has this effect: each adjacent pair of integers differ in their Gray code by a single bit difference (Goldberg 1989). For example, integer 7 is `0111` in binary code and 0100 in the Gray code; integer 8 is 1000 in binary and 1100 in Gray. As a result of this property, Gray codes are frequently used in GA.

19.4.2 Examples: Parameters and Beetles

The literature on GA is massive and growing by leaps and bounds (or, perhaps better: by mutations and crossovers). Many examples can be found in Goldberg (1989), Davis (1991), and the proceedings emanating from the many frequent conferences. Here we describe two from biology that differ radically in their problem definitions.

Model Calibration

Sequeira and Olson (1995) used GA to *calibrate* (Section 7.5) a subset of the free parameters in GOSSYM, a dynamic simulation model of crop growth over a growing season widely applied to cotton, soybeans, and winter wheat. The complete model has over 50 parameters, but Sequeira and Olson (1995) examined the efficiency of GA for estimating five parameters. Each chromosome was partitioned into five genes, and, since the parameters differed in their biologically reasonable ranges, the bit length of genes varied between 8 and 12.

The calibration algorithm was: (1) initialize a population of potentially optimal parameter values based on heuristically obtained values reported in the literature (Reddy et al. 1985); (2) obtain new values by the standard GA algorithm outlined above (binary coding, single crossover); (3) for each member of the population, run the simulation model to obtain predictions at the sampling times; (4) compare model predictions with observations and calculate fitness; (5) repeat step 3 until all potential solutions have been evaluated; and (6) repeat step 2 until the stopping criterion is satisfied. The chance that a given chromosome would survive to the next generation (step 6 in the general evolutionary algorithm, Section 19.3) ws proportional to the ratio of the chromosome's fitness to average population fitness.

Fitness was not a simple sum of squared deviations. Earlier experiments by Sequeira and Olson (1995) showed that absolute differences were more effective for GA. Moreover, complex computer simulation models such as this one produce complex output. In this case, the model predicts both mass accretion and organ generation for several different organ types (e.g., floral buds, immature fruit, etc., see Section 11.4). The model also predicts whole organism measures such as plant height and leaf area. Comparison of all of these measures with observations contributes to the evaluation of model quality. This was implemented in their GA with a fitness function that summed the differences between predictions and data for all of these model outputs.

Sequeira and Olson (1995) found that GA improved model predictions over the heuristic parameters for several of the model outputs. The improvement was most dramatic for dry mass accretion, in which GA improved predictive ability by 25%. For all output quantities, GA resulted in a 15% improvement. Using a population size of 1000, the average error decreased from about 100 to 35 in 30 generations. This application of GA does take some computing resources, however. Sequeira and Olson (1995) ran their GA on a 66 MHz Intel 80486. Some of the calibration experiments took as long as 4.5 days to complete. This helps explain why there are several GA implementations for parallel computers (Goldberg 1989).

Optimal Beetles

The second application used a rather different approach to GA optimization. Since GA was created by drawing an analogy between evolution by natural selection and optimization, it seems reasonable to turn around and apply GA to problems of optimal adaptive traits in evolution. Toquenaga et al. (1994) simulated competition and evolution between two species of beetles that attack beans. *Callosobruchus analis* and *C. phaseoli* lay eggs on the bean surface, the larvae burrow into the interior, develop over a number of days, and emerge as adults. The two species differ in four ecological traits: (1) mode of competition, (2) rate of development, (3) foraging location, and (4) number of eggs laid per bean. Mode of competition is a binary trait and refers to whether the beetle uses *scramble* or *contest* competition. In contest competition, dominant individuals interfere with the foraging of subdominant individuals and thereby acquire more resources. In scramble competition, all individuals compete equally for the resources without interfering with each other directly. *C. analis* uses contest competition, while *C. phaseoli* uses scramble competition. Rate of development refers to the number of days from egg deposition to adult emergence. Foraging location refers to whether the beetles prefer to burrow to the central core of the bean or remain near the surface. There is more and better food in the interior, but emergence rates are higher if the developing larva is near the surface. Number of eggs laid per bean by adults is self-explanatory.

In addition to the behavioral and physiological traits of beetles, the size of the bean is important in evolution. Large beans (i.e., cultivated varieties) should favor scramble competition, while small beans should select for contest competition. To test this hypothesis, Toquenaga et al. (1994) performed laboratory experiments and computer simulations to determine which strategy would out-compete the other on large and small beans. The model was a stochastic individual-based model in which the properties of the individuals evolved using GA. The above four individual traits [(1)–(4)] were encoded as bit strings on a GA chromosome with four genes of length one, five, four, and four bits per gene, respectively.

Toquenaga et al. (1994) constructed an individual-based model (Chapter 13) in which GA was used to determine the phenotypic characteristics of the individuals. The model assumed that individuals having the contest gene [trait (1)] competed with and did not interbreed with individuals having the scramble gene. Starting with five pairs of each species in an arena with either large or small beans, the simulations followed the reproductive fates of individuals over 100 generations. The genetic composition of the next generation was a function of the number of emerging adults produced by each genotype in the previous generation. Consequently, the frequency of genes in the population evolved according to their relative abilities to produce offspring. After emergence, mating occurred randomly among adults of opposite sex that belonged to the same species [based on the competition gene (1)].

The simulations supported the primary hypothesis and agreed with experimental results: small beans favor individuals using contest competition, large beans favor scramble competition (Fig. 19.2). Significantly, the model did not predict extinction of either population, as observed. The evolution of other traits (e.g., developmental rates, etc.) in both populations apparently was able to keep ahead of extinction. Toquenaga et al. (1994) also tracked the evolutionary dynamics of the average life history and behavioral traits in each of the two populations. Surprisingly, they found that contest individuals increased their use of the bean core relative to the peripheral regions, but decreased their developmental rates. This occurred in both large and small beans. Scramble individuals evolved in the opposite direction: they evolved to use the peripheral regions of the bean more than the core and increased their development rate. Both of these results qualitatively agree with experiments.

To summarize, if you are a bean beetle and your strategy is to fight (contest competitor), your best evolvable strategy is to go deep into the bean and out-wait your competitor by developing slowly. If your strategy is to scramble for food, your best strategy is to stay near the surface of the egg, eat the minimum necessary, and get out fast by developing quickly. In essence, Toquenaga's model showed the spontaneous emergence of microhabitat partitioning within a single bean. Neither strategy evolved to produce the maximum number of eggs per female, which you might naively expect to be a winning strategy.

19.5 Genetic Programming (GP)

A computer program, like a sequence of cities visited by a traveling salesperson, is a solution to a problem. It will, no doubt, come as no surprise to learn that there are good and bad computer programs: programs that are more or less efficient, or that give more or less correct answers. As computer users, we tend to think of a computer program as a tool that performs an activity. For example, a word processor is a program that allows us to input text, edit it, format it, and print it. Theoretical computer scientists, however, think of programs as complex objects, which can be studied and classified by their structure. As objects with parts (e.g., `for`

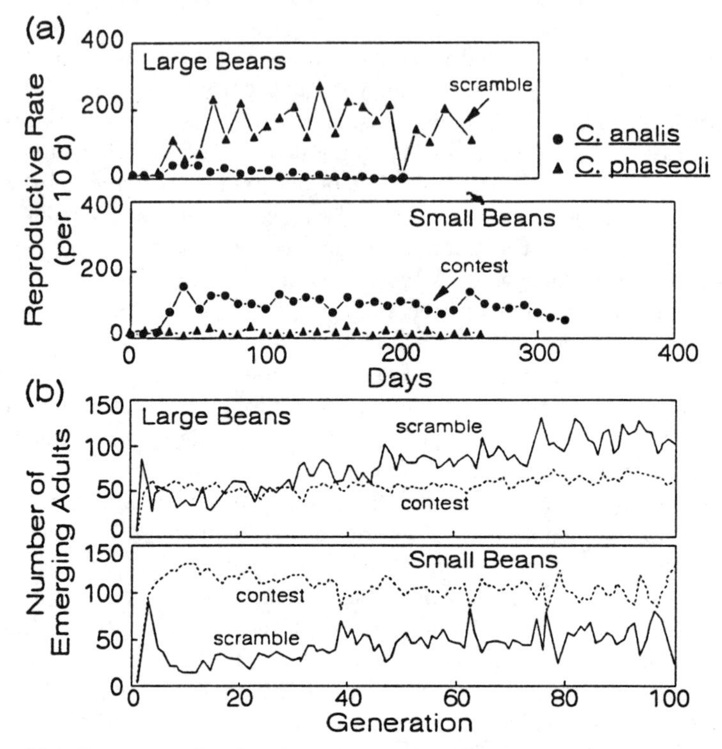

Figure 19.2: Dynamics of real and simulated populations of a contest competitor (*C. analis*) and a scramble competitor (*C. phaseoli*). (a) Dynamics of a laboratory interspecific competition experiment with food renewal at 10-d intervals. In the replicate shown, the scramble strategy dominates in large beans; the contest strategy dominates in small beans. (From Toquenaga and Fujii 1991, Fig. 1; results from controls not shown. © 1991 the Society of Population Ecology. Reprinted by permission of the Society of Population Ecology.) (b) Simulated evolution of competition related life history traits using GA. When the resource is large beans, scramble competitors evolve to dominate; on small beans, contest competitors evolve to dominate. (From Toquenaga et al. 1994, Fig. 9. C. Langton, *Artificial Life III*, © 1994 Addison-Wesley Publishing Company Inc. Reprinted by permission of Addison-Wesley Publishing Company, Inc.)

loops, assignment statements, `if-then` conditionals, etc.), programs can be constructed like any other structure by assembling the available parts. Recently, there has been much research on the use of GA to automatically discover computer programs (Koza 1992b, 1994, Kinnear 1994) by assembling them from parts. Although based on GA, this problem is sufficiently different from other GA applications that it deserves its own name: *genetic programming* (GP).

19.5.1 GP: GAs Applied to S-Expressions

In GP, the search space is defined by the structure of a computer program. Computer programs can be represented as a tree graph that, in turn, can be represented as an *S-expression*. Figure 19.3a shows these two objects

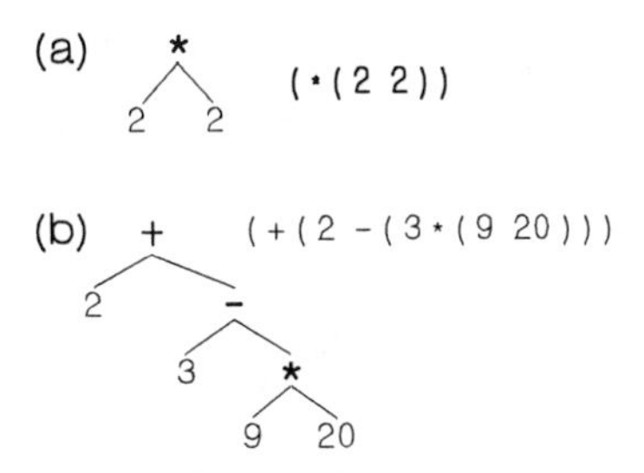

Figure 19.3: Programs represented as trees and S-expressions. (a) Two equivalent representations of a simple program to multiply 2 times 2. The tree structure shows the hierarchical arrangement between the function ('*') and its arguments (2 and 2). The S-expression on the right is a non-graphical representation that uses parentheses to represent the hierarchy. (b) The two representations for a more complicated program.

for a program that multiplies 2 and 2; Fig. 19.3b illustrates a more complicated program. In more typical mathematical notation, the latter program computes the function $y = 2+(3-(9\cdot20))$. Figures 19.1 and 19.3 illustrate a feature of GP that distinguishes it from GA. GA solutions have a fixed size: the number of bits on the chromosome. GP solutions are constructed recursively and are open-ended: GP chromosomes (solutions) can be arbitrarily large. We will discuss below how these structures are created and evolved.

Obviously, GP would not be so wonderful if all it could do was string together arithmetic statements, although, as we will see, this is a very useful idea. GP can also solve combinatorial problems such as discovering a set of moves to be performed by an imaginary ant in following a trail of food (Koza 1992a; also done using GAs by Jefferson et al. 1992). The food is arranged along a contorted trail in a two-dimensional grid, and the task for the ant is to discover as much of the food in the time available. The ant can perform only three simple actions: pivot to the left in the current cell (`LEFT`), pivot right in the cell (`RIGHT`), or move ahead to the next cell in front of the ant (`MOVE`). If the ant moves on to a spatial grid cell containing a food item, then the ant consumes the food. These three activities are analogous to the numbers in the examples of Fig. 19.3; they terminate branches of trees. The arithmetic operations in those examples are analogous to the two functions in the ant system called `IF-FOOD-AHEAD` and `DO-TWO` (we've taken some license with Koza's original terminology). The former function looks in the current facing direction and if food is available in the next cell, a specified activity is performed. If no food is available, another activity is performed. Thus, this function requires two arguments, one for each possible outcome of the test for food. The ant should also be able to move or turn without looking for food and this is the purpose of the second function. It also requires two arguments, but they are performed unconditionally.

Four examples of possible S-expressions are: (1) `(DO-TWO LEFT LEFT)`, (2) `(IF-FOOD-AHEAD MOVE MOVE)`, (3) `(MOVE)`, and (4) `(DO-TWO IF-FOOD-`

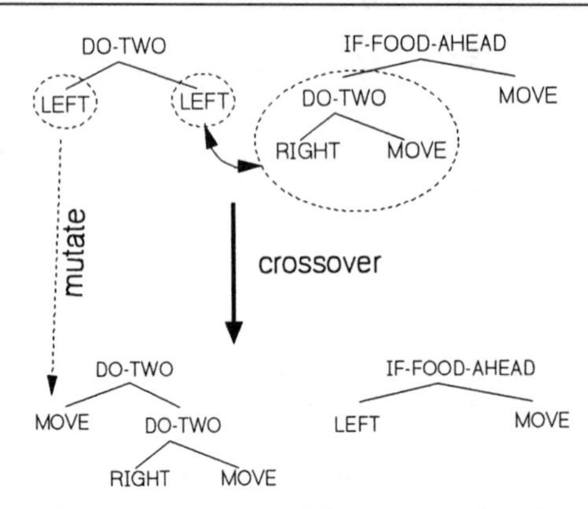

Figure 19.4: GP operations on programs. Crossover: exchanging nodes between two parent programs to form a new offspring. Mutation: randomly altering a single node. Each tree represents one "chromosome" or potential solution.

`AHEAD (IF-FOOD-AHEAD (IF-FOOD-AHEAD DO-TWO (LEFT LEFT) MOVE) RIGHT) DO-TWO (RIGHT MOVE))`. Program (1) causes the ant to spin in place making left turns forever. The ant of program (2) looks for food then moves straight ahead traversing the arena without turning. The ant of (3) does the same thing, but never looks for food. Describing the behavior of the ant of program (4) is left as an exercise for the reader.

As illustrated by this ant example, a GP solution requires that we identify four elements of a problem: (1) a *function set*: a set of functions (e.g., `IF-FOOD-AHEAD`, and `DO-TWO`); (2) a *terminal set*: a set of program elements that do not require arguments (e.g., `LEFT`, `RIGHT`, `MOVE`); (3) a fitness calculation for each possible program; and (4) various GP system control parameters. A computer program that implements a GP system follows a similar structure to that for GA and the general evolutionary algorithm. An initial population of potential solutions is generated at random. Their fitnesses are determined and a subset are chosen for "mating." During the mating process, crossover and mutation can occur. Mutation acts on terminals, giving them new values from the terminal set. Crossover exchanges nodes of a program tree between potential solutions (Fig. 19.4).

To test these small sets of functions and terminals, Jefferson et al. (1992) created a trail called the *Santa Fe Trail* that comprised 89 food items (Fig. 19.5a). GP discovered a program (Fig. 19.5b) that found all of the food before a fixed amount of time had expired (Koza 1992a).

The above solution depends on the initial trail. A different trail will cause a different program to evolve. Indeed, there is some evidence that starting a search for a program to solve a new trail beginning with a population based on the solution to the Santa Fe Trail is actually worse than starting the new search with random programs. Nevertheless, it is a remarkable accomplishment that a "dumb" computer can learn to traverse this trail by means of a relatively simple program.

(a)

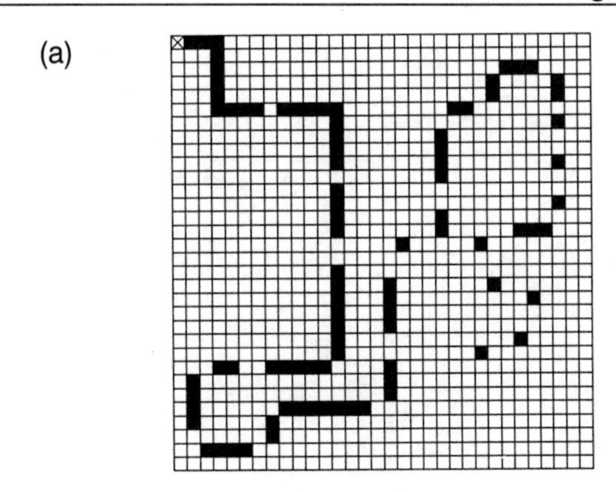

(b)
```
(IF-FOOD-AHEAD MOVE DO-TWO
        (RIGHT IF-FOOD-AHEAD (MOVE LEFT))
        DO-TWO
        (LEFT
        IF-FOOD-AHEAD (LEFT IF-FOOD-AHEAD (MOVE RIGHT))
        MOVE))
```

Figure 19.5: The Santa Fe Trail and its solution. (a) The positions of 89 black food squares. The ant starts in the upper left corner on the square marked with an X. (b) The evolved program discovered by GP that follows the trail exactly. (From Koza 1992, Fig. 5. C. Langton, *Artificial Life: Volume II*, © 1992 Addison-Wesley Publishing Company Inc. Reprinted by permission of Addison-Wesley Publishing Company, Inc.)

19.5.2 Simple Symbolic Regression

Another excellent problem for GP is *symbolic regression*, or the problem to find the best function through a set of datum points that has a single independent variable and a single dependent variable. We have previously discussed parameter estimation techniques to find the best parameters when the function is given. Finding the function *and* the parameters is harder, but GP can help because one of the fundamental methods GP uses to generate potential solutions is the recursive application of mathematical operations. By providing both arithmetic operations shown above in Fig. 19.3 and other fundamental mathematical functions (e.g., log, sine, cosine, etc.), extremely complex functions can be built using recursive applications of the functions.

The function set is the set of all these mathematical functions we care to provide; the terminal set has two elements: a random real number that represents the parameter values (coefficients) and a variable (X) that represents the independent variable. A potential solution's fitness can be calculated using any method that integrates the differences between the data at all values of the independent variable, for example, the sum of the square of the differences. In the following, X can have integer values from 0 to 9. Fitness is calculated by iteratively stepping through each value of X, sub-

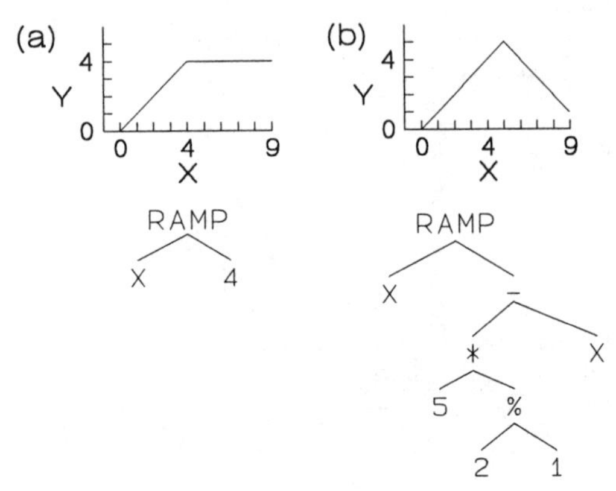

Figure 19.6: GP solution to symbolic regression on a discontinuous function. (a) Above: the input ramp function to fit; below: the program found. (b) Above: the input triangular function to fit; below: the program found. The function "%" is arithmetic division that returns 0 if the denominator is 0. X represents the independent variable and ranges from 0 to 9.

tracting the value calculated by the potential solution from the data value at each X. For example, if the data to fit are generated from the function $Y = AX^3 + BX^2 + CX$, where A, B, and C have definite numerical values, a GP can readily find the S-expression:

```
* ( *(a X) ) (+ ( * ( + ( *(c X) 1 ) *(b X) ) 1 ) ),
```

where a, b, c have definite, evolved values such that $A = abc$, $B = ab$, and $C = a$.

The problem becomes more interesting when we try to fit discontinuous data such as a ramp function that increases linearly to a threshold value of x, and then becomes horizontal (Fig. 19.6a). Simple polynomial equations that result from the combination of multiplication and addition (as above) do not fit this function well. We only obtain a perfect fit if we provide the GP system with a two-argument function RAMP that returns the first argument if it is less than or equal to the second argument, otherwise it returns the second argument. If the first argument is the independent variable (X), then RAMP increases linearly from 0 with slope 1.0 until a threshold (the second argument) is reached [e.g., $X = 4$], for all larger values of X the function is horizontal.

Given this function, the GP system must find a program that, first, uses RAMP, and, second, sets the first argument to X and the second argument to the correct threshold value. This is not hard with moderately large populations (e.g., about 1000). So, the lesson is that the search space (number and nature of elements in the function set) is critical for success using GP. This constraint is not such a great shortcoming, however, since it is only a restatement of the old joke about the drunk who looked for his lost car keys under the street lamp because the lighting was better there than in the place the keys were actually lost. For any optimization method to succeed, the region wherein the solution lies must be searched. The value of GP is that the solution space is automatically generated from the elemental functions.

A slight modification of the present case (D. Neff, *pers. commun.*) provides a good example of this use of elemental functions and illustrates that GP solutions are often not what the designer expected, or even, sometimes, what the designer can fathom. An example of the latter condition is deferred until the next section. The former case occurs when one attempts to extend the application of the simple functions above to fit a dataset that forms a triangular function (Fig. 19.6b, above). One would expect that a perfect fit would require a `TRIANGULAR` function with three arguments analogous to `RAMP`. As Fig. 19.6b illustrates, however, this is not so. The program shown is a perfect fit to the data, but it uses only the `RAMP` function with a complicated, nonconstant second argument. The equivalent S-expression for this program is `( RAMP ( X - ( 10 X ) ) )`. [Because of the complicated method used to compute the number 10 (Fig. 19.6b), this program is not optimized for efficiency.] When this program was evolved, the GP system also had available to it a `TRIANGULAR` function that could have been used, but was not.

Astonishingly, this solution (Fig. 19.6b) implies that the GP system discovered how to count backwards! The solution found is a function that increases as `X` increases from 0 because only the first component of the conditional is invoked (left branch). Then, when the threshold is reached, the second component of the conditional is activated and `(10-X)` is used. This function decreases as `X` increases to 9. The program has discovered how to count backwards from 5 down to 1 using a common trick applied to the loop index long known to experienced programmers. This example of the ability of "mindlessly" created programs to find "new" solutions not originally designed into the system is typical of GP; it sometimes borders on the spooky.

19.5.3 GP Applied to Optimal Foraging

These made-up examples are nice because they illustrate the concepts, but is this system useful for real biological problems? GP is young enough that not many examples exist, but one from optimal foraging theory shows the power of symbolic regression to derive functional relationships.

Anolis lizards are a group of arboreal insectivorous lizards endemic to Central America and the Caribbean islands. They primarily use a foraging strategy called sit-and-wait, which involves sitting on a tree branch until a desirable insect approaches and undertaking a short pursuit followed by a return to the perch. These lizards have excellent binocular vision and can probably detect prey at 8 m. They can eat most species of insect, but these occur in a wide range of sizes and distances from the perch. The lizard's optimization problem is to determine which prey items to pursue. To pursue insects far from the perch risks a lower probability of success and implies a greater time away from the perch (cost of lost opportunity). Large, close insects are clearly worthwhile since the pay-off is high, the risk low, and the opportunity costs are also low. Small, near insects are also probably worthwhile. Distant insects are problematical. Koza et al. (1992) applied GP to this problem, and the following is taken from there.

Formally, the optimization problem is to determine the distance from the perch at which the average energy intake rate is maximized by minimizing the total time required to consume a food item. The time to consume a prey item is a function of four variables: prey abundance, lizard sprint velocity, and the two coordinates of prey location relative to the lizard. From first principles and the assumption that all prey encountered are captured, Roughgarden (cited in Koza et al. 1992) solved the problem analytically for the optimal radius (r_c) beyond which no insect should be pursued. In terms of the pursuit criterion, the lizard should pursue an insect at position x, y (distance r_i from the lizard) only if $r_i < r_c$. Or, the lizard should pursue if $(x^2 + y^2)^{\frac{1}{2}} < (3v/\pi a)^{\frac{1}{3}}$, where (x, y) is the prey location, V is the lizard sprinting velocity, and a is the arrival rate of insects.

To test GP on this problem, Koza et al. (1992) used a five-element terminal set: `X` and `Y` position of insect, `AB` insect abundance, `V` lizard velocity, and a random real number. The function set included the arithmetic operators (`+`, `-`, `%`, `*`), an if-less-than conditional (`IFLTE`), and a special power function (`SREXPT`). The evolved programs calculated the critical distance needed to determine the pursuit decision: if a prey is within the critical distance, pursue and capture, otherwise ignore. Since the arrival of insects is stochastic and lizard sprint velocity unpredictable, each potential solution (program) was tested against several conditions of prey abundance and lizard speed. Program fitness was the average energy intake rate in all these conditions.

The progression of the GP in finding a solution is shown in Fig. 19.7. From the analytical solution, the optimal number of insects captured under the conditions of the GP trial was about 1671 insects (depending on stochastic variables). The number captured by the best program in each generation is shown with the curve of critical distances. The run was terminated at generation 61 and the best program was:

```
(+ (- (+ (- (SREXPT AB -0.9738) (* SREXPT X X) (* X AB)))
(* (+ VEL AB) (% (- VEL (% (- VEL (% AB Y)) (+ 0.7457 0.338898))))
(SREXPT Y X)) (- (- SREXPT AB -0.9738) (SREXPT -0.443604 (- (- (+ (-
(SREXPT AB -0.9738) (SREXPT -0.443604 Y)) (+ AB X)) (SREXPT Y X))
(* (* (+ X 0.0101929) AB) X)))) (* (SREXPT Y Y) (* X AB))))
```

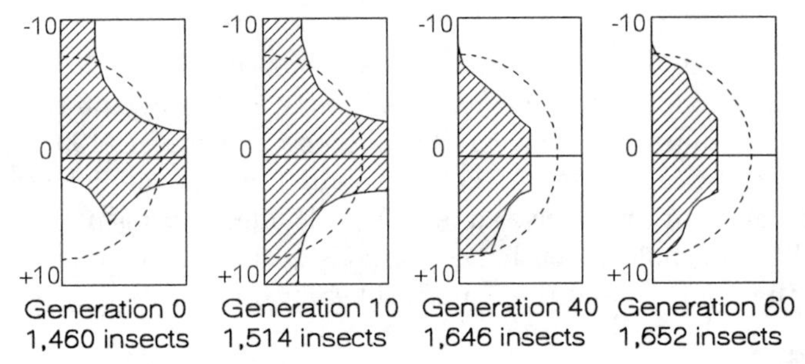

Figure 19.7: Number of captures by best solutions for critical regions to pursue insects at four different stages in the GP search. Insects occurring within the cross-hatched region should be pursued. The dashed semicircle in Generation 60 is the theoretical optimal region. In each panel, axes represent spatial position relative to lizard perch $(0, 0)$. (From Koza et al. 1992, Figs. 8–11. © 1992 the Massachusetts Insitute of Technology. Published by the MIT Press.)

Koza et al. (1992) translated this as:

$$\begin{aligned} C &= -0.44^{(a+x+a^{-0.9738})-(0.44^{y}+y^{x}+ax[x+0.01])} \\ &\quad + 0.922(v+a)(v-a/y)+2a^{-0.97}-y^{x}-ax(x^{x}+y^{y}). \end{aligned}$$

This is a far cry from Roughgarden's simple and elegant analytical result, and it is hard to imagine a lizard keeping track of all those parentheses in deciding to eat or not to eat. Despite the complexity of the answer obtained, there is merit in this application. One value of this exercise was to demonstrate the potential to find discriminating functions without imposing hypotheses of the functional form. Second, the analysis can now be applied to new problems where no analytical solution is known. Koza et al. (1992) applied the GP to a hypothetical lizard that could not see equally well at all angles. In this case, the GP evolved a considerably more complex rule than before (3.4 times as complex) that indicated that regions of poor visibility should be avoided.

19.6 Précis on Evolutionary Computation

Currently, there is much unresolved diversity in approaches to evolutionary computation. You might say that the field of computational optimization has yet to find its global extremum. Since it is not good for scientific disciplines to become trapped in local extrema, the diversity we see is still beneficial. Nevertheless, each of the major approaches have their adherents and sometimes the advocacy is intense and evangelism overblown. There are many, perhaps too many, comparative studies written by an advocate for one of the methods that demonstrate the advantages of the approach favored by the author. For those without strong feelings or great professional investment in one particular approach, the best choice for a given problem is not obvious. A central issue is the relative merits of methods with and without recombination. Moreover, there is great diversity within each approach and many control parameters that must be specified. As a result, users should expect to spend considerable time evaluating alternatives and tuning parameters to obtain the best results. Schwefel (1995) provides discussion and C and FORTRAN code for many function minimization problems with and without constraints. This includes a particularly valuable compendium of multidimensional functions with equations and graphical displays that can be used to compare different optimization approaches.

19.7 Exercises

1. GA and GP software are available from a number of Internet sites. Four of these are:

   `ftp: alife.santafe.edu/pub/user-area/ec`

   `http://www.aic.nrl.navy.mil/galist`

   `http://isl.cps.msu.edu/software`

   `http://alife.santafe.edu/~joke/encore`

   Download a GA package (e.g., SGA). Use it to find the best values for the slope and intercept of a straight line to minimize the sums-of-squares deviations with a hypothetical data set. Compare the results and solution time to a standard parameter estimation package (e.g., SAS linear regression, or simplex).

2. From the above MSU Internet site, download one of the GP packages (e.g., `GPCPP` or `lil-gp`). Compile the symbolic regression package that is used as examples in these systems. When you have been successful, try the following.
   (a) Implement a ramp function (Section 19.5) and test it against the triangle data in Fig. 19.6. Did your system discover counting backwards?
   (b) Using the standard functions supplied with the symbolic regression package you downloaded and a set of table values for a standard mathematical function (e.g., the gamma distribution in the exercises of Chapter 7), compare the solution found by GP with a high order polynomial equation with parameters estimated by standard nonlinear estimation (e.g., simplex).

3. In Fig. 19.1, suppose the ranges of the numerical values represented by the genes from left to right is –10 – 20, 0 – 1.0, and 0 – 10, respectively. If binary coding is used, what values do the three genes contain?

4. Verify that the solution for the Santa Fe Trail works by starting the ant at the "X" in Fig. 19.5 and stepping through the program to find the first 11 food items.

5. Just using your intuition and innate problem-solving skills, try to discover a better program for the Santa Fe Trail. What should "better" mean in this context?

6. Work through the program in Fig. 19.6b to verify that it gives a perfect fit to the input data.
7. In the context of Chapters 1 and 8, what was the objective of Toquenaga's model? How could validation be improved?
8. Can GP be used for *model identification*? In other words, can GP derive a system of ODEs plus parameter values whose solution will fit a given data set. Explore your ideas using Luckinbill's predator-prey data and the functional components taken from Harrison's set of models (Section 13.2.3).

Chapter 20

Complex Adaptive Systems

20.1 What Is Life?

LIVING systems, we all know, are alive, but what *essentially* is it about them that makes them so? What is the nature of living systems that distinguishes them from nonliving objects? How did they come into being? What properties do they possess that permit them to persist in a universe in which order is consistently replaced with disorder? In this final chapter, we will describe a few models whose objective is to examine the fundamental characteristics of living systems that address these basic questions.

To date, an acceptable, one-sentence definition of life has not been stated. But there a number of properties that, while perhaps not unique to living systems, are important in recognizing systems as alive. Living systems are *complex* (they have many parts), but this complexity is *ordered* or organized. This organization comes not from outside the system by means of an intelligent creator, but arises internally from interactions among the parts themselves; living systems are *self-organized* such that high-level coherent behavior or structure *emerges* from local interactions of parts. An example of self-organization is that living systems are able to *replicate* themselves. In addition, living systems that replicate and persist are *goal-directed* in the sense that they are able to satisfy constraints imposed by a a finite, noisy environment. Consequently, living systems can *adapt* to their environments. Adaptation can take the form of *learning* and *problem solving*, *physiological acclimation*, or *genetic evolution*.

Systems that have most or all of these properties are loosely called *complex adaptive systems*. We are interested in studying them in such a way as to gain insight into the properties themselves. Consequently, many models of complex adaptive systems are relatively abstract: they attempt to isolate the property of interest and model it in isolation from other (momentarily) irrelevant properties. Thus, these abstract models do not always appear to be about biology, but the hope is that by studying a simple implementation that has the property of interest, insight will be gained about all biological systems that possess the property.

Over the last 10–15 years, a tangled web of concepts, philosophical perspectives, scientific paradigms, and computational techniques have emerged that together constitute the toolbox used by students of complex adaptive systems. A few of these are: connectionism, neural networks, evolutionary computation, parallel computers, robotics, and nonlinear dynamics (chaos). Recently, the collection of research approaches that use tools such as those listed to study complex adaptive systems have been forged into a new discipline called *Artificial Life* or *ALife*, for short. Many of the examples we discuss below are from this discipline.

20.2 What is ALife?

Artificial life is the study of *man-made systems that exhibit behaviors characteristic of natural living systems* (Langton 1988a). This definition does not distinguish ALife research from standard mathematical modeling based on ODEs (e.g., glucose regulation), which we have already discussed in depth. ALife studies differ from this approach by focusing on the abstract behaviors characteristic of living systems and not (usually) on the compartment dynamics of specific systems.

20.3 Goal Directedness: Robotic Crickets

Robots play a big role in ALife. They are studied in part because they constitute a practical, economically important by-product of this theoretical research. The practical applications are immense for self-reproducing, self-repairing, and self-contained robots that solve problems in novel environments. A few applications are mining and manufacture operations on earth and other planets, toxic waste clean-up, and other menial tasks that require minimal problem-solving abilities. Robots are also studied for theoretical reasons. They are particularly well-suited to validate theoretical models of learning and evolution in the context of two-dimensional movement (Michelsen et al. 1992; Beckers et al. 1994). They are a test-bed for algorithms usually implemented only in computer simulations.

One recent example of the latter type adds a biologically interesting twist: the robot is used as an *analog computer* (Chapter 5). The robot moves about in the environment according to its built-in mechanisms provided by sensors and on-board microprocessors. The robot does not emit digital numbers or voltages on an oscilloscope, but it performs calculations based on its inputs and produces an output in the form of its new spatial position.

In this example, Webb (1994) tested a biological hypothesis explaining how female crickets find a mate. A male cricket adopts a fixed position which is usually in a congregation of males of the same species (sometimes other species are present). The female's problem is to select a mate (based on his call) from a distance and move toward it without getting distracted by competing males or by the call of a different species. The mechanism that accomplishes this task is not known; Schöne (1984) and Weber and Thorson (1989) review the main competing hypotheses.

It is known that biaural location information is provided by the cricket's "ears" when sound waves stimulate a tympanic membrane on each of the two forelegs. These membranes are connected to each other via a trachae. Sound waves cause the membranes to oscillate and stimulate mechanoreceptive neurons. After sufficient stimulation, the neuron's response exceeds a threshold and a chain of motor activities are initiated for walking, turning, etc. Since the tympanic membranes are connected, each is stimulated by pressure waves both directly from the external source as well as from internal pressure waves that originate on the opposite side of the insect's body. The net amplitude of membrane oscillations at each ear is the sum of these two counteracting pressure waves acting on each side of the membrane.

It is also known that a cricket's song is composed of several hierarchical elements. A song is composed of repeated chirps; each chirp is composed of a few (e.g., three) rapidly repeated "syllables." Thus, there is a species specific chirping rate (frequency of chirps); each chirp has a fixed duration; and, there is a species specific duration of each syllable. Finally, if the wrong song is sung (e.g., by a male of another species), the female will respond and attempt to locate the male but will wander more or less at random, unable to pinpoint the source. The primary feature that causes tracking errors is the chirping frequency (Thorson et al. 1982). Incorrect frequencies produce errors in the form of *anomalous phonotaxis.*

One hypothesis for the female's orientation ability and limitations is that the spatial separation of the "ears" results in differences in sound intensity at the two sensors. By comparing relative intensities and turning toward the loudest sound, the cricket will orient to the source. To explain the orientation errors when the wrong song is present, it is hypothesized that each species contains a "filter" that allows it to identify the proper song. This hypothesis is unlikely because the difference in sound intensities at the two ears is small, and the hypothesis does not explain anomalous phonotaxis due to chirping frequency.

An alternative, simpler hypothesis is that the cricket compares, not absolute intensities (which in crickets would be minute differences), but the differences between the two ears. This will give directional information not because of attenuation but because of a phase shift that is transduced into amplified vibrations of the ear nearest the sound source. For a cricket oriented at a right angle to the source, the anatomical distance between the ears will cause a delay or a shift in the phase of the onset of a syllable in the most distant ear. For example, if a sound source is on the cricket's left, it impacts the left tympanic membrane directly, causing it to oscillate. The sound waves also travel the width of the cricket body (b) to enter the right ear and then travel the length of the trachae (t) in the opposite direction to stimulate the left membrane from the inside. If the sum of the distances $b + t$ is one-half the wavelength of the song (chirp), the external oscillation will be 180° out of phase with the internal oscillations. This will cause the two amplitudes to resonate and amplify, producing a strong stimulus to the neurons. The same phenomenon will also occur at the right ear, but the internal and external waves will arrive at the tympanic membrane at approximately the same time (no phase shift). The pressure stimuli from the two sources will be in phase and will cancel each other producing no vibrations of the tympanic membrane at the right ear. The net result is strong stimulation on the side from which the sound emanates and no stimulation on the opposite side. If the wave length of the song is not $2(b + t)$, then only weak resonance will occur on the side of the sound source. As a result, the cricket will have no directional information.

In addition to the phase shift, the distance b will cause a delay in the reception of the chirp on the right ear relative to the left ear. While this time difference is short, it is detectable. Thus, a cricket will turn towards that side for which the two stimulus is first received, and the strength of the turning response will depend on the frequency of the source.

This elegant hypothesis, in one simple stroke, accounts for most of the critical observations: chirping frequency is species specific, females orient to the proper sound, and females become confused when presented with songs of improper frequencies (or if the cricket is the wrong size for the song). There is no need to invoke the ability to detect small differences in sound intensity or the presence of a species-specific filter that turns on or off searching ability when the proper song is presented. Certainly, the actual mechanism is more involved than this and also involves neural activation decay rates, and low- and high-pass filters (Weber and Thorson 1989; Webb 1994). But the central concept is the effect on relative sound intensities of anatomically induced lags resulting in phase shifts and aural delays.

To demonstrate that the frequency-dependent phase shift hypothesis could account for the observations, Webb (1994) built a robot largely from plastic blocks (Fig. 20.1). Using reasonable scaling properties to account for the large robot size and distance between receptors, Webb built circuitry to shift the phase according to a "robotic specific" frequency. She then placed the robot in an arena (Fig. 20.2) with a source capable of playing at different frequencies to test the robot's ability to orient to the source. She found (Fig. 20.2b) that the song syllable frequency that was near the optimum of the robot's phase shift circuitry produced, on the whole, the shortest approaches with little meandering in the longest trial. Frequencies that are too high were the most difficult for the robot to track (Fig. 20.2a). Slow

Figure 20.1: The robotic implementation of the phase cancellation hypothesis of cricket phonotaxis. The phonoreceptors are the paired vertical structures located at the top right. (From Webb 1994, Fig. 3. From *Animals to Animats 3*, © 1994 by the Massachusetts Institute of Technology. Reprinted by permission of the MIT Press, publisher.)

songs could be tracked, but required that the robot ears be perpendicular to the source before a direct approach could be made (see best and median tracks Fig. 20.2c). This is consistent with experiments from real crickets.

Webb also tested if multiple song sources would confuse the robot by presenting it with two equally intense sources separated by 2.5 m but vibrating at the optimum frequency. She found that the robot moved directly to one or other source with little oscillatory movement. Which source was "chosen" appeared to be a random event. This result is qualitatively consistent with real crickets. Alternating the chirps of a song to come from the

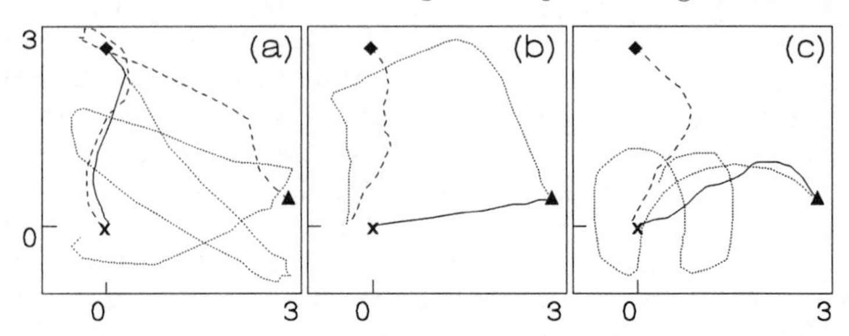

Figure 20.2: Three sets of experimental trials in which the robotic cricket attempts to move to the song source located at 0,0 ("X"). Shown are the x, y positions of the robot (in meters). Ten trials with two starting points each were performed. Solid lines are the shortest tracks in the trials; dashed lines are the medians; dotted lines are the longest tracks. (a) Song syllable frequency too fast: 2.5 Hz, (b) song syllable frequency near optimum: 1.25 Hz, and (c) song syllable frequency too slow: 1.00 Hz. (From Webb 1994, Fig. 5. From *Animals to Animats 3*, © 1994 by the Massachusetts Institute of Technology. Reprinted by permission of the MIT Press, publisher.)

two locations caused both the robot and the real crickets to act as if the source was located between the sources. The robot moves to the midpoint between the sources, then chooses one of the sources to approach. So a novel experimental regimen, which the robot was not explicitly designed to test, was consistent with the data. This further reinforces Webb's belief that the robot embodies the correct mechanism.

Webb (1994), as did Michelsen et al. (1992), uses robots as a simulation tool to test hypotheses on the mechanisms of goal-directedness. Although the method of computation is radically different than digital computer simulation, the philosophy is the same: evaluate competing hypotheses by isolating mechanisms. A natural question to ask at this point is: Since building robots is surely more difficult than programming computers, why not simply simulate the robot's movement where the movement rules adhere to the phase shift hypothesis? First, as Webb (1994) observes, the robot is a subset of cricket aural capabilities, not an abstraction which a mathematical model would have to be. Her model, of course, is an abstraction, since it is larger than real crickets, it uses wires and not nerves, and so on. Nevertheless, it is a real device and is, at least, less abstract than mathematical equations. Second, a computer simulation would have to convince us that the physiological mechanism would work in a realistic aural environment: more realistic than a simple, idealized point source, which we might be tempted to use in a computer simulation. However, simulating the aural environment of the computer robot would not be a trivial task; sound reflections from the floor and sides of the arena, perhaps the entire laboratory setting, would have to be computed. This would be computationally difficult, so much so that it is conceivable that building and running a robot may, in fact, require less time than implementing a computer simulation (see also Cliff et al. 1993). In any case, this is an intriguing application of technology that has emerged from the fields of artificial life and adaptive behavior and has practical importance for future robot design.

20.4 Emergence of Behavior: Evolving Robots

Webb's robot was a tool to test a behavioral and physical hypothesis. The robot was specifically designed to implement the phase cancellation mechanism for cricket orientation. As such, the robot did not adapt or evolve its own structure without human supervision. However, designing self-organizing robots or adaptive robots is useful for three reasons. First, adaptability is one of the signatures of living systems, and any adequate theory of life, including ALife, must propose mechanisms by which this can occur. Second, adaptive robots have obvious practical value. Their designers cannot anticipate all circumstances or problems that the robots will face. Therefore, some built-in ability to solve novel problems will extend the utility of the robot. Three, designing robots to perform real-world, non-trivial activities is very difficult for humans (Harvey et al. 1993). Evolutionary computation lets the system do the hard work.

Although the concept of adaptive robots applies to any programmed machine, we are mainly concerned here with *autonomous* robots: those in which all motive power, sensory inputs, and computation are contained in the robot itself. Usually these are free ranging machines, like that of Webb, that move about their environment performing the task for which they were designed. Many such autonomous robots have been designed to evolve and adapt to new circumstances relative to their designed performance task. Robots have been constructed that learn how to follow walls, to find power stations whereby they can recharge their batteries, collect objects in the environment, and so on. Here, we discuss one approach that combines computer simulation with real robots to design a robot that learns how to explore its environment by wandering around within a defined space.

Miglino et al. (1995) used computer simulation to evolve an abstract representation of a *virtual robot* (Harnard 1994) that would wander. They realized, however, that too often real robots do not behave according to theory, so they also built a *synthetic robot* (Harnard 1994) in which they could implement the abstract, evolved virtual robot and on which they could perform behavioral tests in arenas. In essence, the synthetic robot, which physically resembled Webb's robot (Fig. 20.1), was the laboratory in which the theory was tested.

The test situation that Miglino et al. (1995) used was an arena 2.6 m square inside of which was a 2.0 m square area painted white and surrounded by a large black border 0.6 m in width. The robot was capable of (1) moving forward and backward a specified distance, (2) turning a specified angle to the left or right, and (3) receiving optical input from the front and back of the robot.

The evolutionary process was based on a simple form of genetic algorithms that evolved an *artificial neural network* (ANN) that defined the actions a virtual robot was to take depending on its environment. We do not have the space to delve deeply into ANNs (see Beale and Jackson 1990), but the concept can be simply explained as a decision process. Suppose we wish to assign an unknown object to one of two possible groups (Fig. 20.3a). The two groups can be defined by the values of characteristics x_1 and x_2 as shown. The characteristics might be height and color, for example. The bold line separates the two groups; the other, dashed lines do not separate the groups correctly. Any such line can be represented as $C = w_1x_1 + w_2x_2$, where C is a constant, and w_1 and w_2 are *weighting* factors. For the two sets shown, there are particular values for C, w_1 and w_2 that separate the groups; other values fail. To determine a good "rule" for discriminating the groups, we must select the correct values for the constant and weighting factors. This problem is similar to defining the behavior of a neuron with various inputs (Fig. 20.3b). The neuron (decision unit) is to produce

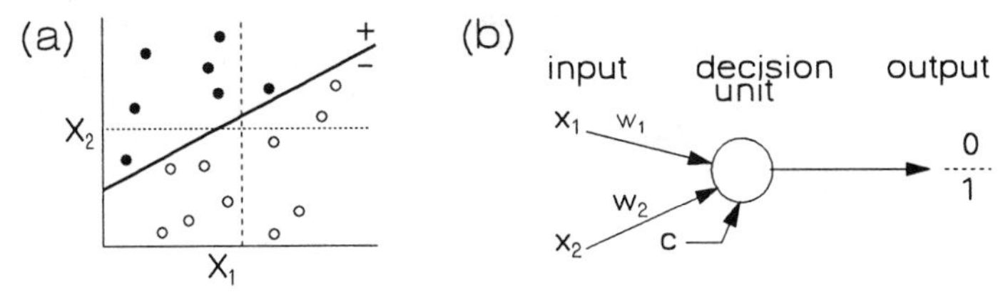

Figure 20.3: Discriminating between sets with neurons. (a) Elements of two sets to be discriminated based on variables x_1 and x_2 and the line that separates: $f(x_1, x_2) = C = w_1x_1 + w_2x_2$. One line (bold) separates the groups; other lines (dashed) do not separate the groups. (b) A neuron with three inputs and a decision unit that produces as output 0 or 1 if an object is an open circle or a filled circle in (a). Neural network algorithms find values for w_i and C so that classification errors are minimized.

a 0 if it is given input values (x_1 and x_2) that fall below the bold line in Fig. 20.3a. The neuron should produce a 1 if the input is above the bold line. In this analogy, teaching a machine to discriminate among groups is equivalent to training a ANN to produce a value (0 or 1) associated with the correct group, given the input. Training such a neuron is equivalent to choosing the correct values for the constant and weighting factors.

Correct robot behavior is similar: do the right thing, given particular inputs. To train a robot to perform a certain task is the same as teaching an ANN to first discriminate among its inputs (current environment), then perform the proper behavior given the type of environment in which it finds itself. For example, if we want to train a robot to wander around in a region demarcated by color, we must give it rules of proper behavior that will cause it to turn when it crosses the boundary. So, when the machine encounters the situation where its front sensor gives no light input (black) and its back sensor gives a positive light input (white), then the robot (ANN) should make a turn in order to remain within the demarcated area. Thus, a working robot has a number of phases: (1) receive optical input, (2) identify what the current environment is, (3) determine the current activity (state) of the robot (e.g., going forward, etc.), (4) define the proper course of action, and (5) activate the motor units to perform the prescribed action.

This definition can be represented as a finite state automaton (FSA, Section 18.2). Figure 20.4 shows one possible FSA for a robot. The states are digits in large circles; arrows are the transitions. Sensory input is represented by the digits in the ellipses, and motor control is represented by the digits on the arrows. Basically, the action of the FSA is interpreted as: "If the sensory input is `ii`, then send control signals `mm` to the motor and change state to `sss`." As the figure illustrates, different sensory inputs will cause different actions, and a robot in one state receiving a given input will behave differently than the robot in a different state but receiving the same input. Action is state- and input-dependent. For example, when a FSA in state `000` receives input `11`, it sends the signal `01` to the motor and changes its state to `011`.

The above is the description of an existing, but not necessarily optimal,

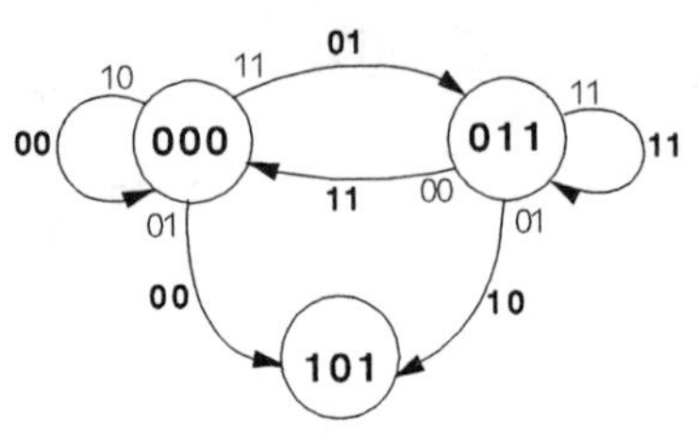

Figure 20.4: A finite state automaton for a robot. Digits in large circles are the states of the FSA; arrows are transitions between states. Digits at the base of the arrows are sensory unit inputs; bold-face digits attached to arrows define motor manipulations.

robot. When we evolve robots, we want to use an automated system that will design a robot of the above form in a way that maximizes a function (robot fitness). The system that Miglino et al. (1995) used both neural networks and genetic algorithms (Chapter 19). The neural network aspect involved using values of the weighting variables and constants to identify when an input should cause a change in state and which motor control output should be produced. Miglino et al. (1995) studied a virtual robot with two sensory inputs and two motor control outputs. The state of the FSA used three digits for eight possible states. The reader should consult the original paper for a detailed description of these states. The system contained a total of 17 weighting values and constants that needed to be specified. Miglino et al. (1995) evolved these 17 unknown values using a genetic algorithm without crossover. Hence, it was closer to evolutionary programming (Section 19.3.2) than genetic algorithms.

The algorithm used by Miglino et al. (1995) was as follows. Each of the 17 values represented a gene. A population of 100 robots were given random initial values for each of the 17 genes. Virtual robot fitness was evaluated in a simulated arena that was marked into discrete grids. Each robot was placed at a random starting point in the arena and allowed to move for 400 steps according to its behavioral rules embodied by the FSA (e.g., Fig. 20.4). The number of different grid points visited was counted. The fitness of the robot was the average number of points visited in 10 trials with different starting points. After each of the 100 robots were tested, those with the 80 lowest fitness values were thrown away and the top 20 were each duplicated four times to replenish the population with the most fit robots. No mating or crossover occurred, but genetic variation was introduced by randomly mutating 10% of the 1700 genes in the population. This procedure was repeated for 600 generations.

The main result from this evolution is summarized in Fig. 20.5. The top panels [(a) and (b)] show the best FSA evolved after 6 (fitness = 0.21) and 207 (fitness = 0.54) generations. Notice the increase in numbers of states with an increase in fitness. The bottom panels (Fig. 20.5c and d) show the behavior of a synthetic robot that uses each of these two FSAs. The synthetic robot did not exactly reproduce the virtual robot movement

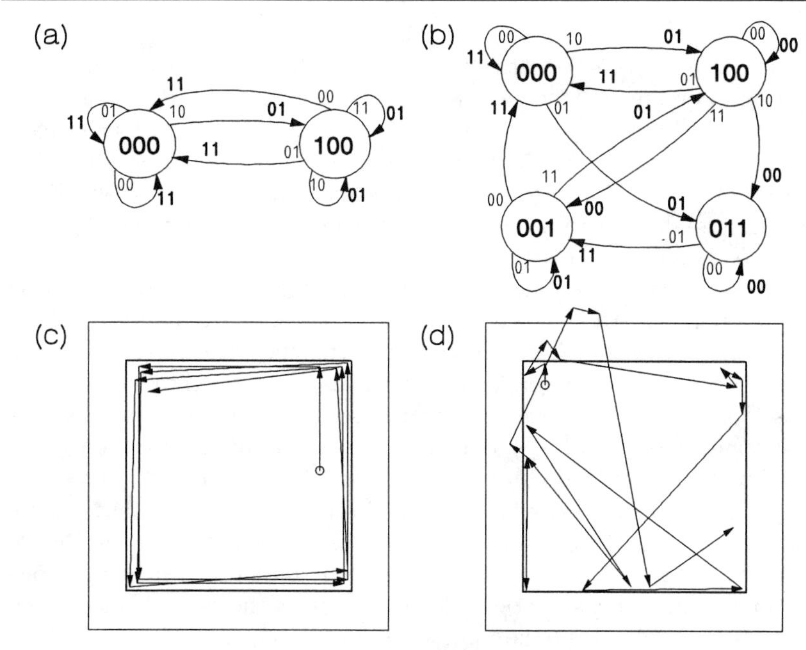

Figure 20.5: Performance of an evolved robot. Numbers in large circles are states, bold numbers on arrows are motor output commands (e.g., forward, turn left, etc.), and other numbers at the base of the arrows are the light sensor inputs (i.e., black or white). (a) The FSA the virtual robot at generation 6. (b) The FSA at generation 207. (c) One movement trial of the synthetic robot based on the FSA in (a). (d) One movement trial based on the FSA in (b). (From Miglino et al. 1995, Figs. 4a, 7b, 8a, and 8c. *Artificial Life, Vol 2*, © 1995 by the Massachusetts Institute of Technology. Reprinted by permission of the MIT Press, publisher.)

simulation due to random variation in traction, turning control, etc. This random noise helped the synthetic robot of generation 6, by breaking out of the cycle of endlessly following the walls. Randomness, however, reduced the fitness of the synthetic robot at generation 207 relative to that of the virtual robot. Figure 20.5 gives a visual demonstration that this approach is an effective method of evolving useful, exploratory behavior in real, autonomous robots. Moreover, the evolutionary approach was clearly superior to a random search. Using GA, Miglino et al. (1995) required 60,000 virtual robots to evolve a fitness of 0.54; random search of 5,000,000 virtual robots achieved only 0.45 fitness. Further, random search was very inefficient: 93% of the random robots had a fitness of less than 0.05.

20.5 Emergence of Structure: Dueling Blockata

Many studies in ALife are concerned with the evolution of organisms. These have stressed two main problem areas: evolution of the morphology of the organism and evolution of the control structures (perceptual and neurophysiological) that govern the use of the physical structure to maximize fitness. Several studies in the former group have emphasized only the structure of organisms. These have been quite elaborate extensions of the basic ideas of Lindenmayer as described in Section 18.5. See, for example, the photographs by Oppenheimer and Lindenmayer and Prusinkiewicz in Langton (1988b). Many studies in the latter group have emphasized the generation of the control structure either through evolution or a learning process (e.g., Koza et al. 1992).

Sims (1994) has broken new ground by performing computer simulations that incorporate both attributes. In Sims' scheme, a virtual creature (VC) is composed of two domains: control and morphology. Morphological structure is created by combining a set of articulating rigid three-dimensional plates ("blocks"). The plates are arranged into a crude body with articulating appendages (see Fig. 20.6). The movements of the appendages are controlled by an evolved control system that fires simulated nerves activating simulated muscles with prescribed strength. Sensors that provide information about the environment to the control system can also evolve. The resultant propulsion is constrained by the physical principles of inertia, momentum, and gravitational forces as well as the range of mechanical movement possible given the number and orientation of the appendages and joints. This "physics-based" approach to evolving organisms, such as that of Niklas (1986b) based on biomechanical principles, represents an important method of providing realism in an otherwise abstract computer simulation.

The control system is set of neurons organized into a network; each neuron computes an output from an input. Inputs are provided by connections to sensors (i.e., angle, contact, or photoreceptive), the output from other neurons, or constant values. Thus, basic feedback mechanisms between nerve firing intensity and the resultant change in a VC's posture can evolve. The number and complexity of neuronal interconnections is evolved. The VC moves by muscular contractions at the joints; the contractions are effected by nerves firing from the control system.

Both control and morphology of a particular VC are generated using recursive computer structures analogous to rewrite systems described for L-systems (Section 18.5). By this mechanism, a wide range of possible VCs are created. Using concepts from genetic algorithms (Chapter 19), a population of VCs are created and tested against each other. The success of each VC in pairwise contests provides a measure of the fitness of the combination of morphology and control system. Successful VCs are mated with each other with crossover so that offspring contain elements of both. Mutations also modify evolving VCs.

This work is significant for its combination of the evolution of neural network-like control systems and physically-based morphological structures. Its use of GA and recursive generation of novel automata demonstrates that complex adaptive systems can be made to evolve from simple

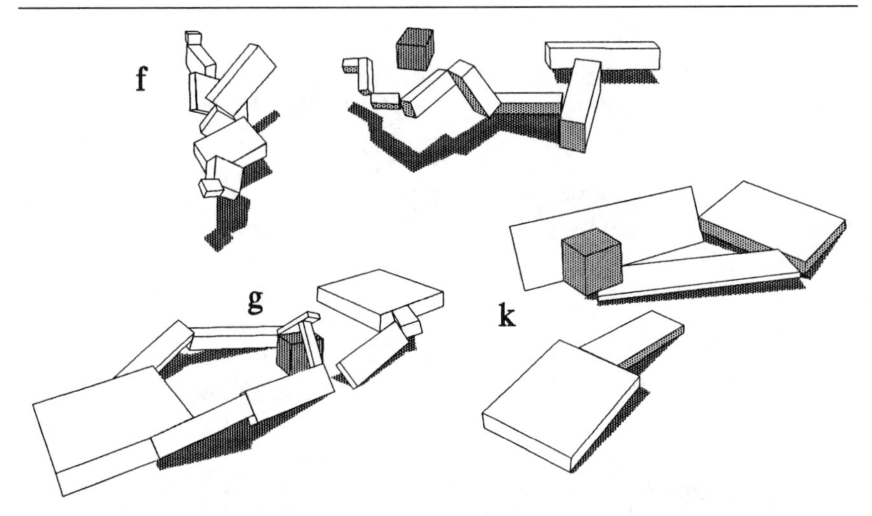

Figure 20.6: Virtual creatures evolved for competition by Karl Sims of Thinking Machines Corporation. Three examples of duels between evolved block-like automata ("virtual creatures", VCs) that are capable of simple movement directed towards capturing the cube that lies between them. VC "f": the VC on the right wins because it can extend its long arm and pursue the cube using sensors on the tip. VC "g": the left VC wins by lunging forward and pins the cube to the ground. VC "k": the right VC is the ultimate winner of all contests because it is fast and able to deflect the cube away from competitors. (From Sims 1994, Figs. 9f, g, and k. *Artificial Life IV*, © 1994 by the Massachusetts Institute of Technology. Reprinted by permission of the MIT Press, publisher.)

components. This is a computationally complex problem and along the way to its solution Sims (1994) provided another use of massively parallel computers in ALife.

20.6 Where Will It All End?

The three examples of simulations of complex adaptive systems were chosen primarily to illustrate the potential practical importance of this approach to biological modeling. There are many other, more abstract studies of complex adaptive systems that equally well show the power of this research area. These studies span a great variety of systems and approaches: the evolution of metabolic systems, epistasis in random Boolean networks, the origin of life, self-replication, the evolution of cooperation, and many others. But robots are physically real entities that have the potential to interact directly with our own bodies and lives. This gives to their study a special poignancy.

The idea of autonomous robots running around adapting to new problems and novel situations may sound like something out of a grade B science fiction movie. But the work of Miglino et al. (1995), Sims (1994), and many others, has demonstrated the power of random variation and selection to rapidly create autonomous entities with the potential to evolve out from under the control of their creators. All computer users are aware of the potential danger of computer viruses. Most of these are static entities that do not evolve; but it is not difficult to imagine the development of an advanced computer virus, more like a real biological virus, that is capable of evolving defenses against the latest virus detection software. [Already computer viruses exist that are capable of mutating their *byte-level appearance*, so as to foil virus detection schemes that search for sequences of bytes uniquely associated with known viruses (Spafford 1994).] It is not hard, under this scenario, to further imagine advanced anti-virus software that itself evolves counter-adaptations to the latest viral tricks. This concept extended into the realm of genetic engineering, nanotechnology, and biocomputers is enough to give even the most complacent computer user technological paranoia. Currently, there are serious research efforts to build robots with *evolvable hardware* (Higuchi et al. 1993). The control program of these robots, unlike those of Miglino et al. (1993) and Sims (1994), will evolve inside the robots themselves, not as an external computer simulation later used in a human-built robot.

This possibility raises social and ethical issues that are new to biological modeling. What are the social benefits of releasing into human society objects (e.g., robots, computer programs) that are, if not *alive*, then, at least, are complex adaptive systems and, therefore, are capable of independently evolving unanticipated actions? These are the same questions that must be addressed when contemplating the release of genetically altered biological material. Tomatoes with genes altered to better resist fungal attack seem to be a simple case compared with more drastic forms of genetic engineering that may appear in the near future. More generally, mathematical models have always had social and ethical implications, a famous example being simulations of the implosion characteristics of different structural designs of the first atomic bomb. But, in some ways, abstract simulations of harmless, wandering virtual robots seem less threatening, and, because of this, may be more dangerous. When the programs become evolving, synthetic robots, we need wisdom. As Farmer and Gelin (1992, p. 836) put it: "...we may be the first species to create its own successors. What will these successors be like?"

We certainly are not yet in a moral or life-threatening crisis induced by ALife and the study of complex adaptive systems, but it is not too soon for those working on this theory, especially in areas such as synthetic robot design, to contemplate the incorporation of fail-safe features that will limit the spread of these potentially self-reproducing entities. Unfortunately, incorporating adaptability into these systems, the very feature that makes them useful, implies that it will be difficult, if not impossible, to prevent the systems from evolving undesirable behavior. It all depends on the fitness function that the entities are attempting to optimize. Isaac Asimov (1950) imagined that future robots would have a fitness function incorporating the dictum: "Never harm a human." It remains to be seen if complex adaptive systems, synthetic or virtual, can be designed that do not have the ability to modify their own fitness function.

References

Abbott, L.C. 1990. *Applying Resource Based Competition Models to Variable Environments.* Ph.D. Thesis, Utah State University, Logan, UT. 155 pages.

Abraham, R. and J.E. Marsden. 1967. *Foundations of Mechanics.* Benjamin Publishers, Reading, MA.

Ackerman, E., Z. Zhuo, M. Altmann, D. Kilis, J-J Yang, S. Seaholm, and L. Gatewood. 1993. Simulation of stochastic micropopulation models—I. The SUMMERS simulation shell. *Computers in Biology and Medicine* 23(3):177–198.

Allen, J.C., W.M. Schaffer, and D. Rosko. 1993. Chaos reduces species extinction by amplifying local population noise. *Nature* 364:229–232.

Adams, S.M. and D.L. DeAngelis. 1987. Indirect effects of early bass-shad interactions on predator population structure and food web dynamics. In: W.C. Kerfoot and A. Sih (eds). *Predation: Direct and Indirect Impacts on Aquatic Communities.* University Press of New England, Hanover, NH. Pages 103–117.

Anderson, S. 1974. Patterns of faunal evolution. *Quarterly Review of Biology* 49:311–332.

Aono, M. and T.L. Kunii. 1984. Botanical tree image generation. *IEEE Computer Graphics and Applications* 4(5):10–34.

Arbib, M.A. 1965. *Brains, Machines, and Mathematics.* McGraw-Hill Book Co., NY.

Asimov, I. 1950. I, robot. New American Library, NY. 192 pages.

Bäck, T. 1994. Evolutionary algorithms: comparison of approaches. In: R. Paton (ed). *Computing with Biological Metaphors.* Chapman and Hall, London. Pages 227–243.

Bäck, T. and H-P. Schwefel. 1993. An overview of evolutionary algorithms for parameter optimization. *Evolutionary Computation* 1(1):1–23.

Baker, G.L. and J.P. Gollub. 1990. *Chaotic Dynamics: An Introduction.* Cambridge University Press, Cambridge, UK. 182 pages.

Balci, O. and R.G. Sargent. 1982. Validation of multivariate response models using Hotelling's two-sample T^2 test. *Simulation* 39(6):185–192.

Ball, J.T., I.E. Woodrow, and J.A. Berry. 1987. A model predicting stomatal conductance and its contribution to the control of photosynthesis under different environmental conditions. In: J. Biggins (ed). *Progress in Photosynthesis Research, Proceedings of the 7th International Congress. Vol. 4.* Kluwer, Boston. Pages 221–224.

Band, L.E., D.L. Peterson, S.W. Running, J. Couglan, R. Lammers, J. Dungan, and R. Nemani. 1991. Forest ecosystem processes at the watershed scale: basis for distributed simulation. *Ecological Modelling* 56:171-196.

Bartell, S.M., J.E. Breck, R.H. Gardner, and A.L. Brenkert. 1986. Individual parameter perturbation and error analysis of fish bioenergetics models. *Canadian Journal of Fisheries and Aquatic Sciences* 43:160–168.

Bartell, S.M., A.L. Brenkert, and S.R. Carpenter. 1988. Parameter uncertainty and the behavior of a size-dependent plankton model. *Ecological Modelling* 40:85–95.

Batschelet, E. 1979. *Introduction to Mathematics for Life Scientists.* Third Edition. Springer-Verlag, Berlin. 643 pages.

Beale, R. and T. Jackson. 1990. *Neural Computing: An Introduction.* Institute of Physics Publishing, Bristol, UK. 240 pages.

Beckers, R., O.E. Holland, and J.L. Deneubourg. 1994. From local actions to global tasks: stigmergy and collective robotics. In: R.A. Brooks and P. Maes (eds). *Artificial Life IV.* The MIT Press, Cambridge, MA. Pages 181–189.

Berg, H.C. 1983. *Random Walks in Biology.* Princeton University Press. Princeton, NJ.

Berne, R.M. and M.N. Levy. (eds). 1993. *Physiology.* 3rd Ed. Mosby-Year Book. St. Louis, MO. 1071 pages.

Berryman, A.A. 1991.Vague notions of density-dependence. *Oikos* 62(2):252–253.

Berryman, A.A. and J.A. Millstein. 1989. Are ecological systems chaotic — And if not, why not? *Trends in Ecology and Evolution* 4:26–28.

Blau, G.E. and W.B. Neely. 1975. Mathematical model building with an application to determine the distribution of Dursban insecticide added to a simulated ecosystem. *Advances in Ecological Research* 9:133–163.

Blumberg, A.A. 1968. Logistic growth rate functions. *Journal of Theoretical Biology* 21:42–44.

Bohachevsky, I.O., M. E. Johnson, and M.L. Stein. 1986. Generalized simulated annealing for function optimization. *Technometrics* 28(3):209–217.

Boote, K.J. and R.S. Loomis. 1991. The prediction of canopy assimilation. In: K.J. Boote and R.S. Loomis (eds). *Modeling Crop Photosynthesis — from Biochemistry to Canopy.* Crop Science Society of America, Madison, WI. Pages 109–140.

Borowski, E.J. and J.M. Borwein. 1991. *The HarperCollins Dictionary of*

Mathematics. HarperCollins Publishers, NY.

Botkin, D.B., J.F. Janak, and J.R. Wallis. 1977. Some ecological consequences of a computer model of forest growth. *Journal of Ecology* 60:849–872.

Boulding, K.E. 1972. Economics and general systems. In: E. Laszlo (ed). *The Relevance of General Systems Theory.* G. Braziller Publisher, NY. Pages 78–92.

Boyce, M.S. 1992. Population viability analysis. *Annual Review of Ecology and Systematics* 23:481-506.

Brackbill, J.U. and B.I. Cohen. (eds). 1985. *Multiple Time Scales.* Academic Press, NY. 442 pages.

Bradley, C.E. and T. Price. 1992. Graduating sample data using generalized Weibull functions. *Applied Mathematics and Computation* 50:115–144.

Bratley, P., B.L. Fox, and L.E. Schrage. 1987. *A Guide to Simulation.* Second Edition. Springer-Verlag, NY.

Brent, R.P. 1973. *Algorithms for Minimization Without Derivatives.* Prentice-Hall, Englewood Cliffs, New Jersey.

Brown, J.H. 1995. *Macroecology.* University of Chicago Press, Chicago, IL. 269 pages.

Brown, R. 1990. Although extremely powerful, polynomial curve fitting springs hidden surprises. *Personal Engineering and Instrumentation News* 7:57–61.

Buis, R. 1991. On the generalization of the logistic law of growth. *Acta Biotheoretica* 39:185–195.

Burmaster, D.E. 1979a. The continuous culture of phytoplankton: mathematical equivalence among three steady-state models. *American Naturalist* 113:123–134.

Burmaster, D.E. 1979b. The unsteady continuous culture of phosphate-limited *Monochrysis lutheri* Droop: experimental and theoretical analysis. *Journal of Experimental Marine Biology and Ecology* 39:167–186.

Burnham, K.P. 1995. Pseudopower and the design of experiments. *Bulletin of the Ecological Society of America* 76(2):35.

Caceci, M. S. and W. P. Cacheris. 1984. Fitting curves to data. *Byte* 9:340–362.

Caldwell, M.M., H.-P. Meister, J.D. Tenhunen, and O.L. Lange. 1986. Canopy structure, light microclimate and leaf gas exchange of *Quercus coccifera* L. in a Portuguese macchia: measurements in different canopy layers and simulations with a canopy model. *Trees: Structure and Function* 1:25–41.

Cale, W.G., R.V. O'Neill, and R.H. Garnder. 1983. Aggregation error in nonlinear ecological models. *Journal of Theoretical Biology* 100:539–550.

Card, O.S. 1982. *Speaker for the Dead.* Tom Doherty Associates, NY.

Cardon, Z.G., K.A. Mott, and J.A. Berry. 1994. Dynamics of patchy stomatal movements, and their contribution to steady-state and oscillating stomatal conductance calculated with gas-exchange techniques. *Plant, Cell and Environment* 17:995–1008.

Carpenter, S.R. 1990. Large-scale perturbations: opportunities for innovation. *Ecology* 71:2038–2043.

Carson, E.R., C. Cobelli, and L. Finkelstein. 1983. *The Mathematical Modeling of Metabolic and Endocrine Systems.* John Wiley and Sons, NY.

Casti, J.L. 1992. *Reality Rules: I. Picturing the World in Mathematics: The Fundamentals.* John Wiley and Sons, NY.

Caswell, H. 1975. The validation problem. In: B.C. Patten (ed). *Systems Analysis and Simulation in Ecology. Vol IV.* Academic Press, NY. Pages 313–325.

Caswell, H. 1976. Community structure: a neutral model analysis. *Ecological Monographs* 46:327–354.

Caswell, H. 1989. *Matrix Population Models.* Sinauer Associates, Sunderland, MA. 328 pages.

Caswell, H., H.E. Koenig, and J.A. Resh. 1972. An introduction to systems science for ecologists. In: B.C. Patten (ed). *Systems Analysis and Simulation in Ecology. Vol II.* Academic Press, NY. Pages 3–78.

Chatfield, C. 1975. *The Analysis of Time Series: Theory and Practice.* Chapman and Hall, London. 263 pages.

Chomsky, N. 1957. *Syntactic Structures.* Moulton. Berlin.

Christini, D.J and J.J. Collins. 1995. Controlling nonchaotic neuronal noise using chaos control techniques. *Physical Review Letters* 75(14):2782–2785.

Cobelli, C., G. Federspil, G. Pacini, A. Salvan, and C. Scandellari. 1982. An integrated mathematical model of the dynamics of blood glucose and its hormonal control. *Mathematical Biosciences* 58:27–60.

Cohen, J. 1988. *Statistical Power Analysis for the Behavioral Sciences.* 2nd Edition. Lawrence Erlbaum, Hillsdale, NJ.

Coleman, T.G. and W.J. Gay. 1990. Simulation of typical physiological systems. In: D.P.F. Möller (ed). *Advanced Simulation in Biomedicine.* Springer-Verlag, NY. Pages 41–69.

Connell, J.H. 1978. Diversity in tropical rain forests and coral reefs. *Science* 199:1304-1310.

Conrad, M. 1986. What is the use of chaos? In: A.V. Holden (ed). *Chaos.* Manchester University Press, Manchester, UK. Pages 3-14.

Copi, I.M. 1954. *Symbolic Logic.* MacMillan Co, NY.

Costantino, R.F., J.M. Cushing, B. Dennis, and R.A. Desharnais. 1995. Experimentally induced transitions in the dynamic behavior of insect populations. *Nature* 375:227–230.

Costanza, R. and F.H. Sklar. 1985. Articulation, accuracy, and effectiveness of mathematical models: a review of freshwater wetland applications. *Ecological Modelling* 27:45–69.

Costanza, R., F.H. Sklar, and M.L. White. 1990. Modeling coastal landscape dynamics. *BioScience* 40(2):91–107.

Crowley, P.H. 1992. Resampling methods for computation-intensive data analysis in ecology and evolution. *Annual Review of Ecology and Systematics* 23:405–447.

Cullinan, V.I. and J.M. Thomas. 1992. A comparison of quantitative methods for examining landscape pattern and scale. *Landscape Ecology* 7(3):211–227.

Cushing, J.M., B. Dennis, R.A. Desharnais, and R.F. Costantino. 1996. An interdisciplinary approach to understanding nonlinear ecological dynamics. *Ecological Modelling* In Press.

Dale, M. 1980. A syntactic basis of classification. *Vegetatio* 42:93–98.

Davis, J.C. 1986. *Statistics and Data Analysis in Geology.* John Wiley and Sons, NY. 646 pages.

Davis, L. (ed) 1991. *Handbook of Genetic Algorithms.* Van Nostrand Reinhold, NY. 385 pages.

Davis, M.B. 1986. Climatic instability, time lags, and community disequilibrium. In: J. Diamond and T.J. Case (eds). *Community Ecology.* Harper and Row, NY. Pages 269–284.

Davis, M.B. and D.B. Botkin. 1985. Sensitivity of cool temperate forests and their fossil pollen record to rapid temperature change. *Quaternary Research* 23:327–340.

Davis, P.J. and R. Hersh. 1981. *The Mathematical Expericence.* Houghton-Mifflin Co., Boston. 440 pages.

DeAngelis, D.L. 1992. *Dynamics of Nutrient Cycling and Food Webs.* Chapman and Hall, London. 270 pages.

DeAngelis, D.L., L. Godbout, B.J. Shuter. 1991. An individual-based approach to predicting density-dependent dynamics in smallmouth bass populations. *Ecological Modelling* 57:91–115.

DeAngelis, D.L. and L.J. Gross. (eds). 1992. *Individual-based Models and Approaches in Ecology: Populations, Communities, and Ecosystems.* Chapman and Hall, NY. 525 pages.

DeAngelis, D.L. and K.A. Rose. 1992. Which individual-based approach is most appropriate for a given problem? In: D.L. DeAngelis and L.J. Gross (eds). *Individual-based Models and Approaches in Ecology: Populations, Communities, and Ecosystems.* Chapman and Hall. NY. Pages 67–87.

Delwiche, M.J. and J.R. Cooke. 1977. An analytical model of the hydraulic aspects of stomatal dynamics. *Journal of Theoretical Biology* 69:113–141.

Dennis, B., R.A. Desharnais, J.M. Cushing, and R.F. Costantino. 1995. Nonlinear demographic dynamics: mathematical models, statistical methods, and biological experiments. *Ecological Monographs* 65(3):261–281.

Dennis, B. and M.L. Taper. 1994. Density dependence in time series observations of natural populations: estimation and testing. *Ecological Monographs* 64(2):205–224.

Dent, J.B. and M.J. Blackie. 1979. *Systems Simulation in Agriculture.* Applied Science Publishers, Ltd, London. 180 pages.

Deutsch, S. and A. Deutsch. 1993. *Understanding the Nervous System: An Engineering Perspective.* IEEE Press, NY. 394 pages.

Dewar, R.C. 1993. A root-shoot partitioning model based on carbon-nitrogen-water interactions and Münch phloem flow. *Functional Ecology* 7:356–368.

Dijkstra, E.W. 1988. "Foreword" in C.A.R. Hoare. 1988. *Communicating Sequential Processes.* Prentice-Hall, Englewood Cliffs, NJ.

DiStefano, J.J., A.R. Stubberud, and I.J. Williams. 1967. *Feedback and Control Systems.* Schaum's Outline of Theory and Problems. McGraw-Hill Book Company, NY.

Dyke, B. and J. MacCluer. (eds). 1973. *Computer Simulation in Human Population Studies.* Academic Press, NY.

Earnshaw, J.C. and D. Haughey. 1993. Lyapunov exponents for pedestrians. *American Journal of Physics* 61(5):401–407.

Ebenman, B. and L. Persson. (eds). 1988. *Size-Structured Populations: Ecology and Evolution.* Springer-Verlag, Berlin. 284 pages.

Edelstein-Keshet, L. 1988. *Mathematical Models in Biology.* Random House, NY. 586 pages.

Efron, B. 1986. Why isn't everyone a Bayesian? *American Statistical Association* 40:1–11.

Ellner, S. 1991. Detecting low-dimensional chaos in population dynamics data: a critical review. In: J. Logan and F.P. Hain (eds). *Chaos and Insect Ecology.* Virginia Agriculture Experiment Station. Information Series 91–3. Virginia Polytechnic Institute and State University. Blacksburg, VA. Pages 63–90.

Ellner, S. and P. Turchin. 1995. Chaos in a noisy world: new methods and evidence from time-series analysis. *American Naturalist* 145(3):343–375.

Evans, J.R. and G.D. Farquhar. 1991. Modeling canopy photosynthesis from the biochemistry of the C_3 chloroplast. In: K.J. Boote and R.S. Loomis (eds). *Modeling Crop Photosynthesis — from Biochemistry to Canopy.* Crop Science Society of America, Madison, WI. Pages 1–15.

Farmer, J.D., E. Ott, and J.A. Yorke. 1983. The dimension of chaotic attractors. *Physica* 7D:153–180.

Farmer, J.D. and J.J. Sidorowich. 1989. Exploiting chaos to predict the future and reduce noise. In: Y.C Lee (ed). *Evolution, Learning and Cognition.* World Scientific Press, NY. Pages 277–304.

Farquhar, G.D. and S. von Caemmerer. 1982. Modelling of photosynthetic response to environmental conditions. In: O.L. Lange, P.S. Nobel, C.B. Osmond, and H. Ziegler (eds). *Encyclopedia of Plant Physiology New*

Series 12B: Physiological Plant Ecology II. Springer-Verlag, Berlin. Pages 549–587.

Feldman, M.W. and J. Roughgarden. 1975. A population's stationary distribution and chance of extinction in a stochastic environment with remarks on the theory of species packing. *Theoretical Population Biology* 7:197–207.

Feldman, R.M., G.L. Curry, and T.E. Wehrly. 1984. Statistical procedure for validating a simple population model. *Environmental Entomology* 13:1446–1451.

Fisher, J.B. 1992. How predictive are computer simulations of tree architecture? *International Journal of Plant Science* 153(3):S137–S146.

Fisher, J.B. and H. Honda. 1979. Branch geometry and effective leaf area: a study of *Terminalia*-branching pattern. I. Theoretical trees. *American Journal of Botany* 66:663–644.

Fleiss, J.L. 1973. *Statistical Methods for Rates and Proportions.* John Wiley and Sons, NY. 223 pages.

Fogel, D.B. 1994a. Evolutionary programming: an introduction and some current directions. *Statistics and Computing* 4:113–129.

Fogel, D.B. 1994b. Applying evolutionary programming to selected control problems. *Computers and Mathematical Applications* 27(11):89-104.

Fogel, D.B. 1994c. On the effectiveness of crossover in simulated evolutionary optimization. *Biosystems* 32:171–182.

Fogel, D.B. and L.C. Stayton. 1994. On the effectiveness of crossover in simulated evolutionary optimization. *BioSystems* 32:171–182.

Fogel, L.J., A.J. Owens, and M.J. Walsh. 1966. *Artificial Intelligence Through Simulated Evolution.* John Wiley and Sons, NY.

Folse, L.J., J.M. Packard, and W.E. Grant. 1989. AI modelling of animal movements in a heterogeneous habitat. *Ecological Modelling* 46:57–72.

Forrester, J.W. 1961. *Industrial Dynamics.* MIT Press, Cambridge, MA.

Forrester, J.W. 1971. *World Dynamics.* Wright-Allen Press. Cambridge, MA.

France, J. and J.H.M. Thornley. 1984. *Mathematical Models in Agriculture.* Butterworths Publishers, London.

Frost, T.M., D.L. DeAngelis, S.M. Bartell, D.J. Hall, and S.H. Hurlbert. 1988. Scale in the design and interpretation of aquatic community research. In: S.R. Carpenter (ed). *Complex Interactions in Lake Communities.* Springer-Verlag, NY. Pages 229–258.

Gardner, M. 1970. Mathematical Games: the fantastic combinations of John Conway's new solitaire game 'life'. *Scientific American* 223(4):120–123.

Gardner, M. 1971. Mathematical games: on cellular automata, self-reproduction, the Garden of Eden, and the game 'life.' *Scientific American* 224(2):112–117.

Gardner, R.H., W.G. Cale, and R.V. O'Neill. 1982. Robust analysis of aggregation error. *Ecology* 63(6):1771–1779.

Gardner, R.H., R.V. O'Neill, J.B. Mankin, and J.H. Carney. 1980. A comparison of sensitivity analysis and error analysis based on a stream ecosystem model. *Ecological Modelling* 12:177–194.

Gatewood, L.C. 1971. *Stochastic Simulation of Influenza A Epidemics Within a Structured Community.* Ph.D. Dissertation, University of Minnesota, Minneapolis, MN.

Gause, G.F. 1934. *The Struggle for Existence.* The Williams and Wilkins Company, Baltimore, MD.

Gershenfeld, N.A. and A.S. Weigend. 1994. The future of time series: Learning and understanding. In: A.S. Weigend and N.A. Gershenfeld (eds). *Time Series Prediction: Forecasting the Future and Understanding the Past.* Santa Fe Institute, Studies in the Sciences of Complexity, Vol. XV. Addison-Wesley, Reading, MA. Pages 1–70.

Gilpin, M.E. 1979. Spiral chaos in a predator-prey model. *American Naturalist* 113(2):306–308.

Glass, L. and M.C. Mackey. 1988. *From Clocks to Chaos: The Rhythms of Life.* Princeton University Press, Princeton, NJ.

Gleick, J. 1987. *Chaos, Making a New Science.* Viking Press, NY.

Goldberg, D.E. 1989. *Genetic Algorithms in Search, Optimization, and Machine Learning.* Addison-Wesley Publishing Co., Reading, MA.

Goldberger, A.L. 1992. Applications of chaos to physiology and medicine. In: J.H. Kim and J. Stringer (eds). *Applied Chaos.* John Wiley and Sons, NY. Pages 321–331.

Goldberger, A.L. and D.R. Rigney. 1991. Nonlinear dynamics at the bedside. In: L. Glass, P. Hunter, and A. McCulloch (eds). *Theory of Heart.* Springer-Verlag, NY. Pages 583–605.

Goldberger, A.L. and B.J. West. 1987. Chaos in physiology: health or disease? In: H. Degn, A.V. Holden, and L.F. Olsen (eds). *Chaos in Biological Systems.* Plenum Press, NY. Pages 233–248.

Goodman, D. 1987. The demography of chance extinction. In: M.E. Soulé (ed). *Viable Populations for Conservation.* Cambridge University Press, Cambridge, UK. Pages 11–34.

Grant, W. E. 1986. *Systems Analysis and Simulation in Wildlife and Fisheries Sciences.* John Wiley and Sons, NY.

Grassberger, P. and I. Procaccia. 1983. Characterization of strange attractors. *Physics Review Letters* 50:346–349.

Green, R.H. 1979. *Sampling Design and Statistical Methods for Environmental Biologists.* John Wiley and Sons, NY.

Grenny, W.J., D.A. Bella, and H.C. Curl, Jr. 1973. A theoretical approach to interspecific competition in phytoplankton communities. *American Naturalist* 107:405–425.

Grimm, V., E. Schmidt, and C. Wissel. 1992. On the application of stability concepts in ecology. *Ecological Modelling* 63:143–161.

Gross, L.J. 1982. Photosynthetic dynamics in varying light environments: a model and its application to whole leaf carbon gain. *Ecology* 63(1):84-93.

Grossman, S.I. and J.E. Turner. 1974. *Mathematics for the Biological Sciences.* Macmillan Publishers, NY.

Grover, J.P. 1990. Resource competition in a variable environment: phytoplankton growing according to Monod's model. *American Naturalist* 136:771–789.

Guckenheimer, J. and P. Holmes. 1990. *Nonlinear Oscillations: Dynamical Systems and Bifurcations of Vector Fields.* 3rd Ed. Springer-Verlag, NY.

Guckenheimer, J., G. Oster, and Ipaktchi, A. 1977. The dynamics of density–dependent population models. *Journal of Mathematical Biology* 4:101–147.

Guyton, A.C. 1986. *Textbook of Medical Physiology.* 7th Ed. W.B. Saunders, Philadelphia, PA.

Haefner, J.W. 1975. *Generative Grammars that Simulate Ecological Systems.* Ph.D. Dissertation. Oregon State University, Corvallis, OR.

Haefner, J.W. 1992. Parallel computers and individual-based models: An overview. In: D.L. DeAngelis and L. Gross (eds). *Individual-based Models and Approaches in Ecology: Populations, Communities, and Ecosystems.* Chapman and Hall, NY. Pages: 126–164.

Haefner, J.W. and T.O. Crist. 1994. Spatial model of movement and foraging in harvester ants (Pogonomyrmex) (I): The roles of memory and communication. *Journal of Theoretical Biology* 166:299–313.

Halfon, E. 1989. Probabilistic validation of computer simulations using the bootstrap. *Ecological Modelling* 46:213–219.

Hall, C.A.S. and D.L. DeAngelis. 1985. Models in ecology: paradigm found or paradigm lost? *Bulletin of the Ecological Society of America* 66:339–346.

Hamming, R.W. 1962. *Numerical Methods for Scientists and Engineers.* McGraw-Hill, NY.

Hansen, S.R. and S.P. Hubbell. 1980. Single-nutrient microbial competition: qualitative agreement between experimental and theoretcially forecast outcomes. *Science* 207:1491–1493.

Hanski, I. 1991. Single-species metapopulation dynamics: concepts, models and observations. *Biological Journal of the Linnean Society* 42:17–38.

Hanson, N.R. 1972. *Patterns of Discovery.* Cambridge University Press, Cambridge, UK.

Harnard, S. 1994. Artificial Life: synthetic *vs* virtual. In: C.G. Langton (ed). *Artificial Life III.* Addison-Wesley Publishing Co., Reading, MA. Pages 539–552.

Harrison, G.W. 1995. Comparing predator-prey models to Luckinbill's experiment with *Didinium* and *Paramecium. Ecology* 76(2):357–374.

Hassell, M.P. 1978. *The Dynamics of Arthropod Predator-Prey Systems.* Princeton University Press. Princeton, New Jersey.

Hastings, A., C.L. Hom, S. Ellner, P. Turchin, and H.C.J. Godfray. 1993. Chaos in ecology: is Mother Nature a strange attractor? *Annual Review of Ecology and Systematics* 24:1–33.

Hayashi, H. and S. Ishizuka. 1987. Chaos in molluscan neuron. In: H. Degn, A.V. Holden, and L.F. Olsen (eds). *Chaos in Biological Systems.* Plenum Press, NY. Pages 157–166.

High Performance Systems Inc. 1992. *Stella II Tutorial and Technical Documentation.* High Performance Systems, Hanover, NH.

Higuchi, T., T. Niwa, T. Tanaka, H. Iba, H. de Garis, and T. Furuya. 1993. Evolving hardware with genetic learning: a first step towards building a Darwin Machine. In: J-A.Meyer, H.L.Roitblat, and S.W. Wilson. (eds). *From Animals to Animats 2: Proceedings of the Second International Conference on Simulation of Adaptive Behavior.* MIT Press, Cambridge, MA. Pages 417–424.

Hilborn, R.C. 1994. *Chaos and Nonlinear Dynamics.* Oxford University Press, Oxford.

Hillier, F.S. and G.J. Lieberman. 1980. *Introduction to Operations Research.* Holden-Day, Inc., San Francisco.

Hodgman, C.D., R.C. Weast, and S.M. Selby. (eds). 1955. *Handbook of Chemistry and Physics.* 37th Ed. Chemical Rubber Publishing, Co. Cleveland, OH.

Hogeweg, P. and B. Hesper. 1974. A model study of biomorphological description. *Pattern Recognition* 6:165–179.

Holland, J.H. 1975. *Adaptation in Natural and Artificial Systems: An Introductory Analysis with Applications to Biology, Control, and Artificial Intelligence.* University of Michigan Press, Ann Arbor, MI. 183 pages.

Holland, J.H. 1992. *Adaptation in Natural and Artificial Systems: An Introductory Analysis with Applications to Biology, Control, and Artificial Intelligence.* The MIT Press, Cambridge, MA. 211 pages. [Re-issue of Holland 1975 with an additional chapter.]

Holling, C.S. 1959. Some characteristics of simple types of predation and parasitism. *Canadian Entomologist* 91:385–398.

Holling, C.S. (ed). 1978. *Adaptive Environmental Assessment and Management.* John Wiley and Sons, NY.

Holton, D. and R.M. May. 1990. Chaos and one-dimensional maps. In: T. Mullin (ed). *The Nature of Chaos.* Clarendon Press, Oxford, UK. Pages 95–119.

Honda, H. 1971. Description of the form of trees by the parameters of the tree-like body: effects of the branching angle and the branching length on the tree-like body. *Journal of Theoretical Biology* 31:331-338.

Honda, H. and J.B. Fisher. 1978. Tree branch angle: maximizing effective leaf area. *Science* 199:888–890.

Hopcroft, J.E. and J.D. Ullman. 1969. *Formal Languages and Their Relation to Automata.* Addison-Wesley Publishing Co., Reading, MA.

Horn, H. 1975. Markovian processes of forest succession. In: M.L. Cody and J. Diamond (eds). *Ecology and Evolution of Communities.* Harvard University Press, Cambridge, MA. Pages 196–213.

Huston, M., D. DeAngelis, and W. Post. 1988. New computer models unify ecological theory. *Bioscience* 38:682–691.

Hyman, J.B., J.B. McAninch, and D.L. DeAngelis. 1991. An individual-based simulation model of herbivory in a heterogeneous landscape. In: M.G. Turner and R.H. Gardner (eds). *Quantitative Methods in Landscape Ecology: The Analysis and Interpretation of Landscape Heterogeneity.* Springer-Verlag, NY. Pages 443–475.

Ingber, L. 1989. Very fast simulated re-annealing. *Mathematical and Computer Modelling* 12(8):967–973.

Ingber, L. and B. Rosen. 1992. Genetic algorithms and very fast simulated reannealing: a comparison. *Mathematical and Computer Modelling* 16(11):87–100.

Innis, G.S. 1975. *Stability Concepts in Mathematics and Ecology and Their Application to Social Systems.* Report No. 1. Regional Analysis of Grassland Environmental Systems. Colorado State University. Fort Collins, CO. 32 pages.

Innis, G.S. (ed). 1978. *Grassland Simulation Model.* Springer-Verlag, NY.

Innis, G.S. 1979. A spiral approach to ecosystem simulation, I. In: G.S. Innis and R.V. O'Neill (eds). *Systems Analysis of Ecosystems.* International Co-operative Publishing House, Fairland, Maryland. Pages 211–386.

Iwasa, Y., V. Andreasen, and S. Levin. 1987. Aggregation in model ecosytems. I. Perfect aggregation. *Ecological Modelling* 37:287–302.

Jaeger, M. and P.H. de Reffye. 1992. Basic concepts of computer simulation of plant growth. *Journal of Bioscience (India)* 17(3):275–291.

Jarvis, P.G. 1993. Prospects for bottom-up models. In: J.R.Ehleringer and C.B. Field (eds). *Scaling Physiological Processes: Leaf to Globe.* Academic Press, San Diego, CA. Pages 115–126.

Jarvis P.G, and K.G. McNaughton. 1986. Stomatal control of transpiration: scaling up from leaf to region. *Advances in Ecological Research* 15:1–49.

Jefferson, D., R. Collins, C. Cooper, M. Dyer, M. Flowers, R. Korf, C. Taylor, and A. Wang. 1992. Evolution as a theme in artificial life: the Genesys/Tracker System. In: C. Langston, C. Taylor, J.D. Farmer and S. Rasmussen (eds). *Artifical Life II.* Addison-Wesley, Redwood City, CA. Pages 549–578.

Johnson, A.R. 1994. Evolution of a size-structured, predator–prey community. In: C.G. Langton (ed). *Artificial Life III.* Addison-Wesley Publishing Co., Reading, MA. Pages 105–129.

Johnson, N.L. and F.C. Leone. 1964. *Statistics and Experimental Design in Engineering and the Physical Sciences. Vol.* 1. John Wiley and Sons,

NY.

Jones, D.D. 1979. The Budworm site model. In: G.A. Norton and C.S. Holling (eds). *Pest Management: Proceedings of an International Conference.* Pergamon Press, Oxford, UK. Pages 91–155.

Jones, H.G. 1992. *Plants and Microclimate: A Quantitative Approach to Environmental Plant Physiology.* 2nd Ed. Cambrige University Press, Cambridge, UK.

Jørgensen, S.E. 1986. *Fundamentals of Ecological Modelling.* Elsevier, Amsterdam.

Judson, O.P. 1994. The rise of the individual-based model in ecology. *Trends in Ecology and Evolution* 9(1):9–14.

Kahaner, D.K., E. Ng, W.E. Schiesser, and S. Thompson. 1992. Experiments with an ordinary differential equation solver in the parallel solution of method of lines problems on a shared memory parallel computer. In: G.D. Byrne and W.E. Schiesser (eds). *Recent Developments in Numerical Methods and Software for ODEs/DAEs/PDEs.* World Scientific Publishers, Singapore. Pages 7–36.

Kareiva, P. and G. Odell. 1987. Swarms of predators exhibit "preytaxis" if individual predators use area-restricted search. *American Naturalist* 130(2):233–270.

Karplus, W.J. 1977. The place of systems ecology models in the spectrum of mathematical models. In: G.S. Innis (ed). *New Directions in the Analysis of Ecological Systems.* Part 2. Simulation Councils Proceedings Series. Volume 5, Number 2, Pages 225–228.

Karplus, W.J. 1983. The spectrum of mathematical models. *Perspectives in Computing* 3:4–14.

Keener, J. 1991. Wave propagation in myocardium. In: L. Glass, P. Hunter, and A. McCulloch (eds). *Theory of Heart: Biomechanics, Biophysics, and Nonlinear Dynamics of Cardiac Function.* Springer-Verlag, NY. Pages 405–436.

Keller, E. and L.A. Segel. 1970. Initiation of slime mold aggregation viewed as an instability. *Journal of Theoretical Biology* 26:399–415.

Kimmel, M. and D.N. Stivers. 1994. Time-continuous branching walk models of unstable gene amplification. *Bulletin of Mathematical Biology* 56(2):337–357.

King, A.W. 1991. Translating models across scales in the landscape. In: M.G. Turner and R.H. Gardner (eds). *Quantitative Methods in Landscape Ecology: The Analysis and Interpretation of Landscape Heterogeneity.* Springer-Verlag, NY. Pages 479–517.

Kinnear, K.E. (ed). 1994 *Advances in Genetic Programming.* The MIT Press, Cambridge, MA. 518 pages.

Kleijnen, J.P.C. and W. van Groenendaal. 1992. *Simulation: A Statistical Perspective.* John Wiley and Sons, Chichester, UK.

Kootsey, J.M., M.C. Kohn, M.D. Freezor, G.R. Mitchell, and P.R. Fletcher.

1986. SCoP: An interactive simulation control program for micro- and minicomputers. *Bulletin of Mathematical Biology* 48:427–441.

Kot, M., G.S. Sayler, and T.W. Schultz. 1992. Complex dynamics in a model microbial system. *Bulletin of Mathematical Biology* 54(4):619–648.

Koza, J.R. 1992a. Genetic evolution and co-evolution of computer programs. In: C. Langston, C. Taylor, J.D. Farmer and S. Rasmussen (eds). *Artificial Life II.* Addison-Wesley, Redwood City, CA. Pages 603–629.

Koza, J.R. 1992b. *Genetic Programming: On the Programming of Computers by Means of Natural Selection.* MIT Press, Cambridge, MA. 819 pages.

Koza, J.R. 1994. *Genetic programming II : automatic discovery of reusable programs.* MIT Press, Cambridge, MA. 746 pages.

Koza, J.R., J.P. Rice, and J. Roughgarden. 1992. Evolution of food foraging strategies for the Caribbean *Anolis* lizard using genetic programming. *Adaptive Behavior* 1(2):47–74.

Lacy, R.C. 1993. VORTEX: a computer simulation model for population viability analysis. *Wildlife Research* 20:45–65.

Lamberson, R.H., K. McKelvey, B.R. Noon, and C. Voss. 1992. A dynamic analysis of Northern Spotted Owl viability in a fragmented forest landscape. *Conservation Biology* 6(4):505–512.

Lande, R. 1987. Extinction thresholds in demographic models of territorial populations. *American Naturalist* 130(4):624–635.

Lande, R. 1988. Demographic models of the Northern Spotted Owl (*Strix occidentalis caurina*). *Oecologia* 75:601–607.

Langton, C. 1988a. Artifical life. In: C. Langton (ed). *Artificial Life.* Addison-Wesley, Redwood City, CA. Pages 1–47.

Langton, C. 1988b (ed). *Artificial Life.* Addison-Wesley, Redwood City, CA.

Langton, C.G. 1992. Life at the edge of chaos. In: C.G. Langton, C. Taylor, J.D. Farmer, and S. Rasmussen (eds). *Artificial Life II.* Addison-Wesley, Redwood City, CA. Pages 41–91.

Lauenroth, W.K., D.L. Urban, D.P. Coffin, W.J. Parton, H.H. Shugart, T.B. Kirchner, and T.M. Smith. 1993. Modeling vegetation structure–ecosystem process interactions across sites and ecosystems. *Ecological Modelling* 67:49–80.

Leslie, P.H. 1945. On the use of matrices in certain population mathematics. *Biometrika* 33:183–212.

Levin, S.A. 1992. The problem of pattern and scale in ecology. *Ecology* 73(6):1943–1967.

Levin, S.A. and L. Buttel. 1987. *Measures of patchiness in ecological systems.* Ecosystems Research Center Report Number ERC–130. Cornell University, Ithaca, NY.

Levins, R. 1966. The strategy of model building in population biology. *American Scientist* 54:421–431.

Levins, R. 1969. Some demographic and genetic consequences of environmental heterogeneity for biological control. *Bulletin of the Entomological Society of America* 15:237–240.

Levins, R. 1970. Complex Systems. In: C.H. Waddington (ed). *Towards a Theoretical Biology 3: Drafts.* Aldine and Atherton, Chicago, IL. Pages 73–88.

Levins, R. 1974. The qualitative analysis of partially specified systems. *Annals of the New York Academy of Sciences* 231:123–138.

Levins, R. 1993. A response to Orzack and Sober: formal analysis and the fluidity of science. *Quarterly Review of Biology* 68(4):547–555.

Lin, C.C. and L.A. Segel. 1988. *Mathematics Applied to Deterministic Problems in the Natural Sciences.* SIAM, Philadelphia, PA. 609 pages.

Lindenmayer, A. 1968a. Mathematical models of cellular interactions in development. Part I. *Journal of Theoretical Biology* 18:280–299.

Lindenmayer, A. 1968b. Mathematical models of cellular interactions in development. Part II. *Journal of Theoretical Biology* 18:300–315.

Lindenmayer, A. 1975. Developmental algorithms for multicellular organisms: a survey of L-systems. *Journal of Theoretical Biology* 54:3–22.

Logan, J.A. 1988. Toward an expert system for development of pest simulation models. *Environmental Entomology* 17:359–376.

Logan, J.A. 1994. In defense of big ugly models. *American Entomologist* 41:202–207.

Logan, J.A. and J.C. Allen. 1992. Nonlinear dynamics and chaos in insect populations. *Annual Review of Entomology* 37:455–477.

Lorenz, E.N. 1963. Deterministic nonperiodic flow. *Journal of Atmospheric Science* 20:130–141.

Luckinbill, L.S. 1973. Coexistence of laboratory populations of *Paramecium aurelia* and its predator *Didinium nasutum. Ecology* 54(6):1320–1327.

Ludwig, D. 1974. *Stochastic Population Theories.* Lecture Notes in Biomathematics. Springer-Verlag. Berlin.

Ludwig, D., D.D. Jones, and C.S. Holling. 1978. Qualitative analysis of insect outbreak systems: the spruce budworm and forest. *Journal of Animal Ecology* 47:315–332.

Mackey, M.C. and L. Glass. 1977. Oscillation and chaos in phyisological control systems. *Science* 197:287–289.

Madenjian, C.P. and S.R. Carpenter. 1991. Individual-based model for growth of young-of-the-year walleye: a piece of the recruitment puzzle. *Ecological Applications* 1(3):268–279.

Madenjian, C.P., S.R. Carpenter, G.W. Eck, and M.A. Miller. 1993. Accumulation of PCBs by Lake Trout (*Salvelinus namaycush*): an individual-based model approach. *Canadian Journal of Fisheries and Aquatic Sciences* 50(1):97–109.

Mandelbrot, B.B. 1977. *Fractals: Form, Chance, and Dimension.* W.H. Freeman, San Francisco, CA.

Mangel, M. and C. Tier. 1993. A simple, direct method for finding persistence times of populations and application to conservation problems. *Proceedings of the National Academy of Sciences (USA)* 90:1083–1086.

Mangel, M. and C. Tier. 1994. Four facts every conservation biologist should know about persistence. *Ecology* 75(3):607–614.

Mankin, J.B., R.V. O'Neill, H.H. Shugart, and B.W. Rust. 1977. The importance of validation in ecosystem analysis. In: G.S. Innis (ed). *New Directions in the Analysis of Ecological Systems.* Part 1. Simulation Councils Proceedings Series. Volume 5, Number 1. Pages 63–71.

Marsili-Libelli, S. 1992. Parameter estimation of ecological models. *Ecological Modelling* 62:233–258.

Matis, J.H., W.E. Grant, and T.H. Miller. 1992. A semi-Markov process model for migration of marine shrimp. *Ecological Modelling* 60:167–184.

Maxwell, T. and R. Costanza. 1993. Spatial ecosystem modeling in a distributed computational environment. In: J. van den Bergh and J. van der Straaten (eds). *Concepts, Methods, and Policy for Sustainable Development.* Island Press, Washington, D.C. Pages 1–26.

May, R.M. 1973. *Stability and Complexity in Model Ecosystems.* Princeton University Press. Princeton, NJ. 235 pages.

May, R.M. 1974. Biological populations with nonoverlapping generations: stable points, stable cycles, and chaos. *Science* 186:645–647.

May, R.M. 1976. Simple mathematical models with very complicated dynamics. *Nature* 261:459–467.

May, R.M. 1977. Thresholds and breakpoints in ecosystems with a multiplicity of stable states. *Nature* 269:471–477.

Mayer, D.G. and D.G. Butler. 1993. Statistical validation. *Ecological Modelling* 68:21–32.

Mayer, D.G., M.A. Stuart, and A.J. Swain. 1994. Regression of real-world data on model output: an appropriate overall test of validity. *Agricultural Systems* 45:93–104.

McCann, K. and P. Yodzis. 1994. Biological conditions for chaos in a three-species food chain. *Ecology* 75(2):561-564.

McKelvey, K., B.R. Noon, and R.H. Lamberson. 1993. Conservation planning for species occupying fragmented landscapes: the case of the Northern Spotted Owl. In: P.M. Kareiva, J.G. Kingsolver, and R.B. Huey (eds). *Biotic Interactions and Global Change.* Sinauer Associates, Sunderland, MA.

Metz, J.A.J. and A.M. de Roos. 1992. The role of physiologically structured population models within a general individual-based modeling perspective. In: D.L. DeAngelis and L.J. Gross (eds). *Individual-based models and Approaches in Ecology: Populations, Communities, and*

Ecosystems. Chapman and Hall, NY. Pages 88–111.

Meyer, S.L. 1975. *Data Analysis for Scientists and Engineers.* John Wiley and Sons, NY.

Michelsen, A., B.B. Andersen, J. Storm, W. H. Kirchner, and M. Lindauer. 1992. How honeybees perceive communication dances, studied by means of a mechanical model. *Behavioral Ecology and Sociobiology* 30:143–150.

Miglino, O., K. Nafashi, and C.E. Taylor. 1995. Selection for wandering behavior in a small robot. *Artificial Life* 2(1):101–116.

Miller, A.R. 1981. *Pascal Programs for Scientists and Engineers.* Sybex, Berkeley, CA.

Milne, B.T. 1991. Lessons from applying fractal models to landscape patterns. In: M.G. Turner and R.H. Gardner (eds). *Quantitative Methods in Landscape Ecology: The Analysis and Interpretation of Landscape Heterogeneity.* Ecological Studies Vol. 82. Springer-Verlag, NY. Pages 199–235.

Mitchell, R.H., A.H. Bailie, and J.M. Anderson. 1992. Cellular automaton model of ventricular conduction. *Medical and Biological Engineering and Computing* 30:481–486.

Moloney, K.A., S.A. Levin, N.R. Chiariello, and L. Buttel. 1992. Pattern and scale in a serpentine grassland. *Theoretical Population Biology* 41:257–276.

Monsi, M. and T. Saeki. (1953). Über den Licht-faktor in den Pflanzengesellschaften und seine Dedeutung für die Stoffproduction. *Japanese Journal of Botany* 14:22–52.

Monsi, M., Z. Uchijima, and T. Oikawa. 1973. Structure of foliage canopies and photosynthesis. *Annual Review of Ecology and Systematics* 4:301–327.

Mott, K.A., Z.G. Cardon, and J.A. Berry. 1993. Asymmetric patchy stomatal closure for the two surfaces of *Xanthium strumarium* L. leaves at low humidity. *Plant, Cell and Environment* 16:25–34.

Mullin, T. 1993a. A dynamical systems approach to time series analysis. In: T. Mullin (ed). *The Nature of Chaos.* Clarendon Press, Oxford, UK. Pages: 23–50.

Mullin, T. 1993b. A multiple bifurcation point as an organizing centre for chaos. In: T. Mullin (ed). *The Nature of Chaos.* Clarendon Press, Oxford, UK. Pages 51–68.

Mullin, T. (ed). 1993c. *The Nature of Chaos.* Clarendon Press, Oxford, UK.

Murray, J.D. 1989. *Mathematical Biology.* Springer-Verlag, Berlin.

Nagel, E. 1961. *The Structure of Science: Problems in the Logic of Scientific Explanations.* Harcourt, Brace, and World, NY.

Nelder, J.A. and R. Mead. 1965. A simplex method for function minimization. *Computer Journal* 7:308–313.

Niklas, K.J. 1986a. Computer-simulated plant evolution. *Scientific Amer-*

ican 254(3):78–86.

Niklas, K.J. 1986b. Computer simulations of branching-patterns and their implications on the evolution of plants. *Lectures on Mathematics in the Life Sciences* 18:1–50.

Niklas, K.J. 1992. *Plant Biomechanics: An Engineering Approach to Plant Form and Function.* University of Chicago Press, Chicago, IL.

Niklas, K.J. and V. Kerchner. 1984. Mechanical and photosynthetic constraints on the evolution of plant shape. *Paleobiology* 10(1):79–101.

Noreen, E.W. 1989. *Computer-Intensive Methods for Testing Hypotheses: An Introduction.* John Wiley and Sons, NY.

Norman, J.M. 1993. Scaling processes between leaf and canopy levels. In: J.R.Ehleringer and C.B. Field (eds). *Scaling Physiological Processes: Leaf to Globe.* Academic Press, Inc. San Diego, CA. Pages 41–76.

Odum, H.T. 1971. *Environment, Power, and Society.* John Wiley and Sons, NY.

O'Neill, R.V., D.L. DeAngelis, J.J. Pastor, B.J. Jackson, and W.M. Post. 1989. Multiple nutrient limitations in ecological models. *Ecological Modelling* 46:147–163.

O'Neill, R.V., D.L. DeAngelis, J.B. Waide, and T.F.H. Allen. 1986. *A Hierarchical Concept of Ecosystems.* Princeton University Press. Princeton, NJ. 253 pages.

O'Neill, R.V. and R.H. Gardner. 1979. Sources of uncertainty in ecological models. In: B.P. Zeigler, M.S. Elizas, G.J. Klir, and T.I. Oren (eds). *Methodology in Systems Modelling and Simulation.* North-Holland Publishing Co., Amsterdam. Pages 447–463.

O'Neill, R.V., R.H. Gardner, and J.B. Mankin. 1980. Analysis of parameter error in a nonlinear model. *Ecological Modelling* 8:297–311.

O'Neill, R.V., R.H. Gardner, B.T. Milne, M.G. Turner, and B. Jackson. 1991. Heterogeneity and spatial hierarchies. In: J. Kolasa and S.T.A. Pickett (eds). *Ecological Heterogeneity.* Springer-Verlag, NY. Pages 85–97.

O'Neill, R.V. and B. Rust. 1979. Aggregation error in ecological models. *Ecological Modelling* 7:91–105.

Orzack, S.H. and E. Sober. 1993. A critical assessment of Levins's *The strategy of model building in population biology (1966). Quarterly Review of Biology* 68(4):533–546.

Ott, E., C. Grebogi, and J.A. Yorke. 1990. Controlling chaos. *Physical Review Letters* 64(11):1196–1199.

Overton, W.S. 1972. Toward a general model structure for a forest ecosystem. In: J.F. Franklin, L.J. Dempster, and R.H. Waring (eds). *Proceedings - Research on Coniferous Forest Ecosystems – A Symposium.* Pacific NW Forest and Range Experiment Station., Portland, OR. Pages 37–47.

Overton, W.S. 1977. A strategy of model construction. In: C.A.S. Hall and J.W. Day (eds). *Ecosystem Modeling in Theory and Practice: An Introduction with Case Histories.* John Wiley and Sons, NY. Pages 49–73.

Pavlov, S. and I.G. Kevrekidis. 1992. Microbial predation in a periodically operated chemostat: a global study of the interaction between natural and externally imposed frequencies. *Mathematical Biosciences* 108:1–55.

Peak, D. and M. Frame. 1994. *Chaos Under Control: The Art and Science of Complexity.* W.H. Freeman, NY. 408 pages.

Percival, I. and D. Richards. 1982. *Introduction to Dynamics.* Cambridge University Press, Cambridge, UK. 228 pages.

Peters, R.H. 1983. *The Ecological Implications of Body Size.* Cambrige University Press, Cambridge, UK. 329 pages.

Pielou, E.C. 1977. *Mathematical Ecology.* John Wiley and Sons, NY. 385 pages.

Platt, J.R. 1964. Strong inference. *Science* 146:347–353.

Platt, T. and K.L. Denman. 1975. Spectral analysis in ecology. *Annual Review of Ecology and Systematics* 6:289–210.

Polya, G. 1973. *How To Solve It: A New Aspect of Mathematical Method.* 2nd Ed. Princeton University Press. Princeton, NJ. 253 pages.

Popper, K. 1968. *The Logic of Scientific Discovery.* Harper Torchbooks, NY.

Power, M. 1993. The predictive validation of ecological and environmental models. *Ecological Modelling* 68:33–50.

Press, W.H., W.A. Teukolsky, W.T. Vetterling, and B.P. Flannery. 1992. *Numerical Recipes in C: The Art of Scientific Computing.* 2nd Ed. Cambridge University Press, Cambridge, UK. 994 pages.

Prusinkiewicz, P. and A. Lindenmayer. 1990. *The Algorithmic Beauty of Plants.* Springer-Verlag, NY.

Purdum, J. 1991. Some thoughts on portability. *C Users Journal* 9:45–52.

Rand, R.H. and J.L. Ellenson. 1986. Dynamics of stomate fields in leaves. *Lectures in Mathematics in the Life Sciences* 18:51–86.

Rand, R.H., S.K. Upadhyaya, J.R. Cooke, and D.W. Storti. 1981. Hopf bifurcation in a stomatal oscillator. *Journal of Mathematical Biology* 12:1–11.

Rastetter, E.B, A.W. King, B.J. Cosby, G.M. Hornberger, R.V. O'Neill, and J.E. Hobbie. 1992. Aggregating fine-scale ecological knowledge to model coarser-scale attributes of ecosystems. *Ecological Applications* 2(1):55-70.

Ratkowsky, D.A. 1983. *Nonlinear Regression Modeling: A Unified Practical Approach.* Marcel Dekker, NY.

Raven, P.H. and G.B. Johnson. 1992. Biology. 3rd Ed. Mosby-Year Book, St. Louis, MO.

Reckhow, K.H. 1990. Bayesian inference in non-replicated ecological studies. *Ecology* 71:2053–2059.

Reckhow, K.H. and S.C. Chapra. 1983a. Confirmation of water quality models. *Ecological Modelling* 20:113–133.

Reckhow, K.H. and S.C. Chapra. 1983b. *Engineering Approaches for Lake Management. Volume 1: Data Analysis and Empirical Modeling.* Butterworth Publishers, Boston, MA. 340 pages.

Reddy, V.R., D.N. Baker,F. Whisler, and J. Lambert. 1985. Validation of GOSSYM: Part II. Mississippi conditions. *Agricultural Systems* 17:133–154.

Reilly, P.M. 1970. Statistical methods in model discrimination. *Canadian Journal of Chemical Engineering* 48:168–173.

Reynolds, J.F., D.W. Hilbert, and P.R. Kemp. 1993. Scaling ecophysiology from the plant to the ecosystem: A conceptual framework. In: J.R.Ehleringer and C.B. Field (eds). *Scaling Physiological Processes: Leaf to Globe.* Academic Press, Inc. San Diego, CA. Pages 127–140.

Rhee, G–Y. 1980. Continuous culture in phytoplankton ecology. In: M.R. Droop and H.W. Jannach (eds). *Advances in Aquatic Microbiology.* Academic Press, NY. Pages 151–203.

Rice, J.R. 1983. *Numerical Methods, Software, and Analysis.* McGraw-Hill, NY. 483 pages.

Rice, J.A. and P.A. Cochran. 1984. Independent evaluation of a bioenergetics model for largemouth bass. *Ecology* 65(3):732-739.

Richards, F.J. 1959. A flexible growth function for empirical use. *Journal of Experimental Botany* 29:290–300.

Richter, O. and D. Söndgerath. 1990. *Parameter Estimation in Ecology: The Link Between Data and Models.* VCH, Weinheim, GDR.

Ricker, W.E. 1973. Linear regression in fishery research. *Journal of the Fisheries Research Board of Canada* 30:409-434.

Rideout, V.C. 1991. *Mathematical and Computer Modeling of Physiological Systems.* Prentice-Hall, Englewood Cliffs, NJ. 261 pages.

Rigney, D.R., A.L. Goldberger, W.C. Ocasio, Y.Ichimaru, G.B. Moody, and R.G. Mark. 1994. Multi-channel physiological data: description and analysis (data set B). In: A.S. Weigend and N.A. Gershenfeld (eds). *Time Series Prediction: Forecasting the Future and Understanding the Past.* Santa Fe Institute, Studies in the Sciences of Complexity, *Vol. XV.* Addison-Wesley, Reading, MA. Pages 105–129.

Romesburg, H.C. 1981. Wildlife science: gaining reliable knowledge. *Journal of Wildlife Management* 45(2):293–313.

Rose, K.A., W.W. Christensen, and D.L. DeAngelis. 1993. Individual-based modeling of populations with high mortality: a new method based on following a fixed number of model individuals. *Ecological Modelling* 68:273–292.

Rosenzweig, M.L. 1971. Paradox of enrichment: destabilization of exploitation ecosystems in ecological time. *Science* 171:385–387.

Royama, T. 1984. Population dynamics of the sruce budworm *Choristoneura fumiferana*. *Ecological Monographs* 54(4):429–462.

Rubinow, S.I. 1975. *Introduction to Mathematical Biology.* John Wiley and Sons, NY.

Rubinow, S.I. and L.A. Segel. 1991. Positive and negative cooperativity. In: L.A. Segel (ed). *Biological Kinetics.* Cambridge University Press, Cambridge. Pages 29–44.

Sargent, R.G. 1984. Simulation model validation. In: T.I. Oren, B.P. Zeigler, and M.S. Elzas. (eds). *Simulation and Model-based Methodologies: An Integrative View.* Springer-Verlag, Berlin. Pages 537–555.

Saxberg, B.E.H. and R.J. Cohen. 1991. Cellular automata models of cardiac conduction. In: L. Glass, P. Hunter, and A. McCulloch (eds). *Theory of Heart: Biomechanics, Biophysics, and Nonlinear Dynamics of Cardiac Function.* Springer-Verlag, NY. Pages 437–476.

Schaffer, W.M. 1987. Chaos in ecology and epidemiology. In: H. Degn, A.V. Holden, and L.F. Olsen (eds). *Chaos in Biological Systems.* Plenum Press, NY. Pages 233–248.

Schaffer, W.M. and M. Kot. 1986. Differential systems in ecology and epidemiology. In: A.V. Holden (ed). *Chaos.* Manchester University Press, Manchester, UK. Pages 158–178.

Schaffer, W.M., L.F. Olsen, G.L. Truty, and S.L. Fulmer. 1990. The case for chaos in childhood epidemics. In: S. Krasner (ed). *The Ubiquity of Chaos.* American Association for the Advancement of Science, Washington, D.C. Pages 138–166.

Schiff, S.J., K. Jerger, D.H. Duong, T. Chang, M.L. Spano, and W.L. Ditto. 1994. Controlling chaos in the brain. *Nature* 370:615–620.

Schmidt-Nielsen, K. 1979. *Animal Physiology: Adaptation and Environment.* 2nd Ed. Cambridge University Press, Cambridge. 560 pages.

Schmidt-Nielsen, K. 1984. *Scaling: Why Animal Size Is so Important.* Cambridge University Press, Cambridge. 241 pages.

Schoener, T.W. 1976. Alternatives to Lotka–Volterra competition: models of intermediate complexity. *Theoretical Population Biology* 10: 309–333.

Schroeder, M. 1991. *Fractals, Chaos, Power Laws: Minutes From an Infinite Paradise.* W.H. Freeman and Co, NY.

Schruben, L.W. 1980. Establishing the credibility of simulations. Simulation 34(3):101-105.

Schwefel, H-P. 1995. *Evolution and Optimum Seeking.* John Wiley and Sons, NY. 444 pages.

Searle, S.R. 1982. *Matrix Algebra Useful for Statistics.* John Wiley and Sons, NY.

Seber, G.A.F. and C.J. Wild. 1989. *Nonlinear Regression.* John Wiley and Sons, NY. 768 pages.

Sequeira, R.A., and R.L. Olson. 1995. Self-correction of simulation models using genetic algorithms. *AI Applications* 9:3–16.

Shannon, R.E. 1975. *Systems Simulation: The Art and Science.* Prentice-Hall, Inc., Englewood Cliffs, NJ. 387 pages.

Shugart, H.H. 1984. *A Theory of Forest Dynamics: The Ecological Implications of Forest Succession Models.* Springer-Verlag, NY. 278 pages.

Shugart, H.H., T.M. Smith, and W.M. Post. 1992. The potential for application of individual-based models for assessing the effects of global change. *Annual Review of Ecology and Systematics* 23:15–38.

Shugart, H.H. and D.C. West. 1977. Development and application of an Appalachian deciduous forest succession model. *Journal of Environmental Management* 5:161–179.

Silvertown, J., S. Holtier, J. Johnson, and P. Dale. 1992. Cellular automaton models of interspecific competition for space — the effect of pattern on process. *Journal of Ecology* 80:527–534.

Sims, K. 1994. Evolving 3D morphology and behavior by competition. In: R.A. Brooks and P. Maes (eds). *Artificial Life IV.* MIT Press, Cambridge, MA. Pages 28–39. [Reprinted in *Artificial Life* 1(4):353–372.]

Sklar, F. and R. Costanza. 1991. The development of dynamic spatial models for landscape ecology: a review and prognosis. In: M.G. Turner and R.H. Gardner (eds). *Quantitative methods in landscape ecology: The analysis and interpretation of landscape heterogeneity.* Ecological Studies *Vol. 82.* Springer-Verlag, NY. Pages 239–288.

Smith, H. and P. Waltman. 1995. *The Theory of the Chemostat.* Cambridge University Press, Cambridge, UK.

Smith, J.M. and R.J. Cohen. 1984. Simple finite-element model accounts for wide range of cardiac dysrhythmias. *Proceedings of the National Academy of Science, USA* 81:233–237.

Sokal, R.R. and F.J. Rohlf. 1981. *Biometry: The Principles and Practice of Statistics in Biological Research.* W.H. Freeman, San Francisco.

Sorenson, H. 1980. *Parameter Estimation: Principles and Problems.* Marcel Dekker, NY. 382 pages.

Spafford, E.H. 1994. Computer viruses as artificial life. *Artificial Life* 1(3):249–266.

Spriet, J.A. and G.C. Vansteenkiste. 1982. *Computer-aided Modelling and Simulation.* Academic Press, London. 490 pages.

Steele, J.H. 1962. Environmental control of photosynthesis in the sea. *Limnology and Oceanography* 7(2):137–150.

Steinhorst, R.K. 1979. Parameter identifiability, validation, and sensitivity analysis of large system models. In: G.S. Innis and R.V. O'Neill (eds). *Systems Analysis of Ecosystems.* Inter-Cooperative Publishing House, Fairland, MD. Pages 33-58.

Stent, A.F. and I.R. McCallum. 1995. Dynamic simulation modeling on a spreadsheet. *Simulation* 64(6):373–379.

Sugihara, G. and R.M. May. 1990. Nonlinear forecasting as a way of distinguishing chaos from measurement error in time series. *Nature* 344:734–741.

Summers, J.K., H.T. Wilson, and J. Kou. 1993. A method for quantifying the prediction uncertainties associated with water quality models. *Ecological Modelling* 65:161–176.

Swartzman, G. 1980. Evaluation of ecological simulation models. In: W. M. Getz (ed). *Mathematical Modeling in Biology and Ecology.* Springer Verlag, Berlin. Pages 230-267.

Swartzman, G.L. and S.P. Kaluzny. 1987. *Ecological Simulation Primer.* MacMillan Publishing, NY.

Sweeney, D.G., D.W. Hand, G. Slack, and J.H.M. Thornley. 1981. Modelling the growth of winter lettuce. In: D.A. Rose and D.A. Charles-Edwards (eds). *Mathematics and Plant Physiology.* Academic Press, London. Pages 217–229.

Taylor, R.J. 1984. *Predation.* Chapman and Hall, London. 166 pages.

Theil, H. 1961. *Economic Forecasts and Policy.* North-Holland Publishing Company, Amsterdam.

Thornley, J.H.M. 1972. A model to describe the partitioning of photosynthate during vegetative plant growth. *Annals of Botany* 36:419–430.

Thornley, J.H.M. and I.R. Johnson. 1990. *Plant and Crop Modelling.* Oxford University Press, Oxford.

Thorson, J., T. Weber, and F. Huber. 1982. Auditory behvaior in the cricket. II. Simplicity of calling-song recognition in *Gryllus* and anomalous phonotaxis at abnormal carrier frequencies. *Journal of Comparative Physiology, A* 146:361–378.

Tilman, D. 1977. Resource competition between algae: an experimental and theoretical approach. *Ecology* 58:338–348.

Timm, N.H. 1975. *Multivariate Analysis with Applications in Education and Psychology.* Brooks/Cole Publishing Co., Monterey, CA.

Toffoli, T. and N. Margolus. 1987. *Cellular Automata Machines: A New Environment for Modeling.* The MIT Press, Cambridge, MA.

Tomovic, R. 1963. *Sensitivity Analysis of Dynamic Systems.* McGraw-Hill, NY.

Toquenaga, Y. and K. Fugii. 1991. Contest and scramble competitions in two bruchid species, *Callosobruchus analis* and *C. phaseoli* (Coleoptera: Bruchidae): III. Multiple-generation competition experiment. *Researches on Population Ecology* 33(2):187–197.

Toquenaga, Y., M. Ichinose, T. Hoshino, and K. Fugii. 1994. Contest and scramble competitions in an artificial world: Genetic analysis with genetic algorithms. In: C.G. Langton (ed). *Artificial Life III.* Addison-Wesley Publishing Co., Reading, MA. Pages 177–199.

Tribus, M. and E.C. McIrvine. 1970. Energy and information. *Scientific American* 224(3):179–188.

Turelli, M. 1977. A reexamination of stability in randomly varying versus deterministic environments with comments on the stochastic theory of limiting similarity. *Theoretical Population Biology.* 13:244–267.

Turelli, M. 1981. Niche overlap and invasion of competitors in random environments. I. Models without demographic stochasticity. *Theoretical Population Biology.* 20(1):1–56.

Turner, M.G., R. Costanza, and F.H. Sklar. 1989. Methods to evaluate the performance of spatial simulation models. *Ecological Modelling* 48:1–18.

Vance, R.R. 1978. Predation and resource partitioning in one predator–two prey model communities. *American Naturalist* 112:797–813.

Vandermeer, J.H. 1981. *Elementary Mathematical Ecology.* John Wiley and Sons, NY.

Van Henten, E.J. 1994. Validation of a dynamic lettuce growth model for greenhouse climate control. *Agricultural Systems* 45:55–72.

van Laarhoven, P.J.M. and E.H.L. Aarts. 1987. *Simulated Annealing: Theory and Applications.* D. Reidel Publishing Co., Dordrecht, Holland. 186 pages.

Walker, J.C. 1992. *Simulated Annealing Applied to the PEANUT growth model for optimization of irrigation scheduling.* Ph.D. Dissertation. Department of Biological and Agricultural Engineering. North Carolina State University, Raleigh, NC.

Walters, C.J. 1986. *Adaptive Management of Renewable Resources.* MacMillan Publishing Company, NY.

Walters, C.J. and F. Bunnell. 1971. A computer management game of land-use in British Columbia. *Journal of Wildlife Management* 35(4):644-357.

Webb, B. 1994. Robotic experiments in cricket phonotaxis. In: D. Cliff, P. Husbands, J-A Meyer, and S.W. Wilson (eds). *From Animals to Animats 3.* The MIT Press, Cambridge, MA. Pages 45–54.

Weber, T. and J. Thorson. 1989. Phonotactic behavior of walking crickets. In: F. Huber, T.E. Moore, and W. Loher. (eds). *Cricket Behavior and Neurobiology.* Cornell University Press, Ithaca, NY. Pages 310–339.

Weigend, A.S. and N.A. Gershenfeld. (eds). 1994. *Time Series Prediction: Forecasting the Future and Understanding the Past.* Santa Fe Institute, Studies in the Sciences of Complexity. Proceedings *Vol. XV.* Addison-Wesley Publishing Co., Reading, MA. 643 pages.

Werner, P.A. and H. Caswell. 1977. Population growth rates and age *versus* state-distribution models for teasel (*Dipsacus sylvestris* HUDS). *Ecology* 58:1102–111.

West, B.J. and M. Shlesinger. 1990. The noise in natural phenomena. *American Scientist* 78:40–45.

White, C. and W.S. Overton. 1974. *Users Manual for the FLEX2 and FLEX3. Model Processors for the FLEX Modelling Paradigm.* Bulletin 15 Review Draft. Forest Research Laboratory, Oregon State University,

Corvallis, OR.
Wilson, E.O. and W.H. Bossert. 1971. *A Primer on Population Biology.* Sinauer and Associates, Sunderland, MA.
Winer, B.J. 1971. *Statistical Principles in Experimental Design.* McGraw-Hill Book Company, NY.
Yodzis, P. 1989. *Introduction to Theoretical Ecology.* Harper and Row, NY.
Zar, J.H. 1984. *Biostatistical Analysis.* Prentice-Hall, Inc. Englewood Cliffs, NJ.
Zeigler, B.P. 1976. The aggregation problem. In: B.C. Patten (ed). *Systems Analysis and Simulation in Ecology. Vol. IV.* Academic Press, NY. Pages 299–311.

Index